SYNOPSIS OF
SLEEP MEDICINE

SYNOPSIS OF SLEEP MEDICINE

Edited by
S. R. Pandi-Perumal

Apple Academic Press Inc.	Apple Academic Press Inc.
3333 Mistwell Crescent	9 Spinnaker Way
Oakville, ON L6L 0A2	Waretown, NJ 08758
Canada	USA

© 2017 by Apple Academic Press, Inc.

First issued in paperback 2021

Exclusive worldwide distribution by CRC Press, a member of Taylor & Francis Group

No claim to original U.S. Government works

ISBN-13: 978-1-77463-613-8 (pbk)
ISBN-13: 978-1-77188-346-7 (hbk)

Library and Archives Canada Cataloguing in Publication

Synopsis of sleep medicine / edited by S.R. Pandi-Perumal.

Includes bibliographical references and index.
Issued in print and electronic formats.
ISBN 978-1-77188-346-7 (hardcover).--ISBN 978-1-77188-347-4 (pdf)

1. Sleep disorders. 2. Sleep disorders--Diagnosis. 3. Sleep disorders--Treatment.
4. Sleep--Physiological aspects. I. Pandi-Perumal, S. R., author, editor

| RC547.S95 2016 | 616.8'498 | C2016-904679-6 | C2016-904680-X |

Library of Congress Cataloging-in-Publication Data

Names: Pandi-Perumal, S. R., editor.
Title: Synopsis of sleep medicine / editor, S.R. Pandi-Perumal.
Description: Toronto ; New Jersey : Apple Academic Press, 2016. | Includes bibliographical references and index.
Identifiers: LCCN 2016030185 (print) | LCCN 2016031120 (ebook) | ISBN 9781771883467 (hardcover : alk. paper) | ISBN 9781771883474 (eBook) | ISBN 9781771883474 ()
Subjects: | MESH: Sleep Wake Disorders | Sleep Disorders, Circadian Rhythm
Classification: LCC RC547 (print) | LCC RC547 (ebook) | NLM WL 108 | DDC 616.8'498--dc23
LC record available at https://lccn.loc.gov/2016030185

Apple Academic Press also publishes its books in a variety of electronic formats. Some content that appears in print may not be available in electronic format. For information about Apple Academic Press products, visit our website at **www.appleacademicpress.com** and the CRC Press website at **www.crcpress.com**

ABOUT THE EDITOR

S. R. Pandi-Perumal, MSc

S. R. Pandi-Perumal is the President and Chief Executive Officer of Somnogen Canada Inc, a Canadian corporation. He holds a Master of Science from the Madurai Kamaraj University, Madurai, India. A well-recognized sleep researcher both nationally and internationally, he has authored over 100 publications in the field of sleep and biological rhythms. His general area of research interest includes sleep and biological rhythms. He has edited nearly 25 volumes related to sleep and biological rhythms research. He had an honorable mention in the *New York Times* in 2004, and the India International Friendship Society awarded him the Bharat Gaurav Award on January 12, 2013. Further details about his academic credentials can be found at: http://pandi-perumal.blogspot.com.

Dedication

To my family....

for their abundant support, for their patience and understanding, and for their everlasting love and affection.

CONTENTS

LIST OF CONTRIBUTORS

K. L. Acosta
University Sleep Disorders Center, College of Medicine, National Plan for Science and Technology, King Saud University, Riyadh, Saudi Arabia

I. M. Ahmed
Department of Neurology, Sleep-Wake Disorders Center, Montefiore Medical Center, Albert Einstein College of Medicine, 3411 Wayne Avenue, Bronx, NY 10467, USA

R. G. Albuquerque
Department of Psychobiology, Universidade Federal de São Paulo, Rua Napoleão de Barros, 925, Vila Clementino, São Paulo, 04024-002, Brazil. E-mail: rac.albuquerque@gmail.com

A. Al-Harbi
Department of Medicine, Pulmonary Division, Sleep Disorders Center, KAMC-KSUHS, Saudi Arabia. E-mail: HarbiA7@NGHA.MED.SA

H. Al-Jahdali
Adjunct Professor - McGill University; Professor Pulmonary and Sleep Medicine, King Saud University for Health Sciences; Head of Pulmonary Division, Medical Director of Sleep Disorders Center, King Abdulaziz Medical City, Riyadh, PO Box 101830*11665, Saudi Arabia. E-mail: Jahdali@yahoo.com

R. N. Al Ismaili
University Sleep Disorders Center, College of Medicine, King Saud University, Riyadh, Saudi Arabia; The Strategic Technologies Program of the National Plan for Sciences and Technology and Innovation in the Kingdom of Saudi Arabia, Saudi Arabia

M. L. Andersen
Department of Psychobiology, Universidade Federal de São Paulo, Rua Napoleão de Barros, 925, Vila Clementino, São Paulo 04024-002, Brazil. E-mail: ml.andersen12@gmail.com

F. Armstrong
Department of Neurology and Neuroscience, Center for Sleep Medicine, Weill Medical College of Cornell University, 525 East 68th Street, Room k-615, New York, NY 10065, USA. E-mail: armstrong.forrest@gmail.com

H. Attarian
Department of Neurology, Circadian Rhythms and Sleep Research Lab, 710 N Lake Shore Drive, Suite 1111, Chicago, IL 60611, USA. E-mail: hattaria@nmff.org

R. Y. Baba
Department of Neurology, Sleep-Wake Disorders Center, Montefiore Medical Center, Albert Einstein College of Medicine, Bronx, New York, NY, USA. E-mail: rbaba@montefiore.org

A. S. BaHammam
Sleep Disorders Center, College of Medicine, King Saud University, Box 225503, Riyadh 11324, Saudi Arabia. E-mail: ashammam2@gmail.com, ashammam@ksu.edu.sa

C. E. Bauer
Center for Neuroscience, West Virginia University, 1 Medical Center Drive, Morgantown, WV 26505, USA. E-mail: cbauer5@mix.wvu.edu

G. M. Brown
Centre for Addiction and Mental Health, University of Toronto, 100 Stokes St., Toronto, ON M6J 1H4, Canada. E-mail: gbrownpn@gmail.com

K. Buttoo
601 Harwood Avenue South, Ajax, ON L1S 2J5, Canada. E-mail: kbuttoo@hotmail.com

D. P. Cardinali
UCA-BIOMED-CONICET, Faculty of Medical Sciences, Pontificia Universidad Católica Argentina, 1107 Buenos Aires, Argentina. E-mail: danielcardinali@uca.edu.ar, danielcardinali@fibertel.com.ar

G. Copinschi
Department of Endocrinology, Université Libre de Bruxelles, Brussels, Belgium. Formerly Chief, Division of Endocrinology, Hôpital Universitaire Saint-Pierre, Université Libre de Bruxelles, Brussels, Belgium. Formerly Chairman, Department of Medicine, Hôpital Universitaire Saint-Pierre, Université Libre de Bruxelles, Brussels, Belgium. E-mail: gcop@ulb.ac.be

M. R. Ebben
Department of Neurology and Neuroscience, Center for
Sleep Medicine, Weill Medical College of Cornell Uni-
versity, 525 East 68th Street, Room k-615, New York,
NY 10065, USA.
E-mail: Mae2001@med.cornell.edu

S. H. Feinsilver
Department of Medicine, Icahn School of Medicine at
Mount Sinai, Division of Pulmonary, Critical Care and
Sleep Medicine, 1 Gustave Levy Place, New York, NY
10029, USA.
E-mail: steven.feinsilver@mssm.edu

D. E. Gacuan
University Sleep Disorders Center, College of Medicine,
National Plan for Science and Technology, King Saud
University, Riyadh, Saudi Arabia

S. George
University Sleep Disorders Center, College of Medicine,
National Plan for Science and Technology, King Saud
University, Riyadh, Saudi Arabia

C. Guilleminault
Stanford Sleep Medicine Center, 450 Broadway St Pavil-
ion C 2nd Fl MC 5704, Redwood City, CA94063, USA.
E-mail: cguil@stanford.edu

R. Gupta
Department of Psychiatry & Sleep Clinic, Himalayan In-
stitute of Medical Sciences, Swami Ram Nagar, Doiwala,
Dehradun, India. E-mail: sleepdoc.ravi@gmail.com

S. F. Harris
Department of Neurology, Sleep-Wake Disorders Cen-
ter, Montefiore Medical Center, Albert Einstein College
of Medicine, 3411 Wayne Avenue, Bronx, NY 10467,
USA.
E-mail: slharris@montefiore.org

A. B. Hernandez
Department of Medicine, Icahn School of Medicine at
Mount Sinai, Division of Pulmonary, Critical Care and
Sleep Medicine, 1 Gustave Levy Place, New York, NY
10029, USA.
E-mail: adam.hernandez@mountsinai.org

C. Hirotsu
Department of Psychobiology, Universidade Federal de
São Paulo, Rua Napoleão de Barros, 925, Vila Clementino, São Paulo, 04024-002, Brazil. E-mail: milahirotsu@
gmail.com

A. Ivanenko
Division of Child and Adolescent Psychiatry, Ann &
Robert H. Lurie Children's Hospital of Chicago, Chi-
cago, IL, USA. E-mail: aivanenko@sbcglobal.net

L. R. Pinto Junior
Department of Neurology and Sleep Medicine, Brazil-
ian Academy of Neurology, Universidade Federal de São
Paulo, Hospital Alemão Oswaldo Cruz, São Paulo (SP),
Brazil. E-mail: lucianoribeiro48@gmail.com

M. Khan
Davison of Pulmonary, KAMC-KSUHS, Saudi Arabia.
E-mail: iyazkhan@aol.com

L. J. Kim
Departamento de Psicobiologia, Universidade Federal
de São Paulo
Rua Napoleão de Barros, 925, Vila Clementino, São
Paulo, 04024-002, Brazil. E-mail: lenisekim@gmail.com

M. H. Lader
Department of Clinical Psychopharmacology, Institute
of Psychiatry, King's College, London SE5 8AF, UK. E-
mail: malcolm.lader@kcl.ac.uk

A. Martin
Universidade Federal do Rio Grande do Sul, Hospital
de Clinicas de Porto Alegre, Rua Ramiro Barcelos, 2350,
Porto Alegre, Brazil. E-mail: adriimartin@gmail.com

E. Medina-Ferret
EUTM, Hospital de Clínicas, Montevideo, Uruguay. E-
mail: emedinaferret@gmail.com

T. Mollayeva
Toronto Rehabilitation Institute-University Health Network, 550 University Avenue, Rm 11120, Toronto, Ontario M5G 2A2, Canada. E-mail: tatyana.mollayeva@
utoronto.ca

H. E. Montgomery-Downs
Department of Psychology, West Virginia University,
Morgantown, WV, USA. E-mail: Hawley.Montgomery-
Downs@mail.wvu.edu

J. M. Monti
Department of Pharmacology and Therapeutics, Department of Physiology, School of Medicine, Universidad de la República, Montevideo, Uruguay. E-mail:
jmonti@mednet.org.uy

D. N. Neubauer
Department of Psychiatry and Behavioral Sciences,
Johns Hopkins Bayview Medical Center, 4940 Eastern
Avenue, Box # 151, Baltimore, MD 21120, USA. E-mail:
neubauer@jhmi.edu

J. F. Pagel
Department of Family Practice, University of Colorado
School of Medicine, Pueblo, CO, USA; Sleep Disorders
Center of Southern Colorado affiliated with Parkview

Neurological Institute, Pueblo, CO, USA. E-mail: pueo34@juno.com

J. Palermo-Neto

Neuroimmunomodulation Research Group, Department of Pathology, School of Veterinary Medicine, University of Sao Paulo (USP), Sao Paulo, Brazil. E-mail: jpalremo@usp.br

S. R. Pandi-Perumal

Somnogen Canada Inc, College Street, Toronto, ON M6H 1C5, Canada. E-mail: pandiperumal2015@gmail.com

Marisa Pedemonte

Facultad de Medicina CLAEH, Punta del Este, Uruguay. E-mail: marisa.pedemonte@gmail.com

S. O. Qasrawi

University Sleep Disorders Center, College of Medicine, King Saud University, Riyadh, Saudi Arabia; The Strategic Technologies Program of the National Plan for Sciences and Technology and Innovation in the Kingdom of Saudi Arabia, Saudi Arabia

A. F. B. Rêgo

Neurology and Sleep Medicine, Brazilian Academy of Neurology, Universidade Federal Do Estado do Rio de Janeiro, Carlos Bacelar Clinic, Rio de Janeiro (RJ), Brazil. E-mail: andreabacelar@uol.com.br

T. Rogula

Bariatric and Metabolic Institute, Cleveland Clinic, 9500 Euclid Ave, M66-06, Cleveland, OH 44118, USA. E-mail: tomrogula@gmail.com

M. M. Schade

Department of Psychology, West Virginia University, Morgantown, WV, USA. E-mail: mmgray@mix.wvu.edu

P. R. Schauer

Bariatric and Metabolic Institute, Cleveland Clinic, 9500 Euclid Ave, M66-06, Cleveland, OH 44118, USA. E-mail: schauep@ccf.org

L. Scrima

Sleep Expert Consultants, LLC, 15011 E Arkansas Dr., Ste. B, Aurora CO 80012, USA. E-mail: scrimasleepdoc@msn.com

C. M. Shapiro

Toronto Western Hospital, 399 Bathurst Street, Rm 7MP421, Toronto, Ontario M5T 2S8, Canada. E-mail: colinshapiro@rogers.com

N. Sherbini

Head of Ethical Research Committee, Medina Region, Certified from Harvard Medical School in Practice of

Clinical Research; Department of Medicine, Pulmonary Division, Sleep Disorders Center, King Saud University for Health Sciences, Riyadh, Saudi Arabia. E-mail: drnahed@yahoo.com

T. J. Swick

American Academy of Neurology, Fellow, American Academy of Sleep Medicine; Department of Neurology, University of Texas Health Sciences Center-Houston; School of Medicine, and Medical Director: 1. Neurology and Sleep Medicine Consultants; 2. North Cypress Medical Center Sleep Disorders Center; 3. Apnix Sleep Diagnostics Laboratories, Neurology and Sleep Medicine Consultants, 7500 San Felipe, Ste. 525, Houston, TX 77063, USA. E-mail: tswick@houstonsleepcenter.com

M. J. Thorpy

Sleep-Wake Disorders Center, Montefiore Medical Center, Albert Einstein College of Medicine, 111 East 210th Street, Bronx, NY 10467, USA. E-mail: michael.thorpy@einstein.yu.edu, thorpy@aecom.yu.edu

P. Torterolo

Departmento de Fisiología, Facultad de Medicina, Universidad de la República, General Flores 2125, 11800 Montevideo, Uruguay

I. Trosman

Sleep Medicine Center, Division of Pulmonary Medicine, Ann & Robert H. Lurie Children's Hospital of Chicago, Chicago IL, USA. E-mail: itrosman@yahoo.com

S. Tufik

Department of Psychobiology, Universidade Federal de São Paulo, Rua Napoleão de Barros, 925, Vila Clementino, São Paulo, 04024-002, Brazil. E-mail: sergio.tufik@unifesp.br

R. A. Velluti

Centro de Medicina del Sueño, Facultad de Medicina CLAEH, Punta del Este, Uruguay. E-mail: ricardo.velluti@gmail.com

C. N. Warren

Department of Psychology, West Virginia University, Morgantown, WV, USA. E-mail: cnwarren@mix.wvu.edu

A. Zager

Neuroimmunomodulation Research Group, Department of Pathology, School of Veterinary Medicine, University of Sao Paulo (USP), Sao Paulo, Brazil. E-mail: adrianozager@gmail.com

LIST OF ABBREVIATIONS

AANAT	Arylalkylamine N-acetyl-transferase	APC	Antigen presenting cells
AAP	American Academy of Pediatrics	APPLES	Apnea positive pressure long-term efficacy study
AASM	American Academy of Sleep Medicine	AR	Allergic rhinitis
AB	Awake bruxism	ArI	Arousal index
ABG	Arterial blood gases	ASP	Advance sleep phase
AC	Alternate current	ASPD	Advance sleep phase disorder
ACTH	Adrenocorticotropic hormone	ASPS	Advanced sleep phase syndrome
AD	Alzheimer's disease	ASV	Adaptive servo-ventilation
ADA	Americans with disabilities	ASWPD	Advanced sleep-wake phase disorder
ADCADN	Autosomal dominant cerebellar ataxia, deafness, and narcolepsy	AVAPS	Average volume-assured pressure-support
ADHD	Attention deficit hyperactivity disorder	Aβ	Amyloid-beta
		BAC	Blood alcohol concentration
ADNOD	Autosomal dominant narcolepsy, obesity, and type 2 diabetes	BAEP	Brainstem auditory evoked potentials
AEs	Adverse experiences	BMI	Body mass index
AHI	Apnea–hypopnea index	BPAP	Bi-level positive airway pressure
AI	Apnea index	BSMI	Benign sleep myoclonus of infancy
ALMA	Alternating leg muscle activation	BZD	Benzodiazepines
AMP	Adenosine monophosphate-activated protein	BZRAs	Benzodiazepine receptor agonists
AMPK	Adenosine monophosphate-activated protein kinase	C4-M1	Central-mastoid1
		CAs	Confusional arousals
AN	Autonomic nervous	CBT	Cognitive behavioral therapy
APA	American Psychiatric Association	CBT-I	Cognitive behavioral therapy for insomnia
APAP	Auto-titrating positive airway pressure	CCGs	Clock-controlled genes
		CD4+	Cluster of differentiation 4

CDC	Centers for Disease Control and Prevention	DHEA	Dehydroepiandrosterone
CH	Chloral hydrate	DISE	Drug-induced sleep endoscopy
CKD	Chronic kidney disease	DLB	Dementia with Lewy bodies
CKId	Casein kinase 1 delta		
CKIe	Casein kinase I epsilon	DLMO	Dim light melatonin onset
CNS	Central nervous system	DNA	Deoxyribonucleic acid
COPD	Chronic obstructive pulmonary disease	DORA	Dual orexin (hypocretin) receptor antagonist
CPAP	Continuous positive airway pressure	DRN	Dorsal raphe nucleus
		DSM-5	Diagnostic and statistical manual of mental disorders, 5th edition
CPS	Cycles per second		
CPSC	Consumer product safety commission		
		DSPS	Delayed sleep phase syndrome
CRH	Corticotropin-releasing hormone		
		DSWPD	Delayed sleep-wake phase disorder
CRSD	Circadian rhythm sleep disorder		
		DZ	Dizygotic
CRSWD	Circadian rhythm sleep-wake disorders	ECG	Electrocardiography
		ECG	Electrocardiogram
Cry1	Cryptochrome 1	EDS	Excessive daytime sleepiness
CSA	Central sleep apnea		
CSAHS	Central sleep apnea-hypopnea syndrome	EEG	Electroencephalogram
		EEG	Electroencephalography
CSAS	Central sleep apnea syndromes	EFM	Excessive fragmentary myoclonus
CSB	Cheyne-Stokes breathing	EFNS	European Federation of Neurological Societies
CSBS	Cheyne-Stokes breathing syndrome		
		EHS	Exploding head syndrome
CSF	Cerebrospinal fluid	EMA	European Medicines Agency
CSF	Colony-stimulating factor		
CSNK-I	Casein kinase-I	EMG	Electromyography
CSR	Cheyne-Stokes respiration	EMG	Electromyogram
CT	Computed tomography	EOG	Electrooculogram
DA	Disorders of arousal	EOG	Electrooculography
DA	Dopamine	EPAP	Expiratory positive airway pressure
DAT	Dopamine transporter		
DC	Direct current	ESS	Epworth sleepiness scale
DEA	Drug Enforcement Agency	E_TCO_2	End-tidal CO2
DEB	Dream-enactment behavior	F4-M1	Frontal4-Mastoid1

FDA	Food and Drug Administration	HPA	Hypothalamus-pituitary-adrenal
FFT	Fast Fourier transformation	hPer2	Human Period2
fMRI	Functional MRI	HRQL	Health-related quality of life
FP	Follicular phase		
FRC	Functional residual capacity	HRT	Hormone replacement therapy
FSH	Follicle-stimulating hormone		
		HTL	Hypothalamus
GABA	Gamma-amino butyric acid	Hz	Hertz
GAD	Generalized anxiety disorder	ICAM	Intercellular adhesion molecule
G-CSF	Granulocyte colony-stimulating factor	ICD	International classification of diseases
GDM	Gestational diabetes mellitus	ICSD	International classification of sleep disorders
GERD	Gastroesophageal reflux disorder	ICSD-3	International classification of sleep disorders, 3rd edition
GH	Growth hormone		
GHB	Gamma-hydroxybutyrate	ICV	Intracerebroventricular
GHRH	GH-releasing hormone	IDO	Indoleamine 2, 3-dioxygenase
GHT	Geniculo-thalamic tract		
GI	Gastrointestinal	IFN	Interferon
GINA	Global initiative for asthma	IH	Idiopathic hypersomnia
GPCR	G protein-coupled receptors	IL	Interleukin
		IL-1β	Interleukin-1 beta
gRLS	Gestational restless leg syndrome	IPAP	Inspiratory positive airway pressure
H1N1	Hemagglutinin Type 1 and Neuraminidase Type 1	IRT	Imagery rehearsal therapy
		ISR	Intensive sleep retraining
HAV	Hepatitis A vaccination	ISWR	Irregular sleep-wake rhythm
Hcrt	Hypocretin		
HDL	High-density lipoprotein	ISWRD	Irregular sleep-wake rhythm disorder
HEENT	Head, eyes, ears, nose, and throat		
		IUGR	Intrauterine growth retardation
HF	Heart failure		
HFF	High frequency filter	KLS	Kleine-Levin syndrome
HFT	Hypnagogic foot tremor	LAEP	Late auditory evoked potentials
HIOMT	Hydroxyindole-O-methyltransferase		
		LAUP	Laser-assisted uvulopalatoplasty
HIV	Human immunodeficiency virus		
		LBW	Low birth weight
HLA	Human leukocyte antigen		

LC	Locus ceruleus	MWT	Maintenance of wakeful-
LC-AN	Locus ceruleus autonomic		ness test
	nervous	MZ	Monozygotic
LD	Light-dark	NA	Nucleus of the amygdale
LDL	Low-density lipoprotein	NADP	Nicotinamide adenine
LDT-PPT	Laterodorsal and pedun-		dinucleotide phosphate
	culopontine tegmental	NADPH	Nicotinamide adenine
	nucleus		dinucleotide phosphate
LFF	Low frequency filter	NASD	Non-apnea sleep disorders
LGN	Lateral geniculate nucleus	NES	Night eating syndrome
LH	Luteinizing hormone	NIH	National Institutes of
LM	Leg movement		Health
LP	Luteal phase	NK	Natural killer
LPS	Lipopolysaccharide	NOS	Not otherwise specified
LSAT	Lowest oxygen desaturation	NREM	Non rapid eye movement
	indices	NSF	National Sleep Foundation
μV	Microvolt	NTSB	National Transportation
MADs	Mandibular advancement		Safety Board
	devices	O2-M1	Occipital-mastoid1
MAO-B	Monoamine oxidase type B	OA	Oral appliances
MAOIs	Monoamine oxidase inhibi-	OCD	Obsessive-compulsive
	tors		disorder
MBSR	Mindfulness-based stress	OCST	Out-of-center sleep testing
	reduction	ODI	Oxygen desaturation index
MCH	Melanin-concentrating	OHS	Obesity hypoventilation
	hormone		syndrome
MCI	Mild cognitive impairment	OR	Odds ratio
MCP-1	Monocyte chemoattractant	OSA	Obstructive sleep apnea
	protein-1	OSAHS	Obstructive sleep apnea-
MEG	Magnetoencephalography		hypopnea syndrome
MES-	Mixed salts/mixed	OTC	Over-the-counter
amphetamine	enantiomers amphetamine	PaCO$_2$	Pressure of carbon dioxide
MHC	Major histocompatibility	PAP	Positive airway pressure
MMA	Maxillomandibular ad-	PAS	p-aminosalicylic acid
	vancement	PCOS	Polycystic ovarian syn-
MRI	Magnetic resonance imag-		drome
	ing	PD	Parkinson's disease
MS	Multiple sclerosis	PD	Panic disorder
MSA	Multiple-system atrophy	PDR	Posterior dominant rhythm
MSLT	Multiple sleep latency test	PDSS	Parkinson's disease sleep
MT$_1$	Melatonin receptor1		scale

PET	Positron emission tomography	REMS	Risk evaluation and mitigation strategy
PFT	Pulmonary function tests	REMw/oA	REM sleep without atonia
PGO	Ponto-geniculo-occipital	RERAs	Respiratory effort-related arousals
PHOX2B	Paired like homeobox 2b		
PLM	Periodic leg movement	RF	Reticular formation
PLMD	Periodic limb movement disorder	RFA	Radiofrequency ablation
		RHT	Retinohypothalamic tract
PLMS	Periodic limb movements of sleep	RIP	Respiratory inductance plethysmography
PMDD	Premenstrual dysphoric disorder	RISP	Recurrent isolated sleep paralysis
PMR	Progressive muscle relaxation	RLS	Restless legs syndrome
		RMD	Sleep-related rhythmic disorder
POA	Preoptic area		
POMC	Pro-opiomelanocortin	RORA	Retinoic acid receptor related orphan receptor-A
PPD	Post partum depression		
PPN	Pedunculopontine nucleus	RORα	Retinoic acid receptor related orphan receptor- alpha
PS	Paradoxical sleep		
PSG	Polysomnogram		
PSG	Polysomnography	RR	Risk ratio
PSM	Propriospinal myoclonus at sleep onset	RRE	Rev response element
		RSWA	REM sleep without atonia
PSP	Progressive supranuclear palsy	RWA	REM sleep without atonia
		SA	Sleep attacks
PSQI	Pittsburgh sleep quality index	SAD	Social anxiety disorder
		SB	Sleep bruxism
PTC	Pressor trigger of cataplexy	SCN	Suprachiasmatic nucleus
PTSD	Post-traumatic stress disorder	SCT	Stimulus control therapy
		SD	Sleep disturbances
PTT	Pulse transit time	SD	Standard deviation
PVN	Paraventricular nucleus	SD	Sleep disordered breathing
QOL	Quality of life	S	Sleep efficiency
RAAS	Reticular ascending activating system	SEM	Slow eye movements
		SGA	Small for gestational age
RBD	REM sleep behavior disorder	SHVS	Sleep hypoventilation syndrome
RDI	Respiratory disturbance index	SIDS	Sudden infant death syndrome
REM	Rapid eye movement	SL	Sleep onset latency
REMOL	REM onset latency	SLD	Sub lateral dorsal nucleus

SN	Substantia nigra pars compacta	TBI	Traumatic brain injury
SNRIs	Selective noradrenergic reuptake inhibitors	TCAs	Tricyclic antidepressants
		Th cells	Thelper cells
SOREM	Sleep onset REM	TMN	Tuberomammillary nucleus
SOREMP	Sleep onset REM period	TNF	Tumor necrosis factor
SRED	Sleep-related eating disorder	TRD	Tongue retaining devices
		TSH	Thyroid-stimulating hormone
SRMDs	Sleep-related movement disorders	TST	Total sleep time
SRT	Sleep restriction therapy	UARS	Upper airway resistance syndrome
SSRIs	Selective serotonin reuptake inhibitors	UPPP	Uvulopalatopharyngoplasty
STs	Sleep terrors	VCAM	Vascular cell adhesion molecule
SUID	Sudden unexpected infant death syndrome	VEGF	Vascular endothelial growth factor
SW	Sleepwalking	VMS	Vasomotor symptoms
SWD	Shift work disorder	VTA	Ventral tegmental area
SWS	Slow wave sleep	W	Wakefulness
SXB	Sodium Oxybate	WASO	Wake after sleep onset
T cell	Thymocytes cell	WED	Willis-Ekbom disease

PREFACE

If anyone were to ask, "why did you decide to edit this volume?", one would immediately think of two answers: First, that they genuinely believed there is a need for a volume of this sort, and, second, that, however pretentious it might sound, they believed that, because of their years of teaching and research experience, he/she is the right person to edit it.

However, I have a third answer. I have, since the beginning of my scientific career, despite my background in botany, been involved in the sleep field. Having edited over 20 volumes along with leading experts in the field of sleep and biological rhythms, I believe that I now have the requisite experience to edit an introductory sleep medicine volume on my own. This first edition of this volume is aimed at residents, fellows, house officers, and physicians of various specialties as well as clinical sleep researchers. The volume will give a basic grounding in sleep medicine to those who are established in related specialties as well as to younger professionals who are considering a future career in sleep medicine. This volume attempts to convey something of the fascinating complexity of the field as well as to separate figure from ground for those who are newcomers to the field and who are seeking guideposts for further research. Sleep medicine encompasses an unusually board spectrum of contributions from biology, technology, and medicine. This volume seeks to summarize the considerable mass of knowledge that has now accumulated in the field and to impart its major findings in a manner that is both comprehensive but not overwhelming.

Inasmuch as sleep problems are frequently co-morbid with other medical conditions, the overt presenting symptoms of many patients may driven by a number of other factors. Disruptions to circadian organization may have a multiple effects of which sleep difficulties are simply the most visible. It is thus in the interest of clinicians to be alert to the ways in which sleep problems interconnect with other pathologies. It is often the case for instance that insomnia is not just insomnia, which is either a symptom or possibly a driver of correlated pathologies. It is thus in the interest of clinicians to be alert to this interconnectedness, and to recognize which difficulties are primary and which are not.

The literature on sleep and sleep medicine is enormous, and expanding rapidly. The objective of the editor has been to make this volume a useful tool for graduate students and newcomers who realistically do not always have time to check original publications. The authors have endeavored to give appropriate references to some of the more recent literature, and at the same time to quote the origins of some of the statements made.

There are often constraints to editing a volume, especially the first edition. For example, it is not always possible to address all the topics that would be desirable for an introductory summary to cover. Additionally it is not always feasible to acquire the best experts in a special area. Nevertheless, for those who are interested in learning more about a specialized area of sleep medicine, the reference sections will represent a rich resource for this purpose. As with all major efforts of this kind we regard this introductory volume and those which will follow as "works in progress," and we anticipate that the content of future editions will evolve to respond to changes in the field as well as to the informational needs of our readers.

We have made every effort to ensure that the dosage recommendations are accurate and in agreement with the standards and collective opinion accepted at the time of publication. The formulations and usage described do not necessarily have specific approval by the regulatory authorities of all countries. Since dosage regimens may be modified as new clinical research accumulates, readers are strongly advised to make note of the most recently recommended prescribing guidelines in their respective countries. Every effort leading up to the creation of this volume has been to make it into a practical and useful introduction to the sleep medicine field. However, as editor I remain responsible for any errors or mistakes which have occurred. This first edition will, I hope, stimulate in you as much excitement and satisfaction as it has in us. I sincerely hope that this volume will serve as a comprehensive guide for diagnostic problems in sleep medicine and it will find its way into the places where the battle against sleep dysfunctions is waged daily in clinics and hospitals around the world.

S. R. Pandi-Perumal
Toronto, Canada

CREDITS AND ACKNOWLEDGEMENTS

I would like to thank some of the many people who have been instrumental in assisting me with the first edition of *Synopsis of Sleep Medicine*. First, I owe an enormous debt to Professor Colin M. Shapiro, Department of Psychiatry, University of Toronto, for getting me started in the field of sleep medicine. Early in my career he transformed me from being an agrostologist (the branch of botany concerned with grasses) to a sleep scientist. I have the deepest gratitude for his mentorship and friendship, and for the great privilege of receiving his contribution to this volume. This meant a lot to me. He is known for the inspiration that he has provided to all who have known him, including me.

At this juncture, I also would like to thank the mentorship of Prof. Martin R. Ralph, Prof. Mircea Steriade, Prof. Michael H. Chase, Prof. Rosalind Cartwright whose invitations to work with them to learn the field of sleep and biological rhythms and were the opportunities of a lifetime and helped to shape my career goals. Additionally, over the decades, I have collaborated with numerous individuals. It will be impossible to list every single one of them. A special mention goes to Prof. Jaime M. Monti, Prof. Daniel P. Cardinali, Prof. Gregory M. Brown, Prof. Ahmed S. BaHammam, Prof. Meera Narasimhan, Prof. Adam Moscovitch, and Mr. Warren Spence. If I am known in this field, they are the reason. All these individuals nurtured and mentored me in some way or other and helped me to become for what I am today.

I would like to thank Prof. Monica Levy Andersen, Department of Psychobiology, Universidade Federal de São Paulo, São Paulo, Brazil, who enthusiastically helped me in identifying some of the authors.

A special thanks to all those who invested time and effort in the compilation of the material that became this book. The many authors who contributed their expertise and perspectives are clearly the backbone of this project and they deserve the lion's share of the commendation. Over 55 biomedical professionals from industry and academia contributed to this work. Most certainly such a distinguished group of authors provided the needed balance and perspective.

I would like to particularly thank Mr. Ashish Kumar, President and Publisher of the Apple Academic Press, Canada, for the solicitation and encouragement in the development of this first edition. He has supported and guided this project from the beginning. I wish to acknowledge the professionalism of the editorial and production staff at the Apple Academic Press, who took on this new project and completed it with remarkable speed and flexibility.

On the top of everything else, I wish to acknowledge my family who provided the encouragement and support that make it possible. They are the reason for my accomplishments. All my books are, in the end, for them.

NEUROANATOMY AND NEUROPHARMACOLOGY OF SLEEP AND WAKEFULNESS

Pablo Torterolo, Jaime M. Monti, and Seithikurippu R. Pandi-Perumal

ABSTRACT

Since the discovery of the ascending reticular activating system more than sixty years ago, the anatomy, electrophysiology, and neurochemistry of the neuronal networks involved in generating and maintaining wakefulness, that is, the activating systems have been characterized in detail. Furthermore, the neural areas critically involved in the generation and maintenance of rapid eye movement (REM) and non-REM (NREM) sleeps, which are called the hypnogenic systems, have also been delineated. The activating and hypnogenic systems deeply interact in order to induce the sleep/wakefulness cycle. These systems are modulated by the suprachiasmatic nucleus (SCN), the circadian rhythm pacemaker, as well as by various somnogenic substances such as adenosine and melatonin.

This chapter is a brief review on the neuroanatomy and functions of the activating and hypnogenic systems. The knowledge of neurobiological basis of these systems is crucial to understand the physiology of wakefulness and sleep, as well as to explain the pathophysiology of conditions such as insomnia, sleepiness, or abnormal behaviors during sleep (parasomnias). Additionally, the chapter highlights the concepts that can be easily applied to understand the neuropharmacology of sleep pathologies.

Pablo Torterolo
Department of Physiology, School of Medicine, Universidad de la República, Montevideo, Uruguay

Jaime M. Monti
Department of Pharmacology and Therapeutics, School of Medicine, Clinics Hospital, Universidad de la República, Montevideo, Uruguay

Seithikurippu Pandi-Perumal
Somnogen Canada Inc., College Street, Toronto, ON, Canada

Corresponding author: Dr. Pablo Torterolo, Departmento de Fisiología, Facultad de Medicina, Universidad de la República. General Flores 2125, 11800 Montevideo, Uruguay. Email: ptortero@fmed.edu.uy. Fax: 598 29243414-3338

1.1. INTRODUCTION

In humans, most mammals, and birds, three behavioral states can be readily distinguished: wakefulness (W), non-Rapid eye movement (NREM) sleep (also called slow wave sleep (SWS)), and rapid eye movement (REM) sleep. Polysomnography (PSG) is the basic tool used to differentiate these states. It consists of the simultaneous recording of various physiological parameters such as the electroencephalogram (EEG), the electromyogram (EMG), and eye movements; other bioelectrical signals can also be recorded in humans or experimental animals (Figure 1.1). The main features of the human PSG are summarized in Table 1.1.

During W, an optimal interaction with the environment allows the development of various behaviors necessary for survival. In humans, W is accompanied by awareness (consciousness) of the environment and internally generated stimuli such as hunger and thirst. The EEG recording of W is marked by the presence of high frequency and low amplitude (cortical activation, Figure 1.1) determined by the activity of thalamic and cortical neurons.

During sleep, there is a marked decrease in the interaction with the environment, an increase of the threshold for the reaction to external stimuli, and a decrease in somatomotor activity. Furthermore, animals adopt a distinctive position to conserve heat.

Presently, three NREM sleep phases (stages N1, N2, and N3 or SWS) are distinguished in humans according to the depth of the state. From stage W, normal adults enter in light NREM sleep (or stage N1). Stage N2 is characterized by the presence of sleep spindles and K-complexes, while the presence of low frequency (0.5–4 Hz) of high-amplitude delta waves characterizes the EEG during N3. Furthermore, tonic parasympathetic activity increases, determining characteristic changes in visceral activity. In the deeper stages of NREM sleep, cognitive activity (that is, dreams) is minimal (Dement and Kleitman, 1957; Pace-Schott, 2005).

REM sleep (or stage R) occurs periodically, and is always preceded by NREM sleep. REM sleep is a deep sleep stage although the EEG is similar to that of stage W; hence, it is also called "paradoxical" sleep. REM sleep is characterized by fast REMs that typically occupy 20–25% of total sleep in human adults. REM sleep occurs ~90 min after sleep onset. There are both *phasic* (episodic) and *tonic* (persistent) components in the stage R. Dreams occur mainly during REM sleep, which is also accompanied by muscle atonia as evidenced in the EMG channel (Figure 1.1), and phasic changes in autonomic activity. A shortened REM onset latency (REMOL; it is the interval between the sleep onset and the appearance of the first REM sleep episode) is a biological marker of primary depression. It is also considered to be a clinically significant pathological feature in other brain diseases.

In rats, a species commonly used in preclinical studies, W and sleep are defined by PSG as follows (Figure 1.1):

1) W, by the presence of low-voltage fast waves in frontal cortex, a mixed theta activity in occipital cortex, and relatively high EMG activity;

2) Light sleep, by the occurrence of high-voltage slow cortical waves

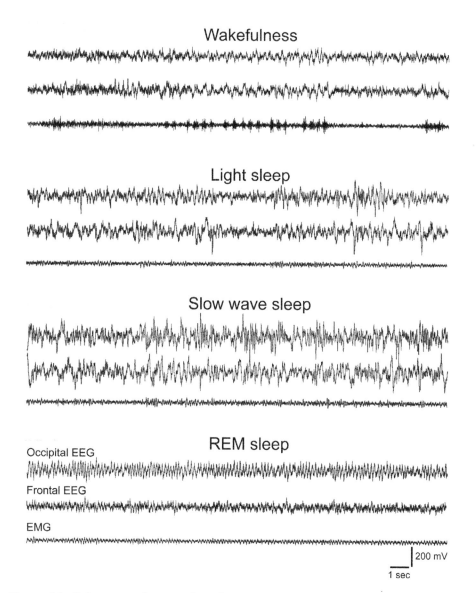

Figure 1.1: Polysomnographic recording during sleep and wakefulness in the rat. EEG, electroencephalogram; EMG, electromyogram.

Table 1.1: Electroencephalographic correlates of sleep stages.

| Sleep stages | TST (%) | Characteristics | | | |
		EEG	EOG	EMG	Other variables
Stage awake (relaxed wakefulness)		Alpha activity (8–12 Hz) or low-amplitude beta (13 35 Hz), mixed-frequency waves	REM (in sync or out of sync deflections), eye blinks	Relatively high tonic EMG activity	Alpha activity in occipital leads compared with central leads, eye opening suppresses alpha activity, movement artifacts
N1, formerly known as stage 1	2–5	Low-voltage, mixed-frequency waves (2–7 Hz range), mainly irregular theta activity, triangular vertex waves	SEMs, waxing and waning of alpha rhythm	Tonic EMG levels typically below range of relaxed wakefulness	Alpha ≤50%, vertex sharp waves in central leads, absence of spindles and K complexes
N2, formerly known as stage 2	45–55	Relatively low-voltage, mixed-frequency waves, some low-amplitude theta and delta activity	No eye movement	Low chin muscle activity	Sleep spindles (7–14 Hz) and K complexes occur intermittently
N3, formerly known as stages 3 and 4	5–20	≥20–50% of epoch consists of delta (0.5–2 Hz) activity	No eye movement	Chin muscle activity is lower than N1 and N2	Sleep spindles may be present
Stage REM	20–25	EEG is relatively low voltage with mixed frequency resembling N1 sleep	REM. Episodic rapid, jerky, and usually lateral eye movements in clusters	EMG tracing almost always reaches its lowest levels owing to muscle atonia	Phasic and tonic components, presence of sawtooth waves, alpha waves are 1–2 Hz slower than waves occurring during wakefulness and non-REM sleep

EEG, electroencephalography; EMG, electromyography; EOG, electrooculography; REM, rapid eye movement; SEMs, slow eye movements; TST, total sleep time.

interrupted by low voltage fast EEG activity;

3) SWS, by the occurrence of continuous high-amplitude slow frontal and occipital waves; light sleep + SWS is called NREM sleep;

4) REM sleep, by the presence of low-voltage fast frontal waves, a regular theta rhythm in the occipital cortex, and a silent EMG except for occasional myoclonic twitching.

Sleep in humans during the night shows four to five NREM-REM sleep cycles (the period from the sleep onset to the end of first REM episode or the period from the end of a REM sleep episode to the subsequent REM sleep episode). The average length of human sleep cycles is about 90–120 min. In contrast, the average sleep cycle duration of the rat is about 10 min (Trachsel et al., 1991).

1.2. COGNITIVE ACTIVITY THROUGH ACTIVATION OF THE THALAMUS AND THE CORTEX

Cognitive activities (consciousness and dreams) and the different EEG rhythms that support these functions are generated by the activity of cortical and thalamic neurons, which are mutually interconnected. Thalamic neurons have a complex electrophysiology that allows them to operate differently according to their level of polarization (Steriade et al., 1993). When hyperpolarized, the thalamic neurons that project to the cortex (thalamocortical neurons) oscillate at low frequency (0.5–4 Hz, delta rhythm), and tend to block the information toward the cortex that goes through the sensory pathways. This "oscillatory mode" of function synchronizes the

cortical neurons and accompanied by other phenomena, generates the slow waves of NREM sleep. On the contrary, when these neurons are relatively depolarized, they enter in the "tonic mode" of function. In this condition, the thalamocortical neurons transmit sensory information toward the cortex in a reliable way. This mode of function occurs during W and REM sleep.

Therefore, the thalamus is critical for the generation of slow waves and spindles that characterize NREM sleep. When the thalamus is lesioned as it occurs in the "fatal familial insomnia," the generation of these electrographic signs is blocked and sleep is prevented (Montagna, 2005).

Neurons that form part of the activating system, that is, the neuronal system that generates and maintains wakefulness are summarized in Figure 1.2. The activating neurons project directly to the thalamus and/or the cortex (Jones, 2005). They depolarize thalamic neurons in order to produce the thalamic tonic mode and desynchronization (activation) of the EEG that accompanies the behavioral awakening. Part of the activating system (the cholinergic nuclei) is also active during REM sleep and activates the corticothalamic system during this behavioral state.

1.3. THE ACTIVATING SYSTEM

Which are the neural mechanisms involved in the generation and maintenance of the behavioral states?

In the 1930s, before REM sleep discovery, Bremer proposed that the baseline state of the brain was sleep (Bremer, 1935). His proposal was based on experimental transections at the level of the intercollicular region of the midbrain, in a prepa-

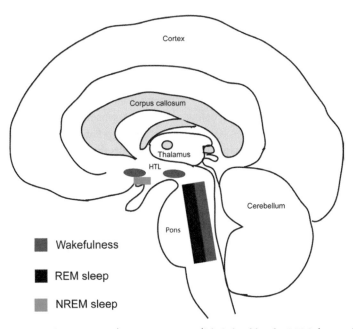

Figure 1.2: Activating (labeled in red) and hypnogenic (labeled in blue for REM sleep and green for NREM sleep) systems are shown. HTL, hypothalamus.

ration known as "cerveau isolé" ("isolated forebrain"). Animals with this injury had an EEG similar to that observed during NREM sleep. Since these animals had lesions in the ascending sensory pathways at the level of the midbrain, at that time it was considered that sleep occurred because sensory inputs to the diencephalon and telencephalon were reduced. In other words, sensory activity was viewed as necessary to maintain W. It was considered that the sensory blockade accompanied by ill-defined neuronal "fatigue," was the cause of sleep. This concept was known as the passive hypothesis of sleep (or deafferentation).

In 1949, Moruzzi and Magoun published their seminal work entitled: *Brainstem reticular formation and activation of the EEG* (Moruzzi and Magoun, 1949).

This work is considered one of the most influential contributions in the field, and it inspired numerous investigations in the following decades. The reticular formation (RF) is located in the central area of the brainstem. Neurons in this region are not grouped in nuclei but are arranged in a complex mesh or network. This region is characterized by its high connectivity, receiving afferents from different sources and sending efferents to different sectors of the central nervous system (CNS) (Jones, 1995). Moruzzi and Magoun (1949) performed electrical stimulation at different rostrocaudal levels of the RF. Their main finding was that electrical stimulation of the RF activates the EEG during anesthesia, that is, from an EEG with high-amplitude and low-frequency waves (SWS-like), the stimuli induced a high frequency, low amplitude EEG as in W. The activation of the EEG was widespread (throughout the neocortex) and in lightly anesthetized animals, it was accompanied by behavioral awakening.

The role of the midbrain and pontine RF in generating and maintaining W was confirmed later making use of different approaches. This area along with its ascending projection was called "reticular ascending activating system" (RAAS). Of note, lesions of this area in patients and experimental animals generate a coma condition (Lindsley et al., 1950; Plum and Posner, 2000).

At the time of its discovery, the RAAS did not contradict the "passive" hypothesis of sleep, but complemented it. It was believed that sleep is initiated by the progressive inactivation of the RAAS, where the decrease in sensory input played a major role. This hypothesis was named "passive inactivation of the RF" or "reticular sleep hypothesis."

The activation of the RAAS promotes the thalamocortical activation (as evidenced by EEG desynchronization) that supports cognitive awakening. The arousal reaction is accompanied by motor, autonomic, and endocrine changes. Then, the RF would also modify directly or indirectly, the activity of motoneurons, autonomic preganglionic neurons, and hypothalamic endocrine-related neurons. In this review, we will emphasize aspects related to the cognitive awakening and EEG activation. Therefore, only the upward or ascending (thalamocortical) component of the RAAS will be considered.

The identification of specific neuronal groups that use different neurotrans-mitters was the beginning of a new era for understanding the RF and the RAAS. Furthermore, different experimental approaches allowed obtaining details of the physiology of the activating system. Nowadays, the following concepts have been established: (i) The RAAS is composed of various neuronal groups that differ in their neurotransmitters; (ii) the neurons from specific regions of the posterior and lateral (perifornical) hypothalamus and basal forebrain, that are considered the rostral extension of the RF of the brainstem, behave as activating nuclei. Thus, the RAAS, the posterior and posterolateral hypothalamus, and the basal forebrain constitute the activating system; (iii) the neuronal groups that make up the activating systems project through a dorsal pathway toward the specific and non-specific thalamic nuclei, and/or a ventral pathway passing through the lateral hypothalamus, and basal forebrain toward the cerebral cortex.

The different components of the activating system are discussed below.

1.3.1. Mesopontine Glutamatergic Neurons

Anatomical and functional studies have shown that the main component of the RAAS is located within the mesopontine RF. With respect to the glutamatergic neurons, they do not form a specific group but are distributed throughout the mesopontine RF intermingled with specific neuronal groups. Regarding their functional activity during the sleep–wakefulness cycle, mesopontine W/REM-on, REM-on, or W-on glutamatergic neurons have been identified (Boucetta et al., 2014).

It has been contended that ketamine, a *N-methyl-D-aspartate* (NMDA) glutamatergic antagonist, inhibits W and produces sedation, hypnosis, or pharmacological coma; part of these effects could be produced by reducing the synaptic effects of mesopontine glutamatergic neurons (Wolff and Winstock, 2006).

1.3.2. Noradrenergic Neurons of the Locus Coeruleus

The locus coeruleus (LC) is a noradrenergic nucleus located in the mesopontine dorsolateral region. The ascending projections of this nucleus are part of the dorsal pathway to the thalamus, and are also included in the ventral pathway that project directly to the cerebral cortex. LC noradrenergic neurons show their maximum firing rate during W; it decreases during NREM sleep and is minimal during REM sleep (Aston-Jones and Bloom, 1981). This profile of activity is in agreement with the pattern of release of noradrenaline in the cerebral cortex as measured by microdialysis (Berridge and Abercrombie, 1999). It should be mentioned that during W, these neurons markedly increase their firing rate following a new stimulus, but the response is reduced after habituation, which led to the proposal that this neuronal group regulates attention (Foote et al., 1991). Interestingly, a1 antagonists including prazosin facilitate the generation of sleep, while a2 agonists such as dexmedetomidine inhibit the activity of LC neurons and are used as sedatives (Nishino and Mignot, 1997; Nelson et al., 2003).

1.3.3. Midbrain Dopaminergic Neurons

The substantia nigra pars compacta (SN) and the ventral tegmental area (VTA) are located in the midbrain. Both regions are characterized by the presence of dopaminergic neurons; while dopaminergic neurons of the SN project to the dorsal striatum, VTA neurons project to the prefrontal cortex and nucleus accumbens (ventral striatum) (Oades and Halliday, 1987). The firing rate of dopaminergic neurons and extracellular concentration of dopamine in the prefrontal cortex increases during reward-related stimuli (Mirenowicz and Schultz, 1996; Feenstra, 2000). Dopamine agonists and antagonists increase and decrease W, respectively (Monti and Monti, 2007; Monti and Jantos, 2008). Presently available evidence tends to indicate that these dopaminergic neurons are involved in the arousal that accompanies reward and motivation. Cocaine and amphetamines inhibit dopamine reuptake and induce its release, respectively. As expected, their administration produces an increase of W. In addition, these drugs are strong positive reinforcers, which is because of their addictive power. Notwithstanding this, drugs that increase synaptic dopamine levels are in the first-line for the treatment of hypersomnia (Nishino and Mignot, 1997).

1.3.4. Serotonergic Neurons of the Rostral Raphe Nuclei

Serotonergic neurons of the dorsal (Figure 1.3A) and median raphe nuclei are located within the mesopontine midline (Jacobs and Azmitia, 1992; Monti, 2010b, a). These neurons project toward the thalamus and cortex. Serotonergic neurons discharge more frequently during W, decrease their activity during NREM sleep, and virtually turn off during REM sleep (McGinty and Harper, 1976). A similar pattern of

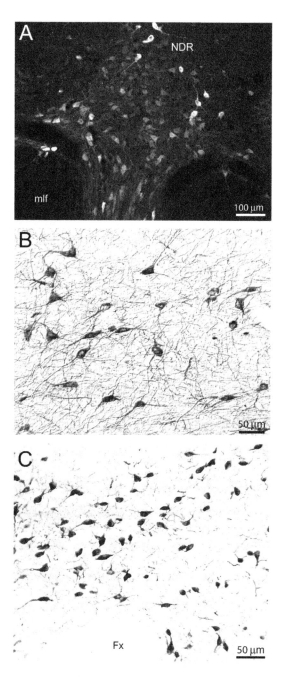

Figure 1.3. (A) Photomicrographs illustrating serotonergic neurons of the dorsal raphe nucleus of the rat; (B) cholinergic neurons of basal forebrain of the cat, and; (C) hypocretinergic neurons of the guinea pig. These neurons were revealed following immunohistochemistry procedures.

activity has been observed with respect to the release of serotonin as measured by microdialysis (Portas et al., 2000). Subgroups of these serotonergic neurons are activated during stereotyped movements that take place when the experimental animal is moving or grooming (Jacobs and Fornal, 2008). In turn, electrical stimulation of the dorsal raphe nucleus produces a marked EEG activation (Dringenberg and Vanderwolf, 1997). It has been proposed that serotonergic neurons play a permissive role in the generation of REM sleep such that they must be inhibited for REM sleep to occur (McCarley, 2007). Local GABAergic neurons would be involved in this inhibition (Torterolo et al., 2000). Since there are several types of receptors for serotonin, the effect of serotonergic drugs on sleep is complex (Monti and Jantos, 2008).

1.3.5. Cholinergic Neurons of the LDT-PPT

Mesopontine cholinergic neurons are located in the laterodorsal and pedunculopontine tegmental nucleus (LDT-PPT). These neurons project directly to the thalamus (Satoh and Fibiger, 1986). Cholinergic neurons are activated during W in close relation to the cortical activation. They are inhibited during NREM sleep and re-activated during REM sleep (Boucetta et al., 2014). In the thalamus, acetylcholine acts on muscarinic and nicotinic receptors in order to produce cortical activation (Curro Dossi et al., 1991). In humans, increasing synaptic levels of acetylcholine by acetylcholinesterase inhibitors produces W and cortical activation, while REM sleep precipitates if this drug is applied during NREM sleep (Gillin and Sitaram, 1984). These data suggest a bimodal role of cholinergic neurons, promoting both the generation of W and REM sleep. Of note, drugs that increase synaptic levels of acetylcholine, such as physostigmine, reverse the state of general anesthesia produced by *sevoflurane* in humans (Plourde et al., 2003).

1.3.6. Mesopontine GABAergic Neurons

GABAergic neurons, terminals, and receptors are distributed throughout the mesopontine region. In contrast to the effects of hypnotics that enhance GABAergic neurotransmission and facilitate sleep, the application of GABAergic receptor agonists into the NPO generates W (Xi et al., 1999). Furthermore, local increase of GABA levels in the NPO prolongs the time necessary to induce general anesthesia, while isoflurane anesthesia reduces GABA levels within the NPO (Vanini et al., 2008).

1.3.7. Histaminergic Neurons of the Posterior Hypothalamus

Neurons using histamine as a neurotransmitter are located only in the tuberomammillary nucleus of the posterior hypothalamus, and project to the thalamus and cortex (Monti, 2011). The firing rate of the histaminergic neurons decreases when passing from W to NREM sleep and is minimal during REM sleep (Takahashi et al., 2006).

The information provided by "knockout" mice lacking histidine decarboxylase (enzyme involved in the synthesis of histamine) is revealing; these animals are unable to stay awake when they are placed in a new environment (Parmentier et al., 2002). Drugs that increase synaptic levels of histamine augment cortical activation and W (Kalivas, 1982). In humans, drugs that antagonize the H1 receptor including pyrilamine and diphenhydramine and have been prescribed as anti-allergic, cause drowsiness as a side effect (Roth et al., 1987).

1.3.8. Hypocretinergic Neurons of the Posterior Hypothalamus

In 1998, two independent research groups identified hypocretins almost simultaneously by different techniques (de Lecea et al., 1998; Sakurai et al., 1998). Hypocretin 1 and 2 (also called orexin A and B) neuropeptides are synthesized by a small group of neurons located exclusively in the dorsal, posterior, and lateral hypothalamic region (de Lecea et al., 1998; Sakurai et al.,

1998) (Figure 1.3C). These neurons use the hypocretins as neurotransmitters and project diffusely throughout the CNS, including mesopontine areas critical for waking and sleep generation (Figure 1.4). Hypocretins act on two types of metabotropic receptors exerting presynaptic and postsynaptic excitatory effects.

The intracerebral or intraventricular administration of hypocretins facilitates the generation of W (Piper et al., 2000). In turn, several experimental approaches have shown that these neurons are primarily activated during motivated W; their activity is reduced during NREM sleep and is almost absent during tonic REM sleep; however, hypocretinergic neuronal activity seems to increase in the presence of the phasic components of REM sleep.

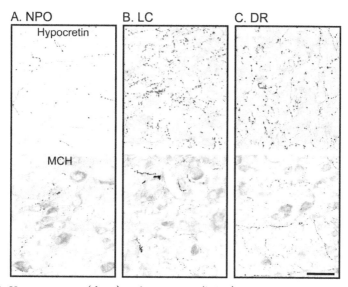

Figure 1.4. Hypocretinergic (above) and MCHergic (below) fibers in the nucleus pontis oralis (A. NPO), locus coeruleus (B, LC), and dorsal raphe nucleus (C, DR) of the cat. The sections were prepared for immunohistochemistry to detect hypocretin-1 and MCH. The sections treated with MCH antibodies (below) were also counterstained with pyronin-y. Calibration bar, 100 mm. This figure highlights the strong interconnection among the neurons that are critically involved in the generation of sleep and wakefulness.

(Torterolo et al., 2001b; Kiyashchenko et al., 2002; Torterolo et al., 2003; Lee et al., 2005b; Mileykovskiy et al., 2005; Torterolo et al., 2009c; Torterolo and Chase, 2014). The medical importance of this system boosted when Nishino et al., (2000) showed the absence of hypocretin 1 in the cerebrospinal fluid of narcoleptic patients; degeneration of these neurons is the pathological basis of narcolepsy-cataplexy.

1.3.9. Basal Forebrain Cholinergic Neurons

These cholinergic neurons are located in the area known as basal forebrain (anterior to the hypothalamus), which includes the nucleus basalis of Meynert (Figure 1.3B). The main projections of these neurons are to the neocortex, hippocampus, and reticular thalamic nucleus (Semba, 2000). Chemical and electrical stimulation of this region generates cortical activation and W, whereas its inactivation produces NREM sleep (Belardetti et al., 1977; Cape and Jones, 2000). During W and REMS, there is an increase in the firing rate of basal forebrain cholinergic neurons which is correlated with EEG activation and an increase in the release of the acetylcholine at cortical levels (Marrosu et al., 1995; Lee et al., 2005a). During W, these neurons regulate sensory information processing, attention, and learning. Of note, cognitive disorders characteristic of Alzheimer's disease are related to lesions of this neuronal group

(Coyle et al., 1983; Vitiello and Borson, 2001).

1.3.10. Role of Wake-Promoting Neurons in the Different Types of Wakefulness

There is an important anatomical and functional relationship between the activating neuronal groups (Figure 1.4), which tends to indicate that these neurons act in tandem to generate and maintain W. W is a heterogeneous process. Thus, it is not the same state of W when caused by nociceptive stimulation, by intense motor activity, or just during relaxed activity. There is evidence showing that the relative activity of the different components of the activating system varies with the type or level of W. For example, experimental studies using Fos protein as an index of neuronal activity have shown that the hypocretinergic neuronal activity increases during W with motor activity related to the motivation to explore a new environment, but not during quiet wakefulness or forced locomotion (Figure 1.5) (Torterolo et al., 2001b; Torterolo et al., 2003; Torterolo et al., 2009c). Moreover, serotonergic neurons are active during W related to stereotyped and automatic motor activity, while LC noradrenergic neurons would be critical in the increased surveillance that occurs following a new stimulus (Foote et al., 1991; Jacobs and Fornal, 2008).

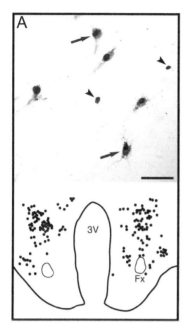
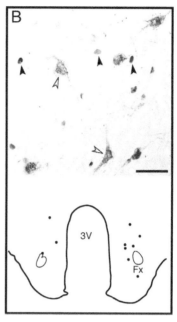

Figure 1.5. Photomicrographs illustrating hypocretin and Fos (a marker of neuronal activity) immunoreactive neurons from the posterolateral area of the hypothalamus. (A) Above, hypocretinergic neurons that express c-Fos (arrows) during active wakefulness with motor exploratory activity. Hypocretinergic neurons are stained in brown, Fos immunoreactivity, which is black, is restricted to nuclei. Hcrt- Fos+ neurons (arrowheads) are also intermingled with Hcrt+ Fos+ neurons. Below, camera lucida drawing of a representative hypothalamic sections of the same animal. The distribution of Hcrt+ Fos+ neurons is represented. Each mark indicates one labeled neuron (3V, third ventricle; Fx, fornix). (B) Above, group of hypocretinergic neurons during quiet wakefulness. The hypocretinergic neurons shown did not express c-Fos (unfilled arrowheads, i.e., not active), although Hcrt- Fos+ neurons are intermingled with these neurons (filled arrowheads). Below, the distribution of Hcrt+ Fos+ neurons in a representative section for the same animal is shown. This figure highlights the fact that hypocretinergic neurons are active during active wakefulness, but not during quiet (relaxed) wakefulness. Modified from Torterolo et al. (2001b).

1.4. HYPNOGENIC SYSTEMS

1.4.1. NREM Sleep: Preoptic Area

The neuronal groups critical in the generation and maintenance of NREM sleep are located in the preoptic area (POA) of the hypothalamus (Kumar, 2004; Szymusiak et al., 2007; Torterolo et al., 2009a; Benedetto et al., 2012) (Figure 1.2). These neurons, increase their firing rate during NREM sleep, and have been identified in the medial, median, and ventrolateral region of the POA. Electrical stimulation of the POA and adjacent basal forebrain induces NREM sleep and inhibits the activating system; in fact, GABAergic neurons of the POA project toward the activating system (McGinty and Szymusiak, 2005).

In turn, neurons from the activating system inhibit hypnogenic regions (Gallopin et al., 2000). This reciprocal inhibition between activating and hypnogenic neurons is critical for the transition between sleep and W.

1.4.2. REM Sleep Generation

The necessary and sufficient neuronal networks critical for the generation and maintenance of REM sleep are located in the mesopontine RF, where the RAAS is located (Figure 1.2) (Siegel, 2011). In fact, the LC noradrenergic neurons, and the dorsal raphe nucleus serotonergic neuron are active during W but turn off their firing during REM sleep (REM off neurons). Conversely, cholinergic neurons of the LDT-PPT increase their firing rate during REM sleep, thus contributing to the cortical activation of this state (McCarley, 2007; Boucetta et al., 2014). These cholinergic neurons also project to the NPO

that is the executive area for REM sleep generation. Then, acetylcholine is released within this area to induce REM sleep. This effect is mimicked by microinjection of carbachol, a mixed (nicotinic and muscarinic) cholinergic agonist, into the NPO of a cat. Carbachol generates a long duration (up to 2 h) REM sleep episode with a very short latency (down to 30 s) (Figure 1.6). Physiologically, it is considered that cholinergic neurons of the LDT-PPT activate glutamatergic neurons of the NPO. In turn, these neurons activate different groups of neurons that execute REM sleep functions such as atonia, REMs, EEG activation, autonomic unsteadiness, and so forth. For example, REM sleep atonia depends upon glutamatergic neurons of the NPO that project to the magnocellular medullary RF and depolarize premotor glycinergic neurons (Chase, 2013). These neurons produce the postsynaptic inhibition of motoneurons during REM sleep.

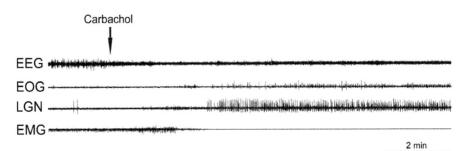

Figure 1.6. Polygraphic recording during the onset of an episode of REM sleep induced by a microinjection of carbachol into the nucleus pontis oralis of the cat. REM sleep is signaled by the appearance of ponto-geniculo-occipital (PGO) waves in the lateral geniculate nucleus (LGN), a decrease in muscle tone, REMs, and EEG desynchronization. This episode of REM sleep was maintained for approximately 2 h. An arrow signals the beginning of the microinjection of carbachol into the nucleus pontis oralis. EEG, electroencephalogram; EOG, electrooculogram; EMG, electromyogram; NGL, electrogram of the lateral geniculate nucleus (visual thalamus).

GABAergic neurons of the mesopontine area (dorsal raphe nucleus, LDT-PPT, ventrolateral periaqueductal gray, and so forth) also play an important role in the generation of REM sleep (Torterolo et al., 2000, 2001a; Torterolo et al., 2002a; Torterolo et al., 2002b; Vanini et al., 2007; Torterolo and Vanini, 2010). In fact, there are models which propose that REM sleep generation depends upon the activity of mesopontine GABAergic neurons (Lu et al., 2006; Luppi et al., 2007).

The hypothalamus also participates in the generation of REM sleep. As mentioned before, histaminergic neurons are REM-off. Moreover, hypocretinergic neurons decrease their firing rate during "tonic" REM sleep; however, Fos protein and microdialysis studies conducted in cats suggest that these neurons may be active during the "phasic" components of REM sleep, and may contribute to the induction of twitches, REMs, autonomic instability, and so forth (Torterolo and Chase, 2014). The role of the hypothalamic melanin-concentrating hormone (MCH) will be described in the next section.

1.4.3. Melanin-Concentrating Hormone: A Sleep Promoting Factor

There are neurons in the posterolateral hypothalamus (intermingled with the hypocretinergic neurons) and incertohypothalamic area that utilize the neuropeptide MCH as a neuromodulator (Torterolo et al., 2006; Torterolo et al., 2011a; Monti et al., 2013), and project throughout the CNS, including the mesopontine RF, where the RAAS is located (Figure 1.4) (Bittencourt et al., 1992). These neurons fire scarcely during W, increase their

firing rate during NREM sleep, and reach a maximum during REM sleep. Since, the administration of MCH into the cerebral ventricles, preoptic area, basal forebrain, dorsal raphe nucleus, LC, and NPO facilitates the generation of NREM sleep and/or REM sleep, it is possible that MCHergic neurons inhibit the activating systems and/or activate the hypnogenic nucleus in order to promote sleep (Verret et al., 2003; Lagos et al., 2009; Torterolo et al., 2009b; Lagos et al., 2012; Benedetto et al., 2013; Monti et al., 2015a). Recent studies have demonstrated that MCH inhibits dorsal raphe nucleus serotonergic neurons (Devera et al., 2015); this finding may explain, at least in part, the promotion of REM sleep induced by MCH (Lagos et al., 2009). The suppression of the serotonergic activity by MCH could explain also the pro-depressive effect of this neuropeptide (Lagos et al., 2011; Lopez Hill et al., 2013; Urbanavicius et al., 2014).

1.5. TRANSITION FROM WAKEFULNESS TO NREM SLEEP

The physiological transition between W and sleep is regulated by a circadian and a homeostatic component (Borbely, 1982). Like all circadian rhythms, sleep and W are regulated by commands from the suprachiasmatic nucleus (SCN) of the hypothalamus. The SCN receives photic information directly from the retina, and regulates the activity of both the hypnogenic and activating systems (Mistlberger, 2005). Furthermore, through indirect modulation of the sympathetic system, the SCN regulates the release of melatonin from the pineal gland during the night (Pandi-Perumal et al., 2008). Melatonin has a weak sleep-promoting effect.

The homeostatic component also regulates the sleep–wakefulness cycle, that is, prolonged W facilitates the generation of sleep. Different lines of research have shown that sleep-promoting substances including adenosine, are released and accumulated during W (Basheer et al., 2004; Huang et al., 2011). Adenosine promotes sleep by inhibiting the activating systems and stimulating the hypnogenic systems. Of interest, caffeine promotes W by blocking the receptors for adenosine (Nishino and Mignot, 2005).

1.6. TRANSITION FROM NREM SLEEP TO REM SLEEP

There is limited knowledge about the neuronal basis involved in the transition from NREM sleep to REM sleep. However, a role for the caudolateral peribrachial region during this transition has been proposed (Torterolo et al., 2011b).

1.7. BRIEF SYNOPSIS OF THE NEUROPHARMACOLOGY IN SLEEP PATHOLOGY

Knowing the neurobiological basis of W and sleep provides the clinician with the frame to understand sleep pathologies and the pharmacological approaches to their treatment.

Sleep pathological conditions are characterized by either a lack of the necessary amount (or quality) of sleep that is called insomnia, or by an excess of sleep, that is hypersomnia. Another type of syndromes is caused by the appearance of abnormal behaviors during sleep and is called parasomnias.

The chronic insomnia disorder in adult patients occurs no less than three times per week, for at least three months and is characterized by an inability to fall or stay asleep, and daytime complaints such as somnolence and fatigue (American Academy of Sleep Medicine, 2005). Medications approved for this disorder include benzodiazepine (BZD) receptor allosteric modulators, BZD (triazolam, temazepam, and flurazepam) or non-BZD (zolpidem, eszopiclone, and zaleplon) agents. These drugs promote sleep, at least in part, by reducing the activity of the activating systems. Melatonin and the melatonin receptor agonist ramelteon are also used for the treatment of an insomnia disorder; these drugs make the most of the sleep-promoting effect of natural melatonin. Low-dose doxepin (a tricyclic antidepressant) is also employed for the treatment of sleep disorders. Its mechanism of action is mainly related to the blockade of histamine H1 receptor. Finally, the dual orexin (hypocretin) receptor antagonist (DORA) suvorexant that blocks the effect of endogenous hypocretin (a wake-promoting neuromodulator), has been recently approved by the FDA for the treatment of insomnia disorders.

Briefly, drugs currently used for the treatment of chronic primary insomnia address sleep onset latency (zolpidem immediate-release, zaleplon, and ramelteon) and/or sleep maintenance (temazepam, flurazepam, zolpidem extended-release, eszopiclone, and low-dose doxepin). However, during their administration, N3 sleep and REM sleep do not regain normal levels or can be even further reduced. With respect to suvorexant, the compound increases N2 sleep and REM sleep in patients with insomnia disorder (Monti et al., 2015b).

Hypersomnia disorder is a term used for a group of disorders in which the primary characteristic is excessive daytime sleepiness in the presence of normal or longer than normal nocturnal sleep (Larson-Prior et al., 2014). Among these disorders, the most common is narcolepsy. The management of this pathology includes several behavioral approaches and pharmacological treatment. For excessive daytime sleepiness, modafinil is in the first line pharmacological treatment. This drug blocks the dopamine transporter (DAT) and increases dopamine synaptic levels, which is central in its wake-promoting effect; in fact, genetic ablation of the DAT abolishes the wake-promoting effect of modafinil. However, there are other possible sites of action of this drug (Wisor, 2013). Amphetamine-like drugs such as methylphenidate are also used to reduce sleepiness.

Parasomnias are unpleasant or undesirable behavioral phenomena that occur during the sleep period. There are different types of parasomnias (Pandi-Perumal et al., 2014). One of them is the REM sleep behavioral disorder (RBD). During RBD, the REM sleep atonia does not occur and the patients act out their dreams. Severe lesions can occur during the REM without atonia episodes. About 90% of patients with chronic RBD respond well to clonazepam (0.5–2 mg) administered half an hour before sleep time (Mahowald and Schneck, 2009). Clonazepam is a BDZ whose mechanism of action for the RBD is still unknown.

1.8. CONCLUSIONS AND FUTURE DIRECTIONS

A detailed knowledge of the anatomy and physiology of the activating and hypnogenic systems is important to understand and treat sleep pathologies. A recent achievement in relation to the activating systems has been the unveiling of the pathogenesis of narcolepsy. This pathology is caused by the degeneration of hypocretinergic neurons (Mignot, 2011), which prompted paraclinical studies such as the titration of hypocretin-1 in the cerebrospinal fluid for diagnostic confirmation of narcolepsy, and therapeutic advances such as intranasal hypocretin-1 administration for the treatment of some aspects of the disease (Baier et al., 2008).

KEYWORDS

- **REM**
- **reticular formation**
- **hypothalamus**
- **basal forebrain**
- **sleep**
- **MCH**
- **acetylcholine**
- **dopamine**
- **hypocretin**
- **histamine**
- **norepinephrine**
- **serotonin**

REFERENCES

American Academy of Sleep Medicine. *International Classification of Sleep Disorders: Diagnostic and Coding Manual*; Westchester, IL, 2005.

Aston-Jones, G.; Bloom, F. E. Activity of Norepinephrine-Containing Locus Coeruleus Neurons in Behaving Rats Anticipates Fluctuations in the Sleep-Waking Cycle. *J. Neurosci.* **1981**, *1*, 876–886.

Baier, P. C.; Weinhold, S. L.; Huth, V.; Gottwald, B.; Ferstl, R.; Hinze-Selch, D. Olfactory Dysfunction in Patients with Narcolepsy with Cataplexy is Restored by Intranasal Orexin A (Hypocretin-1). *Brain.* **2008**, *131*, 2734–2741.

Basheer, R.; Strecker, R. E.; Thakkar, M. M.; McCarley, R. W. Adenosine and Sleep-Wake Regulation. *Prog. Neurobiol.* **2004**, *73*, 379–396.

Belardetti, F.; Borgia, R.; Mancia, M. Prosencephalic Mechanisms of ECoG Desynchronization in Cerveau Isole Cats. *Electroencephalogr. Clin. Neurophysiol.* **1977**, *42*, 213–225.

Benedetto, L.; Chase, M.; Torterolo, P. GABAergic Processes Within the Median Preoptic Nucleus Promote NREM Sleep. *Behav. Brain Res.* **2012**, *232*, 60–65.

Benedetto, L.; Rodriguez-Servetti, Z.; Lagos, P.; D'Almeida, V.; Monti, J. M.; Torterolo, P. Microinjection of Melanin Concentrating Hormone into the Lateral Preoptic Area Promotes non-REM Sleep in the Rat. *Peptides.* **2013**, *39C*, 11–15.

Berridge, C. W.; Abercrombie, E. D. Relationship between Locus Coeruleus Discharge Rates and Rates of Norepinephrine Release within Neocortex as Assessed by in vivo Microdialysis. *Neuroscience.* **1999**, *93*, 1263–1270.

Bittencourt, J. C.; Presse, F.; Arias, C.; Peto, C.; Vaughan, J.; Nahon, J. L. Vale, W. Sawchenko, P. E. The Melanin-Concentrating Hormone System of the Rat Brain: An Immuno- and Hybridization Histochemical Characterization. *J. Comp. Neurol.* **1992**, *319*, 218–245.

Borbely, A. A. A Two Process Model of Sleep Regulation. *Hum. Neurobiol.* **1982**, *1*, 195–204.

Boucetta, S.; Cisse, Y.; Mainville, L.; Morales, M.; Jones, B. E. Discharge Profiles across the Sleep-Waking Cycle of Identified Cholinergic, GABAergic, and Glutamatergic Neurons in the Pontomesencephalic Tegmentum of the Rat. *J. Neurosci.* **2014**, *34*, 4708–4727.

Bremer, F. Cerveau "Isole" et Physiologie du Sommeil. *C. R. Soc. Biol.* **1935**, *118*, 1235–1241.

Cape, E. G.; Jones, B. E. Effects of Glutamate Agonist Versus Procaine Microinjections into the Basal Forebrain Cholinergic Cell Area Upon Gamma and Theta Eeg Activity and Sleep-Wake State. *Eur. J. Neurosci.* **2000**, *12*, 2166–2184.

Chase, M. H. Motor Control during Sleep and Wakefulness: Clarifying Controversies and Resolving Paradoxes. *Sleep Med. Rev.* **2013**, *17*, 299–312.

Coyle, J. T.; Price, D. L.; DeLong, M. R. Alzheimer's Disease: A Disorder of Cortical Cholinergic Innervation. *Science.* **1983**, *219*, 1184–1190.

Curro Dossi, R.; Pare, D.; Steriade, M. Short-Lasting Nicotinic and Long-lasting Muscarinic Depolarizing Responses of Thalamocortical Neurons to Stimulation of Mesopontine Cholinergic Nuclei. *J. Neurophysiol.* **1991**, *65*, 393–406.

de Lecea, L.; Kilduff, T. S.; Peyron, C.; Gao, X.; Foye, P. E.; Danielson, P. E.; Fukuhara, C.; Battenberg, E. L.; Gautvik, V. T.; Bartlett, F. S.; 2nd, Frankel, W. N.; van den Pol, A. N.; Bloom, F. E.; Gautvik, K. M.; Sutcliffe, J. G. The Hypocretins: Hypothalamus-Specific Peptides with Neuroexcitatory Activity. *Proc. Natl Acad. Sci. U S A.* **1998**, *95*, 322–327.

Dement, W.; Kleitman, N. The Relation of Eye Movements during Sleep to Dream Activity: An Objective Method for the Study of Dreaming. *J. Exp. Psychol.* **1957**, *53*, 339–346.

Devera, A.; Pascovich, C.; Lagos, P.; Falconi, A.; Sampogna, S.; Chase, M. H.; Torterolo, P. Melanin-Concentrating Hormone (MCH) Modulates the Activity of Dorsal Raphe Neurons. *Brain Res.* **2015**, *1598*, 114–128.

Dringenberg, H. C.; Vanderwolf, C. H. Neocortical Activation: Modulation by Multiple Pathways Acting on Central Cholinergic and Serotonergic Systems. *Exp. Brain. Res.* **1997**, *116*, 160–174.

Feenstra, M. G. Dopamine and Noradrenaline Release in the Prefrontal Cortex in Relation to Unconditioned and Conditioned Stress and Reward. *Prog. Brain Res.* **2000**, *126*, 133–163.

Foote, S. L.; Berridge, C. W.; Adams, L. M.; Pineda, J. A. Electrophysiological Evidence for the Involvement of the Locus Coeruleus in the Alerting, Orienting, and Attending. In *Progress in Brain Research;* Barnes, C. D., Pompeiano, O., Eds.; Elsevier: Amsterdam, 1991; vol.88, pp 521–531.

Gallopin, T.; Fort, P.; Eggermann, E.; Cauli, B.; Luppi, P. H.; Rossier, J.; Audinat, E.; Muhlethaler, M.; Serafin, M. Identification of Sleep-Promoting Neurons in vitro. *Nature.* **2000**, *404*, 992–995.

Gillin, J. C.; Sitaram, N. Rapid Eye Movement (REM) Sleep: Cholinergic Mechanisms. *Psychol. Med.* **1984**, *14*, 501–506.

Huang, Z. L.; Urade, Y.; Hayaishi, O. The Role of Adenosine in the Regulation of Sleep. *Curr. Top. Med. Chem.* **2011**, *11*, 1047–1057.

Jacobs, B. L.; Azmitia, E. C. Structure and Function of the Brain Serotonin System. *Physiol. Rev.* **1992**, 72, 165–229.

Jacobs, B. L.; Fornal, C. A. Brain Serotonergic Neuronal Activity in Behaving Cats. In *Serotonin and Sleep: Molecular, Functional and Clinical Aspects*; Monti, J. M. et al., Eds.; Birkhauser, Basel: Boston, Berlin, 2008.

Jones, B. Reticular Formation: Cytoarchitecture, Transmitter and Projections. In *The Rat Nervous System*; Paxinos, G., Ed.; Academic Press: San Diego, CA, 1995; 2nd ed., pp 155–171.

Jones, B. Basic Mechanisms of Sleep-Wake States. In *Principles and Practices of Sleep Medicine*; Kryger, M. H. et al., Eds.; Elsevier-Saunders: Philadelphia, 2005, pp 136–153.

Kalivas, P. W. Histamine-Induced Arousal in the Conscious and Pentobarbital-Pretreated Rat. *J. Pharmacol. Exp. Ther.* **1982**, 222, 37–42.

Kiyashchenko, L. I.; Mileykovskiy, B. Y.; Maidment, N.; Lam, H. A.; Wu, M. F.; John, J.; Peever, J.; Siegel, J. M. Release of Hypocretin (Orexin) during Waking and Sleep States. *J. Neurosci.* **2002**, 22, 5282–5286.

Kumar, V. M. Why the Medial Preoptic Area is Important for Sleep Regulation. *Indian J. Physiol. Pharmacol.* **2004**, 48, 137–149.

Lagos, P.; Monti, J. M.; Jantos, H.; Torterolo, P. Microinjection of the Melanin-Concentrating Hormone into the Lateral Basal Forebrain Increases REM Sleep and Reduces Wakefulness in the Rat. *Life Sci.* **2012**, 90, 895–899.

Lagos, P.; Torterolo, P.; Jantos, H.; Chase, M. H.; Monti, J. M. Effects on Sleep of Melanin-Concentrating Hormone Microinjections into the Dorsal Raphe Nucleus. *Brain Res.* **2009**, 1265, 103–110.

Lagos, P.; Urbanavicius, J.; Scorza, C.; Miraballes, R.; Torterolo, P. Depressive-Like Profile Induced by MCH Microinjections into the Dorsal Raphe Nucleus Evaluated in the Forced Swim Test. *Behav. Brain Res.* **2011**, 218, 259–66.

Larson-Prior, L. J.; Ju, Y. E.; Galvin, J. E. Cortical-Subcortical Interactions in Hypersomnia Disorders: Mechanisms Underlying Cognitive and Behavioral Aspects of the Sleep-Wake Cycle. *Front. Neurol.* **2014**, 5, 165.

Lee, M. G.; Hassani, O. K.; Alonso, A.; Jones, B. E. Cholinergic Basal Forebrain Neurons Burst with Theta During Waking and Paradoxical Sleep. *J. Neurosci.* **2005a**, 25, 4365–4369.

Lee, M. G.; Hassani, O. K.; Jones, B. E. Discharge of Identified Orexin/Hypocretin Neurons across the Sleep-Waking Cycle. *J. Neurosci.* **2005b**, 25, 6716–6720.

Lindsley, D. B.; Schreiner, L. H.; Knowles, W. B.; Magoun, H. W. Behavioral and Eeg Changes Following Chronic Brain Stem Lesions in the Cat. *Electroencephalogr. Clin. Neurophysiol.* **1950**, 2, 483–498.

Lopez Hill, X.; Pascovich, C.; Urbanavicius, J.; Torterolo, P., Scorza, C. The Median Raphe Nucleus Participates in the Depressive-Like Behavior Induced by MCH: Differences with the Dorsal Raphe Nucleus. *Peptides.* **2013**, 50C, 96–99.

Lu, J.; Sherman, D.; Devor, M.; Saper, C. B. A Putative Flip-Flop Switch for Control of REM Sleep. *Nature.* **2006**, 441, 589–594.

Luppi, P. H.; Gervasoni, D.; Verret, L.; Goutagny, R.; Peyron, C.; Salvert, D.; Leger, L.; Fort, P. Paradoxical (REM) Sleep Genesis: The Switch from an Aminergic-Cholinergic to a GABAergic-Glutamatergic Hypothesis. *J. Physiol. Paris.* **2007**, 100, 271–283.

Mahowald, M. W.; Schneck, C. REM Sleep Parasomnias. In *Principles and Practices of Sleep Medicine*; Kryger, M. H. et al., Eds.; Elsevier-Saunders: Philadelphia, 2009, pp 1083–1097.

Marrosu, F.; Portas, C.; Mascia, M. S.; Casu, M. A.; Fa, M.; Giagheddu, M.; Imperato, A.; Gessa, G. L. Microdialysis Measurement of Cortical and Hippocampal Acetylcholine Release during Sleep-Wake Cycle in Freely Moving Cats. *Brain Res.* **1995**, 671, 329–332.

McCarley, R. W. Neurobiology of REM and NREM Sleep. *Sleep Med.* **2007**, 8, 302–330.

McGinty, D.; Szymusiak, R. Sleep-Promoting Mechanisms in Mammals. In *Principles and Practices of Sleep Medicine*; Kryger, M. H. et al., Eds.; Elsevier-Saunders: Philadelphia, 2005, pp 169–184.

McGinty, D. J.; Harper, R. M. Dorsal Raphe Neurons: Depression of Firing during Sleep in Cats. *Brain Res.* **1976**, 101, 569–575.

Mignot, E. Narcolepsy: Pathophysiology and Genetic Predisposition. In *Principles and Practices of Sleep Medicine*; Krieger, M. H. et al., Eds.; Saunders: Philadelphia, 2011, pp 938–956.

Mileykovskiy, B. Y.; Kiyashchenko, L. I.; Siegel, J. M. Behavioral Correlates of Activity in Identified Hypocretin/Orexin Neurons. *Neuron.* **2005**, 46, 787–798.

Mirenowicz, J.; Schultz, W. Preferential Activation of Midbrain Dopamine Neurons by Appetitive Rather Than Aversive Stimuli. *Nature.* **1996**, 379, 449–451.

Mistlberger, R. E. Circadian Regulation of Sleep in Mammals: Role of the Suprachiasmatic Nucleus. *Brain Res. Brain Res. Rev.* **2005**, *49*, 429–454.

Montagna, P. Fatal Familial Insomnia: A Model Disease in Sleep Physiopathology. *Sleep Med. Rev.* **2005**, *9*, 339–353.

Monti, J. M. The Role of Dorsal Raphe Nucleus Serotonergic and Non-Serotonergic Neurons, and of Their Receptors, in Regulating Waking and Rapid Eye Movement (REM) Sleep. *Sleep Med. Rev.* **2010a**, *14*, 319–327.

Monti, J. M. The Structure of the Dorsal Raphe Nucleus and Its Relevance to the Regulation of Sleep and Wakefulness. *Sleep Med. Rev.* **2010b**, *14*, 307–317.

Monti, J. M. The Role of Tuberomammillary Nucleus Histaminergic Neurons, and of Their Receptors, in the Regulation of Sleep and Waking. In *REM Sleep: Regulation and Function;* Mallick B.N., P.-P.S., McCarley R.W., Morrison A.R., Eds.; Cambridge University Press: Cambridge, UK, 2011, pp 223–233.

Monti, J. M.; Jantos, H. The Roles of Dopamine and Serotonin, And of Their Receptors, In Regulating Sleep and Waking. *Prog. Brain Res.* **2008**, *172*, 625–646.

Monti, J. M.; Lagos, P.; Jantos, H.; Torterolo, P. Increased REM Sleep After Intra-Locus Coeruleus Nucleus Microinjection of Melanin-Concentrating Hormone (MCH) in the Rat. *Prog. Neuropsychopharmacol. Biol. Psychiatry.* **2015a**, *56*, 185–188.

Monti, J. M.; Monti, D. The Involvement of Dopamine in the Modulation of Sleep and Waking. *Sleep Med. Rev.* **2007**, *11*, 113–133.

Monti, J. M.; Torterolo, P.; Lagos, P. Melanin-Concentrating Hormone Control of Sleep-Wake Behavior. *Sleep Med. Rev.* **2013**, *17*, 293–298.

Monti, J. M.; Torterolo, P.; Pandi-Perumal, S. R. The Effects of Selective Serotonin 5-HT2A Receptor Antagonists and Inverse Agonists on Sleep and Wakefulness. Submitted for publication, **2015b.**

Moruzzi, G.; Magoun, H. W. Brain Stem Reticular Formation and Activation of the EEG. *Electroencephalogr. Clin. Neurophysiol.* **1949**, *1*, 455–473.

Nelson, L. E.; Lu, J., Guo, T.; Saper, C. B.; Franks, N. P.; Maze, M. The Alpha2-Adrenoceptor Agonist Dexmedetomidine Converges on an Endogenous Sleep-Promoting Pathway to Exert its Sedative Effects. *Anesthesiology.* **2003**, *98*, 428–436.

Nishino, S.; Mignot, E. Pharmacological Aspects of Human and Canine Narcolepsy. *Prog. Neurobiol.* **1997**, *52*, 27–78.

Nishino, S.; Mignot, E. Wake-Promoting Medications: Basic Mechanisms and Pharmacology. In *Principles and Practices of Sleep Medicine;* Kryger, M. H. et al. Eds.; Elsevier-Saunders: Philadelphia, 2005, pp 468–483.

Nishino, S.; Ripley, B.; Overeem, S.; Lammers, G. J.; Mignot, E. Hypocretin (Orexin) Deficiency in Human Narcolepsy. *Lancet.* **2000** *355*, 39–40.

Oades, R. D.; Halliday, G. M. Ventral Tegmental (A10) System: Neurobiology. 1. Anatomy and Connectivity. *Brain Res.* **1987** *434*, 117–165.

Pace-Schott, E. The Neurobiology of Dreaming. In *Principles and Practices of Sleep Medicine;* Kryger, M. H. et al., Eds.; Elsevier-Saunders: Philadelphia, 2005, pp 551–564.

Pandi-Perumal, S. R.; BaHammam, A. S.; Shapiro, C. M. Parasomnias. In *Encyclopedia of Psychopharmacology;* Stolerman, I. P., Price, L. H., Eds.; Springer-Verlag: Berlin, Heidelberg, 2014.

Pandi-Perumal, S. R.; Trakht, I.; Srinivasan, V.; Spence, D. W.; Maestroni, G. J.; Zisapel, N.; Cardinali, D. P. Physiological Effects of Melatonin: Role of Melatonin Receptors and Signal Transduction Pathways. *Prog. Neurobiol.* **2008**, *85*, 335–353.

Parmentier, R.; Ohtsu, H.; Djebbara-Hannas, Z.; Valatx, J. L.; Watanabe, T.; Lin, J. S. Anatomical, Physiological, and Pharmacological Characteristics of Histidine Decarboxylase Knock-Out Mice: Evidence for the Role of Brain Histamine in Behavioral and Sleep-Wake Control. *J. Neurosci.* **2002**, *22*, 7695–7711.

Piper, D. C.; Upton, N.; Smith, M. I.; Hunter, A. J. The Novel Brain Neuropeptide, Orexin-A, Modulates the Sleep-Wake Cycle of Rats. *Eur. J. Neurosci.* **2000**, *12*, 726–730.

Plourde, G.; Chartrand, D.; Fiset, P.; Font, S.; Backman, S. B. Antagonism of Sevoflurane Anaesthesia by Physostigmine: Effects on the Auditory Steady-State Response and Bispectral Index. *Br. J. Anaesth.* **2003**, *91*, 583–586.

Plum, F.; Posner, J. *The Diagnosis of Stupor and Coma.* Oxford University Press: New York, 2000.

Portas, C. M.; Bjorvatn, B.; Ursin, R. Serotonin and the Sleep/Wake Cycle: Special Emphasis on Microdialysis Studies. *Prog. Neurobiol.* **2000**, *60*, 13–35.

Roth, T.; Roehrs, T.; Koshorek, G.; Sicklesteel, J.; Zorick, F. Sedative Effects of Antihistamines. *J. Allergy Clin. Immunol.* **1987**, *80*, 94–98.

Sakurai, T.; Amemiya, A.; Ishii, M.; Matsuzaki, I.; Chemelli, R. M.; Tanaka, H.; Williams, S. C.; Richardson, J. A.; Kozlowski, G. P.; Wilson, S.; Arch, J. R.; Buckingham, R. E.; Haynes, A. C.; Carr, S. A.; Annan, R. S.; McNulty, D. E.; Liu, W. S.; Terrett, J. A.; Elshourbagy, N. A.; Bergsma, D. J.; Yanagisawa, M. Orexins and Orexin Receptors: A Family of Hypothalamic Neuropeptides and G Protein-Coupled Receptors That Regulate Feeding Behavior. *Cell.* **1998,** *92,* 573–585.

Satoh, K.; Fibiger, H. C. Cholinergic Neurons of the Laterodorsal Tegmental Nucleus: Efferent and Afferent Connections. *J. Comp. Neurol.* **1986,** *253,* 277–302.

Semba, K. Multiple Output Pathways of the Basal Forebrain: Organization, Chemical Heterogeneity, and Roles in Vigilance. *Behav. Brain Res.* **2000,** *115,* 117–141.

Siegel, J. M. REM Sleep. In *Principles and Practices of Sleep Medicine*; Kryger, M. H. et al., Eds.; Elsevier-Saunders: Philadelphia, 2011, pp 92–111.

Steriade, M.; McCormick, D. A.; Sejnowski, T. J. Thalamocortical Oscillations in the Sleeping and Aroused Brain. *Science.* **1993,** *262,* 679–685.

Szymusiak, R., Gvilia, I., McGinty, D. Hypothalamic Control of Sleep. *Sleep Med.* **2007,** *8,* 291–301.

Takahashi, K.; Lin, J. S.; Sakai, K. Neuronal Activity of Histaminergic Tuberomammillary Neurons during Wake-Sleep States in the Mouse. *J. Neurosci.* **2006,** *26,* 10292–10298.

Torterolo, P.; Benedetto, L.; Lagos, P.; Sampogna, S.; Chase, M. H. State-Dependent Pattern of Fos Protein Expression in Regionally-Specific Sites Within the Preoptic Area of the Cat. *Brain Res.* **2009a,** *1267,* 44–56.

Torterolo, P.; Chase, M. H. The Hypocretins (Orexins) Mediate the "Phasic" Components of REM sleep: A new hypothesis. *Sleep Sci.* **2014,** *7,* 19–29.

Torterolo, P.; Lagos, P.; Monti, J. M. Melanin-Concentrating Hormone (MCH): A New Sleep Factor? *Front. Neurol.* **2011a,** *2,* 1–12.

Torterolo, P.; Morales, F. R.; Chase, M. H. GABAergic Mechanisms in the Pedunculopontine Tegmental Nucleus of the Cat Promote Active (REM) Sleep. *Brain Res.* **2002a,** *944,* 1–9.

Torterolo, P.; Sampogna, S.; Chase, M. H. MCHergic Projections to the Nucleus Pontis Oralis Participate in the Control of Active (REM) Sleep. *Brain Res.* **2009b,** *1268,* 76–87.

Torterolo, P.; Sampogna, S.; Chase, M. H. A Restricted Parabrachial Pontine Region is Active During Non-Rapid Eye Movement Sleep. *Neuroscience.* **2011b,** *190,* 184–193.

Torterolo, P.; Sampogna, S.; Morales, F. R.; Chase, M. H. Gudden's Dorsal Tegmental Nucleus is Activated in Carbachol Induced Active (REM) Sleep and Active Wakefulness. *Brain Res.* **2002b,** *944,* 184–189.

Torterolo, P.; Sampogna, S.; Morales, F. R.; Chase, M. H. MCH-Containing Neurons in the Hypothalamus of the Cat: Searching For a Role in the Control of Sleep and Wakefulness. *Brain Res.* **2006,** *1119,* 101–114.

Torterolo, P.; Vanini, G. Involvement of GABAergic Mechanisms in the Laterodorsal and Pedunculopontine Tegmental Nuclei (LDT-PPT) in the Promotion of REM Sleep. In *GABA and Sleep: Molecular, Functional and Clinical Aspects*; Monti, J. et al., Eds.; Springer: Basel, 2010, pp 213-231.

Torterolo, P.; Vanini, G.; Cabrera, G.; Chase, M.; Falconi, A. Hypocretins (Orexins) in the Inferior Colliculus: Anatomy and Physiology. *Sleep Med.* **2009c,** *10,* S61.

Torterolo, P.; Yamuy, J.; Sampogna, S.; Morales, F. R.; Chase, M. H. GABAergic Neurons of the Cat Dorsal Raphe Nucleus Express C-Fos During Carbachol-Induced Active Sleep. *Brain Res.* **2000,** *884,* 68–76.

Torterolo, P.; Yamuy, J.; Sampogna, S.; Morales, F. R.; Chase, M. H. GABAergic Neurons of the Laterodorsal and Pedunculopontine Tegmental Nuclei of the Cat Express C-Fos During Carbachol-Induced Active Sleep. *Brain Res.* **2001a,** *892,* 309–319.

Torterolo, P.; Yamuy, J.; Sampogna, S.; Morales, F. R.; Chase, M. H. Hypothalamic Neurons that Contain Hypocretin (Orexin) Express C-Fos during Active Wakefulness and Carbachol-Induced Active Sleep. *Sleep Res.* **2001b,** Online 4, 25–32. http://www.sro.org/2001/Torterolo/2025.

Torterolo, P.; Yamuy, J.; Sampogna, S.; Morales, F. R.; Chase, M. H. Hypocretinergic Neurons are Primarily Involved in Activation of the Somatomotor System. *Sleep.* **2003,** *1,* 25–28.

Trachsel, L.; Tobler, I.; Achermann, P.; Borbely, A. A. Sleep Continuity and the REM-NonREM Cycle in the Rat under Baseline Conditions and After Sleep Deprivation. *Physiol. Behav.* **1991,** *49,* 575–580.

Urbanavicius, J.; Lagos, P.; Torterolo, P.; Scorza, C. Pro-Depressive Effect Induced by Microinjections of MCH into the Dorsal Raphe: Time-Course, Dose-Dependence, Effects on Anxiety-Related

Behaviors and Reversion by Nortriptyline. *Behav. Pharmacol.* **2014,** *25,* 316–324.

Vanini, G.; Torterolo, P.; McGregor, R.; Chase, M. H.; Morales, F. R. GABAergic Processes in the Mesencephalic Tegmentum Modulate the Occurrence of Active (Rapid Eye Movement) Sleep in Guinea Pigs. *Neuroscience.* **2007,** *145,* 1157–1167.

Vanini, G.; Watson, C. J.; Lydic, R.; Baghdoyan, H. A. Gamma-Aminobutyric Acid-Mediated Neurotransmission in the Pontine Reticular Formation Modulates Hypnosis, Immobility, and Breathing during Isoflurane Anesthesia. *Anesthesiology.* **2008,** *109,* 978–988.

Verret, L.; Goutagny, R.; Fort, P.; Cagnon, L.; Salvert, D.; Leger, L.; Boissard, R.; Salin, P.; Peyron, C.; Luppi, P. H. A Role of Melanin-Concentrating Hormone Producing Neurons in the Central Regulation of Paradoxical Sleep. *BMC Neurosci.* **2003,** *4,* 19.

Vitiello, M. V.; Borson, S. Sleep Disturbances in Patients with Alzheimer's Disease: Epidemiology, Pathophysiology and Treatment. *CNS Drugs.* **2001,** *15,* 777–796.

Wisor, J. Modafinil as a Catecholaminergic Agent: Empirical Evidence and Unanswered Questions. *Front. Neurol.* **2013,** *4,* 139.

Wolff, K.; Winstock, A. R. Ketamine : From Medicine to Misuse. *CNS Drugs.* **2006,** *20,* 199–218.

Xi, M. C.; Morales, F. R.; Chase, M. H. Evidence that Wakefulness and REM Sleep are Controlled by a GABAergic Pontine Mechanism. *J. Neurophysiol.* **1999,** *82,* 2015–2019.

2

NEUROENDOCRINOLOGY OF SLEEP AND WAKEFULNESS

Georges Copinschi*

2.1. INTRODUCTION

Sleep involves two states of distinct brain activity: rapid eye movement (REM) sleep, and non-REM sleep. Sleep is normally initiated by light non-REM stages (stages 1 and 2), followed by a deeper non-REM stage (stage 3, also called slow-wave sleep, SWS), then by REM sleep. In healthy young subjects, this cycle, which lasts around 90 min, is generally repeated 4–6 times per night. As the night progresses, non-REM sleep becomes shallower, the duration of REM episodes becomes longer, and the number and the duration of transient awakenings increase. Aging is rapidly associated with marked alterations in sleep architecture. SWS is already markedly reduced in healthy subjects 35 years old. Thereafter, REM sleep progressively declines, in mirror image of an equivalent increase in time spent awake.[1]

Circulating hormonal levels undergo pronounced temporal oscillations, providing the endocrine system with remarkable flexibility. The temporal organization of hormonal secretions results from the interaction of two time-keeping mechanisms in the central nervous system, the *circadian pacemaker* and the *sleep–wake homeostasis*. The circadian pacemaker, or circadian clock, located in the suprachiasmatic nucleus of the hypothalamus, totally or partially controls the timing and the intensity of most hormonal circadian rhythms. In addition, it regulates the timings of sleep onset and sleep offset and the distribution

Georges Copinschi
Laboratory of Physiology and Pharmacology, Université Libre de Bruxelles, Brussels, Belgium

Corresponding author: Laboratory of Physiology and Pharmacology, Faculty of Medicine, Université Libre de Bruxelles, CP 604, 808 route de Lennik, B-1070 Brussels, Belgium. E-mail: gcop@ulb.ac.be; Tel.: +32 478 40 44 90

*Professor Emeritus of Endocrinology, Université Libre de Bruxelles, Brussels, Belgium. Formerly Chief, Division of Endocrinology, Hôpital Universitaire Saint-Pierre, Université Libre de Bruxelles, Brussels, Belgium. Formerly Chairman, Department of Medicine, Hôpital Universitaire Saint-Pierre, Université Libre de Bruxelles, Brussels, Belgium.

of REM sleep. The sleep–wake homeostasis, a mechanism relating the timing and the architecture of sleep to the duration of prior wakefulness, controls the non-REM sleep, in particular SWS, and regulates, at least partially, the timing and the intensity of virtually all endocrine–metabolic systems. Conversely, hormones may modulate sleep architecture. Thus, bidirectional interrelations are operative between sleep and endocrine function.[2] The mechanisms of those sleep–endocrine interactions, which markedly differ from one hormone to another, will be reviewed in the following sections.

2.2. GROWTH HORMONE
Hormone Primarily Controlled by the Sleep–Wake Homeostasis

Pituitary growth hormone (GH) secretion is stimulated by hypothalamic GH-releasing hormone (GHRH) and inhibited by somatostatin. In addition, the acylated form of ghrelin (a peptide principally produced by the stomach) also stimulates GH secretion. In adults, the normal 24 h profile of circulating GH levels consists of stable low levels abruptly interrupted by secretory pulses. In young healthy adults, the most reproducible GH secretory pulse is consistently observed shortly after sleep onset and is temporally and quantitatively related to SWS. As illustrated in Figure 2.1, this sleep-onset pulse is generally the largest GH pulse over the 24 h span in adult males, while high amplitude daytime pulses are frequent in young adult females.[3,4]

A sleep-onset associated GH pulse is still observed when sleep is delayed or advanced.[5–8] Therefore, shifts of the sleep–wake cycle are always associated with parallel shifts of GH secretion (Fig. 2.2). Conversely, GH secretion during the sleep period is inhibited by transitional awakenings and sleep fragmentation.[9] In night workers, the most important GH secretory pulse is generally observed during the first part of the shifted sleep period.[10]

In healthy young adults, GHRH injection, when given at a time of decreased sleep propensity at a dose eliciting a GH secretory response within the physiological range, was shown to dramatically reduce the duration of awakenings, to markedly stimulate SWS (Fig. 2.3), and possibly, though to lesser extent, to stimulate REM sleep.[11] Animal studies indicate that the effects of GHRH on SWS are not mediated by GH, while GHRH effects on REM sleep are probably mediated by GH.[12–14] In addition, administration of ritanserin (a selective 5HT2 antagonist) or of gamma-hydroxybutyrate (GHB; a naturally occurring metabolite of GABA used in the treatment of narcolepsy) results in highly correlated simultaneous stimulations of nocturnal GH secretion and of SWS.[6,15] Thus, a robust, though complex, interaction exists between sleep and GH secretion, and sleep is clearly the major determinant of GH secretion. However, there is also some evidence for the existence of a minor circadian modulation of the occurrence and magnitude of GH secretory pulses.[3]

Several studies have shown that in normally cycling young women the amplitude of GH pulses is positively correlated with circulating estradiol levels.[4,16] In addition, daytime GH secretion was found to be higher during the luteal than during the follicular phase (Fig. 2.1), and this elevation correlated positively with circulating

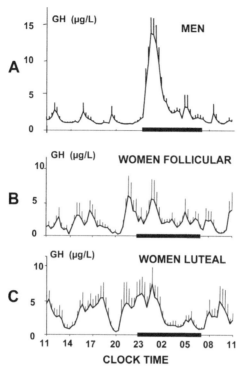

Figure 2.1: A. Mean (+ SEM) 24 h plasma growth hormone profiles in a group of 9 healthy men, aged 18–30 years. (Adapted from Van Cauter et al.[3]).
B. and C. Mean (+ SEM) 24 h plasma growth hormone profiles in a group of 10 normally cycling women, age 21–36 years, during follicular (B) and luteal (C) phase. (Data from Caufriez et al.[17]).

PLASMA GROWTH HORMONE (μg/L)

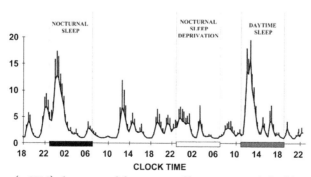

Figure 2.2: Mean (+ SEM) plasma growth hormone profiles in a group of 8 healthy men, aged 20–27 years, investigated during a 53 h period that included 8 h of nocturnal sleep, 28 h of sleep deprivation, and 8 h of daytime sleep. The black bar denotes the nocturnal sleep period, the open bar denotes the period of nocturnal sleep deprivation, and the grey bar denotes the period of daytime sleep. Blood samples were obtained at 20 min intervals. (Adapted from Van Cauter and Spiegel[100]).

EFFECTS OF GHRH ON SLEEP

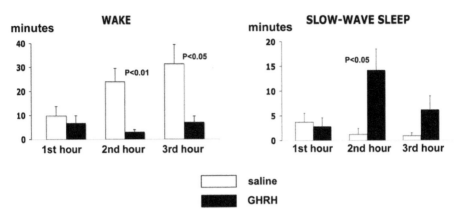

Figure 2.3: Effects of GHRH on mean (+ SEM) duration of wake and SWS in 8 healthy men, aged 24–31 years. GHRH (0.3 μg/kg body weight) was injected intravenously after 60 s of the third REM period of the sleep cycle. (Data from Kerkhofs et al.[11]).

progesterone levels (but not with estradiol levels).[17]

As illustrated in Figure 2. 4, aging is associated with dramatic parallel decreases in SWS and in GH secretion during sleep. Those age-related decreases occur in an exponential fashion between young adulthood and midlife, despite the persistence of high sex steroids levels. Thereafter, minor progressive decreases occur from midlife to old age.[1,18–20] In healthy old subjects, GH secretory profiles are similar in women and in men.[4] In postmenopausal women, progesterone administration was shown to stimulate nocturnal GH secretion, and also to prevent sleep disturbances.[21]

During pregnancy, a placental GH variant substitutes for pituitary GH to regulate maternal insulin-like growth factor-1 (IGF-1).[22,23] Interestingly, this GH variant is released by the placenta in a tonic, rather than pulsatile, fashion.[24] In acromegaly, elevated pituitary GH levels

throughout the 24-h span result from a highly irregular pulsatile pattern superimposed on elevated basal levels, indicating the existence of tonic secretion.[25,26]

Because of the pulsatile nature of GH secretion, determinations of random circulating GH levels are of little or no value in the diagnosis of growth hormone disorders. Appropriate determinations of GH secretion necessarily involve multiple blood sampling collected at 15–20 min intervals for 24 h, together with polygraphic sleep recordings.

2.3. THE CORTICOTROPIC AXIS
Hormones Primarily Controlled by the Circadian Clock

Outputs from the hypothalamic suprachiasmatic nucleus activate the release of corticotropin-releasing hormone (CRH). CRH in turn stimulates pituitary secretion

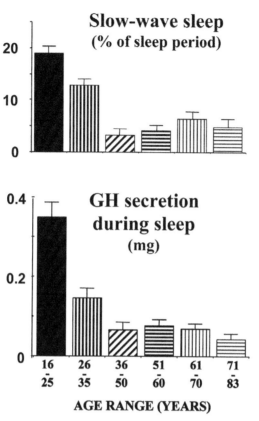

Figure 2.4: Slow-wave sleep and growth hormone secretion during sleep as a function of age in 149 healthy men, 16–83 years old. For each age group, values shown are mean + SEM. Note the temporal concomitance between the decrease in SWS and the decrease in nocturnal GH secretion. (Data from Van Cauter et al.[1]).

of adrenocorticotropic hormone (ACTH). The diurnal patterns of adrenocortical hormones are primarily dependent on the 24 h rhythm of ACTH release. Circulating profiles of ACTH, cortisol, and dehydroepiandrosterone (DHEA) show an early morning maximum, declining levels during daytime, a period of minimal levels centered around midnight (referred to as the *quiescent period*) followed during late sleep by an abrupt elevation (referred to as the *circadian rise*) toward the morning peak. Similar diurnal variations are observed for other adrenocortical steroids. The 24 h cortisol profile results from a succession of secretory pulses of magnitude modulated by the circadian rhythmicity, without any evidence of tonic secretion.[27,28] Diurnal ACTH and cortisol patterns are largely unaffected when sleep is advanced or delayed (Fig. 2.5). Thus, these rhythms appear to be primarily driven by the circadian pacemaker.

However, cortisol rhythmicity is clearly modulated by the sleep–wake homeostasis, as illustrated in Figure 2.6. Sleep

PLASMA CORTISOL (μg/L)

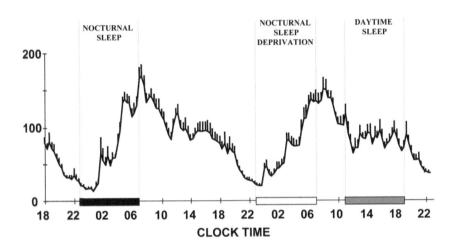

Figure 2.5: Mean (+ SEM) cortisol profiles in a group of 8 healthy men, aged 20–27 years, investigated during a 53 h period that included 8 h of nocturnal sleep, 28 h of sleep deprivation, and 8 h of daytime sleep. The black bar denotes the nocturnal sleep period, the open bar denotes the period of nocturnal sleep deprivation, and the grey bar denotes the period of daytime sleep. Blood samples were obtained at 20 min intervals. (Adapted from Van Cauter and Spiegel[100]).

onset (and more precisely SWS) is consistently associated with an immediate drop in circulating cortisol levels (this drop may be undetectable if sleep is initiated in early morning, at the peak of corticotropic activity).[29–35] Conversely, final awakening at the end of the sleep period as well as transient awakenings interrupting the sleep period are associated with cortisol secretory pulses.[30–32,36] In normal sleepers, 24 h circulating cortisol concentrations are positively correlated with the duration of nocturnal wake.[37]

From 50 years of age onward, aging is associated with a progressive increase in evening cortisol levels, which occurs as a mirror image of the decline in REM sleep (Fig. 2.7).[38,39] In the elderly, the cortisol

quiescent period is markedly shortened and fragmented. Interestingly, this elevation of evening cortisol in healthy elderly subjects with normal sleep schedules is strikingly similar to that observed in young subjects submitted to sleep deprivation. Indeed, in healthy young subjects, acute (one night) or semi-chronic (six nights) total or partial (4 h in bed per night) sleep curtailment results on the following day in abnormal elevations of circulating cortisol levels in late afternoon and evening,[40, 41] indicating that sleep curtailment might accelerate the senescence process of the corticotropic axis and therefore induce a number of deleterious effects. Hyperactivity of the corticotropic axis inhibits SWS and promotes nocturnal awakenings, initiating a feed-for-

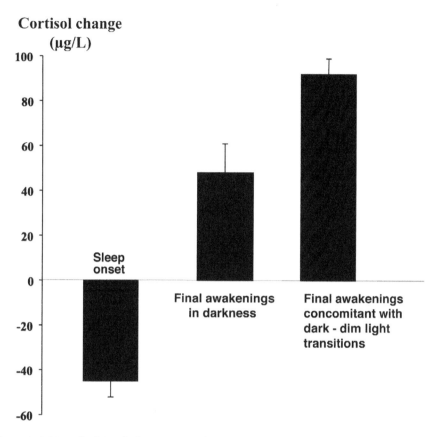

Figure 2.6: Mean (and SEM) changes in circulating cortisol levels observed in eight healthy young subjects aged 21–33 years: within 120 min following sleep onset; within 20 min following final awakening in darkness; within 20 min following final awakening concomitant with transition from darkness to dim light. (Data from Caufriez et al.[30]).

ward cascade of negative events generated by both sleep and hormonal disruptions.[2] Moreover, animal and human studies indicate that hyperactivity of the corticotropic axis at a time of the day when cortisol levels are normally low could favor, among others, the development of insulin resistance, memory deficits and osteoporosis.[42–45]

In rats and in rabbits, intracerebral injections of CRH inhibit SWS, while blockade of CRH receptors reduces wakefulness.[46–48] In humans, intravenous injections of CRH may inhibit SWS and sleep efficiency,[49–51] while cortisol administration enhances SWS,[52–54] an effect presumably mediated by a negative feedback action of cortisol on endogenous CRH secretion.

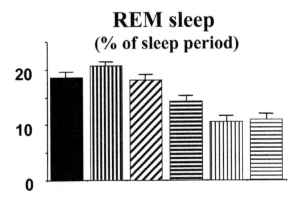

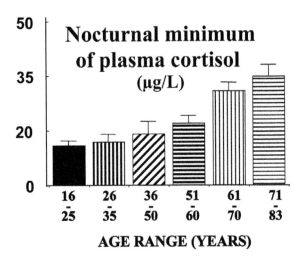

Figure 2.7: REM sleep and nocturnal minimum of plasma cortisol as a function of age in 149 healthy men, 16–83 years old. For each age group, values shown are mean + SEM. Note the temporal concomitance between the decline in REM sleep and a mirror increase in cortisol levels after 50 years of age. (Data from Van Cauter et al.[1])

Thus, interactions clearly occur between sleep and the corticotropic axis.

In pituitary-dependent Cushing's disease, cortisol circadian variations are markedly dampened and cortisol pulsatility is blunted in the majority of patients,[2] suggesting an autonomous tonic secretion of ACTH by the pituitary tumor.

Because of the amplitude of circadian variations and the magnitude of secretory pulses, isolated random determinations of circulating cortisol levels are of little value in the diagnosis of adrenocortical disorders.

2.4. THYROTROPIN
Hormone Controlled by Both the Circadian Clock and the Sleep–Wake Homeostasis

During daytime, circulating thyrotropin (TSH) levels are low and fairly stable. A rise in TSH levels starts in late afternoon or early evening, well before sleep onset, and is considered to reflect a circadian control. Maximal levels are reached around the beginning of the sleep period. Thereafter, TSH levels progressively decline during the sleep period to return to daytime values after morning awakening.[55] This nocturnal decline results from an inhibitory action of sleep (and more precisely of SWS) on TSH secretion.[56,57] Indeed, this nocturnal decrease is not observed under conditions of sleep deprivation,[55,58] and transient nocturnal awakenings are frequently associated with TSH peaks.[59] Of note, daytime sleep does not result in a decrease of TSH below normal daytime levels,[59] suggesting a modulation by the circadian clock of the sleep-associated TSH inhibition.

Early morning TSH levels may occasionally be elevated, even in healthy subjects with regular normal sleep schedules.

2.5. GONADOTROPINS AND SEX STEROIDS
Modulation by the Sleep–Wake Homeostasis

2.5.1. Women

Gonadotropin secretion is modulated by sleep. In prepubertal girls, an association can be observed between sleep and pulsatile gonadotropin secretion.[60] During puberty, pulse amplitude of luteinizing hormone (LH) and of follicle-stimulating hormone (FSH) is enhanced during sleep.[60] In normally cycling adult women, gonadotropin pulse frequency is reduced during the night in the early follicular phase, but not in the luteal phase.[61, 62] This slowing is also observed during daytime sleep, but not during nighttime wakefulness,[63] indicating that it is specifically related to sleep, rather than to the circadian pacemaker. Moreover, LH pulses during the sleep period are preferentially associated with transient awakenings.[63] The circadian clock plays no significant role in diurnal gonadotropin variations.

Since shift and night work is consistently associated with shorter and fragmented sleep, altered menstrual cycles frequently observed in those workers are likely to partially result from altered sleep patterns.

Sleep may be severely impaired in postmenopausal women. It has been shown in a randomized, double-blind crossover study that oral administration of progesterone may improve sleep duration and quality in postmenopausal women when sleep is disturbed by environmental conditions.[21]

2.5.2. Men

In contrast to gonadotropins, circulating testosterone levels exhibit a clear diurnal rhythm, with maximal and minimal levels in the early morning and in the late evening, respectively.[64] This rhythm appears to be primarily driven by the sleep–wake cycle, since both nighttime and daytime sleep are associated with increases in testosterone levels.[65] However, a blunted testosterone increase is observed during nocturnal wakefulness, indicating the existence of a circadian component, possibly reflecting the adrenal secretion of testosterone.[65] In young healthy men, one week of sleep restriction (5 h in bed per night) was shown to result in a significant decrease in 24 h testosterone levels.[66] In older men, morning testosterone levels are positively associated with the duration of sleep the night before.[67]

2.6. GLUCOSE REGULATION AND INSULIN SECRETION
Modulation by the Sleep–Wake Homeostasis

Glucose and insulin homeostasis primarily depend on the balance between the production of glucose by the liver on the one hand, and the glucose utilization by insulin dependent tissues (such as adipose tissue and muscles) and non-insulin dependent

tissues (principally the brain) on the other hand. However, under basal conditions in healthy subjects, glucose utilization varies with time of the day,[68] as illustrated in Figure 2.8. Circulating glucose levels progressively decline during prolonged daytime wakefulness and fasting, while they remain fairly stable during nocturnal sleep despite the overnight fast. Conversely, glucose levels decrease during enforced nighttime wakefulness, while they remain stable during daytime sleep.[33] Thus, some sleep-associated mechanisms are normally operative in healthy subjects to maintain stable circulating glucose concentrations despite prolonged sleep-associated fasting.

This is not surprising since brain glucose utilization is estimated to represent at least 20% of total body glucose utilization.[69]

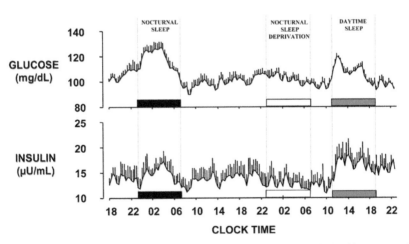

GLUCOSE AND INSULIN PROFILES
under continuous glucose infusion at a constant rate

Figure 2.8: Mean (+ SEM) plasma glucose and insulin profiles in a group of 8 healthy men, aged 20–27 years, investigated, under continuous glucose infusion at a constant rate, during a 53 h period that included 8 h of nocturnal sleep, 28 h of sleep deprivation, and 8 h of daytime sleep. The black bars denote the nocturnal sleep period, the open bars denote the period of nocturnal sleep deprivation, and the grey bars denote the period of daytime sleep. Blood samples were obtained at 20 min intervals. (Adapted from Van Cauter and Spiegel[100]).

Therefore, any modification in cerebral glucose utilization will markedly impact glucose tolerance. Experimental protocols involving constant intravenous glucose infusion or constant parenteral nutrition have shown that in healthy subjects, glucose tolerance deteriorates during the first part of the night to reach a minimum around mid-sleep. Thereafter, glucose tolerance improves to return to the daytime levels.[33,70,71] This nocturnal decrease in glucose tolerance results from a reduction in both peripheral and cerebral glucose utilization.[70,72] On the one hand, the sleep-onset associated GH release is likely to inhibit glucose uptake by insulin-dependent tissues. On the other hand, it was evidenced by positron emission tomography studies that glucose utilization by the brain is markedly lower during SWS (a dominant sleep stage during the first part of the sleep period in healthy young subjects) than during REM sleep or during wake.[70,72,73] Thus, glucose regulation is markedly modulated by the sleep–wake homeostasis.

Glucose regulation is markedly disturbed by sleep restriction. Multiple epidemiologic studies indicate a robust association between short sleep duration and the risk of developing diabetes.[74] Moreover, individuals with a parental history of type 2 diabetes have an increased risk of developing diabetes if they are short sleepers.

Those observations are consistent with data obtained in several well-controlled laboratory studies. In a *princeps* experimental landmark study, published in 1999, Spiegel et al. investigated, under strictly controlled conditions of caloric intake and physical activity, normal young men submitted to 6 days of partial sleep deprivation without circadian disruption (4 h in bed, centered around 03 h), followed by 6 days of sleep extension (12 h in bed, centered around 03 h).[41] Sleep restriction induced important decreases in the duration of light non-REM (stages 1 and 2) and of REM sleep, but SWS was largely preserved. Sleep curtailment was associated with marked alterations of glucose metabolism, as shown in Figure 2.9: increased insulin resistance, decreased insulin response to intravenous glucose infusion, decreased glucose tolerance. The disposition index (product of acute insulin response to glucose and insulin sensitivity), a good marker of diabetes risk, was reduced by more than 30%.[75] Actually, glucose tolerance fell to values observed in healthy elderly subjects with normal sleep schedule, suggesting that chronic sleep restriction might accelerate the age-related deterioration of glucose metabolism.

Deleterious effects of sleep curtailment on glucose metabolism were confirmed in a number of subsequent experimental studies which found that partial sleep curtailment for 1–7 days is likely to result in the development of insulin resistance without compensatory increase in insulin release, resulting in impaired glucose tolerance.[75–77] In a randomized crossover study performed in healthy volunteers, cellular insulin sensitivity in subcutaneous fat was reduced after sleep restriction (4.5 h in bed) compared with the normal sleep duration (8.5 h in bed).[78] Recently, sleep curtailment was shown to be associated with nocturnal and early morning elevations in circulating levels of free fatty acids, which may contribute to insulin resistance and increased diabetes risk.[79]

Conversely, a recent study has demonstrated a beneficial effect of sleep extension

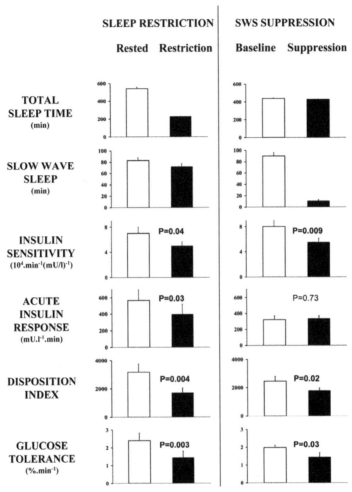

Figure 2.9: Effects of sleep restriction and of slow-wave sleep suppression on total sleep time, SWS, insulin sensitivity, acute insulin response to intravenous glucose infusion, disposition index and glucose tolerance in healthy young subjects. Values shown are mean + SEM. (Data from Spiegel et al.[41,75] and Tasali et al.[81]).

on fasting insulin sensitivity in adults with habitual sleep restriction.[80]

Even alterations in sleep quality, without any alteration in sleep duration, may have deleterious effects on glucose metabolism: in a very elegant study performed in nine healthy volunteers, an all-night selective SWS suppression, with preservation of REM sleep and of total sleep duration, was found to result in a decrease in insulin sensitivity without adequate compensatory increase in insulin levels, resulting in a reduction of the disposition index and of glucose tolerance (Fig. 2.9). A robust correlation was evidenced between the decrease in SWS and the decrease in insulin sensitivity.[81]

Thus, both a reduction in sleep duration with preservation of SWS, and a reduction of SWS with preservation of total sleep duration, may exert negative effects on glucose metabolism and enhance the risk of type 2 diabetes. A recent study provides evidence that this risk is further increased when sleep loss is associated with circadian misalignment, as it is the case among the shift workers.[82]

Obstructive Sleep Apnea and Glucose Metabolism

Obstructive sleep apnea (OSA), the most common type of sleep apnea, characterized by repetitive 20 to 40 s pauses in breathing during sleep, results in intermittent hypoxia and in marked sleep disturbances: fragmented and shallow sleep with reduced SWS and reduced total sleep duration. OSA is more frequently observed in obese men.[74] Large population-based studies have evidenced robust associations between OSA and insulin resistance, resulting in reduced glucose tolerance. These observations were confirmed by several well-controlled experimental investigations. Healthy volunteers were submitted, while awake, to 5 h per day of intermittent hypoxia, resulting in episodes of desaturation similar to those observed in moderate OSA.[83] Glucose effectiveness and insulin sensitivity were reduced, without any compensatory increase in insulin release. Overall, both experimental and epidemiologic studies strongly suggest that OSA-associated hypoxia may increase diabetes risk.[74]

2.7. LEPTIN AND GHRELIN: HORMONES INVOLVED IN APPETITE REGULATION
Interaction of the Circadian Clock, the Sleep–Wake Homeostasis and the Food Intake Schedule

2.7.1. Leptin

Leptin, an anorexigenic hormone mainly released by the adipocytes, is a key component of the energy balance. Overall, circulating leptin levels reflect cumulative energy balance. Leptin levels decrease in response to underfeeding, together with an increase in hunger. Conversely, they increase in response to overfeeding, together with a decrease in hunger.[84, 85] However, in healthy lean subjects, 24 h leptin profiles undergo important diurnal variations, with minimum values during the daytime and maximum values during the early or midnocturnal sleep.[86,87] This rhythm persists, albeit with smaller amplitude, in subjects receiving continuous enteral nutrition[71] and in subjects submitted to 88 consecutive hours of wakefulness.[88,89] Following an abrupt shift of the sleep–wake schedule, nocturnal leptin levels rise in spite of the absence of sleep,[71] indicating that diurnal leptin variations are, at least partially, under the control of the circadian clock. However, after the shift of the sleep period, another rise in leptin levels is also observed following the onset of daytime recovery sleep,[71] indicating an effect of the sleep–wake cycle. Thus, leptin diurnal variations appear to result from the interaction of the food intake schedule, the circadian clock and the sleep–wake homeostasis.

2.7.2. Ghrelin

The acylated form of ghrelin, an orexigenic hormone, which also stimulates the pituitary GH secretion, is another key component of the energy balance.[90] Ghrelin is known to reduce energy expenditure, increase glucose production by the liver and promote fat retention. In healthy subjects, daytime profiles are primarily dependent on the food intake schedule: circulating total and acylated ghrelin levels rapidly decrease after each meal to reach minimum levels 90–120 min after meal presentation, followed by a rebound until the following meal, irrespective of time of day. However, after evening dinner, this rebound only persists until the first part of the sleep period. Thereafter, ghrelin levels progressively decline despite prolonged fasting. Moreover, the ratio of acylated (the active form of the hormone) to total ghrelin is higher during wake than during sleep.[90] These findings strongly suggest a sleep-associated inhibition of the orexigenic signal. In another experimental investigation involving three consecutive days of fasting, ghrelin profiles showed diurnal variations, with minimum levels in early morning, peak levels in the afternoon, followed by a progressive nocturnal decrease, in mirror image of cortisol variations. Thus, ghrelin diurnal variations also appear to result from the interaction of the food intake schedule, the circadian clock and the sleep–wake homeostasis.

2.7.3. Effects of Sleep Deprivation on Leptin, Ghrelin and Hunger

Sleep deprivation exerts major effects on leptin, ghrelin, and hunger regulation. In another pioneering study published in 2004, Spiegel et al. investigated healthy normal-weight young men who were included in a randomized crossover investigation, performed under strictly controlled conditions of caloric intake and physical activity and without any circadian disruption. All subjects participated in two sessions spaced at least six weeks apart. One session involved two consecutive nights with 10 h in bed (sleep extension), the other session involved two consecutive nights with only 4 h in bed (sleep restriction). The day after the two consecutive nights of sleep extension or sleep restriction, hunger and appetite were rated every hour, and circulating ghrelin and leptin levels were measured at frequent intervals from 8 h to 21 h. After sleep restriction, hunger ratings, appetite ratings and ghrelin levels were consistently higher, and leptin levels consistently lower, than after sleep extension. The increase in hunger rating was proportional to the increase in the ghrelin-to-leptin ratio (Fig. 2.10).[91]

Similar results were obtained in subsequent studies performed in healthy normal-weight men investigated under strictly controlled conditions of caloric intake. A dose-response relationship between sleep duration and 24 h leptin profiles was evidenced in a study performed in healthy young subjects, comparing leptin levels after six consecutive nights with 4 h in bed with levels observed after six consecutive nights with 8 h in bed and with levels recorded after six consecutive nights with 12 h in bed: circulating leptin levels progressively decreased from the 12 h in bed to the 4 h in bed condition.[75] Between the 12 h and the 4 h in bed conditions, the decrease in the maximum value of the 24 h leptin profile was similar to changes observed under normal sleep conditions in

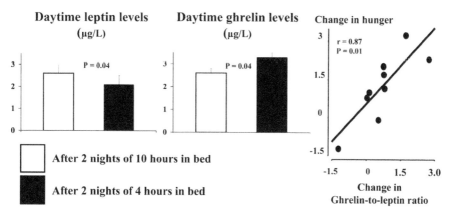

Figure 2.10. Effects of sleep restriction on daytime circulating levels of leptin and ghrelin (left panels). Correlation (Spearman coefficient) between the changes in hunger ratings and the changes in ghrelin-to-leptin ratio after sleep restriction (4 h in bed) compared with sleep extension (10 h in bed) (right panel). (Data from Spiegel et al.[91]).

healthy subjects fed only 70% of their energy requirements during three consecutive nights.

On the other hand, in middle-aged overweight sedentary subjects, sleep restriction (14 consecutive nights with only 5.5 h in bed vs 14 consecutive nights with 8.5 h in bed), with ad libitum access to food, was found to result in an increased intake of calories from snacks, while regular meal intake was not affected.[92] Thus, in a obesity-promoting environment, recurrent sleep curtailment may induce excessive consumption of calorie-dense food.

Using a similar protocol, but with identical moderate caloric restriction in both conditions, total weight loss (~ 3 kg) was quite similar in both conditions. However, sleep restriction, compared with the normal sleep condition, resulted in a 55% reduction in fat loss, together with a 60% increase in lean mass loss (Fig. 2.11), suggesting the existence of mechanistic relationships between energy metabolism

and sleep. Indeed, despite similar energy consumption, sleep curtailment was associated with increased 24 h levels of acylated ghrelin, resulting in a reduction in fat oxidation and in resting metabolic rate, and an increase in respiratory quotient values.[93]

The effects of sleep restriction on energy intake and energy expenditure were investigated in healthy normal-weight young volunteers enrolled in a crossover study involving five consecutive nights of sleep restriction without circadian disruption and five consecutive nights of normal sleep duration, with ad libitum access to food. As expected, sleep restriction was associated with a ~ 5% increase in total daily energy expenditure, but food intake (especially at night, following dinner) increased beyond what was required to maintain energy balance. This imbalance resulted in a weight gain averaging ~ 0.8 kg after five days.[94]

Brain functional magnetic resonance was used to investigate the impact of sleep deprivation on central brain mechanisms

WEIGHT LOSS
14 days of moderate caloric restriction, similar in both conditions
Effects of sleep restriction

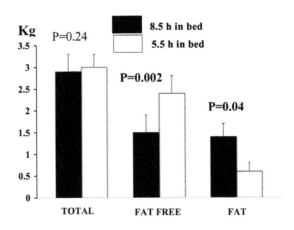

Figure 2.11: Effects of sleep restriction on weight loss in middle-aged overweight subjects submitted to moderate caloric restriction during 14 consecutive days. Values shown are mean + SEM. (Data from Nedeltcheva et al.[93]).

regulating food desire. In a first random-ized crossover study, healthy normal-weight subjects were assigned, four weeks apart, to either six consecutive nights of ha-bitual sleep (9 h per night in bed) or to six consecutive nights of restricted sleep (4 h per night in bed). The volunteers received a controlled diet during the first four days and had ad libitum access to food the last two days of each session. On the morn-ing of day 6, after an overnight fast, sleep restriction was found to be associated with increased stimulation of cerebral regions sensitive to food stimuli, as evidenced by functional magnetic resonance.[95] In anoth-er study, healthy normal-weight subjects

completed two experimental sessions: one night of normal sleep, one night of total sleep deprivation. Experimental sessions, placed at least seven days apart, were coun-terbalanced in order across participants. The morning following sleep depriva-tion, the desire for high-calorie food was increased and brain functional magnetic resonance showed increased activity with-in the amygdala (leading to stimulate de-sire for high-calorie food), contrasting with decreased activity in appetitive evaluation cortical regions (leading to improper food choice selection). Furthermore, subjec-tive sleepiness following sleep deprivation correlated positively with the desire for high-calorie food.[96] It was also reported in another study that healthy normal-weight men, when given a fixed budget to get food at their convenience, purchased more calories and more grams on the morning following one night of total sleep deprivation than after one night of normal sleep.[97]

Conversely, it was shown in a recent study that in overweight young subjects who habitually curtail their sleep, sleep extension is associated with a decrease in overall appetite and in desire for salty and sweet foods, without any effect on desire for vegetables, fruits and protein-rich nutrients.[98]

A large population study, involving more than 1000 pairs of twins, suggests that sleep restriction may also provide a permissive environment for the expression of genes that promote obesity. As expected, it was found that shorter sleep duration was associated with increased body mass index (BMI). But in addition, it was also evidenced that the heritability of BMI was more than twice as large when sleep duration was less than 7 h per night than when it was more than 9 h per night.[99]

2.8. CONCLUSIONS

Sleep–wake homeostasis and the endocrine–metabolic system closely interact with each other. Although a progressive adaptation to minor partial sleep deprivation cannot be definitively excluded, a bulk of epidemiologic and experimental laboratory studies provides consistent and convincing evidence that recurrent sleep curtailment is likely to accelerate the senescence of endocrine–metabolic function, increase the risk of diabetes and of obesity and compromise the efficacy of dietary

strategies in overweight patients. Considering the current epidemics of obesity and diabetes and the fact that recurrent sleep curtailment, whether voluntary or not, is a hallmark of modern society, strategies to optimize sleep duration and quality should be considered for the prevention and the treatment of obesity and diabetes.

KEY WORDS

- **growth hormone**
- **ACTH**
- **cortisol**
- **thyrotropin**
- **gonadotropins**
- **LH**
- **FSH**
- **testosterone**
- **glucose**
- **insulin**
- **obstructive sleep apnea**
- **leptin**
- **ghrelin**
- **sleep**

REFERENCES

1. Van Cauter, E.; Leproult, R.; Plat, L. Age-Related Changes in Slow Wave Sleep and REM Sleep and Relationship with Growth Hormone and Cortisol Levels in Healthy Men. *Jama.* **2000**, *284*, 861–868.
2. Copinschi, G.; Challet, E. Endocrine Rhythms, the Sleep-Wake Cycle, and Biological Clocks. In *Endocrinology: Adult and Pediatric*, Jameson, L. J.; DeGroot, L. J., Eds.; Elsevier -Saunders: Philadelphia, 2015; Vol. 1; 147–173.
3. Van Cauter, E.; Plat, L.; Copinschi, G. Interrelations Between Sleep and the Somatotropic Axis. *Sleep* **1998**, *21*, 553–566.

4. Ho, K. Y.; Evans, W. S.; Blizzard, R. M.; Veld-huis, J. D.; Merriam, G. R.; Samojlik, E.; Furla-netto, R.; Rogol, A. D.; Kaiser, D. L.; Thorner, M. O. Effects of Sex and Age on the 24-Hour Profile of Growth Hormone Secretion in Man: Importance of Endogenous Estradiol Concen-trations. *J. Clin. Endocrinol. Metab.* **1987**, *64*, 51–58.

5. Van Cauter, E. Spiegel, K, Circadian and Sleep Control of Endocrine Secretions. In *Regulation of sleep and circadian rhythms*, F.W. Turek, F. W.; Zee, P. C. Eds.; Marcel Dekker, Inc: New york, 1999; pp 397–426.

6. Gronfier, C.; Luthringer, R.; Follenius, M.; Schaltenbrand, N.; Macher, J. P.; Muzet, A.; Brandenberger, G. A Quantitative Evaluation of the Relationships Between Growth Hormone Secretion and Delta Wave Electroencephalo-graphic Activity During Normal Sleep and after Enrichment in Delta Waves. *Sleep* **1996**, *19*, 817–824.

7. Holl, R. W.; Hartman, M. L.; Veldhuis, J. D.; Taylor, W. M.; Thorner, M. O. Thirty-Second Sampling of Plasma Growth Hormone in Man: Correlation with Sleep Stages. *J. Clin. Endocri-nol. Metab.* **1991**, *72*, 854–861.

8. Van Cauter, E.; Kerkhofs, M.; Caufriez, A.; Van Onderbergen, A.; Thorner, M. O.; Copinschi, G. A Quantitative Estimation of Growth Hor-mone Secretion in Normal Man: Reproduc-ibility and Relation to Sleep and Time of Day. *J. Clin. Endocrinol. Metab.* **1992**, *74*, 1441–1450.

9. Van Cauter, E.; Caufriez, A.; Kerkhofs, M.; Van Onderbergen, A.; Thorner, M. O.; Copinschi, G. Sleep, Awakenings, and Insulin-like Growth Factor-I Modulate the Growth Hormone (GH) Secretory Response to GH-Releasing Hormone. *J. Clin. Endocrinol. Metab.* **1992**, *74*, 1451–1459.

10. Weibel, L.; Spiegel, K.; Gronfier, C.; Follenius, M.; Brandenberger, G. Twenty-Four-Hour Mel-atonin and Core Body Temperature Rhythms: Their Adaptation in Night Workers. *Am. J. Physiol.* **1997**, *272*, R948–954.

11. Kerkhofs, M.; Van Cauter, E.; Van Onderber-gen, A.; Caufriez, A.; Thorner, M. O.; Copin-schi, G. Sleep-Promoting Effects of Growth Hormone-Releasing Hormone in Normal Men. *Am. J. Physiol.* **1993**, *264*, E594–598.

12. Obal, F., Jr.; Alfoldi, P.; Cady, A. B.; Johannsen, L.; Sary, G.; Krueger, J. M. Growth Hormone-Releasing Factor Enhances Sleep in Rats and Rabbits. *Am. J. Physiol.* **1988**, *255*, R310–316.

13. Obal, F., Jr.; Floyd, R.; Kapas, L.; Bodosi, B.; Krueger, J. M. Effects of Systemic GHRH on Sleep in Intact and Hypophysectomized Rats. *Am. J. Physiol.* **1996**, *270*, E230–237.

14. Ocampo-Lim, B.; Guo, W.; DeMott-Friberg, R.; Barkan, A. L.; Jaffe, C. A. Nocturnal Growth Hormone (GH) Secretion Is Eliminated by In-fusion of GH-Releasing Hormone Antagonist. *J. Clin. Endocrinol. Metab.* **1996**, *81*, 4396–4399.

15. Van Cauter, E.; Plat, L.; Scharf, M. B.; Lep-roult, R.; Cespedes, S.; L'Hermite-Baleriaux, M.; Copinschi, G. Simultaneous Stimulation of Slow-Wave Sleep and Growth Hormone Secre-tion by Gamma-Hydroxybutyrate in Normal Young Men. *J. Clin. Invest.* **1997**, *100*, 745–53.

16. Shah, N.; Evans, W. S.; Veldhuis, J. D. Actions of Estrogen on Pulsatile, Nyctohemeral, and En-tropic Modes of Growth Hormone Secretion. *Am. J. Physiol.* **1999**, *276*, R1351–1358.

17. Caufriez, A.; Leproult, R.; L'Hermite-Baleriaux, M.; Moreno-Reyes, R.; Copinschi, G. A Poten-tial Role of Endogenous Progesterone in Modu-lation of GH, Prolactin and Thyrotrophin Se-cretion During Normal Menstrual Cycle. *Clin. Endocrinol. (Oxf)* **2009**, *71*, 535–542.

18. Van Coevorden, A.; Mockel, J.; Laurent, E.; Kerkhofs, M.; L'Hermite-Baleriaux, M.; De-coster, C.; Neve, P.; Van Cauter, E. Neuroendo-crine Rhythms and Sleep in Aging Men. *Am. J. Physiol.* **1991**, *260*, E651–661.

19. Veldhuis, J. D.; Liem, A. Y.; South, S.; Weltman, A.; Weltman, J.; Clemmons, D. A.; Abbott, R.; Mulligan, T.; Johnson, M. L.; Pincus, S.; et al. Differential Impact of Age, Sex Steroid Hor-mones, and Obesity on Basal Versus Pulsatile Growth Hormone Secretion in Men as Assessed in an Ultrasensitive Chemiluminescence Assay. *J. Clin. Endocrinol. Metab.* **1995**, *80*, 3209–3222.

20. Vermeulen, A. Nyctohemeral Growth Hor-mone Profiles in Young and Aged Men: Cor-relation with Somatomedin-C Levels. *J. Clin. Endocrinol. Metab.* **1987**, *64*, 884–888.

21. Caufriez, A.; Leproult, R.; L'Hermite-Baleriaux, M.; Kerkhofs, M.; Copinschi, G. Progesterone Prevents Sleep Disturbances and Modulates GH, TSH, and Melatonin Secretion in Post-menopausal Women. *J. Clin. Endocrinol. Metab.* **2011**, *96*, E614–623.

22. Caufriez, A.; Frankenne, F.; Englert, Y.; Gol-stein, J.; Cantraine, F.; Hennen, G.; Copinschi,

G. Placental Growth Hormone as a Potential Regulator of Maternal IGF-I During Human Pregnancy. *Am. J. Physiol.* **1990**, *258*, E1014–1019.

23. Caufriez, A.; Frankenne, F.; Hennen, G.; Copinschi, G. Regulation of Maternal IGF-I by Placental GH in Normal and Abnormal Human Pregnancies. *Am. J. Physiol.* **1993**, *265*, E572–577.

24. Eriksson, L.; Frankenne, F.; Eden, S.; Hennen, G.; Von Schoultz, B. Growth Hormone 24-h Serum Profiles During Pregnancy--Lack of Pulsatility for the Secretion of the Placental Variant. *Br. J. Obstet. Gynaecol.* **1989**, *96*, 949–953.

25. Hartman, M. L.; Veldhuis, J. D.; Vance, M. L.; Faria, A. C.; Furlanetto, R. W.; Thorner, M. O. Somatotropin Pulse Frequency and Basal Concentrations are Increased in Acromegaly and Are Reduced by Successful Therapy. *J. Clin. Endocrinol. Metab.* **1990**, *70*, 1375–1384.

26. Hartman, M. L.; Pincus, S. M.; Johnson, M. L.; Matthews, D. H.; Faunt, L. M.; Vance, M. L.; Thorner, M. O.; Veldhuis, J. D. Enhanced Basal and Disorderly Growth Hormone Secretion Distinguish Acromegalic from Normal Pulsatile Growth Hormone Release. *J. Clin. Invest.* **1994**, *94*, 1277–1288.

27. Van Cauter, E. v. C., A. Blackman, JD, Modulation of Neuroendocrine Release by Sleep and Circadian Rhythmicity. In *Advances in neuroendocrine regulation of reproduction*, S. Yen, W. V. Ed.; Serono symposium: Norwell, MA, 1990; pp 113–122.

28. Veldhuis, J. D.; Iranmanesh, A.; Johnson, M. L.; Lizarralde, G. Amplitude, But not Frequency, Modulation of Adrenocorticotropin Secretory Bursts Gives Rise to the Nyctohemeral Rhythm of the Corticotropic Axis in Man. *J. Clin. Endocrinol. Metab.* **1989**, *71*, 452–463.

29. Bierwolf, C.; Struve, K.; Marshall, L.; Born, J.; Fehm, H. L. Slow Wave Sleep Drives Inhibition of Pituitary-Adrenal Secretion in Humans. *J. Neuroendocrinol.* **1997**, *9*, 479–484.

30. Caufriez, A.; Moreno-Reyes, R.; Leproult, R.; Vertongen, F.; Van Cauter, E.; Copinschi, G. Immediate Effects of an 8-h Advance Shift of the Rest-Activity Cycle on 24-h Profiles of Cortisol. *Am. J. Physiol. Endocrinol. Metab.* **2002**, *282*, E1147–1153.

31. Follenius, M.; Brandenberger, G.; Bandesapt, J. J.; Libert, J. P.; Ehrhart, J. Nocturnal Cortisol Release in Relation to Sleep Structure. *Sleep* **1992**, *15*, 21–27.

32. Gronfier, C.; Luthringer, R.; Follenius, M.; Schaltenbrand, N.; Macher, J. P.; Muzet, A.; Brandenberger, G. Temporal Relationships between Pulsatile Cortisol Secretion and Electroencephalographic Activity During Sleep in Man. *Electroencephalogr. Clin. Neurophysiol.* **1997**, *103*, 405–408.

33. Van Cauter, E.; Blackman, J. D.; Roland, D.; Spire, J. P.; Refetoff, S.; Polonsky, K. S. Modulation of Glucose Regulation and Insulin Secretion by Circadian Rhythmicity and Sleep. *J. Clin. Invest.* **1991**, *88*, 934–942.

34. Weibel, L.; Follenius, M.; Spiegel, K.; Ehrhart, J.; Brandenberger, G. Comparative Effect of Night and Daytime Sleep on the 24-hour Cortisol Secretory Profile. *Sleep* **1995**, *18*, 549–556.

35. Weitzman, E. D.; Zimmerman, J. C.; Czeisler, C. A.; Ronda, J. Cortisol Secretion is Inhibited During Sleep in Normal Man. *J. Clin. Endocrinol. Metab.* **1983**, *56*, 352–358.

36. Spath-Schwalbe, E.; Gofferje, M.; Kern, W.; Born, J.; Fehm, H. L. Sleep Disruption Alters Nocturnal ACTH and Cortisol Secretory Patterns. *Biol Psychiatry* **1991**, *29*, 575–584.

37. Vgontzas, A. N.; Zoumakis, M.; Bixler, E. O.; Lin, H. M.; Prolo, P.; Vela-Bueno, A.; Kales, A.; Chrousos, G. P. Impaired Nighttime Sleep in Healthy Old Versus Young Adults is Associated with Elevated Plasma Interleukin-6 and Cortisol Levels: Physiologic and Therapeutic Implications. *J. Clin. Endocrinol. Metab.* **2003**, *88*, 2087–2095.

38. Harman, S. M.; Metter, E. J.; Tobin, J. D.; Pearson, J.; Blackman, M. R. Longitudinal Effects of Aging on Serum Total and Free Testosterone Levels in Healthy Men. Baltimore Longitudinal Study of Aging. *J. Clin. Endocrinol. Metab.* **2001**, *86*, 724–731.

39. Van Cauter, E.; Leproult, R.; Kupfer, D. J. Effects of Gender and Age on the Levels and Circadian Rhythmicity of Plasma Cortisol. *J. Clin. Endocrinol. Metab.* **1996**, *81*, 2468–2473.

40. Leproult, R.; Copinschi, G.; Buxton, O.; Van Cauter, E. Sleep Loss Results in an Elevation of Cortisol Levels the Next Evening. *Sleep* **1997**, *20*, 865–870.

41. Spiegel, K.; Leproult, R.; Van Cauter, E. Impact of Sleep Debt on Metabolic and Endocrine Function. *Lancet* **1999**, *354*, 1435–1439.

42. McEwen, B. S. Stress, Adaptation, and Disease. Allostasis and Allostatic Load. *Ann. N. Y. Acad. Sci.* **1998**, *840*, 33–44.

43. Dallman, M. F.; Strack, A. M.; Akana, S. F.; Bradbury, M. J.; Hanson, E. S.; Scribner, K. A.; Smith, M. Feast and Famine: Critical Role of Glucocorticoids with Insulin in Daily Energy Flow. *Front. Neuroendocrinol.* **1993**, *14*, 303–347.

44. Dennison, E.; Hindmarsh, P.; Fall, C.; Kellingray, S.; Barker, D.; Phillips, D.; Cooper, C. Profiles of Endogenous Circulating Cortisol and Bone Mineral Density in Healthy Elderly Men. *J. Clin. Endocrinol. Metab.* **1999**, *84*, 3058–3063.

45. Plat, L.; Leproult, R; L'Hermite-Balériaux, M.; Féry, F.; Mockel, J.; Polonsky, K.S.; Van Cauter, E. Metabolic Effects of Short-Term Physiological Elevations of Plasma Cortisol Are More Pronounced in the Evening Than in the Morning. *J. Clin. Endocrinol. Metab.* **1999**, *84*, 3082–3092.

46. Ehlers, C. L.; Reed, T. K.; Henriksen, S. J. Effects of Corticotropin-Releasing Factor and Growth Hormone-Releasing Factor on Sleep and Activity in Rats. *Neuroendocrinology* **1986**, *42*, 467–474.

47. Opp, M.; Obal, F., Jr.; Krueger, J. M. Corticotropin-Releasing Factor Attenuates Interleukin 1-Induced Sleep and Fever in Rabbits. *Am. J. Physiol.* **1989**, *257*, R528–535.

48. Chang, F. C.; Opp, M. R. Blockade of Corticotropin-Releasing Hormone Receptors Reduces Spontaneous Waking in the Rat. *Am. J. Physiol.* **1998**, *275*, R793–802.

49. Holsboer, F.; von Bardeleben, U.; Steiger, A. Effects of Intravenous Corticotropin-Releasing Hormone upon Sleep-Related Growth Hormone Surge and Sleep EEG in Man. *Neuroendocrinology* **1988**, *48*, 32–38.

50. Tsuchiyama, Y.; Uchimura, N.; Sakamoto, T.; Maeda, H.; Kotorii, T. Effects of hCRH on Sleep and Body Temperature Rhythms. *Psychiatry. Clin. Neurosci.* **1995**, *49*, 299–304.

51. Vgontzas, A. N.; Bixler, E. O.; Wittman, A. M.; Zachman, K.; Lin, H. M.; Vela-Bueno, A.; Kales, A.; Chrousos, G. P. Middle-Aged Men Show Higher Sensitivity of Sleep to the Arousing Effects of Corticotropin-Releasing Hormone Than Young Men: Clinical Implications. *J. Clin. Endocrinol. Metab.* **2001**, *86*, 1489–1495.

52. Bohlhalter, S.; Murck, H.; Holsboer, F.; Steiger, A. Cortisol Enhances non-REM Sleep and Growth Hormone Secretion in Elderly Subjects. *Neurobiol. Aging* **1997**, *18*, 423–429.

53. Born, J.; DeKloet, E. R.; Wenz, H.; Kern, W.; Fehm, H. L. Gluco- and Antimineralocorticoid Effects on Human Sleep: A Role of Central Corticosteroid Receptors. *Am. J. Physiol.* **1991**, *260*, E183–188.

54. Friess, E.; U, V. B.; Wiedemann, K.; Lauer, C. J.; Holsboer, F. Effects of Pulsatile Cortisol Infusion on Sleep-EEG and Nocturnal Growth Hormone Release in Healthy Men. *J. Sleep Res.* **1994**, *3*, 73–79.

55. Brabant, G.; Prank, K.; Ranft, U.; Schuermeyer, T.; Wagner, T. O.; Hauser, H.; Kummer, B.; Feistner, H.; Hesch, R. D.; von zur Muhlen, A. Physiological Regulation of Circadian and Pulsatile Thyrotropin Secretion in Normal Man and Woman. *J. Clin. Endocrinol. Metab.* **1990**, *70*, 403–409.

56. Goichot, B.; Brandenberger, G.; Saini, J.; Wittersheim, G.; Follenius, M. Nocturnal Plasma Thyrotropin Variations are Related to Slow-Wave Sleep. *J. Sleep Res.* **1992**, *1*, 186–190.

57. Gronfier, C.; Luthringer, R.; Follenius, M.; Schaltenbrand, N.; Macher, J. P.; Muzet, A.; Brandenberger, G. Temporal Link Between Plasma Thyrotropin Levels and Electroencephalographic Activity in Man. *Neurosci. Lett.* **1995**, *200*, 97–100.

58. Parker, D. C.; Pekary, A. E.; Hershman, J. M. Effect of Normal and Reversed Sleep-Wake Cycles upon Nyctohemeral Rhythmicity of Plasma Thyrotropin: Evidence Suggestive of an Inhibitory Influence in Sleep. *J. Clin. Endocrinol. Metab.* **1976**, *43*, 318–329.

59. Hirschfeld, U.; Moreno-Reyes, R.; Akseki, E.; L'Hermite-Baleriaux, M.; Leproult, R.; Copinschi, G.; Van Cauter, E. Progressive Elevation of Plasma Thyrotropin During Adaptation to Simulated Jet Lag: Effects of Treatment with Bright Light or Zolpidem. *J. Clin. Endocrinol. Metab.* **1996**, *81*, 3270–3277.

60. Apter, D.; Butzow, T. L.; Laughlin, G. A.; Yen, S. S. Gonadotropin-Releasing Hormone Pulse Generator Activity During Pubertal Transition in Girls: Pulsatile and Diurnal Patterns of Circulating Gonadotropins. *J. Clin. Endocrinol. Metab.* **1993**, *76*, 940–949.

61. Filicori, M.; Santoro, N.; Merriam, G. R.; Crowley, W. F., Jr. Characterization of the Physiological Pattern of Episodic Gonadotropin Secretion

Throughout the Human Menstrual Cycle. *J. Clin. Endocrinol. Metab.* **1986**, *62*, 1136–1144.

62. Reame, N.; Sauder, S. E.; Kelch, R. P.; Marshall, J. C. Pulsatile Gonadotropin Secretion During the Human Menstrual Cycle: Evidence for Altered Frequency of Gonadotropin-Releasing Hormone Secretion. *J. Clin. Endocrinol. Metab.* **1984**, *59*, 328–337.

63. Hall, J. E.; Sullivan, J. P.; Richardson, G. S. Brief Wake Episodes Modulate Sleep-Inhibited Luteinizing Hormone Secretion in the Early Follicular Phase. *J. Clin. Endocrinol. Metab.* **2005**, *90*, 2050–2055.

64. Lejeune-Lenain, C.; Van Cauter, E.; Desir, D.; Beyloos, M.; Franckson, J. R. Control of Circadian and Episodic Variations of Adrenal Androgens Secretion in Man. *J. Endocrinol. Invest.* **1987**, *10*, 267–276.

65. Axelsson, J.; Ingre, M.; Akerstedt, T.; Holmback, U. Effects of Acutely Displaced Sleep on Testosterone. *J. Clin. Endocrinol. Metab.* **2005**, *90*, 4530–4535.

66. Leproult, R.; Van Cauter, E. Effect of 1 Week of Sleep Restriction on Testosterone Levels in Young Healthy Men. *JAMA* **2011**, *305*, 2173–2174.

67. Penev, P. D. Association Between Sleep and Morning Testosterone Levels in Older Men. *Sleep* **2007**, *30*, 427–432.

68. Van Cauter, E.; Polonsky, K. S.; Scheen, A. J. Roles of Circadian Rhythmicity and Sleep in Human Glucose Regulation. *Endocr. Rev.* **1997**, *18*, 716–738.

69. Magistretti, P. J. Neuron-glia Metabolic Coupling and Plasticity. *J. Exp. Biol.* **2006**, *209*, 2304–2311.

70. Scheen, A. J.; Byrne, M. M.; Plat, L.; Leproult, R.; Van Cauter, E. Relationships Between Sleep Quality and Glucose Regulation in Normal Humans. *Am. J. Physiol.* **1996**, *271*, E261–270.

71. Simon, C.; Gronfier, C.; Schlienger, J. L.; Brandenberger, G. Circadian and Ultradian Variations of Leptin in Normal Man Under Continuous Enteral Nutrition: Relationship to Sleep and Body Temperature. *J. Clin. Endocrinol. Metab.* **1998**, *83*, 1893–1899.

72. Maquet, P. Functional Neuroimaging of Normal Human Sleep by Positron Emission Tomography. *J. Sleep Res.* **2000**, *9*, 207–231.

73. Maquet, P. Positron Emission Tomography Studies of Sleep and Sleep Disorders. *J. Neurol.* **1997**, *244*, S23–28.

74. Reutrakul, S.; VanCauter, E. Interactions Between Sleep, Circadian Function and Glucose Metabolism: Implications for Risk and Severity of Diabetes. *Annals of the New York Academy of Sciences* **2014**, *1311*, 151–173.

75. Spiegel, K.; Leproult, R.; L'Hermite-Baleriaux, M.; Copinschi, G.; Penev, P. D.; Van Cauter, E. Leptin Levels Are Dependent on Sleep Duration: Relationships with Sympathovagal Balance, Carbohydrate Regulation, Cortisol, and Thyrotropin. *J. Clin. Endocrinol. Metab.* **2004**, *89*, 5762–5771.

76. Buxton, O. M.; Pavlova, M.; Reid, E. W.; Wang, W.; Simonson, D. C.; Adler, G. K. Sleep Restriction for 1 Week Reduces Insulin Sensitivity in Healthy Men. *Diabetes* **2010**, *59*, 2126–2133.

77. Spiegel, K.; Knutson, K.; Leproult, R.; Tasali, E.; Van Cauter, E. Sleep Loss: A Novel Risk Factor for Insulin Resistance and Type 2 Diabetes. *J. Appl. Physiol.* **2005**, *99*, 2008–2019.

78. Broussard, J. L.; Ehrmann, D. A.; Van Cauter, E.; Tasali, E.; Brady, M. J. Impaired Insulin Signaling in Human Adipocytes after Experimental Sleep Restriction: A Randomized, Crossover Study. *Ann. Intern. Med.* **2012**, *157*, 549–557.

79. Broussard, J. L.; Chapotot, F.; Abraham, V.; Day, A.; Delebecque, F.; Whitmore, H. R.; Tasali, E. Sleep Restriction Increases Free Fatty Acids in Healthy Men. *Diabetologia.* **2015**, *58*, 791–798.

80. Leproult, R.; Deliens, G.; Gilson, M.; Peigneux, P. Beneficial Impact of Sleep Extension on Fasting Insulin Sensitivity in Adults with Habitual Sleep Restriction. *Sleep* **2014**.

81. Tasali, E.; Leproult, R.; Ehrmann, D. A.; Van Cauter, E. Slow-Wave Sleep and the Risk of Type 2 Diabetes in Humans. *Proc. Natl. Acad. Sci. U.S.A.* **2008**, *105*, 1044–1049.

82. Leproult, R.; Holmback, U.; Van Cauter, E. Circadian Misalignment Augments Markers of Insulin Resistance and Inflammation, Independently of Sleep Loss. *Diabetes* **2014**, *63*, 1860–1869.

83. Louis, M.; Punjabi, N. M. Effects of Acute Intermittent Hypoxia on Glucose Metabolism in Awake Healthy Volunteers. *J. Appl. Physiol. (1985)* **2009**, *106*, 1538–1544.

84. Kershaw, E. E.; Flier, J. S. Adipose Tissue as an Endocrine Organ. *J. Clin. Endocrinol. Metab.* **2004**, *89*, 2548–2556.

85. Flier, J. S. Obesity Wars: Molecular Progress Confronts an Expanding Epidemic. *Cell* **2004**, *116*, 337–350.

86. Sinha, M. K.; Ohannesian, J. P.; Heiman, M. L.; Kriauciunas, A.; Stephens, T. W.; Magosin, S.; Marco, C.; Caro, J. F. Nocturnal Rise of Leptin in Lean, Obese, and Non-insulin-Dependent Diabetes Mellitus Subjects. *J. Clin. Invest.* **1996**, *97*, 1344–1347.

87. Saad, M. F.; Riad-Gabriel, M. G.; Khan, A.; Sharma, A.; Michael, R.; Jinagouda, S. D.; Boyadjian, R.; Steil, G. M. Diurnal and Ultradian Rhythmicity of Plasma Leptin: Effects of Gender and Adiposity. *J. Clin. Endocrinol. Metab.* **1998**, *83*, 453–459.

88. Mullington, J. M.; Chan, J. L.; Van Dongen, H. P.; Szuba, M. P.; Samaras, J.; Price, N. J.; Meier-Ewert, H. K.; Dinges, D. F.; Mantzoros, C. S. Sleep Loss Reduces Diurnal Rhythm Amplitude of Leptin in Healthy Men. *J. Neuroendocrinol.* **2003**, *15*, 851–854.

89. Shea, S. A.; Hilton, M. F.; Orlova, C.; Ayers, R. T.; Mantzoros, C. S. Independent Circadian and Sleep/Wake Regulation of Adipokines and Glucose in Humans. *J. Clin. Endocrinol. Metab.* **2005**, *90*, 2537–2544.

90. Spiegel, K.; Tasali, E.; Leproult, R.; Scherberg, N.; Van Cauter, E. Twenty-Four-Hour Profiles of Acylated and Total Ghrelin: Relationship with Glucose Levels and Impact of Time of Day and Sleep. *J. Clin. Endocrinol. Metab.* **2011**, *96*, 486–493.

91. Spiegel, K.; Tasali, E.; Penev, P.; Van Cauter, E. Brief Communication: Sleep Curtailment in Healthy Young Men Is Associated with Decreased Leptin Levels, Elevated Ghrelin Levels, and Increased Hunger and Appetite. *Ann. Intern. Med.* **2004**, *141*, 846–850.

92. Nedeltcheva, A. V.; Kilkus, J. M.; Imperial, J.; Kasza, K.; Schoeller, D. A.; Penev, P. D. Sleep Curtailment Is Accompanied by Increased Intake of Calories from Snacks. *Am. J. Clin. Nutr.* **2009**, *89*, 126–133.

93. Nedeltcheva, A. V.; Kilkus, J. M.; Imperial, J.; Schoeller, D. A.; Penev, P. D. Insufficient Sleep Undermines Dietary Efforts to Reduce Adiposity. *Ann. Intern. Med.* **2010**, *153*, 435–441.

94. Markwald, R. R.; Melanson, E. L.; Smith, M. R.; Higgins, J.; Perreault, L.; Eckel, R. H.; Wright, K. P. Jr. Impact of Insufficient Sleep on Total Daily Energy Expenditure, Food Intake, and Weight Gain. *Proc. Natl. Acad. Sci. U.S.A.* **2013**, *110*, 5695–5700.

95. St-Onge, M. P.; McReynolds, A.; Trivedi, Z. B.; Roberts, A. L.; Sy, M.; Hirsch, J. Sleep Restriction Leads to Increased Activation of Brain Regions Sensitive to Food Stimuli. *Am. J. Clin. Nutr.* **2012**, *95*, 818–824.

96. Greer, S. M.; Goldstein, A. N.; Walker, M. P. The Impact of Sleep Deprivation on Food Desire in the Human Brain. *Nat. Commun.* **2013**, *4*, 2259.

97. Chapman, C. D.; Nilsson, E. K.; Nilsson, V. C.; Cedernaes, J.; Rangtell, F. H.; Vogel, H.; Dickson, S. L.; Broman, J. E.; Hogenkamp, P. S.; Schioth, H. B.; Benedict, C. Acute Sleep Deprivation Increases Food Purchasing in Men. *Obesity (Silver Spring)* **2013**, *21*, E555–560.

98. Tasali, E.; Chapotot, F.; Wroblewski, K.; Schoeller, D. The Effects of Extended Bedtimes on Sleep Duration and Food Desire in Overweight Young Adults: A Home-Based Intervention. *Appetite* **2014**, *80*, 220–224.

99. Watson, N. F.; Harden, K. P.; Buchwald, D.; Vitiello, M. V.; Pack, A. I.; Weigle, D. S.; Goldberg, J. Sleep Duration and Body Mass Index in Twins: A Gene-Environment Interaction. *Sleep* **2012**, *35*, 597–603.

100. Van Cauter, E.; Spiegel, K. Circadian and Sleep Control of Endocrine Secretions. In *Neurobiology of sleep and circadian rhythms*, Turek, F. W.; Zee, P. C., Eds.; Marcel Dekker: New York, 1999; Vol. 133, pp 397–426.

NEUROIMMUNE ASPECTS OF SLEEP AND WAKEFULNESS

Adriano Zager and João Palermo-Neto

3.1. BIDIRECTIONAL INTERACTIONS BETWEEN THE NERVOUS AND IMMUNE SYSTEM

Neuroscience and immunology are two disciplines that have developed independently from each other. For this reason, hypotheses on how the brain communicates with the immune system remained anecdotal until recently. Claudius Galen, the second century Greek physician and philosopher was not only among the first to describe the scientific method and use experimentation, but also proposed a link between the emotionality (or "temperament") of a person and their susceptibility to particular diseases, such as infection and cancer.

Adriano Zager
Neuroimmunomodulation Research Group, Department of Pathology, School of Veterinary Medicine, University of Sao Paulo (USP), Sao Paulo, Brazil

João Palermo-Neto
Neuroimmunomodulation Research Group, Department of Pathology, School of Veterinary Medicine, University of Sao Paulo (USP), Sao Paulo, Brazil

Many centuries passed until the depiction of sympathetic nerves and adrenal glands by Bartolomeo Eustachius in 1552 helped physicians to later understand the role and function of cortisol and adrenalin, two hormones produced by the adrenal glands, and their relation with physiological stress. In the 19th century, the works of Claude Bernard and Louis Pasteur consolidated the concept of homeostasis and how it is influenced by mental states.

After the first description of the general adaptation syndrome, made by Hans Selye in 1936, and the introduction of the term "stress" in medicine, it became clear that mental states affect the physiology of an organism, especially the immune system. Growing evidence indicated that this interaction between the nervous and immune systems is reciprocal, becoming the basis for *Neuroimmunomodulation* or *Psychoneuroimmunology* Research.

This bidirectional interaction between the nervous and the immune systems is manifested by a dynamic equilibrium in both physiological and pathological conditions. On the one hand, mediators produced by immune cells, such as cytokines and prostaglandins, act on the central nervous system (CNS) by modulating the activity of neurons and altering behavior and neurotransmission. On the other hand, activation of the sympathetic nervous system and the hypothalamus-pituitary-adrenal axis (HPA) modulates immune cell activity in a manner dependent on the intensity and duration of the stressing stimuli. Stressors, when applied acutely, stimulate the activity of certain cells and improve the immune response. Paradoxically, chronic stressors impair immunity, especially the cell-mediated immunity (Th1 profile). However, this phenomenon can also be modulated by receptors of other neurotransmitters present in immune cells, for example, glutamate, cholinergic, and adrenergic receptors.

Among the physiological brain states that influence and are influenced by the immune system, sleep is one of the most controversial and poorly understood. This subject has recently gained particular attention as modern society is becoming progressively sleep-deprived due to social and economic pressures and because of the important part sleep plays in our lives given the fact that we spend almost one third of our time asleep.

Sleep comprises two major phases, as measured by electroencephalogram (EEG) changes: non-rapid eye movement sleep (NREM) and REM sleep. In human sleep research, NREM sleep is further divided into four stages, which can be correlated with depth of sleep. Stage 1 is considered to be a transitional state between wakefulness and sleep, whereas stages 2–4 are characterized by greater depths of sleep. The human stages 3 and 4 are known as slow-wave sleep (SWS), characterized by low-frequency and high-amplitude components in the EEG, and a deeper level of unconsciousness. In contrast, REM sleep is predominant in the second half of nighttime sleep and is characterized by high-frequency and low-amplitude EEG components, and a higher level of consciousness in relation to NREM sleep.

The first documented observations that infectious diseases could increase sleepiness were made by Aristotle in the fourth century B.C. In his text *On Sleep and Sleeplessness*, Aristotle notes that fever is commonly associated with the sensation of fatigue and tiredness, and introduced the concept that "vapors" were responsible for the sleep induction.

The general notion that sleep increases during an infectious disease, along with the common belief that poor sleep leads to increased susceptibility to pathogens, have supported the hypothesis that sleep has a restorative function. Over the past century, there has been a dramatic increase in the scientific interest in the interaction between sleep and immunity, which has led to recent advances in the field.

This chapter aims to contribute to this growing area of research by exploring the reciprocal relations between sleep and the immune system by means of two basic experimental approaches that are currently being adopted in research of the field: one is the increased sleep induced either by spontaneous or experimental inflammation/infection, and the other is how sleep loss modulates immune responses.

3.2. SLEEP PATTERN ALTERATIONS AFTER AN IMMUNE-INFLAMMATORY STIMULUS

Growing evidence has shown a close link between infections and sleepiness.

The vast majority of studies about this subject come from animal data, using different inflammatory stimuli. Bacterial fragments, such as lipopolysaccharide (LPS, a fragment of the cell wall of Gram-negative bacteria) and muramyl peptides (fragments of peptidoglycans from the cell walls of Gram-positive bacteria), as well as influenza virus infection, have been extensively used to examine the impact of immune activation on sleep pattern, with consistent results. Overall, experimental infection in animals resulted in altered sleep pattern, characterized by increased sleepiness, increased non-REM sleep (NREM), and decreased REM sleep (Krueger et al., 1986; Lancel et al., 1995). Despite the limited evidence linking the inflammatory response to physiological REM sleep modulation, EEG findings have shown that LPS administration in mice greatly impairs REM sleep increase following REM sleep deprivation (REM sleep rebound) in a dose-dependent manner, with no effect on NREM sleep response (Zager et al., 2009).

Although there were minor microorganism-specific differences, these common results suggest that inflammatory mediators released during the acute phase response might be implicated in the increased sleepiness. Pro-inflammatory cytokines, such as interleukin (IL)-1β, IL-6, and tumor necrosis factor (TNF), released during acute infections, are known to increase NREM sleep and induce sleepiness. The administration of TNF in rodents was shown to increase sleepiness in a dose-dependent manner (Kapas and Krueger, 1992), whereas the administration of an anti-TNF antibody suppresses sleep (Takahashi et al., 1995).

Opp et al. carried out a range of experiments in which IL-1β was directly injected into the dorsal raphe nucleus (DRN) of rats, and suggested that IL-1β-induced enhancement of NREM sleep could be due to the inhibition of DRN wake-promoting serotonergic neurons (Alam et al., 2004; Brambilla et al., 2007; Manfridi et al., 2003). The inhibition of the cleavage of biologically active IL-1β in the brain significantly reduced physiological NREM sleep time and prevented NREM sleep enhancement induced by peripheral LPS administration, supporting the role of IL-1β as a mediator of both physiological and pathological sleep (Imeri et al., 2006).

As elegantly reviewed by Imeri and Opp (2009), serotonin is involved in the mechanism of increased sleep pressure after an immune response activation. On one hand, serotonin metabolism by the serotonergic wake-promoting neurons during wakefulness is linked to melatonin secretion during night and initiation of sleep. On the other hand, pro-inflammatory cytokine expression in the CNS negatively impact tryptophan metabolism, and consequently, serotonin synthesis, via Indoleamine 2,3-dioxygenase (IDO) activation (Catena-Dell'Osso et al., 2013; Maes et al., 2011). Neurons immunoreactive to IL-1β and TNF, and their receptors are located in brain regions implicated in the regulation of sleep-wake behavior, such as the hypothalamus, hippocampus and brain stem, and the functioning of these serotonergic wake-promoting neurons is negatively influenced by an inflammatory environment, such as peripheral inflammation during acute infection.

Recent studies have also indicated a role for IL-6 as a modulator of sleep during pathological conditions such as narco-lepsy, sleep apnea, and obesity (Alberti et al., 2003; Okun et al., 2004; Vgontzas et al., 1997). In this sense, ICV administration of IL-6 increases NREM and induces fever in rats. However, in contrast to IL-1β and TNF, IL-6 administration does not affect REM sleep (Hogan et al., 2003).

Anti-inflammatory cytokines, such as IL-10 and IL-4, antagonize many effects of pro-inflammatory cytokines during the immune response. Therefore, it is reason-able to expect that anti-inflammatory cy-tokines have the opposite effect on sleep than pro-inflammatory cytokines. In fact, central administration of IL-10 reduces the amount of NREM sleep in rats and rabbits (Kushikata et al., 1999; Opp et al., 1995), and mice lacking IL-10 spend more time in NREM sleep (Toth and Opp, 2001). In addition, IL-4 reduces the amount of spon-taneous NREM sleep in rabbits (Kushikata et al., 1998).

3.2.1. Human Data

Although there is not as much human data as animal data, human experimentation have been accumulating over the past 30 years and points in the same direction re-garding increased sleepiness during infec-tions. However, this trend is less clear in humans than in animals, which probably reflects the difficulties of studying spon-taneous human infections and/or the dif-ferences in responses to the various patho-gens.

Since it would be unethical to ex-perimentally infect humans, to date, few studies have investigated the influence of infection on human sleep. Smith (1992) studied sleep time during influenza infec-tion. Sleep duration was described as being decreased during the incubation period, but increased during the symptomatic pe-riod of infection.

Using experimental immune challeng-es, the pioneering studies of Trachsel and colleagues have demonstrated in humans that a single low-dose endotoxin (LPS) administered at the beginning of the dark period was able to increase the time spent in NREM sleep and EEG spectral power (an indicator of enhanced slow wave ac-tivity), as well as to suppress REM sleep, which was accompanied by a rise in body temperature and heart rate (Trachsel et al., 1994). In addition, Mullington and colleagues (2000) have shown that the ef-fect of endotoxin treatment on circulating cytokines and sleep pattern is dose-depen-dent. The mild inflammatory activation caused by a low-dose endotoxin increased NREM sleep time and intensity, whereas a high-dose of endotoxin disrupts sleep by increasing wakefulness, demonstrating how sensitive the modulation of sleep is to activation of host defense.

In contrast, when the endotoxin was administered during the morning (that is, given at 9 a.m.), a higher dose of endotoxin was necessary to elicit an immune response comparable to that induced by a low-dose endotoxin given during the dark period. In addition, the administration of endotoxin during the day provoked a suppression of REM sleep without affecting the NREM sleep during monitored naps across the day (Korth et al., 1996). These discrepan-cies evidence a clear circadian trend on the effects of infections on human sleep, and highlight the suppression of REM sleep as a common outcome, regardless of the time of antigen exposure.

However, enhancing the daytime endotoxin-induced host response by pre-treating patients with granulocyte colony-stimulating factor (G-CSF) reveals that a more robust immune activation reduces daytime NREM sleep and increases sleepiness, suggesting that disruption of regular sleep pattern might be prior to increased sleep pressure following a high-dose endotoxin treatment (Hermann et al., 1998).

In humans, the modulatory effect of cytokines on sleep is supported by clinical observations of increased sleepiness in the course of infections. However, compared to the available animal data, the role of specific cytokines in human sleep is less elucidated. In contrast with animal studies, human studies are unclear as to which cytokines of interest induce sleep, as well as whether these cytokines have a function on sleep only during infections or also under normal physiological conditions. In this regard, Vgontzas et al. (2004) have demonstrated that neutralizing TNF activity caused a significant reduction of sleepiness and sleep apnea/hypopnea in obese patients.

Cytokines released peripherally during the host response relay messages to the CNS via active transport mechanisms, vagal afferents, and/or circumventricular organs, leading to neuroendocrine activation and fever (Hopkins and Rothwell, 1995; Rothwell and Hopkins, 1995). The infiltration of peripheral macrophages into the CNS, as well as neuron and glial activation, contribute to the *de novo* cytokine synthesis in the CNS (Breder et al., 1988; Garden and Moller, 2006; Marz et al., 1998). In addition, the presence of lymphatic vessels in the brain parenchyma has recently been described for the first time. These vessels are connected to the deep cervical lymph nodes and carry immune cells from the cerebrospinal fluid (Louveau et al., 2015).

In the following section, we will focus on the many inflammatory diseases that are known to disrupt regular sleep pattern by means of inflammatory mediators released during the progression of the disease and how they can increase sleep pressure and NREM sleep or impair the quality of sleep.

3.2.2. Sleep in Infectious and Inflammatory Conditions

Spontaneous infections such as African trypanosomiasis (sleeping sickness) and HIV are characterized by sleep-related problems. Sleeping sickness is often accompanied by meningoencephalitis. Polysomnographic recordings indicate a disruption of circadian rhythm in patients with this condition, rather than increased sleep time (Buguet et al., 1993; Buguet et al., 2005; Buguet et al., 1999).

HIV infections are related to characteristic alterations in sleep structure. For instance, asymptomatic HIV patients are reported to present increased daytime fatigue and changes in nighttime sleep, such as increased SWS, which is accompanied by enhanced plasmatic TNF levels (Darko et al., 1995; Norman et al., 1990; White et al., 1995).

In addition to infections, cytokines are released in the CNS in many distinct autoimmune and neuroinflammatory conditions, such as narcolepsy, multiple sclerosis (MS), Parkinson's, and Alzheimer's disease (AD). The onset of these diseases is correlated with neuroinflammation and is frequently linked with sleep disturbances. In this regard, the best example of a neuroimmune interaction in sleep is type 1 narcolepsy, which is a sleep disorder caused by loss of hypothalamic hypocretin/orexin (HCRT) neurons, characterized by sleep attacks and increased daytime sleepiness. The exact cause of narcolepsy is not known. The most accepted hypothesis as to the etiology of the disease is that narcolepsy is driven by an autoimmune reaction, characterized by infiltration of autoreactive T cells in the CNS, glial activation, and destruction of HCRT neurons (Kornum et al., 2011). To date, some factors are known to elicit or trigger the autoimmunity in narcolepsy, including previous exposure to pathogens and infections.

There is increasing evidence that previous infection (Aran et al., 2009) and vaccination against the H1N1 virus (Jacob et al., 2015, Lecendreux et al., 2015) might trigger autoimmunity and narcolepsy with cataplexy due to cross-reactivity with previous antigens. It is believed that molecular mimicry with previous antigens causes a robust immune response against the HCRT neurons, which, in turn, promotes sleep attacks, increased sleepiness, and loss of muscle tone.

MS progression is also related to sleep disturbances. The etiology of MS is not completely understood, but is considered to be a T-cell driven autoimmune disease, causing immune response to myelin antigens, and axonal loss. Among the symptoms, fatigue is the most commonly described, but sleep disorders are frequent outcomes of MS, present in 25–54% of symptomatic patients (Brass et al., 2010; Fleming and Pollak, 2005; Lobentanz et al., 2004). EEG findings show that MS patients present reduced sleep efficiency and increased nocturnal awakenings compared to normal controls (Ferini-Strambi et al., 1994). In addition, a higher prevalence of insomnia complaints (Stanton et al., 2006), obstructive sleep apnea (Kaminska et al., 2012), and restless leg syndrome (Auger et al., 2005; Deriu et al., 2009; Manconi et al., 2008a; Manconi et al., 2008b) are correlated to MS.

In addition, some studies have revealed that MS is the fourth most common cause of narcolepsy, with 12% of the cases of secondary narcolepsy being due to MS, indicating a similar mechanism and cross-reactivity between the two conditions (Nishino and Kanbayashi, 2005; Poirier et al., 1987). The MS-related narcoleptic symptoms are possibly caused by MS hypothalamic lesions that result in low CSF HCRT levels and hypersomnia (Oka et al., 2004). Both MS and narcolepsy are related to human leukocyte antigen DQB1*0602, which indicates that both diseases might have a similar autoimmune etiology (Barun, 2013). Paradoxically, MS lesions on the proximity of the pedunculopontine nucleus (PPN) are described to be related to REM sleep behavior disorder (RBD), a parasomnia characterized by loss of muscle atonia during REM sleep, either as a first symptom of disease (Plazzi and Montagna, 2002), or during the course of MS (Tippmann-Peikert et al., 2006).

Sleep disturbances are one of the most common non-motor symptoms of Parkinson's disease (PD). Among sleep complaints, RBD, obstructive sleep apnea (OSA), restless legs syndrome (RLS), and periodic limb movements of sleep (PLMS) are often reported in PD patients (Ondo, 2014). In this sense, the PPN seems to be involved in the sleep disruption associated with PD, since it modulates the sleep cycle and REM sleep via cholinergic ascending thalamic pathways (Alessandro et al., 2010; Monderer and Thorpy, 2009).

AD progression is also accompanied by multiple sleep disturbances. Poor sleep quality, with excessive daytime sleepiness, and insomnia at night; affects approximately 25–40% of AD patients with dementia (Moran et al., 2005). Aggregation of amyloid-β (Aβ) into extracellular plaques in the brain, and the consequent local inflammatory reaction, is a key step in AD pathogenesis and recent advances in the field have indicated in both humans (Huang et al., 2012) and mouse models (Kang et al., 2009) that Aβ concentrations rise during wakefulness and fall during sleep, suggesting an Aβ diurnal pattern (Lucey and Bateman, 2014). In transgenic APP mice, when the Aβ plaque formation occurs, there is a loss in this Aβ diurnal pattern, which coincides with disruption of sleep-wake pattern with increased wakefulness and decreased sleep during the light phase (Kang et al., 2009). Yet, if the plaque formation is inhibited by active immunization of the mice with Aβ42, the sleep disturbances are completely prevented, suggesting the Aβ aggregation is the cause of sleep disruption (Roh et al., 2012).

It is noteworthy that all of the aforementioned neurodegenerative conditions that alter sleep pattern have neuroinflammation as a pathophysiological hallmark (Glass et al., 2010). In situ production of cytokines following astrocytic and microglial activation is one of the main perpetrators of neuronal loss and sleep disruption, sometimes accompanied by infiltration of peripheral immune cells, such as lymphocytes and macrophages.

Based on the evidence presented in this part of the chapter, we can conclude that immune system activation, either by experimental challenges or spontaneous infectious/inflammatory conditions, elicits the release of pro-inflammatory cytokines, which cross the Blood-brain barrier, and act on multiple neurotransmitter systems, disrupting regular sleep. In some diseases such as narcolepsy and MS, peripheral immune cells infiltrate the brain, and together with glial activation, help to propagate the inflammation and consequent sleep disturbances.

3.3. SLEEP DEPRIVATION, SLEEP RESTRICTION, AND IMMUNE FUNCTION

Based on our previous discussion, it is evident that cytokines, such as IL-1β and TNF, are mediators of both immune response and physiological sleep. During a pathological condition, such as infection and experimental immune challenge, the exaggerated release of these cytokines increases sleepiness, and NREM sleep. The elevated sleep pressure during acute immune response indicates that sleep has a pivotal role in proper immune function.

It is therefore reasonable to argue that subjecting the organism to sleep curtailment/deprivation might have immunosuppressive consequences. Sleep loss and consequent circadian disruption are stressful conditions that have many neurological, cardiovascular, and immunological consequences. Animal data regarding the effects of sleep deprivation on immune function have been accumulating over the past 30 years. Animal experiments have shown that sleep deprivation can lead to immune suppression (Brown et al., 1989; Everson, 2005; Zager et al., 2007; Zager et al., 2012), or even to immune stimulation (Renegar et al., 2000). These contradictory findings suggest that the effects of sleep deprivation on immunity depends on many factors such as type of sleep deprivation (total versus selective REM sleep deprivation), duration (acute versus chronic), species, and even strain of mice used. Overall, sleep deprivation is described as decreasing circulating lymphocytes (Zager et al., 2007; Zager et al., 2012), as well as inducing endotoxemia in rodents (Everson, 2005). Although the majority of experimental data points to sleep loss having a negative impact on immunity, the discrepancies between animal studies make it extremely difficult to draw any firm conclusions.

3.3.1. Circadian Rhythm of Immune Cells

There is growing evidence that indicates a significant diurnal rhythm for all blood cell types during normal condition. Circulating lymphocytes show peak levels during the night, whereas granulocytes and NK/dendritic cells peak during the late afternoon. In addition, circulating granulocyte rhythm is severely affected by 29 h of prolonged wakefulness, with higher circulating levels, lower amplitude, and loss of rhythmicity (Ackermann et al., 2012). Supporting this hypothesis, additional data indicates a shift in the Th1/Th2 cytokine balance during nocturnal sleep in humans, tracked by IFN-γ/IL-4 producing CD4$^+$ cells. Interestingly, there is a shift toward Th1 cytokines (cell-mediated immunity) during early sleep, when SWS is predominant, favoring cellular aspects of adaptive immune responses over activity of anti-inflammato-

ry, or Th2 cytokines (humoral immunity). However, the effect is reversed during the late part of sleep, when REM sleep is dominant (Dimitrov et al., 2004b).

The regulatory T cells (Treg), a subtype of CD4+ T cells responsible for the control of inflammation by producing IL-10 and TGF-β, also follow a rhythm across the 24 h period. Interestingly, Treg number follows a circadian rhythm independent of sleep. In turn, Treg function follows a sleep-dependent rhythm with highest suppressive activity during sleep, with nocturnal sleep deprivation dampening T cell proliferation, possibly resulting in impaired immune responses (Bollinger et al., 2009).

3.3.2 Sleep Deprivation in Humans

As in animal research, human experiments have been performed using partial sleep deprivation, total sleep deprivation, and sleep restriction in order to understand the fine regulation of immune function by sleep. In most instances, the role of sleep was studied by subjecting human volunteers to either restricted nighttime sleep (that is, 4 h/night), or forced wakefulness for one night. Some studies have applied prolonged periods of sleep deprivation to address this issue. However, submitting the human volunteer to more than one night of prolonged wakefulness represents a stressful event which, intrinsically, might influence the immune function.

Partial sleep deprivation has been associated with decreased natural killer (NK) cells function (Irwin et al., 1994) and number, decreased IL-2 production (Irwin et al., 1996) and impaired lymphocyte proliferation (Wilder-Smith et al., 2013), suggesting impairment in NK morning response. However, when NK cells are measured during sleep, there is a reduction in NK cell number compared to waking condition (Born et al., 1997), indicating that these discrepancies could be due to the time of blood collection (sleeping vs after awakening).

In contrast, one night of total sleep deprivation strikingly decreases the number of dendritic cells producing IL-12, which is a main inducer of the Th1 response (Dimitrov et al., 2007). In addition, nocturnal wakefulness suppresses the IL-12-producing monocytes to approximately 40% of sleep levels and induces an enhancement of IL-10-producing monocytes (Lange et al., 2006; Redwine et al., 2003), indicating a shift of antigen presenting cells (APC) function toward type 1 cytokines and cell-mediated immunity (Th1) during sleep and representing a key target for the regulatory effect of sleep on immune function. Regarding other cytokines, IL-2 is increased in sleeping subjects, compared to wake condition (Born et al., 1997; Irwin et al., 1996), further supporting the notion that sleep facilitates the type 1 cytokine activity, positively influencing a cell-mediated immune response (Th1 immunity).

One interesting method of measuring immune competence and its relation to sleep in humans is by analyzing the effectiveness of a vaccination. Lange and colleagues (2003) have demonstrated that, when healthy volunteers are subjected to sleep deprivation in the night following hepatitis A vaccination, the antibody titers measured 28 days later were reduced by more than 50%, when compared to volunteers who slept the night following vaccination.

More recently, the same authors monitored for a year the vaccination-induced T cell and antibody responses in volunteers that were or were not sleep-deprived following the three doses of Hepatitis A vaccination. Compared to the volunteers who slept regularly after the vaccine, sleep deprivation caused a decline in the frequency of Ag-specific T helper (Th) cells and in the fraction of IFN-producing T cells. This effect was correlated with the HAV-specific IgG1 levels (Lange et al., 2011). Corroborating this finding, Prather and colleagues have demonstrated an association between shorter sleep duration and lower antibody response to hepatitis B vaccination, which is independent of age, sex, and body mass index (Prather et al., 2012). These findings indicate that sleep should be considered an important factor when determining the effectiveness of a vaccination program.

In contrast, a number of human studies have examined the effects of prolonged sleep deprivation and restriction (for days rather than hours), with interesting results. Reduced sleep duration for a longer period (across 10 days) caused increased IL-6 levels, which is correlated with increased pain sensitivity (Haack et al., 2007). Corroborating this finding, cross-sectional analyses indicated that shorter sleep is associated with higher levels of inflammatory markers, such as C-reactive protein, and IL-6 (Ferrie et al., 2013). Interestingly, it has been shown that a daytime 2 h nap is able to reverse the effects of one night sleep deprivation, particularly the increased IL-6 and cortisol levels, and consequently improve alertness and performance in the sleep-deprived subjects. This study indicates that short naps may counterbalance sleep deprivation and its stress-related effects by restoring the immune and hormonal *milieu* necessary for proper immune response (Faraut et al., 2015; Vgontzas et al., 2007).

Prolonged sleep deprivation studies have consistently shown a progressive decrease in the NK number and activity and in the number of major lymphocyte populations (Boyum et al., 1996; Dinges et al., 1994; Ozturk et al., 1999). Extended periods of partial or total sleep deprivation are associated with alterations in many aspects of immunity, such as increased counts of white blood cells, granulocytes, and monocytes (Boyum et al., 1996; Dinges et al., 1994; Heiser et al., 2000; Lasselin et al., 2015).

The analysis of cytokine activity indicates activation of innate immunity after extended sleep deprivation. A pioneering study of Moldofsky et al. (1989) reported increased plasma IL-1β and IL-2 activity after 40 h of forced wakefulness. In addition, soluble TNF receptor 1 and IL-6 levels were also increased after four days of total sleep deprivation (Shearer et al., 2001). Consistent with the notion of innate immune activation, prolonged sleep restriction has been associated to higher levels of TNF and MCP-1 produced by mitogen-stimulated cells, and a shift toward Th2 activity (Axelsson et al., 2013).

This enhanced inflammatory response to chronic sleep loss is clinically relevant because it is prevalent in several epidemic diseases, such as obstructive sleep apnea, atherosclerosis, obesity, metabolic syndrome, diabetes, and depression. However, the role of sleep as a modulator of inflammation in the aforementioned diseases has been extensively addressed elsewhere (Gozal and Kheirandish-Gozal, 2008; Irwin et al., 2013; Mehra and Redline, 2008; Miller, 2011; Nadeem et al., 2013) and is not the focus of this chapter.

Overall, the alterations in immune parameters after acute periods of sleep deprivation suggest an immunosuppressive effect and further support the role of sleep in the adaptive immune response. On the contrary, extended periods of sleep loss, either by total or partial sleep deprivation is characterized by stress-related leukocytosis and increased activity of pro-inflammatory cytokines, supporting the notion of a continuous inflammatory state induced by chronic sleep loss.

3.4. POSSIBLE LINKS BETWEEN SLEEP AND IMMUNITY

Several humoral, endocrine, and neural mediators are candidates for modulating the bidirectional interactions of sleep and immunity. The endocrine and autonomic nervous system convey efferent information from the CNS to the periphery ("brain to immune"), but also afferent information from the periphery to the brain ("immune to brain"). In addition, this bidirectional communication is critically modulated by inflammatory mediators, which influence and are influenced by sleep.

Pro-inflammatory cytokines, such as IL-1β, TNF, and IL-6, are the main immune mediators that are robustly released during an immune response and, by acting on the CNS, change neurotransmission, elevate body temperature, and increase sleepiness. On the other hand, chronic sleep loss increases inflammatory cytokines, possibly creating a sustained inflammatory state in sleep-deprived subjects, which per se, increases sleep pressure and necessity. For many years, it was believed that the effects of specific cytokines on increased sleepiness and NREM sleep was due, at least in part, to the rise in core body temperature. However, recent data have indicated that cytokine-mediated increase in NREM sleep time occurs independently from the febrile response. For instance, low doses of IL-1β administered in rodents and endotoxin injected in human volunteers both increase NREM sleep, with no effect on brain and body temperature (Mullington et al., 2000; Opp et al., 1991). Corroborating this finding, inhibition of brain IL-1β and TNF response suppress the cytokine-induced increase in NREM sleep without affecting the febrile response of rabbits (Takahashi et al., 1999). Although sleep and thermoregulation are indeed closely linked, the influence of cytokines on human sleep can be distinguished from their effect on body temperature, suggesting that other hormonal and autonomic mediators might be responsible for the immunomodulatory action of sleep.

Nocturnal sleep elicits a unique endocrine *milieu* of a number of hormones that are known to have immunomodulatory effects. During the early period of sleep, which is dominated by SWS in humans, there is a marked secretory activity of the somatotropic hormones. For example, hypothalamic secretion of growth hormone-releasing hormone (GHRH) and, consequently, pituitary release of growth hormone (GH), peak during SWS, and decrease to a minimum in late nocturnal sleep, when REM sleep predominates. In addition, other hormones such as prolactin are also increased during sleep, particularly SWS. Both GH and prolactin are secreted during physiological sleep, and when this

secretory pattern is disrupted, either by acute sleep deprivation or chronic sleep disturbances, the immune response mediated by these hormones is impaired.

The studies of Lange et al. (Lange et al., 2011; Lange et al., 2003) address this issue by investigating the effects of one night sleep deprivation immediately after vaccination to hepatitis A. As previously cited, the T helper and antibody responses to the vaccine were dramatically decreased in the sleep-deprived subjects, even one year after the first shot of the vaccine. More interestingly, the sleep pattern and hormonal levels of the control group was monitored during the nights following the vaccination and showed a robust increase in slow wave activity, which is consistent with the immune response elicited by vaccine. However, the increased slow wave activity in the sleeping subjects is positively correlated to a surge of immunostimulating hormones such as GH and prolactin released during the nighttime sleep. The surge of GH and prolactin elicited by SWS is suppressed in sleep-deprived volunteers. The same disrupted hormonal pattern was shown in other studies employing one night of sleep deprivation (Dimitrov et al., 2015; Dimitrov et al., 2004a; Lange et al., 2003). This data indicates that SWS elicits a unique hormonal milieu that exerts adjuvant activity by stimulating the production of proinflammatory cytokines and the development of adaptive immunity, and suggesting that sleep quality is determinant to the effectiveness of a vaccination.

While GH and prolactin seem to positively influence the immune response, glucocorticoids are known to negatively impact immune response following sleep loss. In fact, the activity of the HPA axis shows a temporal pattern that is opposite in relation to GH and prolactin. Corticotropin releasing hormone (CRH), adrenocorticotropic hormone (ACTH) and corticosteroids reaches their minimum levels during SWS, and increase during REM sleep, peaking at about the time of morning awakening. Glucocorticoids, particularly cortisol, are known to elicit a marked decrease in circulating lymphocytes, due to enhanced trafficking to the bone marrow (Ottaway and Husband, 1992; Sackstein and Borenstein, 1995). In addition, glucocorticoids suppress the production of proinflammatory cytokines such as IL-1β, TNF, and IL-6 (Besedovsky and del Rey, 1996; Petrovsky and Harrison, 1998), and might affect the immunity of sleep-deprived subjects. Collectively, the present data further supports the hypothesis that the GH and prolactin secretion during SWS induces a shift toward type 1 cytokine activity, whereas the concomitant suppression of cortisol during early night facilitates the production of type 1 and type 2 cytokines.

In addition, sympathetic activation is also possibly involved in the immunomodulation by sleep. Lymphoid tissue, such as the spleen and lymph nodes, is highly enervated by the autonomic nervous system, and leukocytes constitutively express adrenergic and cholinergic receptors. Moreover, a decrease in sympathetic activity, as well as an increase in parasympathetic tone, is reported during nocturnal sleep in humans (Burgess et al., 1997). Studies on pharmacological manipulation of adrenergic receptors have shown that sympathetic activation induces leukocytosis, characterized by an increased number of lymphocytes and mobilization of granulocytes, and NK cells to the circulation (Benschop et al., 1996; Ottaway and Husband, 1992, 1994). However, the influence of the autonomic nervous system on immunity in the context of sleep remains to be fully understood. It is believed that the marked sleep-associated suppression of sympathetic tone, accompanied by the enhanced parasympathetic tone, influences the adaptive T cell response by increasing type 1 cytokines and by enhancing the homing of lymphocytes in lymphoid tissues (Abo and Kawamura, 2002; Besedovsky and del Rey, 1996; Ottaway and Husband, 1994; Petrovsky, 2001; Sanders, 1995).

More recently, the relation between sleep and immunity has been linked with immune cell aging, as indicated by the telomere length of leukocytes. Telomeres are DNA-protein complexes at the end of chromosomes that protect DNA. Telomeres shorten after each cell cycle, and critically short telomeres are associated to cell senescence, malfunction, and apoptosis. In the studies of Prather et al., poorer global sleep quality is associated with shorter telomere length in leukocytes of midlife women (Prather et al., 2011), as well as in CD8+ and CD4+ T lymphocytes of obese subjects (Prather et al., 2015). Interestingly, poorer sleep predicts shorter telomere length in CD8+ cells in individuals reporting higher perceived psychological stress, but not in low stress participants, suggesting that self-perceived stress might be a potentiating factor and a link between poor sleep effects and immune cell aging (Prather et al., 2015). It was recently proposed that telomere length could be a marker of sleep loss and sleep disturbances (as reviewed by Tempaku et al., 2015).

3.5. CONCLUDING REMARKS

In this chapter, we have reviewed the reciprocal interaction between sleep and the immune system in humans. On one hand, sleep pressure (or sleep demand) and SWS are increased during an experimental immune challenge, suggesting that inflammatory mediators released during the immune response are modulators of both physiological and pathological sleep, by acting on neurotransmission systems. To date, IL-1β and TNF effects on the brain (produced either in periphery or the CNS) have been extensively studied and are implicated in the modulation of sleep response both in animals and humans.

On the other hand, sleep is required for a proper immune response, possibly due to the endocrine *milieu* elicited by SWS during nighttime sleep. Hormones such as GH and prolactin peak during SWS and have facilitatory effects in the adaptive immune response, particularly on Th1 immunity. Short and prolonged periods of sleep loss negatively impact SWS and disrupt this hormonal pattern, preventing these immunosupportive actions and leading to immune suppression. Interestingly, during an ongoing immune response pro-inflammatory and Th1 cytokines feed back to the brain to enhance SWS.

Therefore, it is possible to speculate that sleep acts as an adjuvant to the optimal immune response, creating a hormonal *milieu* that favors type 1 cytokines and Th1 immunity (Fig. 3.1). In turn, the minimum sleep requirement (or sleep pressure) is

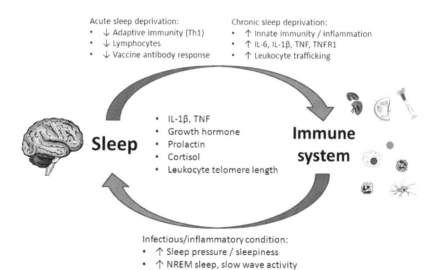

Acute sleep deprivation:
- ↓ Adaptive immunity (Th1)
- ↓ Lymphocytes
- ↓ Vaccine antibody response

Chronic sleep deprivation:
- ↑ Innate immunity / inflammation
- ↑ IL-6, IL-1β, TNF, TNFR1
- ↑ Leukocyte trafficking

Sleep
- IL-1β, TNF
- Growth hormone
- Prolactin
- Cortisol
- Leukocyte telomere length

Immune system

Infectious/inflammatory condition:
- ↑ Sleep pressure / sleepiness
- ↑ NREM sleep, slow wave activity
- ↓ REM sleep

Figure 3.1. Brain-immune interactions in sleep. This figure challenges the traditional view of medicine, with the CNS above (suggesting a more important role) other physiological systems. Here, the immune system is presented alongside the brain, emphasizing that the link between sleep and the immune system is bidirectional. This occurs via inflammatory cytokines (which increase sleep pressure and SWS), GH, and prolactin (immunostimulatory hormones released during nighttime SWS), cortisol and leukocyte telomere length (a marker of immune cell senescence in sleep-deprived subjects).

increased during an infection or immune response, possibly resulting in immune suppression if this minimum requirement is not reached. These outcomes create a vicious cycle between sleep loss, immune modulation, and infectious-inflammatory diseases. Future basic and clinical research is warranted to further understand this phenomenon from the clinical perspective in order to develop novel strategies for patients with sleep disturbances.

KEYWORDS

- **sleep**
- **immune system**
- **cytokines**
- **slow-wave sleep**
- **sleep deprivation**
- **Th1 immunity**
- **prolactin**

REFERENCES

Abo, T.; Kawamura, T. Immunomodulation by The Autonomic Nervous System: Therapeutic Approach for Cancer, Collagen Diseases, and Inflammatory Bowel Diseases. *Ther. Apher.* **2002**, *6*, 348–357.

Ackermann, K.; Revell, V. L.; Lao, O.; Rombouts, E. J.; Skene, D. J.; Kayser, M. Diurnal Rhythms in Blood Cell Populations and the Effect of Acute Sleep Deprivation in Healthy Young Men. *Sleep.* **2012**, *35*, 933–940.

Alam, M. N.; McGinty, D.; Bashir, T.; Kumar, S.; Imeri, L.; Opp, M. R.; Szymusiak, R. Interleukin-1beta Modulates State-Dependent Discharge Activity of Preoptic Area and Basal Forebrain Neurons: Role in Sleep Regulation. *Eur. J. Neurosci.* **2004**, *20*, 207–216.

Alberti, A.; Sarchielli, P.; Gallinella, E.; Floridi, A.; Mazzotta, G.; Gallai, V. Plasma Cytokine Levels in Patients with Obstructive Sleep Apnea Syndrome: A Preliminary Study. *J. Sleep Res.* **2003**, *12*, 305–311.

Alessandro, S.; Ceravolo, R.; Brusa, L.; Pierantozzi, M.; Costa, A.; Galati, S.; Placidi, F.; Romigi, A.; Iani, C.; Marzetti, F.; Peppe, A. Non-Motor Functions in Parkinsonian Patients Implanted in the Pedunculopontine Nucleus: Focus on Sleep and Cognitive Domains. *J. Neurol. Sci.* **2010**, *289*, 44–48.

Aran, A.; Lin, L.; Nevsimalova, S.; Plazzi, G.; Hong, S. C.; Weiner, K.; Zeitzer, J.; Mignot, E. Elevated Anti-Streptococcal Antibodies in Patients with Recent Narcolepsy Onset. *Sleep.* **2009**, *32*, 979–983.

Auger, C.; Montplaisir, J.; Duquette, P. Increased Frequency of Restless Legs Syndrome in a French-Canadian Population with Multiple Sclerosis. *Neurology.* **2005**, *65*, 1652–1653.

Axelsson, J.; Rehman, J. U.; Akerstedt, T.; Ekman, R.; Miller, G. E.; Hoglund, C. O.; Lekander, M. Effects of Sustained Sleep Restriction on Mitogen-Stimulated Cytokines, Chemokines, and T Helper 1/T Helper 2 Balance in Humans. *PLoS ONE.* **2013**, *8*.

Barun, B. Pathophysiological Background and Clinical Characteristics of Sleep Disorders in Multiple Sclerosis. *Clin. Neurol. Neurosurg.* **2013**, *115*, Suppl 1, S82-S85.

Benschop, R. J.; Rodriguez-Feuerhahn, M.; Schedlowski, M. Catecholamine-Induced Leukocytosis: Early Observations, Current Research, and Future Directions. *Brain Behav. Immun.* **1996**, *10*, 77–91.

Besedovsky, H. O.; del Rey, A. Immune-Neuro-Endocrine Interactions: Facts and Hypotheses. *Endocr. Rev.* **1996**, *17*, 64–102.

Bollinger, T.; Bollinger, A.; Skrum, L.; Dimitrov, S.; Lange, T.; Solbach, W. Sleep-Dependent Activity of T Cells and Regulatory T Cells. *Clin. Exp. Immunol.* **2009**, *155*, 231–238.

Born, J.; Lange, T.; Hansen, K.; Molle, M.; Fehm, H. L. Effects of Sleep and Circadian Rhythm on Human Circulating Immune Cells. *J. Immunol.* **1997**, *158*, 4454–4464.

Boyum, A.; Wiik, P.; Gustavsson, E.; Veiby, O. P.; Reseland, J.; Haugen, A. H.; Opstad, P. K. The Effect Of Strenuous Exercise, Calorie Deficiency and Sleep Deprivation on White Blood Cells, Plasma Immunoglobulins, and Cytokines. *Scand. J. Immunol.* **1996**, *43*, 228235.

Brambilla, D.; Franciosi, S.; Opp, M. R.; Imeri, L. Interleukin-1 Inhibits Firing of Serotonergic Neurons in the Dorsal Raphe Nucleus and Enhances

Gabaergic Inhibitory Post-Synaptic Potentials. *Eur. J. Neurosci.* **2007**, *26*, 1862–1869.

Brass, S. D.; Duquette, P.; Proulx-Therrien, J.; Auerbach, S. Sleep Disorders in Patients with Multiple Sclerosis. *Sleep Med. Rev.* **2010**, *14*, 121–129.

Breder, C. D.; Dinarello, C. A.; Saper, C. B. Interleukin-1 Immunoreactive Innervation of the Human Hypothalamus. *Science.* **1988**, *240*, 321–324.

Brown, R.; Pang, G.; Husband, A. J.; King, M. G. Suppression of Immunity to Influenza Virus Infection in the Respiratory Tract Following Sleep Disturbance. *Reg. Immunol.* **1989**, *2*, 321–325.

Buguet, A.; Bert, J.; Tapie, P.; Tabaraud, F.; Doua, F.; Lonsdorfer, J.; Bogui, P.; Dumas, M. Sleep-Wake Cycle in Human African Trypanosomiasis. *J. Clin. Neurophysiol.* **1993**, *10*, 190–196.

Buguet, A.; Bisser, S.; Josenando, T.; Chapotot, F.; Cespuglio, R. Sleep Structure: A New Diagnostic Tool for Stage Determination in Sleeping Sickness. *Acta Trop.* **2005**, *93*, 107–117.

Buguet, A.; Tapie, P.; Bert, J. Reversal of the Sleep/Wake Cycle Disorder of Sleeping Sickness after Trypanosomicide Treatment. *J. Sleep. Res.* **1999**, *8*, 225–235.

Burgess, H. J.; Trinder, J.; Kim, Y.; Luke, D. Sleep and Circadian Influences on Cardiac Autonomic Nervous System Activity. *Am. J. Physiol.* **1997**, *273*, H1761–H1768.

Catena-Dell'Osso, M.; Rotella, F.; Dell'Osso, A.; Fagiolini, A.; Marazziti, D. Inflammation, Serotonin, and Major Depression. *Curr. Drug Targets.* **2013**, *14*, 571–577.

Darko, D. F.; Miller, J. C.; Gallen, C.; White, J.; Koziol, J.; Brown, S. J.; Hayduk, R.; Atkinson, J. H.; Assmus, J.; Munnell, D. T.; Naitoh, P.; McCutchan, J. A.; Mitler, M. M. Sleep Electroencephalogram Delta-Frequency Amplitude, Night Plasma Levels of Tumor Necrosis Factor Alpha, and Human Immunodeficiency Virus Infection. *Pro. Natl Acad. Sci. U S A* **1995**, *92*, 12080–12084.

Deriu, M.; Cossu, G.; Molari, A.; Murgia, D.; Mereu, A.; Ferrigno, P.; Manca, D.; Contu, P.; Melis, M. Restless Legs Syndrome in Multiple Sclerosis: A Case-Control Study. *Mov. Disord.* **2009**, *24*, 697–701.

Dimitrov, S.; Besedovsky, L.; Born, J.; Lange, T. Differential Acute Effects of Sleep on Spontaneous and Stimulated Production of Tumor Necrosis Factor in Men. *Brain Behav. Immun.* **2015**, *47*, 201–210.

Dimitrov, S.; Lange, T.; Fehm, H. L.; Born, J. A Regulatory Role of Prolactin, Growth Hormone, and Corticosteroids for Human T-Cell Production of Cytokines. *Brain Behav. Immun.* **2004a**, *18*, 368–374.

Dimitrov, S.; Lange, T.; Nohroudi, K.; Born, J. Number and Function of Circulating Human Antigen Presenting Cells Regulated by Sleep. *Sleep.* **2007**, *30*, 401–411.

Dimitrov, S.; Lange, T.; Tieken, S.; Fehm, H. L.; Born, J. Sleep Associated Regulation of T Helper 1/T Helper 2 Cytokine Balance in Humans. *Brain Behav. Immun.* **2004b**, *18*, 341–348.

Dinges, D. F.; Douglas, S. D.; Zaugg, L.; Campbell, D. E.; McMann, J. M.; Whitehouse, W. G.; Orne, E. C.; Kapoor, S. C.; Icaza, E.; Orne, M. T. Leukocytosis and Natural Killer Cell Function Parallel Neurobehavioral Fatigue Induced by 64 Hours of Sleep Deprivation. *J. Clin. Invest.* **1994**, *93*, 1930–1939.

Everson, C. A. Clinical Assessment of Blood Leukocytes, Serum Cytokines, and Serum Immunoglobulins as Responses to Sleep Deprivation in Laboratory Rats. *Am. J. Physiol. Regul. Integr. Comp. Physiol.* **2005**, *289*, R1054–R1063.

Faraut, B.; Nakib, S.; Drogou, C.; Elbaz, M.; Sauvet, F.; De Bandt, J. P.; Leger, D. Napping Reverses the Salivary Interleukin-6 and Urinary Norepinephrine Changes Induced by Sleep Restriction. *J. Clin. Endocrinol. Metab.* **2015**, *100*, E416–E426.

Ferini-Strambi, L.; Filippi, M.; Martinelli, V.; Oldani, A.; Rovaris, M.; Zucconi, M.; Comi, G.; Smirne, S. Nocturnal Sleep Study in Multiple Sclerosis: Correlations with Clinical and Brain Magnetic Resonance Imaging Findings. *J. Neurol. Sci.* **1994**, *125*, 194–197.

Ferrie, J. E.; Kivimaki, M.; Akbaraly, T. N.; Singh-Manoux, A.; Miller, M. A.; Gimeno, D.; Kumari, M.; Smith, G. D.; Shipley, M. J. Associations Between Change in Sleep Duration and Inflammation: Findings on C-reactive Protein and Interleukin 6 in the Whitehall II Study. *Am. J. Epidemiol.* **2013**, *178*, 956–961.

Fleming, W. E.; Pollak, C. P. Sleep Disorders in Multiple Sclerosis. *Semin. Neurol.* **2005**, *25*, 64–68.

Garden, G. A.; Moller, T. Microglia Biology in Health and Disease. *J. Neuroimmune Pharmacol.* **2006**, *1*, 127–137.

Glass, C. K.; Saijo, K.; Winner, B.; Marchetto, M. C.; Gage, F. H. Mechanisms Underlying Inflammation in Neurodegeneration. *Cell.* **2010**, *140*, 918–934.

Gozal, D., Kheirandish-Gozal, L. Cardiovascular Morbidity in Obstructive Sleep Apnea: Oxidative Stress, Inflammation, and Much More. *Am. J. Respir. Crit. Care Med.* **2008**, *177*, 369–375.

Haack, M.; Sanchez, E.; Mullington, J. M. Elevated Inflammatory Markers in Response to Prolonged Sleep Restriction are Associated with Increased Pain Experience in Healthy Volunteers. *Sleep.* **2007**, *30*, 1145–1152.

Heiser, P.; Dickhaus, B.; Schreiber, W.; Clement, H. W.; Hasse, C.; Hennig, J.; Remschmidt, H.; Krieg, J. C.; Wesemann, W.; Opper, C. White Blood Cells and Cortisol After Sleep Deprivation and Recovery Sleep in Humans. *Eur. Arch. Psychiatry Clin. Neurosci.* **2000**, *250*, 16–23.

Hermann, D. M.; Mullington, J.; Hinze-Selch, D.; Schreiber, W.; Galanos, C.; Pollmacher, T. Endotoxin-Induced Changes in Sleep and Sleepiness During the Day. *Psychoneuroendocrinology.* **1998**, *23*, 427–437.

Hogan, D.; Morrow, J. D.; Smith, E. M.; Opp, M. R. Interleukin-6 Alters Sleep of Rats. *J. Neuroimmunol.* **2003**, *137*, 59–66.

Hopkins, S. J.; Rothwell, N. J. Cytokines and the Nervous System. I: Expression and Recognition. *Trends Neurosci.* **1995**, *18*, 83–88.

Huang, Y.; Potter, R.; Sigurdson, W.; Santacruz, A.; Shih, S.; Ju, Y. E.; Kasten, T.; Morris, J. C.; Mintun, M.; Duntley, S.; Bateman, R. J. Effects of Age and Amyloid Deposition on Abeta Dynamics in the Human Central Nervous System. *Arch. Neurol.* **2012**, *69*, 51–58.

Imeri, L.; Bianchi, S.; Opp, M. R. Inhibition of Caspase-1 in Rat Brain Reduces Spontaneous Non-Rapid Eye Movement Sleep And Non-Rapid Eye Movement Sleep Enhancement Induced by Lipopolysaccharide. *Am. J. Physiol. Regul. Integr. Comp. Physiol.* **2006**, *291*, R197–R204.

Imeri, L.; Opp, M. R. How (and why) the Immune System Makes us Sleep. *Nat. Rev. Neurosci.* **2009**, *10*, 199–210.

Irwin, M.; Mascovich, A.; Gillin, J. C.; Willoughby, R.; Pike, J.; Smith, T. L. Partial Sleep Deprivation Reduces Natural Killer Cell Activity in Humans. *Psychosom. Med.* **1994**, *56*, 493–498.

Irwin, M.; McClintick, J.; Costlow, C.; Fortner, M.; White, J.; Gillin, J. C. Partial Night Sleep Deprivation Reduces Natural Killer and Cellular Immune Responses in Humans. *Faseb J* **1996**, *10*, 643–653.

Irwin, M. R.; Olmstead, R. E.; Ganz, P. A.; Haque, R. Sleep Disturbance, Inflammation and Depression Risk in Cancer Survivors. *Brain Behav. Immun.* **2013**, *30 Suppl*, S58–S67.

Jacob, L.; Leib, R.; Ollila, H. M.; Bonvalet, M.; Adams, C. M.; Mignot, E. Comparison of Pandemrix and Arepanrix, two Ph1n1 As03-Adjuvanted Vaccines Differentially Associated with Narcolepsy Development. *Brain Behav. Immun.* **2015**, *47*, 44–57.

Kaminska, M.; Kimoff, R. J.; Benedetti, A.; Robinson, A.; Bar-Or, A.; Lapierre, Y.; Schwartzman, K.; Trojan, D. A. Obstructive Sleep Apnea is Associated with Fatigue in Multiple Sclerosis. *Mult. Scler.* **2012**, *18*, 1159–1169.

Kang, J. E.; Lim, M. M.; Bateman, R. J.; Lee, J. J.; Smyth, L. P.; Cirrito, J. R.; Fujiki, N.; Nishino, S.; Holtzman, D. M. Amyloid-Beta Dynamics are Regulated by Orexin and the Sleep-Wake Cycle. *Science.* **2009**, *326*, 1005–1007.

Kapas, L.; Krueger, J. M. Tumor Necrosis Factor-Beta Induces Sleep, Fever, and Anorexia. *Am J Physiol.* **1992**, *263*, R703–R707.

Kornum, B. R.; Faraco, J.; Mignot, E. Narcolepsy with Hypocretin/Orexin Deficiency, Infections and Autoimmunity of the Brain. *Curr. Opin. Neurobiol.* **2011**, *21*, 897–903.

Korth, C.; Mullington, J.; Schreiber, W.; Pollmacher, T. Influence of Endotoxin on Daytime Sleep in Humans. *Infect. Immun.* **1996**, *64*, 1110–1115.

Krueger, J. M.; Kubillus, S.; Shoham, S.; Davenne, D. Enhancement of Slow-Wave Sleep by Endotoxin and Lipid A. *Am. J. Physiol.* **1986**, *251*, R591–R597.

Kushikata, T.; Fang, J.; Krueger, J. M. Interleukin-10 Inhibits Spontaneous Sleep in Rabbits. *J. Interferon Cytokine Res.* **1999**, *19*, 1025–1030.

Kushikata, T.; Fang, J.; Wang, Y.; Krueger, J. M. Interleukin-4 Inhibits Spontaneous Sleep in Rabbits. *Am. J. Physiol.* **1998**, *275*, R1185–R1191.

Lancel, M.; Cronlein, J.; Muller-Preuss, P.; Holsboer, F. Lipopolysaccharide Increases Eeg Delta Activity Within Non-Rem Sleep and Disrupts Sleep Continuity in Rats. *Am. J. Physiol.* **1995**, *268*, R1310–R1318.

Lange, T.; Dimitrov, S.; Bollinger, T.; Diekelmann, S.; Born, J. Sleep After Vaccination Boosts Immunological Memory. *J. Immunol.* **2011**, *187*, 283–290.

Lange, T.; Dimitrov, S.; Fehm, H. L.; Westermann, J.; Born, J. Shift of Monocyte Function Toward Cellular Immunity During Sleep. *Arch. Intern. Med.* **2006**, *166*, 1695–1700.

Lange, T.; Perras, B.; Fehm, H. L.; Born, J. Sleep Enhances the Human Antibody Response to Hepatitis A Vaccination. *Psychosom. Med.* **2003**, *65*, 831–835.

Lasselin, J.; Rehman, J. U.; Akerstedt, T.; Lekander, M.; Axelsson, J. Effect of Long-Term Sleep Restriction and Subsequent Recovery Sleep on the Diurnal Rhythms of White Blood Cell Subpopulations. *Brain Behav. Immun.* **2015**, *47*, 93–99.

Lecendreux, M.; Libri, V.; Jaussent, I.; Mottez, E.; Lopez, R.; Lavault, S.; Regnault, A.; Arnulf, I.; Dauvilliers, Y. Impact of Cytokine in Type 1 Narcolepsy: Role of Pandemic H1n1 Vaccination ? *J. Autoimmun.* **2015**, *60*, 20–31.

Lobentanz, I. S.; Asenbaum, S.; Vass, K.; Sauter, C.; Klosch, G.; Kollegger, H.; Kristoferitsch, W.; Zeitlhofer, J. Factors Influencing Quality of Life in Multiple Sclerosis Patients: Disability, Depressive Mood, Fatigue, and Sleep Quality. *Acta Neurol Scand.* **2004**, *110*, 6–13.

Louveau, A.; Smirnov, I.; Keyes, T. J.; Eccles, J. D.; Rouhani, S. J.; Peske, J. D.; Derecki, N. C.; Castle, D.; Mandell, J. W.; Lee, K. S.; Harris, T. H.; Kipnis, J. Structural and Functional Features of Central Nervous System Lymphatic Vessels. *Nature.* **2015**, *523*, 337–341.

Lucey, B. P.; Bateman, R. J. Amyloid-Beta Diurnal Pattern: Possible Role of Sleep in Alzheimer's Disease Pathogenesis. *Neurobiol. Aging.* **2014**, *35 Suppl 2*, S29–S34.

Maes, M.; Leonard, B. E.; Myint, A. M.; Kubera, M.; Verkerk, R. The New '5-Ht' Hypothesis of Depression: Cell-Mediated Immune Activation Induces Indoleamine 2,3-Dioxygenase, Which Leads to Lower Plasma Tryptophan and an Increased Synthesis of Detrimental Tryptophan Catabolites (Trycats), Both of Which Contribute to the Onset of Depression. *Prog. Neuro-Psychoph.* **2011**, *35*, 702–721.

Manconi, M.; Ferini-Strambi, L.; Filippi, M.; Bonanni, E.; Iudice, A.; Murri, L.; Gigli, G. L.; Fratticci, L.; Merlino, G.; Terzano, G.; Granella, F.; Parrino, L.; Silvestri, R.; Arico, I.; Dattola, V.; Russo, G.; Luongo, C.; Cicolin, A.; Tribolo, A.; Cavalla, P.; Savarese, M.; Trojano, M.; Ottaviano, S.; Cirignotta, F.; Simioni, V.; Salvi, F.; Mondino, F.; Perla, F.; Chinaglia, G.; Zuliani, C.; Cesnik, E.; Granieri, E.; Placidi, F.; Palmieri, M. G.; Manni, R.; Terzaghi, M.; Bergamaschi, R.; Rocchi, R.; Ulivelli, M.; Bartalini, S.; Ferri, R.; Lo Fermo, S.; Ubiali, E.; Viscardi, M.; Rottoli, M.; Nobili, L.; Protti, A.; Ferrillo, F.; Allena, M.; Mancardi, G.; Guarnieri, B.; Londrillo, F. Multicenter Case-Control Study on Restless Legs Syndrome in Multiple Sclerosis: The Rems Study. *Sleep.* **2008a**, *31*, 944–952.

Manconi, M.; Rocca, M. A.; Ferini-Strambi, L.; Tortorella, P.; Agosta, F.; Comi, G.; Filippi, M. Restless Legs Syndrome is a Common Finding in Multiple Sclerosis and Correlates with Cervical Cord Damage. *Mult. Scler.* **2008h**, *14*, 86–93.

Manfridi, A.; Brambilla, D.; Bianchi, S.; Mariotti, M.; Opp, M. R.; Imeri, L. Interleukin-1beta Enhances Non-Rapid Eye Movement Sleep When Microinjected into the Dorsal Raphe Nucleus and Inhibits Serotonergic Neurons in Vitro. *Eur. J. Neurosci.* **2003**, *18*, 1041–1049.

Marz, P.; Cheng, J. G.; Gadient, R. A.; Patterson, P. H.; Stoyan, T.; Otten, U.; Rose-John, S. Sympathetic Neurons can Produce and Respond to Interleukin 6. *Proc. Natl Acad. Sci. USA.* **1998**, *95*, 3251–3256.

Mehra, R.; Redline, S. Sleep apnea: A Proinflammatory Disorder that Coaggregates with Obesity. *J. Allergy Clin. Immun.* **2008**, *121*, 1096–1102.

Miller, M. A. Association of Inflammatory Markers with Cardiovascular Risk and Sleepiness. *J. Clin. Sleep Med.* **2011**, *7*, S31–S33.

Moldofsky, H.; Lue, F. A.; Davidson, J. R.; Gorczynski, R. Effects of Sleep Deprivation on Human Immune Functions. *Faseb J.* **1989**, *3*, 1972–1977.

Monderer, R.; Thorpy, M. Sleep Disorders and Daytime Sleepiness in Parkinson's Disease. *Curr. Neurol. Neurosci. Rep.* **2009**, *9*, 173–180.

Moran, M.; Lynch, C. A.; Walsh, C.; Coen, R.; Coakley, D.; Lawlor, B. A. Sleep Disturbance in Mild to Moderate Alzheimer's Disease. *Sleep Med.* **2005**, *6*, 347–352.

Mullington, J.; Korth, C.; Hermann, D. M.; Orth, A.; Galanos, C.; Holsboer, F.; Pollmacher, T. Dose-Dependent Effects of Endotoxin on Human Sleep. *Am. J. Physiol. Regul. Integr. Comp. Physiol.* **2000**, *278*, R947–R955.

Nadeem, R.; Molnar, J.; Madbouly, E. M.; Nida, M.; Aggarwal, S.; Sajid, H.; Naseem, J.; Loomba, R. Serum Inflammatory Markers in Obstructive Sleep Apnea: A Meta-Analysis. *J Clin Sleep Med.* **2013**, *9*, 1003–1012.

Nishino, S.; Kanbayashi, T. Symptomatic Narcolepsy, Cataplexy and Hypersomnia, and Their Implications in the Hypothalamic Hypocretin/Orexin System. *Sleep Med. Rev.* **2005**, *9*, 269–310.

Norman, S. E.; Chediak, A. D.; Kiel, M.; Cohn, M. A. Sleep Disturbances in HIV-Infected Homosexual Men. *Aids* **1990**, *4*, 775–781.

Oka, Y.; Kanbayashi, T.; Mezaki, T.; Iseki, K.; Matsubayashi, J.; Murakami, G.; Matsui, M.; Shimizu, T.; Shibasaki, H. Low Csf Hypocretin-1/Orexin-A

Associated with Hypersomnia Secondary to Hypothalamic Lesion in a Case of Multiple Sclerosis. *J. Neurol.* **2004**, *251*, 885–886.

Okun, M. L.; Giese, S.; Lin, L.; Einen, M.; Mignot, E.; Coussons-Read, M. E. Exploring the Cytokine and Endocrine Involvement in Narcolepsy. *Brain Behav. Immun.* **2004**, *18*, 326–332.

Ondo, W. G. Sleep/Wake Problems in Parkinson's Disease: Pathophysiology and Clinicopathologic Correlations. *J. Neural. Transm.* **2014**, *121 Suppl 1*, S3–S13.

Opp, M. R.; Obal, F., Jr.; Krueger, J. M. Interleukin 1 Alters Rat Sleep: Temporal and Dose-Related Effects. *Am. J. Physiol.* **1991**, *260*, R52–R58.

Opp, M. R.; Smith, E. M.; Hughes, T. K., Jr. Interleukin-10 (Cytokine Synthesis Inhibitory Factor) Acts in the Central Nervous System of Rats to Reduce Sleep. *J. Neuroimmunol.* **1995**, *60*, 165–168.

Ottaway, C. A.; Husband, A. J. Central Nervous System Influences on Lymphocyte Migration. *Brain Behav. Immun.* **1992**, *6*, 97–116.

Ottaway, C. A.; Husband, A. J. The Influence of Neuroendocrine Pathways on Lymphocyte Migration. *Immunol. Today.* **1994**, *15*, 511–517.

Ozturk, L.; Pelin, Z.; Karadeniz, D.; Kaynak, H.; Cakar, L.; Gozukirmizi, E. Effects of 48 Hours Sleep Deprivation on Human Immune Profile. *Sleep Res. Online.* **1999**, *2*, 107–111.

Petrovsky, N. Towards a Unified Model of Neuroendocrine-Immune Interaction. *Immunol. Cell Biol.* **2001**, *79*, 350–357.

Petrovsky, N.; Harrison, L. C. The Chronobiology of Human Cytokine Production. *Int. Rev. Immunol.* **1998**, *16*, 635–649.

Plazzi, G.; Montagna, P. Remitting Rem Sleep Behavior Disorder as the Initial Sign of Multiple Sclerosis. *Sleep Med.* **2002**, *3*, 437–439.

Poirier, G.; Montplaisir, J.; Dumont, M.; Duquette, P.; Decary, F.; Pleines, J.; Lamoureux, G. Clinical and Sleep Laboratory Study of Narcoleptic Symptoms in Multiple Sclerosis. *Neurology.* **1987**, *37*, 693–695.

Prather, A. A.; Gurfein, B.; Moran, P.; Daubenmier, J.; Acree, M.; Bacchetti, P.; Sinclair, E.; Lin, J.; Blackburn, E.; Hecht, F. M.; Epel, E. S. Tired Telomeres: Poor Global Sleep Quality, Perceived Stress, and Telomere Length In Immune Cell Subsets in Obese Men and Women. *Brain Behav. Immun.* **2015**, *47*, 155–162.

Prather, A. A.; Hall, M.; Fury, J. M.; Ross, D. C.; Muldoon, M. F.; Cohen, S.; Marsland, A. L. Sleep and Antibody Response to Hepatitis B Vaccination. *Sleep.* **2012**, *35*, 1063–1069.

Prather, A. A.; Puterman, E.; Lin, J.; O'Donovan, A.; Krauss, J.; Tomiyama, A. J.; Epel, E. S.; Blackburn, E. H. Shorter Leukocyte Telomere Length in Midlife Women with Poor Sleep Quality. *J. Aging Res.* **2011**, *2011*, 721390.

Redwine, L.; Dang, J.; Hall, M.; Irwin, M. Disordered Sleep, Nocturnal Cytokines, and Immunity in Alcoholics. *Psychosom. Med.* **2003**, *65*, 75–85.

Renegar, K. B.; Crouse, D.; Floyd, R. A.; Krueger, J. Progression of Influenza Viral Infection Through the Murine Respiratory Tract: The Protective Role of Sleep Deprivation. *Sleep.* **2000**, *23*, 859–863.

Roh, J. H.; Huang, Y. F.; Bero, A. W.; Kasten, T.; Stewart, F. R.; Bateman, R. J.; Holtzman, D. M. Disruption of the Sleep-Wake Cycle and Diurnal Fluctuation of Amyloid-beta in Mice with Alzheimer's Disease Pathology. *Sci. Transl. Med.* **2012**, *4*.

Rothwell, N. J.; Hopkins, S. J. Cytokines and the Nervous System Ii: Actions and Mechanisms of Action. *Trends Neurosci.* **1995**, *18*, 130–136.

Sackstein, R.; Borenstein, M. The Effects of Corticosteroids on Lymphocyte Recirculation in Humans: Analysis of the Mechanism of Impaired Lymphocyte Migration to Lymph Node Following Methylprednisolone Administration. *J. Investig. Med.* **1995**, *43*, 68–77.

Sanders, V. M. The Role of Adrenoceptor-Mediated Signals in the Modulation of Lymphocyte Function. *Adv. Neuroimmunol.* **1995**, *5*, 283–298.

Shearer, W. T.; Reuben, J. M.; Mullington, J. M.; Price, N. J.; Lee, B. N.; Smith, E. O.; Szuba, M. P.; Van Dongen, H. P.; Dinges, D. F. Soluble Tnf-Alpha Receptor 1 and Il-6 Plasma Levels in Humans Subjected to the Sleep Deprivation Model of Spaceflight. *J Allergy Clin. Immunol.* **2001**, *107*, 165–170.

Smith, A. *Sleep, Arousal and Performance*; Birkhauser: Boston, 1992.

Stanton, B. R.; Barnes, F.; Silber, E. Sleep and Fatigue in Multiple Sclerosis. *Mult. Scler.* **2006**, *12*, 481–486.

Takahashi, S.; Kapas, L.; Fang, J., Krueger, J. M. An Anti-Tumor Necrosis Factor Antibody Suppresses Sleep in Rats and Rabbits. *Brain Res.* **1995**, *690*, 241–244.

Takahashi, S.; Kapas, L.; Fang, J. D.; Krueger, J. M. Somnogenic Relationships Between Tumor Necrosis Factor and Interleukin-1. *Am. J. Physiol-Reg I.* **1999**, *276*, R1132–R1140.

Tempaku, P. F.; Mazzotti, D. R., Tufik, S. Telomere Length as a Marker of Sleep Loss and Sleep Disturbances: A Potential Link Between Sleep and Cellular Senescence. *Sleep Med.* **2015**, *16*, 559–563.

Tippmann-Peikert, M.; Boeve, B. F.; Keegan, B. M. Rem Sleep Behavior Disorder Initiated by Acute Brainstem Multiple Sclerosis. *Neurology.* **2006**, *66*, 1277–1279.

Toth, L. A.; Opp, M. R. Cytokine- and Microbially Induced Sleep Responses of Interleukin-10 Deficient Mice. *Am. J. Physiol. Regul. Integr. Comp. Physiol.* **2001**, *280*, R1806–R1814.

Trachsel, L.; Schreiber, W.; Holsboer, F.; Pollmacher, T. Endotoxin Enhances Eeg Alpha and Beta Power in Human Sleep. *Sleep.* **1994**, *17*, 132–139.

Vgontzas, A. N.; Papanicolaou, D. A.; Bixler, E. O.; Kales, A.; Tyson, K.; Chrousos, G. P. Elevation of Plasma Cytokines in Disorders of Excessive Daytime Sleepiness: Role of Sleep Disturbance and Obesity. *J. Clin. Endocrinol. Metab.* **1997**, *82*, 1313–1316.

Vgontzas, A. N.; Pejovic, S.; Zoumakis, E.; Lin, H. M.; Bixler, E. O.; Basta, M.; Fang, J.; Sarrigiannidis, A.; Chrousos, G. P. Daytime Napping After a Night of Sleep Loss Decreases Sleepiness, Improves Performance, and Causes Beneficial Changes in Cortisol and Interleukin-6 Secretion. *Am. J. Physiol. Endocrinol. Metab.* **2007**, *292*, E253–E261.

Vgontzas, A. N.; Zoumakis, E.; Lin, H. M.; Bixler, E. O.; Trakada, G.; Chrousos, G. P. Marked Decrease in Sleepiness in Patients with Sleep Apnea by Etanercept, a Tumor Necrosis Factor-Alpha Antagonist. *J. Clin. Endocrinol. Metab.* **2004**, *89*, 4409–4413.

White, J. L.; Darko, D. F.; Brown, S. J.; Miller, J. C.; Hayduk, R.; Kelly, T.; Mitler, M. M. Early Central Nervous System Response to HIV Infection: Sleep Distortion and Cognitive-Motor Decrements. *Aids.* **1995**, *9*, 1043–1050.

Wilder-Smith, A.; Mustafa, F. B.; Earnest, A.; Gen, L.; Macary, P. A. Impact of Partial Sleep Deprivation on Immune Markers. *Sleep Med.* **2013**, *14*, 1031–1034.

Zager, A.; Andersen, M. L.; Lima, M. M.; Reksidler, A. B.; Machado, R. B., Tufik, S. Modulation of Sickness Behavior by Sleep: The Role of Neurochemical and Neuroinflammatory Pathways in Mice. *Eur. Neuropsychopharmacol.* **2009**, *19*, 589–602.

Zager, A.; Andersen, M. L.; Ruiz, F. S.; Antunes, I. B.; Tufik, S. Effects of Acute and Chronic Sleep Loss on Immune Modulation of Rats. *Am. J. Physiol. Regul. Integr. Comp. Physiol.* **2007**, *293*, R504–R509.

Zager, A.; Ruiz, F. S.; Tufik, S.; Andersen, M. L. Immune Outcomes of Paradoxical Sleep Deprivation on Cellular Distribution in Naive and Lipopolysaccharide-Stimulated Mice. *Neuroimmunomodulation.* **2012**, *19*, 79–87.

4

EPIDEMIOLOGY OF SLEEP DISORDERS

Irina Trosman and Anna Ivanenko

ABSTRACT

Sleep disorders are common in the general population. Between 10 and 30 % of adults experience insomnia at least few nights per month. Daytime sleepiness has been reported by 4–21% of the population. Sleep disturbances are associated with significant medical and mental health morbidities. Despite their high prevalence rate and impact on general health and wellbeing sleep disorders remain underrecognized and undertreated. This chapter provides an overview of the epidemiology of the most common sleep disorders.

4.1. INTRODUCTION

Sleep disorders are common in the general population and are highly prevalent across

Irina Trosman
Sleep Medicine Center, Division of Pulmonary Medicine, Ann & Robert H. Lurie Children's Hospital of Chicago, Chicago, IL, USA

Anna Ivanenko
Division of Child and Adolescent Psychiatry, Ann & Robert H. Lurie Children's Hospital of Chicago, Chicago, IL, USA

Corresponding author: Anna Ivanenko, E-mail: a-ivanenko@northwestern.edu

the lifespan. Untreated sleep disorders present a significant public health problem. As many as 10% of the population suffer from at least one clinically significant sleep disorder that impacts their quality of life and increases economical burden and health care costs.

The most prevalent sleep disorders are insomnia, sleep disordered breathing, and restless legs syndrome. Excessive daytime sleepiness related to sleep-wakefulness disorders are due to insufficient sleep and posses a high public health hazard as it directly correlates with the rate of motor vehicle accidents, work-related injuries, and high utilization of healthcare resources.

Although present at all ages, frequency of sleep disorders increases with age. Numerous epidemiological studies have been conducted over the past several decades. However, prevalence estimates of sleep disorders vary widely across studies due to differences in sample characteristics, assessment procedures, and clinical definitions of disorders.

There have been recent revisions made to several classification systems available for sleep disorders: the Diagnostic and Statistical Manual of Mental Disorders, 5th edition (DSM-5), and the International

Classification of Sleep Disorders, 3rd edition (ICSD-3).

This chapter aims to review the epidemiology of common sleep disorders based on the most recent classification criteria outlined in the ICSD-3.

4.2. SLEEP-RELATED BREATHING DISORDERS

Sleep disordered breathing (SDB) represents a group of disorders characterized by abnormal respiratory patterns during sleep. SDB includes at least three distinct clinical syndromes - obstructive sleep apnea-hypopnea syndrome (OSAHS), central sleep apnea-hypopnea syndrome (CSAHS) including Cheyne-Stokes breathing syndrome (CSBS), and sleep hypoventilation syndrome (SHVS). Each syndrome has its own specific diagnostic criteria.

4.3. EPIDEMIOLOGY OF OBSTRUCTIVE SLEEP APNEA SYNDROME IN ADULTS

4.3.1. Obstructive Sleep Apnea Hypopnea Syndrome (OSAHS)

OSAHS is a potentially disabling condition characterized by excessive daytime sleepiness, disruptive snoring, repeated episodes of upper airway obstruction during sleep, and nocturnal hypoxemia. These acute physiologic disruptions evolve into long-term sequelae, such as hypertension and cardiovascular morbidities (1,2), cognitive impairment, decreased mood and quality of life, and premature death. Daytime symptoms may include excessive sleepiness, headaches, fatigue, and decreased attention.

OSAHS is an extremely common chronic medical condition, with an estimated U.S. prevalence of 17% of women and 34% of men ages 30–70 years (3), when OSAHS is defined broadly as an apnea hypopnea index (AHI) greater than five events per hour as measured by a polysomnography. When more stringent definitions are used, either combining an AHI of more than five episodes per hour with report of at least one symptom of disturbed sleep (e.g., daytime sleepiness) or using an AHI of 15 or event per hour, the estimated prevalence is approximately 15% in males and 5% in females (4–6).

The important risk factors for OSAHS are advanced age, male gender, obesity, and craniofacial, or upper airway soft tissue abnormalities. Additional risk factors identified in some studies include smoking, nasal congestion, menopause, and family history. Rates of OSAHS are also increased in association with certain medical conditions, such as pregnancy, end-stage renal disease, congestive heart failure, chronic lung disease, and stroke.

Longitudinal data collected by the Wisconsin Sleep Cohort Study, also confirmed by the Cleveland Family Study, showed that incidence of OSAHS is independently determined by **body weight, gender, and age.**

Multiple mechanisms likely explain the associations of **excess body mass** and OSAHS, including increased upper airway collapsibility and impaired neuromuscular control of upper airway patency due to local fat deposition (7). The escalation in the prevalence of obesity in the U.S. has subsequently lead to increased prevalence of SDB among both adults and children.

Gender: OSAHS is approximately two to three times more common in males than females, although the gap narrows around the age of menopause in women (8). The mechanisms for this difference are not well understood. One possible reason for the lower prevalence of OSAHS in females may be reluctance on the part of many women to report symptoms mostly considered inappropriate, like snoring; this reluctance may cause a clinical underestimation of the problem in females (9). Analyses from different referral centers show that women with OSAHS have a greater tendency to report symptoms of fatigue and lack of energy than men. Thus, it's possible that health care providers do not recognize and refer women presenting with fatigue and lack of energy for OSAHS evaluation. It is also possible that health care provider have a lower index of suspicion for considering OSAHS in men than women given the general expectation that the disorder predominantly affects men. The role of sex hormones in OSAHS pathogenesis has been hypothesized to account for the disparity seen (10). Clear sex differences in upper airway shape and genioglossal muscle activity during the awake state, in craniofacial morphology, and in pattern of fat deposition have been proposed to account for a higher male OSAHS risk as well (11). OSAHS prevalence is higher in post- versus pre-menopausal women. Furthermore, hormone replacement therapy in post-menopausal women has been associated with a lower OSAHS prevalence (12). Women with polycystic ovarian syndrome (**PCOS**) have a higher prevalence of OSAHS (17%–70% depending on the definition of OSAHS used) probably due to obesity and increased androgen production (13,14).

Pregnancy: It is another condition in which women are at higher risk for OSAHS. The proposed mechanisms are excess weight gain, diffuse pharyngeal edema of pregnancy, and/or the effect of sleep deprivation on pharyngeal dilator muscle activity.

Age: The prevalence of OSA increases from young adulthood through the sixth to seventh decade, then appears to plateau (4,15). Epidemiologic studies tend to find a peak prevalence of clinically significant OSHS in middle age, but population-based studies typically report increasing levels of OSAHS with aging; this occurs even though obesity is less prevalent with aging.

Race: Prevalence of OSAHS is a least as high in African American population and may be higher than in Caucasian populations. Interestingly, OSAHS is common in Asia where obesity is much less common. Moreover, for a given age, sex, and BMI, Asian have greater disease severity than whites. Difference in craniofacial features between Asians and whites have been demonstrated and thought to be responsible for these findings (16).

Exposure to tobacco smoke: It causes upper airway irritation and the airway narrowing. OSAHS is more prevalent in current smokers than in non-smoker or ex-smokers (17).

Comorbid medical conditions: Certain medical conditions also increase the risk for OSAHS. These conditions include nasal obstruction, rhinitis, hypothyroidism, PCOS, acromegaly and syndromes associated with craniofacial abnormalities, and airway obstruction including Trisomy 21, Pierre Robin, Apert, and Treacher-Collins.

Craniofacial and upper airway abnormalities: Craniofacial and upper airway soft tissue abnormalities each increase the likelihood of having or developing OSA (18). These factors are well recognized in Asian patients. Examples of such abnormalities include an abnormal maxillary or short mandibular size, a wide craniofacial base, and adenotonsillar hypertrophy.

Risk factor for OSAHS
1. Obesity (BMI above 30 kg/m2)
2. Male gender (until women reach menopause)
3. Age
4. Race
5. Consumption of alcohol
6. Smoking
7. Comorbid conditions (hypothyroidism, acromegaly, craniofacial abnormalities)

4.3.2. Epidemiology of Central Sleep Apnea Syndromes (CSA)

Central sleep apnea (CSA) is a disorder characterized by repetitive cessation or decrease of both airflow and ventilatory effort during sleep. Although CSA is less common than OSAHS, the two syndromes frequently overlap. The condition can be primary (that is, idiopathic CSA) or secondary. Because CSA includes heterogeneous group of disorder, it is impossible to evaluate its prevalence.

Primary CSA is a rare condition. It is, by definition, a diagnosis of exclusion. It probably represents less than 5% of all cases of sleep apnea.

Secondary CSA can arise in association with Cheyne-Stokes breathing, congestive heart failure, narcotics use, high-altitude periodic breathing (18), and neurologic disorders. It is further categorized into hypercapnic CSA and hypocapnic/eucapnic CSA.

Risk factors: The prevalence of symptomatic CSA (that is, CSA syndrome) appears to be higher among individuals who are elderly, male, or have certain comorbid conditions.

Age: The prevalence of CSA syndrome is greater amongst adults who are older than 65 years than amongst younger adults (1.1 versus 0.4% (19). The increased prevalence of CSA syndrome in older adults may reflect the higher frequency of comorbid conditions that influence breathing during sleep in this population, including heart failure (20), atrial fibrillation (21), and cerebrovascular diseases (22,23).

Sex: CSA syndrome is more prevalent among men than women (24). One study found a lower apneic threshold among women than men due to the interaction of sleep state and chemical stimuli in sustaining rhythmic ventilation (25). $PaCO_2$ has to be reduced a greater amount to induce central apnea. Therefore, a higher apneic threshold is indicative of greater susceptibility to central apnea.

The difference between women and men appears to also be related to lower testosterone levels in women. Administration of testosterone to healthy premenopausal women for 12 days increases the apneic threshold during non-rapid eye movement (NREM) sleep (26). Conversely, suppression of testosterone with leuprolide acetate in healthy males decreases the apneic threshold (27).

Heart failure: Both CSA associated with Cheyne-Stokes breathing, and obstructive sleep apnea (OSA) are prevalent among patients with heart failure. The former is particularly common in individuals with heart failure who are male, older than 60 years, have atrial fibrillation, or have daytime hypocapnia ($PaCO_2 < 38$ mmHg) (28).

4.4. SLEEP-RELATED MOVEMENT DISORDERS

According to the ICSD-3 (2014), sleep-related movement disorders category includes:

- Restless legs syndrome/Willis-Ekbom disease
- Periodic limb movement disorder
- Sleep-related leg cramps
- Sleep-related bruxism (teeth grinding)
- Sleep-related rhythmic movement disorder
- Benign sleep myoclonus of infancy
- Propriospinal myoclonus at sleep onset
- Sleep-related movement disorder due to a medical disorder
- Sleep-related movement disorder due to a medication or substance

- Sleep-related movement disorder, unspecified

4.4.1. Restless Legs Syndrome (RLS) Epidemiology

RLS is a sensorimotor condition characterized by a strong urge to move the legs accompanied by other disagreeable sensations or parestheseas relieved by walking or movements. The symptoms have a strong circadian pattern and occur mostly in the evening or at night. RLS is divided into primary (typically idiopathic) or secondary RLS (observed in association with multiple disorders such as end-stage renal disease, iron deficiency, pregnancy, venous insufficiency, and so forth).

The diagnosis of RLS is clinical. It is based on fulfillment of four essential criteria and relies on the patient to report an accurate history. Establishing the diagnosis strictly based on these criteria yields a specificity of 84% with an estimated positive predictive value of 40% (29). Thus, use of only minimal (four) symptoms does not exclude "mimicking conditions like, leg cramps or arthritic pain in the lower limbs." In addition, the prevalence of RLS in the general population can be overestimated when these criteria are used.

RLS epidemiological studies have often been limited by lack of differentiation between primary and secondary RLS, lack of exclusion of mimicking conditions, variable population characteristics and inconsistencies in RLS diagnostic criteria, and procedures. Thus, reported RLS prevalence estimates in epidemiological studies have varied widely. When a differential diagnostic approach was taken into consideration in addition to the essential four diagnostic criteria, RLS prevalence was estimated at 1.9–4.6% of the European and North American general adult population (30).The most recent synthesis of the literature reveals that RLS prevalence rates are linked to:

Sex of the individuals: In most of studies, RLS prevalence is about twice as high in women as in men (30).

Age: RLS prevalence increases with age in North America and Europe. However, the same trend is not observed in Asian countries (30).

Race: Data on race is still too fragmentary, but Asian countries appear to have much lower prevalence of RLS (30).

Comorbid conditions: Insomnia, excessive daytime sleepiness, depression, and anxiety have been consistently associated with RLS (30).

Risk factors: End stage renal disease, pregnancy, frequent blood donations, iron deficiency, and neuropathy are considered to be risk factors for RLS (31).

4.4.2. Epidemiology of Periodic Limb Movement Disorder

Periodic leg movements in sleep (PLMS) are a frequent finding in polysomnography. In sleep studies, PLMS are found most frequently in restless legs syndrome (RLS). PLMS also often occur in narcolepsy, sleep apnea syndrome, and REM sleep behavior disorder. PLMS are also found in various medical and neurological disorders that do not primarily affect sleep.

Some patients with otherwise unexplained insomnia, fatigue, or hypersomnia have polysomnography results that reveal an elevated number of PLMS, a condition defined as periodic limb movement of sleep. Periodic limb movement disorder (PLMD) is defined by the International Classification of Sleep Disorders as periodic leg movement > 15 times/h accompanied by a clinical sleep disturbance or a complaint of daytime fatigue.

The prevalence of PLMS is estimated to be 4–11% in adults (32).There are few PSG studies examining the prevalence of PLMS and PLMD in the general population. One population-based survey obtained respondent reports of PLMD and estimated the prevalence of PLMD at approximately 3.9%.While no racial difference were examined in this study, the factors found to be associated with PLMD were shift work, daily coffee intake, and stress (33).

Another population-based study using standardized PSG criteria found a PLMD prevalence rate of 7.6%. This study also had shown that racial differences in PLMD exist, with African American being less likely to have a PLMS (34).

4.4.3. Epidemiology of Nocturnal Bruxism

Nocturnal Bruxism: It is defined as nocturnal activity that includes unconscious clenching, grinding, or bracing of the teeth. The cause of bruxism is generally accepted to be multifactorial though largely unknown. Studies have reported that awake bruxism (AB) affects females more commonly than males (35) while in sleep bruxism (SB), males are as equally affected as females (36).

A 2013 systematic review of the epidemiologic reports of bruxism showed a prevalence of about 9.7–15.9% for SB, and a prevalence of about 8–31.4% for bruxism in general. The review concluded that bruxism affects males and females equally and affects elderly people less commonly (36).

4.5. CIRCADIAN RHYTHM SLEEP-WAKE DISORDERS

4.5.1. Epidemiology of Circadian Rhythm Sleep Disorders (CRSD)

CRSD are characterized by complaints of insomnia and excessive sleepiness that are primarily due to alterations in the internal circadian timing system or a misalignment between the timing of sleep and the 24 h social and physical environment. In addition to physiological and environmental factors, maladaptive behaviors often play an important role in the development of many circadian rhythm sleep disorders. The prevalence of circadian rhythm sleep disorders in the general population is largely unknown.

Delayed sleep phase disorder (DSPD) is characterized by sleep times that are delayed three to six hours relative to the desired or socially acceptable sleep-wake schedules. It has been estimated that the prevalence of DSPD in the general population is between 0.13 and 0.17% (37,38). DSPD is more common in adolescents, with a reported prevalence of 7–16%, and may represent approximately 7% of patients presenting to sleep clinics with complaints of chronic insomnia (39).

There is no consistent data on gender differences. A British study of adolescents from 1988 found a male predominance of 10:1 (39). In contrast, a Norwegian epidemiological found no evidence for gender differences in delayed sleep phase (40). Generally adolescent boys seem to have a somewhat more delayed sleep phase than adolescent girls.

Advance sleep phase disorder (**ASPD**) is characterized by habitual and involuntary sleep times (6–9 p.m.) and wake times (2–5 a.m.) that are several hours early relative to conventional and desired times. Patients with ASPD typically present with complaints of sleep maintenance insomnia, early morning awakenings and sleepiness in the late afternoon, or early evening. In general, individuals with ASPD tend to have less difficulty adjusting to their preferred earlier schedules than those with DSPD. Non-age-related ASPD is believed to be rare, and there are only a few reported cases in the literature. However, the prevalence tends to increase with age, and has been estimated to be about 1% in middle age and older adults (41).

Non-24-hour sleep-wake syndrome is characterized by a steady daily drift of the major sleep and wake times. Because the endogenous circadian period in humans is usually slightly longer than 24 h, patients will report a progressive delay in sleep and wake times. It has been estimated that approximately 50% of legally blind individuals have non-entrained circadian rhythms (42) and about 70% have complaints of chronic sleep disturbances (43,44). In addition to non-entrained rhythms, advanced, or delayed circadian rhythms with a period of 24 h have also been reported in the blind population (45). Although rare, non-entrained sleep and wake patterns have also been reported in sighted individuals (46).

Jet lag is the result of the external environment being temporarily altered in relation to the timing of the endogenous circadian rhythm by rapid traveling across time zones. It is characterized by symptoms such as daytime fatigue and sleepiness, nighttime insomnia, mood changes, difficulty concentrating, general malaise, and gastrointestinal problems. Though considered to be a common condition, the true prevalence is unknown.

Shift work disorder (SWD) typically presents with complaints of unrefreshing sleep, excessive sleepiness, and insomnia that vary depending on the work schedule. One study suggests that the prevalence of shift work sleep disorder is approximately 10% in night and rotating shift workers (47).

Irregular sleep-wake rhythm disorder (ISWRD) is characterized by the lack of a clearly identifiable circadian pattern of sleep and wake times. Although total sleep time over 24 h may be normal for age, sleep and wake periods occur in short bouts throughout the day, and night. Although the prevalence of ISWRD increases in later life, age itself is not an independent risk factor for ISWRD. Rather, the age-associated increases in medical, neurologic, and psychiatric disorders have been shown to be the greatest contributors to the development of ISWRD. The disorder is seen more commonly in institutionalized older adults and most commonly in patients who have Alzheimer's disease. Other disorders of the central nervous system, including traumatic brain injury, and mental retardation, also can lead to an ISWR pattern (48).

Circadian rhythm disorder summary

- ASP is more common as people age, occurring in about one percent of middle-aged and older adults.
- DSPS young adults.

- Irregular sleep-wake rhythm may occur in nursing home residents and other people who have little exposure to time cues such as light, activity, and social schedules.
- Free-running (non-entrained) type occurs in more than half of all people who are totally blind.
- Jet lag can affect anyone who travels by air, but symptoms may be more severe and may last longer in older people, and when anyone travels in an eastward direction.
- Shift work disorder is most common in people who work night shifts and early morning shifts.

4.6. PARASOMNIAS

Parasomnias are characterized as abnormal behaviors that may occur during non-rapid eye movement sleep (NREM), rapid eye movement sleep (REM), or transitions between sleep and wakefulness. Parasomnias may result in arousal or partial arousal of the patient. Parasomnias can also cause physical and/or psychosocial consequences to the patient and cohabitants therefore classifying them as clinical disorders.

According to the ICSD- 3 parasomnias include:

NREM-related parasomnias:
- Confusional arousals (CAs),
- Sleepwalking (SW)
- Sleep terrors (STs)
- Sleep-related eating disorder

REM-related parasomnias:
- REM sleep behavior disorder (RBD)
- Nightmare disorder
- Recurrent isolated sleep paralysis

Other parasomnias:

- Sleep-related hallucinations
- Sleep enuresis
- Exploding head syndrome
- Parasomnia due to medical disorder
- Parasomnia due to medication and substance
- Parasomnia, unspecified

4.6.1. Epidemiology of NREM-Related Parasomnias

Sleepwalking (Somnambulism) is a parasomnia, which begins as a partial arousal typically from NREM 3 sleep. SW is one of a NREM arousal that also includes confusional arousals, sleep terrors, sleep-related eating disorder, and sexual behaviors during sleep. Other complex motor behaviors may occur in association with sleep walking such as sleep driving, highly agitated, inappropriate, and even violent behaviors.

SW is most common in children and adolescents. A longitudinal study revealed the prevalence of SW decreases with age with 14.5% frequency in children two and one half to six years of age (49). By age 13 years majority of children stop sleepwalking. However, SW persists into adulthood in approximately 24% of those with frequent sleepwalking episodes (50). In a large Finnish Twin Cohort study SW was reported in 3.9% of men and 3.1% of women (51). The vast majority of the adults who did sleepwalk reported sleepwalking in childhood.

A recent prospective cross-sectional study of 19,136 U.S. adults found that 29% reported a lifetime prevalence of "nocturnal wandering" (52). Nocturnal wandering decreased with age and was not related to gender. A Cross-sectional Norwegian study conducted among 1,000 adults

showed a history of SW in 22.4% of participants with only 1.7% reported SW within the past three months (53).

SW is more common among individuals with psychiatric disorders. Presence of bipolar disorder, major depressive, or obsessive compulsive disorders were major risk factors for SW, sleep terrors, and nocturnal wandering in U.S. adults (52,54). Also, panic or anxiety disorders, were more frequently reported in adolescents with SW and/or sleep terrors compared to healthy controls (55).

Risk factors include family history of SW or other parasomnias, other sleep disorders such as OSA, RLS, PLMD, sleep deprivation, and use certain medications (neuroleptics, non-benzodiazepine hypnotics, antidepressants, beta-blockers) in predisposed individuals.

Confusional arousals are characterized by partial awakening in a disoriented state, while the patient is in bed. They can be accompanied by vocalization or behavioral agitation. There is often no memory of the previous arousal. They usually occur in the early part of the night.

Sleep terrors, like confusional arousals, are characterized by partial arousal and are accompanied by autonomic nervous system and behavioral manifestation of intense fear. They are typically initiated by a cry or loud scream and may result in violent behavior or injuries followed by complete amnesia of the event.

Sleep terrors are most common in young children with peak prevalence at 5–7 years of age and spontaneous remission in adolescents (56,57). Depending on the population sample and methods used in the studies the prevalence rates vary between 1 and 6.5% in children with relatively stable rates of 2.3–2.6 % after 15 years of age (57). One Canadian study reported a 40% overall prevalence of STs and CAs in children with a prevalence of 20% in 2.5 years old dropping to 11% at six years of age (49). There are no genders differences have been reported across the studies.

Risk factors for STs and CAs include genetic predisposition, sleep deprivation, poor sleep hygiene, shift work, other circadian sleep disturbances, and presence of other sleep disorders.

4.6.2. Epidemiology of REM-Related Parasomnias

REM sleep behavior disorder (RBD): RBD is characterized by partial arousal and abnormal behaviors emerging during REM sleep that may cause injury. The exact prevalence of RBD is unknown.

Studies based on phone surveys of violent nighttime behavior suggest a prevalence of less than one percent. In a study of nearly 5,000 individuals conducted in Great Britain, 2% of subjects reported violent behaviors during sleep (58). Only 0.5% of those surveyed had possibly met behavioral criteria for RBD based on potential for harm, actual injury, dreams that appeared to be "acted out," and disrupted sleep. There were no PSG performed to confirm the diagnosis. Other similar survey of 1,034 individuals in Hong Kong over the age of 70 found that 0.8% reported sleep-related injury. When evaluated using PSG only four individuals met diagnostic criteria for RBD leading to the prevalence of 0.38% (59). Most studies based on the phone surveys screen for sleep-related injuries and therefore underestimate an actual prevalence of RBD since a lot of cases

may not involve physical injury and are underreported. There appears to be a male predominance of the RBD with some studies showing up to 80% of cases occurring in males. Females with RBD are less likely to have violent reenactment dreams and are probably underreported.

RBD is more common in individuals with neurodegenerative disorders. It has been shown that 50–70% of patients presenting with idiopathic RBD go on to develop a synucleinopathy over 10–15 years (60–65).

Clinical manifestations in idiopathic RBD are identical to those seen in cases secondary to Parkinson's disease, multiple system atrophy, Lewy body disease, Alzheimer's disease, and has been associated with mitochondrial disorders, brain tumors and many other diseases, and clinical conditions that may damage brainstem mechanisms involved in generating REM Sleep atonia.

Risk factor for RBD include male sex, age over 50 years, neurological disorders, narcolepsy, PTSD, family history of RBD, use of medications, particularly SSRIs and SNRIs, beta-blockers, and anticholinesterase inhibitors.

Nightmares disorder Surveys of college students and general adult population indicate that between 8 and 30% having one or more nightmares per month (66–70). Other studies conducted across different countries find that 2–6% of people report experiencing one or more nightmares per week (68,71). An overall prevalence of nightmares is estimated at 5–8% of the general population (72,73). Nightmares are common in children. Epidemiological surveys showed occurrence of occasional nightmares in 60–75% of children at different ages. The predictor of nightmare disorder at an older age is a history of recurrent nightmares in childhood.

Frequent nightmares and nightmare disorder shown to have high prevalence among patients with psychiatric disturbances, especially in those with PTSD, or those exposed to a wide range of traumatic experiences (74–76).

However, frequent nightmares can be experienced by relatively well-functioning individuals, who do not present clinically significant psychiatric symptoms.

Risk factors for nightmare disorder include genetic predisposition, younger age, psychiatric disorders, especially PTSD, certain pharmacological agents affecting neurotransmitters norepinephrine, serotonin, and dopamine.

4.7. INSOMNIA

This category of sleep disorders is characterized by persistent difficulty with sleep initiation, duration, consolidation, and quality that results in daytime impairment. There are several types of insomnia that include:

- Chronic insomnia disorder
- Short-term insomnia disorder
- Other insomnia disorder
- Isolative symptoms and normal variants:
  - Excessive time in bed
  - Short sleeper

4.7.1. Epidemiology of Insomnia

Numerous epidemiological studies on insomnia have been conducted with prevalence estimates varying widely across studies due differences in the definitions,

assessments tools, sample characteristics and type of study design, and duration of assessments.

Insomnia is highly prevalent in the general population and in primary care settings with approximately 40% of patients reporting clinically significant difficulty sleeping (77).

In general population date-base about 1/3 of adults reports at least one symptoms of insomnia associated with onset or sleep maintenance. However, this rate decreases to 10–15% when daytime consequences are added to case definition (78).

There has been a strong association reported between insomnia and psychiatric disorders and psychological stress. Insomnia has been also associated with lower socioeconomic status and with being separated or living alone.

Studies of insomnia conducted across different ages demonstrated that sleep initiation problems are more common among younger individuals, where sleep maintenance insomnia is more prevalent among middle-ages and older adults.

A meta-analysis study revealed a gender risk ratio of 1.41 for women versus men (79).

Among older adults hormonal therapy was found to be protective against insomnia (80). Cross-cultural studies across different countries demonstrated the prevalence rate ranging from as high as 79% in Brazil to as low as 6.6% in Japan (81). The prevalence rate in the U.S. was estimated to be 27% (82). A recent Swedish population-based survey demonstrated a 24.6% prevalence of insomnia symptoms. Insomnia disorder with daytime consequences accounted for 10.5% with women ages 40–

49 demonstrating higher rates if insomnia of 21.6% (83).

Cultural, ethnic and religious believes influence the way people perceive sleep experience and what is normal or abnormal sleep. These differences affect reported rates of insomnia and make it difficult to conduct epidemiological studies of insomnia. Because incidence rates of insomnia vary extensively across the studies, longitudinal studies of insomnia found incidence rates of 2.8% in Sweden (84), 6% in the U.S. (85), and 15% in U.K. (86).

Pediatric insomnia is generally defined as "a repeated difficulty with sleep initiation, duration, consolidation, or quality that occurs despite age-appropriate time and opportunity for sleep and results in daytime functional impairment for the child and/or family" (87).

The prevalence of pediatric insomnia is estimated to be about 1–6% among general pediatric populations with a lot higher prevalence among children with neurodevelopmental and chronic medical and psychiatric conditions (88,89).

Risk factors for insomnia include female gender, older age, family history of insomnia, presence of psychiatric, and chronic medical conditions.

4.8. CENTRAL DISORDERS OF HYPRSOMNOLENCE

According to the ICSD-3 daytime sleepiness is defined as "the inability to stay awake and alert during the major waking episodes of the day" and results in the periods of irrepressible sleep or lapses in to drowsiness. Central disorders of hypersomnolence include the following clinical conditions:

o   Narcolepsy type 1
o   Narcolepsy type 2
o   Idiopathic hypersomnia (IH)
o   Kleine-Levin syndrome (KLS)
o   Hypersomnia due to medical condition
o   Hypersomnia due to medication or substance
o   Hypersomnia associated with a psychiatric disorder
o   Insufficient sleep syndrome\
o   Isolated symptoms and normal variants
    • Long sleeper

4.8.1. Epidemiology of Narcolepsy

Narcolepsy type 1 is characterized by excessive daytime sleepiness, signs of REM sleep intrusion into wakefulness, and most specific cataplexy. It has been associated with hypothalamic hypocretin deficiency.

Narcolepsy with cataplexy is a rare disorder with estimated prevalence of 0.02–0.18% in the U.S. and Western Europe (90–92). There are more significant differences in the distribution of narcolepsy type 1 across the world with the highest prevalence reported in Japan (93) and the lowest prevalence in Israel (94).

There have been fewer studies on the incidence of narcolepsy with some estimates at around 1.4 per 100,000 person-years (95). Due to difficulties diagnosing this condition, narcolepsy remains underdiagnosed and undertreated illness.

There seem to be two peaks in the onset of narcolepsy, first during adolescence and second at approximately 35 years of age. Both sexes seem to be affected equally, however, a slightly higher prevalence has been reported in males.

Narcolepsy usually manifests with symptoms of excessive daytime sleepiness. Cataplexy develops in the following several years. Only 10–15% of patients present with a full spectrum picture of narcolepsy type 1 that includes a classical tetrad of symptoms.

Narcolepsy type 2 is characterized by excessive daytime sleepiness, manifestations of REM sleep on polysomnography and MSLT and absence of cataplexy. Other associated features like hypnagogic hallucinations, sleep paralysis, or automatic behaviors may be present. Although the exact prevalence of narcolepsy type 2 is unknown, it is estimated at approximately 15–25% of entire population of patients with narcolepsy. Some studies have shown a prevalence of 20.5 cases per 100,000 populations.

Risk factors and precipitating factors include head injury, viral illness, sustained sleep deprivation/sleep loss, association with HLA subtypes DR2/DRB 1*1501, and DQB1*0602.

4.8.2. Epidemiology of Idiopathic Hypersomnia

Idiopathic hypersomnia is characterized by excessive daytime sleepiness in the absence of cataplexy and is associated with less than one SOREM on MSLT and preceding polysomnography combined. There is a lack of systematic epidemiological studies of IH and the exact prevalence of IH is unknown. The existing literature indicates an approximate prevalence of IH among general population between 2 and 8 per 100, 000 individuals (96,97).

No consistent predisposing or risk factors have been identifies for IH, and it is

not known to be associated with a certain HLA subtypes.

4.8.3.Epidemiology of Kleine-Levin Syndrome

Kleine-Levin syndrome (KLS) is characterized by recurring episodes of severe hypersomnolence in association with cognitive, behavioral, and psychiatric disturbances. KLS is a rare disorder and according to one of the recent study of multiple cases from different countries the male to female ratio was approximately 4:1 with the median onset of 15 years of age (98).

There seems to be overrepresentation of Ashkenazi Jewish patients among those diagnosed with KLS (99,100). Familial cases have also been reported (101).

Risk factors and precipitating factors for KLS include upper airway infection, flulike illness, head trauma, emotional stress, alcohol intake, and anesthesia.

4.9. CONCLUSION

Sleep disorders are highly prevalent in general population and across all age groups. They have a significant impact on cognitive, emotional, behavioral, and occupational functioning. Individuals with chronic medical, neurodevelopmental and psychiatric disorders, and the elderly have a significantly higher rate of reported sleep-wake disorders that if not treated properly can persist over time. High public awareness of sleep disorders should be promoted to reduce the prevalence and minimize the impact of chronic sleep loss among individuals and within the entire society.

KEY WORDS

- **sleep**
- **disorders**
- **epidemiology**
- **prevalence**
- **incidence**

REFERENCES

1. Somers, V. K.; White, D. P.; Amin, R., et al. Sleep Apnea and Cardiovascular Disease: An American Heart Association/ America College of Cardiology Foundation Scientific Statement from the American Heart Association Council for High Blood Pressure Research Professional Education Committee, Council on Clinical Cardiology, Stroke Council, and Council on Cardiovascular Nursing. *J Am Coll Cardiol.* **2008**, *52* (8), 686–717.

2. Shamsuzzaman, A. S.; Gersh, B. J.; Somers, V. K. Obstructive Sleep Apnea: Implications for Cardiac and Vascular Disease. *JAMA.* **2003**, *290* (14), 1906–1914; Peppard, P. E.; Young, T.; Palta, M., et al. Prospective Study of the Association between Sleep-Disordered Breathing and Hypertension. *N. Engl J Med.* **2000**, *342* (19), 1378–1384.

3. Barnet, J. H.; Palta, M.; Hagen, E. W., et al. Increased Prevalence of Sleep Disordered Breathing in Adults. *Am J Epidemiol.* **2013**, *177*, 1006–1014.

4. Young, T.; Palta, M.; Dempsey, J.; Peppard, P. E.; Nieto, F. J.; Hla, K. M. Burden of Sleep Apnea: Rationale, Design, and Major Findings of the Wisconsin Sleep Cohort Study. *WMJ.* **2009**, *108* (5), 246–249.

5. Peppard, P. E.; Young, T.; Barnet, J. H.; Palta, M.; Hagen, E. W.; Hla, K. M. Increased Prevalence of Sleep-Disordered Breathing in Adults. *Am J Epidemiol.* **2013**, *177* (9), 1006–1014

6. Dempsey, J. A.; Veasey, S. C.; Morgan, B. J.; O'Donnell, C. P. Pathophysiology of Sleep Apnea. *Physiol Rev.* **2010**, *90*, 47–112.

7. Schwartz, A. R.; Patil, S. P., et al. Obesity and Obstructive Sleep Apnea: Pathogenic Mechanisms and Therapeutic Approaches. *Proc Am Thorac Soc.* **2008**, *5*(2), 185–192.

8. Young, T.; Skatrud, J.; Peppard, P. E. Risk Factors for Obstructive Sleep Apnea in Adults. *JAMA.* **2004**, *291*(16), 2013–2016.

9. Punjabi, N. The Epidemiology of Adult Obstructive Sleep Apnea. Proc Am Thorac Soc Vol 5. 2008, pp136-143

10. Krystal, A. D., et al. Sleep in Peri-Menopausal and Post-Menopausal Women. *Sleep Med Rev.* **1997**, *2* (4), 243–253.

11. Young, T.; Peppard, P. E.; Gottlieb, D. J. Epidemiology of Obstructive Sleep Apnea: A Population Health Perspective. *Am J Crit Care Med.* **2002**, *165* (9), 1217–1239.

12. Bixler, E. O., et al. Prevalence of SDB in Women: Effect of Gender. *Am J Respir Cri Care Med.* **2001**, *163*, 608–613.; Shahar, E., et al. Hormone replacement therapy and sleep -disordered breathing. *Am J Respir Crit Care Med.* **2003**, *167*, 1186–1192

13. Gopal, M. J., et al. The Role of Obesity in the Increased Prevalence of OSA in Patient with PCOS. *Sleep Med.* **2002**, *3* (5), 401–404.

14. Fogel, R. B., et al. Increased Prevalence of OSA in Obese Women with PCOS. *J Clin Endocrinol Metab.* **2001**, *86* (3), 1175–1180.

15. Tufik, S.; Santos-Silva, R.; Taddei, J. A.; Bittencourt, L. R. Obstructive Sleep Apnea Syndrome in the Sao Paulo Epidemiologic Sleep Study. *Sleep Med.* **2010** May, *11* (5), 441–446. Epub 2010 Apr 1.

16. Ong, K. C., et al. Comparison of the Severity of Sleep-Disordered Breathing in Asian and Caucasian Patient Seen at a Sleep Disorder Center. *Respir Med.* **1998**, *92*, 843–848.

17. Wetter, D.W.; Young, T. B.; Bidweel, T. R., et al. Smoking as a Risk Factor for Sleep-Disordered Breathing. *Arch Intern Med.* **1994**, *154*, 2219–2224.

18. International Classification of Sleep Disorders, 3rd Ed, American Academy of Sleep Medicine, Darien, IL **2014**. CSA associated with Cheyne-Stokes breathing is particularly common, especially among patients who have heart failure or have had a stroke.

19. Bixler, E. O.; Vgontzas, A. N.; Ten Have, T., et al. Effects of Age on Sleep Apnea in Men: I. Prevalence and Severity. *Am J Respir Crit Care Med.* **1998**, *157*, 144–148.

20. Bradley, T. D.; Floras, J. S. Sleep Apnea and Heart Failure: Part II: Central Sleep Apnea. *Circulation.* **2003**, *107*, 1822–1826.

21. Leung, R. S.; Huber, M. A.; Rogge, T., et al. Association between Atrial Fibrillation and Central Sleep Apnea. *Sleep.* **2005,** *28,* 1543–1546

22. Bassetti, C.; Aldrich, M. S. Sleep Apnea in Acute Cerebrovascular Diseases: Final Report on 128 Patients. *Sleep.* **1999,** *22,* 217–223.

23. Johnson, K. G.; Johnson, D. C. Frequency of Sleep Apnea in Stroke and Tia Patients: A Meta-Analysis. *J Clin Sleep Med.* **2010,** *6,* 131–137.

24. Bixler, E. O.; Vgontzas, A. N.; Lin, H. M., et al. Prevalence of Sleep-Disordered Breathing in Women: Effects of Gender. *Am J Respir Crit Care Med.* **2001,** *163,* 608–613.

25. Skatrud, J. B.; Dempsey, J. A. *J Appl Physiol Respir Environ Exerc Physiol.* **1983,** 55 (3), 813–822.

26. Zhou, X. S.; Rowley, J. A.; Demirovic, F., et al. Effect of Testosterone on the Apneic Threshold in Women during NREM Sleep. *J Appl Physiol (1985).* **2003,** *94,* 101–107

27. Mateika, J. H.; Omran, Q.; Rowley, J. A., et al. Treatment with Leuprolide Acetate Decreases the Threshold of the Ventilatory Response to Carbon Dioxide in Healthy Males. *J Physiol.* **2004,** *561,* 637–646.

28. Sin, D. D.; Fitzgerald, F.; Parker, J. D.; Newton, G.; Floras, J. S.; Bradley, T. D. Risk Factors for Central and Obstructive Sleep Apnea in 450 Men and Women with Congestive Heart Failure. *Am J Respir Crit Care Med.* **1999,** *160* (4), 1101–1106

29. Hening, W. A.; Allen, R. P.; Washburn, M., et al. *Sleep Med.* **2009** Oct, *10* (9), 976–981.

30. Ohayon, M.; O'Hara, R.; Vitiello, M. Epidemiology of Restless Legs Syndrome: A Synthesis of the Literature. *Sleep Med Rev.* **2012** August, *16* (4), 283–295.

31. Garcia-Borreguero, D.; Winkelmann, J.; Berger, K. Epidemiology of Restless Legs Syndrome; The Current Status. *Sleep Med Rev.* **2006** Jun, *10* (3), 153–167

32. Hornyak, M.; Feige, B.; Riemann, D.; Voderholzer, U. *Sleep Med Rev,* **2006** Jun, *10* (3), 169–177.

33. Thomas Roth; Ohayon, M. Prevalence of Restless Legs Syndrome and Periodic Limb Movement Disorder in the General Population. *J Psychosom Res.* **2002** July, *53,* 547–554.

34. Holly Scofield, Thom Roth. Periodic Limb Movement during Sleep: Population Prevalence, Clinical Correlates, and Racial Difference. *Sleep.* **2008,** *31* (9), 1221–1227.

35. American Academy of Sleep Medicine, **2001.** American Academy of Sleep Medicine. International Classification of Sleep Disorders, Revised: Diagnostic and Coding Manual. Chicago, Illinois: AASM.

36. Manfredini, D.; Winocur, E.; Guarda-Nardini, L.; Paesani, D.; Lobbezoo, F. Epidemiology of Bruxism in Adults: A Systematic Review of the Literature. *J Orofac Pain.* **2013** Spring, *27* (2), 99–110.

37. Schrader, H.; Bovim, G.; Sand, T. The Prevalence of Delayed and Advanced Sleep Phase Syndromes. *J Sleep Res.* **1993,** *13,* 51–55; Yazaki, M.; Shirakawa, S.; Okawa, M.; Takahashi, K. Demography of Sleep Disturbances Associated with Circadian Rhythm Disorders in Japan. *Psychiatry Clin Neurosci.* **1999,** *13,* 267–268.

38. Barion, A.; Zee, P. C. A Clinical Approach to Circadian Rhythm Sleep Disorders. *Sleep Med.* **2007** Sep, *8* (6), 566–577.

39. Thorpy, M. J.; Korman, E.; Spielman, A. J.; Glovinsky, P. B. Delayed Sleep Phase Syndrome in Adolescents. *J Adolesc Health Care.* **1988** Jan, *9* (1), 22–27.

40. Saxvig, I. W.; Pallesen, S.; Wilhelmsen-Langeland, A.; Molde, H.; Bjorvatn, B. Prevalence and Correlates of Delayed Sleep Phase in High School Students. *Sleep Med.* **2012** Feb, *13* (2), 193–199.

41. Ando, K.; Kripke, D. F.; Ancoli-Israel, S. Estimated Prevalence of Delayed and Advanced Sleep Phase Syndromes. *Sleep Res.* **1995,** *24,* 509.

42. Sack, R. L.; Lewy, A. J.; Blood, M. L.; Keith, L. D.; Nakagawa, H. J. Circadian Rhythm Abnormalities in Totally Blind People: Incidence and Clinical Significance. *J Clin Endocrinol Metab.* **1992** Jul, 75 (1), 127–134.

43. Miles, L. E.; Raynal, D. M.; Wilson, M. A. Blind Man Living in Normal Society has Circadian Rhythms of 24.9 Hours. *Science.* **1977,** *198* (4315), 421–423.

44. Martens, H.; Endlich, H.; Hildebrandt, G. Sleep/Wake Distribution in Blind Subjects with and without Sleep Complaints. *Sleep Res.* **1990,** 9, 398.

45. Lockley, S. W.; Skene, D. J.; Butler, L. J.; Arendt, J. Sleep and Activity Rhythms are Related to Circadian Phase in the Blind. *Sleep.* **1999,** *22* (5), 616–623.

46. McArthur, A. J.; Lewy, A. J.; Sack, R. L. Non-24-h Sleep-Wake Syndrome in a Sighted Man:

Circadian Rhythm Studies and Efficacy of Melatonin Treatment. *Sleep.* **1996**, *19* (7), 544–553.

47. Drake, C. L.; Roehrs, T.; Richardson, G.; Walsh, J. K.; Roth, T. Shift Work Sleep Disorder: Prevalence and Consequences beyond that of Symptomatic Day Workers. *Sleep.* **2004**, *27* (8), 1453–1462.

48. Rhythm Phyllis C. Zee, MD; Michael V. Vitiello. Circadian Rhythm Sleep Disorder: Irregular Sleep-Wake. *Sleep Med Clin.* **2009**, *4*, 213–218.

49. Petit, D.; Touchette, E.; Tremblay, R. E.; Boivin, M.; Montplaisir, J. Dyssomnias and Parasomnias in Early Childhood. *Pediatrics.* **2007** May, *119* (5), e1016–e10 25. Pub Med PMID: 17438080.

50. Klackenberg, G. Somnambulism in Childhood - Prevalence, Course and Behavioral Correlations. A Prospective Longitudinal Study (6–16 years). *Acta Paediatr Scand.* **1982** May, *71* (3), 495–499.

51. Hublin, C.; Kaprio, J.; Partinen, M.; Heikkila, K.; Koskenvuo, M. Prevalence and Genetics of Sleepwalking: A Population-Based Twin Study. *Neurology.* **1997** Jan, *48* (1), 177–181.

52. Ohayon, M. M.; Mahowald, M. W.; Dauvilliers, Y.; Krystal, A. D.; Leger, D. Prevalence and Comorbidity of Nocturnal Wandering in the U.S. Adult General Population. *Neurology.* **2012** May 15, *78* (20), 1583–1589.

53. Bjorvatn, B.; Gronli, J.; Pallesen, S. Prevalence of Different Parasomnias in the General Population. *Sleep Med.* **2010** Dec, *11* (10), 1031–1034.

54. Ohayon, M. M.; Guilleminault, C.; Priest, R. G. Night Terrors, Sleepwalking, and Confusional Arousals in the General Population: Their Frequency and Relationship to Other Sleep and Mental Disorders. *J Clin Psychiatry.* **1999** Apr, *60* (4), 268–276.

55. Gau, S. F.; Soong, W. T. Psychiatric Comorbidity of Adolescents with Sleep Terrors or Sleepwalking: A Case-Control Study. *Aust N Z J Psychiatry.* **1999** Oct, *33* (5), 734–739.

56. Mason, T. B. 2nd; Pack, A. I. Pediatric Parasomnias. *Sleep.* **2007** Feb, *30* (2), 141–151.

57. American Academy of Sleep Medicine. International Classification of Sleep Disorders. 3rd Edition. American Academy of Sleep Medicine, Darien, IL **2014**.

58. Ohayon, M. M.; Caulet M, Priest RG. Violent Behavior during Sleep. *J Clin Psychiatry.* **1997** Aug, *58* (8), 369–376.

59. Chiu, H. F.; Wing, Y. K.; Lam, L. C.; Li, S. W.; Lum, C. M.; Leung, T., et al. Sleep-Related Injury in the Elderly–an Epidemiological Study in Hong Kong. *Sleep.* **2000** Jun 15, *23* (4), 513–517.

60. Schenck, C. H.; Hurwitz, T. D.; Mahowald, M. W. Symposium: Normal And Abnormal REM Sleep Regulation: REM Sleep Behaviour Disorder: An Update on a Series of 96 Patients and a Review of the World Literature. *J Sleep Res.* **1993** Dec, *2* (4), 224–231.

61. Olson, E. J.; Boeve, B. F.; Silber, M. H. Rapid Eye Movement Sleep Behaviour Disorder: Demographic, Clinical and Laboratory Findings in 93 Cases. *Brain.* **2000** Feb, *123* (Pt 2), 331–339.

62. Schenck, C. H.; Bundlie, S. R.; Mahowald, M. W. Delayed Emergence of a Parkinsonian Disorder in 38% of 29 Older Men Initially Diagnosed with Idiopathic Rapid Eye Movement Sleep Behaviour Disorder. *Neurology.* **1996** Feb, *46* (2), 388–393.

63. Britton, T. C.; Chaudhuri, K. R. REM Sleep Behavior Disorder and the Risk of Developing Parkinson Disease or Dementia. *Neurology.* **2009** Apr 14, *72* (15), 1294–1295.

64. Postuma, R. B.; Gagnon, J. F.; Vendette, M.; Montplaisir, J. Y. Idiopathic REM Sleep Behavior Disorder in the Transition to Degenerative Disease. *Mov Disord.* **2009** Nov 15, *24* (15), 2225–2232.

65. Postuma, R. B.; Gagnon, J. F.; Vendette, M.; Fantini, M. L.; Massicotte-Marquez, J.; Montplaisir, J. Quantifying the Risk of Neurodegenerative Disease in Idiopathic REM Sleep Behavior Disorder. *Neurology.* **2009** Apr 14, *72* (15), 1296–1300.

66. Haynes, S. N.; Mooney, D. K. Nightmares: Etiological, Theoretical, and Behavioral Treatment Considerations. *Psychol Rec.* **1975**, *25* (2), 225–236.

67. Belicki. K.; Belicki, D. Predisposition for Nightmares: A Study of Hypnotic Ability, Vividness of Imagery, and Absorption. *J Clin Psychol.* **1986**, *42* (5), 714–718.

68. Feldman, M. J.; Hersen, M. Attitudes toward Death in Nightmare Subjects. *J Abnorm Psychol.* **1967**, *72* (5), 421–425.

69. Wood, J. M.; Bootzin, R. R. The Prevalence of Nightmares and their Independence from Anxiety. *J Abnorm Psychol.* **1990**, *99* (1), 64–68.

70. Zadra, A.; Donderi, D. C. Nightmares and Bad Dreams: Their Prevalence and Relationship to

Well-Being. *J Abnorm Psychol.* **2000**, **109** (2), 273–281.

71. Janson, C.; Gislason, T.; De Backer, W.; Plaschke, P.; Bjornsson, E.; Hetta, J., et al. Prevalence of Sleep Disturbances among Young Adults in Three European Countries. *Sleep.* **1995**, *18* (7), 589–597.

72. Bixler, E.O., et al. Prevalence of Sleep Disorders in the Los Angeles Metropolitan Area. *Am J Psychiatry.* **1979**, *136* (10), 1257–1262.

73. Klink, M.; Quan, S. F. Prevalence of Reported Sleep Disturbances in a General Adult Population and their Relationship to Obstructive Airways Diseases. *Chest.* **1987**, *91* (4), 540–546.

74. Krakow, B.; Schrader, R.; Tandberg, D.; Hollifield, M.; Koss, M. P.; Yau, C. L., et al. Nightmare Frequency in Sexual Assault Survivors with PTSD. *J Anxiety Disord.* **2002**, *16* (2), 175–190.

75. Phelps, A. J.; Forbes, D.; Creamer, M. Understanding Posttraumatic Nightmares: An Empirical and Conceptual Review. *Clin Psychol Rev.* **2008**, *28* (2), 338–355.

76. Wittmann, L.; Schredl, M.; Kramer, M. Dreaming in Posttraumatic Stress Disorder: A Critical Review of Phenomenology, Psychophysiology and Treatment. *Psychother Psychosom.* **2006**, *76* (1), 25–39.

77. Simon, G. E.; VonKorff, M. Prevalence, Burden, and Treatment of Insomnia in Primary Care. *Am J Psychiatry.* **1997**, *154*, 1417–1423.

78. Ohayon, M. M.; Reynolds, C. F. 3rd. Epidemiological and Clinical Relevance of Insomnia Diagnosis Algorithms According to the DSM-IV and the International Classification of Sleep Disorders (ICSD). *Sleep Med.* **2009**, *10*, 952–960.

79. Zhang, B.; Wing, Y. K. Sex Differences in Insomnia: A Meta-Analysis. *Sleep.* **2006**, *29*, 85–93.

80. Jaussent, A.; Dauvillierrs, Y.; Ancelin, M. L., et al. Insomnia Symptoms in Older Adults: Associated Factors and Gender Differences. *Am J Geriatr Psychiatry.* **2011**, *19*, 88–97.

81. Soldatos, C.; Allaert, F.; Ohta, T., et al. How do Individuals Sleep around the World? Results from a Single-Day Survey in Ten Countries. *Sleep Med.* **2005**, *6*, 5–13.

82. Leger, D.; Poursain, B. An International Survey of Insomnia: Under-Recognition and Under-Treatment of a Polysymptomatic Condition. *Curr Med Res Opin.* **2005**, *21*, 1785–1792.

83. Mallon, L.; Broman, J. E.; Akerstedt, T.; Hetta, J. Insomnia in Sweden: A Population-Based Survey. *Sleep Disord.* **2014**, *2014*, 843126. doi: 10.1155/2014/843126. Epub 2014 May 12. PMID: 24955254.

84. Jansson-Frojmark, M.; Linton, S. J. The Course of Insomnia over one Year: A Longitudinal Study in the General Population in Sweden. *Sleep.* **2008**, *31*, 881–886.

85. Ford, D. E.; Kamerow, D. B. Epidemiologic Study of Sleep Disturbances and Psychiatric Disorders. An Opportunity for Prevention? *JAMA.* **1989**, *262*, 1479–1484.

86. Morphy, H.; Dunn, K. M.; Lewis, M., et al. Epidemiology of Insomnia: A Longitudinal Study in a UK Population. *Sleep.* **2007**, *30*, 274–280.

87. Mindell, J. A.; Emslie, G.; Blumer, J., et al. *Pharmacological Management of Insomnia in Children and Adolescents: Consensus Statement. Pediatrics.* **2006**, *117* (6), e1223–e1232.

88. Stein, M. A.; Mendelsohn. J; Obermeyer, W. H., et al. *Sleep and Behavior Problems in School-Aged Children. Pediatrics.* **2001**, *107* (4), E60.

89. Ivanenko, A.; Crabtree, V. M.; Gozal, D. *Sleep in Children with Psychiatric Disorders. Pediatr Clin North Am.* **2004**, *51* (1), 51–68.

90. Longstreth, W.T. Jr; Koepsell, T. D.; Ton, T. G., et al. The Epidemiology of Narcolepsy. *Sleep.* **2007**, *30*, 13–26.

91. Ohayon, M. M.; Priest, R. G.; Zulley, J., et al. Prevalence of Narcolepsy Symptomatology and Diagnosis in the European General Population. *Neurology.* **2002**, *58*, 1826–1833.

92. Hublin, C.; Kaprio, J.; Partinen, M., et al. The Prevalence of Narcolepsy: An Epidemiological Study of the Finnish Twin Cohort. *Ann Neurol.* **1994**, *35*, 709–716.

93. Honda, Y. Census of Narcolepsy, Cataplexy and Sleep Life among Teenagers in Fujisawa City. *Sleep Res.* **1979**, *8*, 191.

94. Lavie, P.; Peled, R. Narcolepsy is a Rare Disease in Israel. *Sleep.* **1987**, *10*, 608–609.

95. Silber, M. H.; Krahn, L. E.; Olson, E. J., et al. The Epidemiology of Narcolepsy in Olmsted County, Minnesota: A Population-Based Study. *Sleep.* **2002**, *25*, 197–202.

96. Bassetti, C.; Aldrich, M. S. Idiopathic Hypersomnia. A Series of 42 Patients. *Brain.* **1997**, *120* (Pt 8), 1423–1435.

97. Billiard, M.; Dauvilliers, Y. Idiopathic Hypersomnia. *Sleep Med Rev.* **2001**, *5* (5), 351–360.

98. Billiard, M.; Jaussent, I.; Dauvilliers, Y., et al. Recurrent Hypersomnias: A Review of 339 Cases. *Sleep Med Rev.* **2011**, 15, 247–257.

99. Arnulf, I.; Zeitzer J. M.; File, J., et al. Kleine-Levin Syndrome: A Systematic Review of 186 Cases in the Literature. *Brain.* **2005**, 128, 2763–2776.

100. Arnulf, I.; Lin, L.; Gadoth, N., et al. Kleine-Levin Syndrome: A Systematic Study of 108 Patients. *Ann Neurol.* **2008**, 63, 482–492.

101. BaHammam, A. S.; GadElRab, M. O.; Owais, S. M., et al. Clinical Characteristics and HLA Typing of a Family with Kleine-Levin Syndrome. *Sleep Med.* **2008**, 9, 575–578.

5

CURRENT CLASSIFICATIONS OF SLEEP DISORDERS

Michael Thorpy

The first formal classification of sleep disorders was in 1979, when the Association of Sleep Disorders Centers published the Classification of Sleep and Arousal Disorders in the journal Sleep.[1] In 2013, the *Diagnostic and Statistical Manual of Mental Disorders fifth edition (DSM-V)* included a section entitled "Sleep-Wake Disorders."[2] This classification differs from that of the *International Classification of Sleep Disorders* (ICSD-3) that was produced by the American Academy of Sleep Medicine in 2014.[3] *The International Classification of Diseases* (ICD)-modified version, the ICD-10-CM, will be adopted in the USA in 2015, and contains classification that more closely conforms to the ICSD-3 (Table 5.1).[4]

Michael Thorpy
Sleep-Wake Disorders Center, Montefiore Medical Center, and Albert Einstein College of Medicine, Bronx, NY, USA

Corresponding author: Michael Thorpy, Sleep-Wake Disorders Center, Montefiore Medical Center, Bronx, NY 1046, USA. Email: thorpy@aecom.yu.edu. Tel.: +1 718-920-4841

5.1. DIAGNOSTIC AND STATISTICAL MANUAL OF MENTAL DISORDERS 5TH EDITION (DSM-V)

Insomnia disorder, the first entry in the DSM-V, is a diagnostic entry that requires the presence of at least one sleep complaint, such as difficulty initiating sleep that must be present at least three nights per week for at least three months. The diagnosis is coded along with other mental, medical, and sleep disorders. The diagnosis can be specified as being episodic if it occurs for at least one month; however, acute and short-term insomnia which has symptoms of less than three months should be coded as "other specified insomnia disorder. "

Hypersomnolence disorder requires a three-month history of excessive sleepiness in the presence of significant distress or other impairment. Objective documentation by electrophysiological tests, such as the multiple sleep latency test (MSLT), is not required. This diagnosis is coded along with any other concurrent mental, medical, and sleep disorder. Narcolepsy is defined as recurrent episodes of sleep that occur for at least three months along

Table 5.1. DSM-V.

Sleep–wake disorders
- Insomnia disorder
- Hypersomnolence disorder
- Narcolepsy
  - Subtypes
    - Presence or absence of hypocretin deficiency
    - Presence or absence of hypocretin deficiency
    - Autosomal dominant cerebellar ataxia
    - Deafness and narcolepsy (ADCADN)
    - Autosomal dominant narcolepsy
    - Obesity and type 2 diabetes (ADNOD)
    - Secondary to another medical condition
- Obstructive sleep apnea syndrome
- Central sleep apnea
- Sleep-related hypoventilation
  - Subtypes:
    - Idiopathic hyperventilation
    - Congenital central alveolar hypoventilation
    - Comorbid sleep-related hypoventilation
- Circadian rhythm sleep disorder
    - Delayed sleep phase type
    - Advance sleep phase type
    - Irregular sleep–wake type
    - Non-24 h sleep–wake type
    - Shift work disorder
    - Unspecified type
- Parasomnias
    - Non-rapid eye movement sleep arousal disorder
      - Subtypes
        - o Sleepwalking type
        - o Sleep terror type
    - Nightmare disorder
    - Rapid eye movement sleep behavior disorder
- Restless legs syndrome
- Substance/medication-induced sleep disorder
- Other specified insomnia disorder
- Other specified hypersomnolence disorder
- Unspecified sleep–wake disorder
- Unspecified insomnia disorder
- Unspecified hypersomnolence disorder
Unspecified sleep–wake disorder

Adapted from: American Psychiatric Association. *Diagnostic and Statistical Manual of Mental Disorders 5th Edition DSM-V.* Washington DC, 2013.

with one of three additional features, such as cataplexy, hypocretin deficiency, or polysomnographic features, either a sleep onset rapid eye movement (REM) period (SOREMP) on a nighttime polysomnogram (PSG) or an MSLT that shows a mean sleep latency of 8 min or less and two or more SOREMPs. So narcolepsy can be diagnosed in DSM-V if just sleepiness occurs for three months and there is a SOREMP on the nocturnal PSG. This has the potential of leading to errors in diagnosis as other disorders including obstructive sleep apnea syndrome (OSA) can produce similar features. Five subtypes of narcolepsy are specified according to the presence or absence of hypocretin deficiency, autosomal dominant cerebellar ataxia, deafness, and narcolepsy (ADCADN), autosomal dominant narcolepsy, obesity, and type 2 diabetes (ADNOD), or secondary to another medical condition.

OSA is an apnea hypopnea index (AHI) of at least five per hour along with typical nocturnal respiratory symptoms, or daytime excessive sleepiness or fatigue. Alternatively, the diagnosis requires an AHI of at least 15 regardless of accompanying symptoms. Mild is regarded as an AHI of less than 15; moderate, 15–30; and severe, greater than 30. Central sleep apnea (CSA) requires the presence of five or more central apneas per hour of sleep. Sleep-related hypoventilation has PSG evidence of decreased ventilation with either elevated CO_2 levels or persistent oxygen desaturation unassociated with apneic/hypopneic events. Subtypes of sleep-related hypoventilation include idiopathic hypoventilation, congenital central alveolar hypoventilation, and comorbid sleep-related hypoventilation.

Circadian rhythm sleep–wake disorders (CRSWDs) with five subtypes is caused by a persistent recurrent pattern of sleep disruption due to an alteration or misalignment of the endogenous circadian rhythm and the individual's required sleep–wake schedule, along with symptoms of either insomnia or excessive sleepiness or both. Subtypes are delayed sleep phase type, advanced sleep phase type, irregular sleep–wake type, non-24 h sleep–wake type, and shift work type none of which has specific diagnostic criteria.

The parasomnias are subdivided into five disorders: non-rapid eye movement (NREM) sleep arousal disorder, nightmare disorder, REM sleep behavior disorder (RBD), restless legs syndrome, and substance/medication-induced sleep disorder. NREM sleep arousal disorder consists of two types with the typical features of either sleepwalking or sleep terrors. The sleepwalking type can be subtyped into sleep-related eating or sleep-related sexual behavior (sexsomnia). Nightmare disorder is due to repeated occurrences of extended, dysphoric, and well-remembered dreams that threaten the individual. Rapid orientation and alertness follows the episode, but it causes significant distress. RBD is recurrent episodes of arousal with vocalization and /or complex movements from REM sleep that is documented by either PSG or a history suggesting a synucleinopathy. Restless legs syndrome, an urge to move the legs is accompanied by uncomfortable sensations in the legs with the typical features that occur at least three times per week for at least three months. Sleep/medication-induced sleep disorder is a sleep disturbance during substance intoxication, soon after substance intoxication or after withdrawal, or when the substance is known to cause sleep disturbance.

Other specified insomnia disorders are diagnosed when the insomnia does not meet the criteria for insomnia disorder, and other hypersomnolence disorder when the excessive sleepiness does not meet the criteria for hypersomnolence disorder. Similarly, unspecified sleep–wake disorder is diagnosed when it does not meet the full criteria for the specified sleep–wake disorders. Unspecified forms consist of insomnia disorder, hypersomnolence disorder, and sleep–wake disorder.

5.2. INTERNATIONAL CLASSIFICATION OF SLEEP DISORDERS 3RD REVISION (ICSD-3)

The ICSD-3 (Table 5.2) has six major disorder sections and a related isolated symptoms and normal variants section that contains conditions that are not believed to be specific disorders at this time.

The insomnia section has three disorders, and the isolated symptoms and normal variants section has two conditions. The sleep-related breathing disorders section has four subsections: obstructive sleep

Table 5.2. ICSD-3.

ICD-9-CM code	ICD-10-CM code	
Insomnia disorders		
Chronic insomnia disorder	342	F51.01
Short-term insomnia disorder	307.41	F51.02
Other insomnia disorder	307.49	F51.09
Isolated symptoms and normal variants		
Excessive time in bed		
Short sleeper		
Sleep-related breathing disorders		
Obstructive sleep apnea disorders		
Obstructive sleep apnea, adult	327.23	G47.33
Obstructive sleep apnea, pediatric	327.23	G47.33
Central sleep apnea syndromes		
Central sleep apnea with Cheyne-Stokes breathing	786.04	R06.3
Central apnea due to a medical disorder without Cheyne-Stokes		
Breathing	327.27	G47.37
Central sleep apnea due to high-altitude periodic breathing	327.22	G47.32
Central sleep apnea due to a medication or substance	327.29	G47.39
Primary central sleep apnea	327.21	G47.31
Primary central sleep apnea of infancy	770.81	P28.3
Primary central sleep apnea of prematurity	770.82	P28.4
Treatment–emergent central sleep apnea	327.29	G47.39
Sleep-related hypoventilation disorders		
Obesity hypoventilation syndrome	278.03	E66.2
Congenital central alveolar hypoventilation syndrome	327.25	G47.35
Late-onset central hypoventilation with hypothalamic dysfunction	327.26	G47.36
Idiopathic central alveolar hypoventilation	327.24	G47.34
Sleep-related hypoventilation due to a medication or substance	327.26	G47.36

Sleep-related hypoventilation due to a medical disorder	327.26	G47.36
Sleep-related hypoxemia disorder		
Sleep-related hypoxemia	327.26	G47.36
Isolated symptoms and normal variants		
Snoring		
Catathrenia		
Central disorders of hypersomnolence		
Narcolepsy type 1	347.01	G47.411
Narcolepsy type 2	347.00	G47.419
Idiopathic hypersomnia	327.11	G47.11
Kleine-Levin syndrome	327.13	G47.13
Hypersomnia due to a medical disorder	327.14	G47.14
Hypersomnia due to a medication or substance	292.85 (Drug-induced) F11-F19 291.82 (Alcohol-induced)	
Hypersomnia associated with a psychiatric disorder	327.15	F51.13
Insufficient sleep syndrome	307.44	F51.12
Isolated symptoms and normal variants		
Long sleeper		
Circadian rhythm sleep–wake disorders		
Delayed sleep–wake phase disorder	327.31	G47.21
Advanced sleep–wake phase disorder	327.32	G47.22
Irregular sleep–wake rhythm disorder	327.33	G47.23
Non-24 h sleep–wake rhythm disorder	327.34	G47.24
Shift work disorder	327.36	G47.26
Jet lag disorder	327.35	G47.25
Circadian sleep–wake disorder not otherwise specified (NOS)	327.30	G47.20
Parasomnias		
NREM-related parasomnias		
Disorders of arousal (from NREM sleep)		
Confusional arousals	327.41	G47.51
Sleepwalking	307.46	F51.3
Sleep terrors	307.46	F51.4
Sleep-related eating disorder	327.40	G47.59
REM-related parasomnias		
REM sleep behavior disorder:	327.42	G47.52
Recurrent isolated sleep paralysis	327.43	G47.51
Nightmare disorder	307.47	F51.5
Other parasomnias		
Exploding head syndrome	327.49	G47.59
Sleep-related hallucinations	368.16	H53.16
Sleep enuresis 788.36	N39.44	
Parasomnia due to a medical disorder	327.44	G47.54
Parasomnia due to a medication or substance	292.85 (Drug-induced) F11-F19 291.82 (Alcohol-induced)	
Parasomnia, unspecified	327.40	G47.50

Isolated symptoms and normal
    Sleep talking
Sleep-related movement disorders

Restless legs syndrome	333.94	G25.81	
Periodic limb movement disorder		327.51	G47.61
Sleep-related leg cramps	327.52	G47.62	
Sleep-related bruxism	327.53	G47.63	
Sleep-related rhythmic movement disorder		327.59	G47.69
Benign sleep myoclonus of infancy		327.59	G47.69
Propriospinal myoclonus at sleep onset		327.59	G47.69
Sleep-related movement disorder due to a medical disorder		327.59	G47.69
Sleep-related movement disorder due to a medication or substance		292.85 (Drug-induced)F11-F19 291.82 (Alcohol-induced)	
Sleep-related movement disorder, unspecified		327.59	G47.69

Isolated symptoms and normal variants
    Excessive fragmentary myoclonus
    Hypnagogic foot tremor and alternating leg muscle activation
    Sleep starts (hypnic jerks)

Other sleep disorder		327.8	G47.8

Adapted from: American Academy of Sleep Medicine. *International classification of sleep disorders, 3rd ed.* Darien, IL: American Academy of Sleep Medicine, 2014.

APPENDIX A

Fatal familial insomnia	046.8	A81.83
Sleep-related epilepsy	345	G40.5
Sleep-related headaches	784.0	R51
Sleep-related laryngospasm	787.2	J38.5
Sleep-related gastroesophageal reflux	530.1	K21.9
Sleep-related myocardial ischemia	411.8	I25.6

APPENDIX B

ICD-10-CM coding for substance-induced sleep disorders	F10–F19

apnea disorders section has two disorders, CSA syndromes section has eight disorders, sleep-related hypoventilation disorders section has six disorders, sleep-related hypoxemia disorder section has one disorder, and isolated symptoms and normal variants section has two disorders. The central disorders of hypersomnolence have eight disorders and the isolated symptoms and normal variants have one condition.

CRSWDs have seven disorders. Parasomnias have three subdivisions: NREM-related parasomnias with four specific disorders, REM-related parasomnias with three disorders, other parasomnias with six disorders, and isolated symptoms and normal variants with one condition. The sleep-related movement disorders (SRDMs) have 10 disorders, and the isolated symptoms and normal variants section have three conditions.

5.2.1. Insomnia Disorders

Insomnia is defined as a persistent difficulty with sleep initiation, duration, consolidation, or quality that occurs despite adequate opportunity and circumstances for sleep, and results in some form of daytime impairment. The insomnia disorders consist mainly of one major disorder termed chronic insomnia disorder. The clinical features of insomnia are believed to be the result of a primary or secondary process, but the consequences are similar no matter what the etiology.[5] The diagnosis rests upon a sleep symptom of difficulty initiating sleep, difficulty maintaining sleep, early-morning awakening, and mainly for pediatric age groups, resistance to going to bed, and difficulty in sleeping without a caregiver intervention. The symptoms occur three times per week for at least three months and have daytime consequences, such as fatigue/malaise, cognitive impairment, mood lability, daytime sleepiness or social, family, academic, or occupational impairment. Psychophysiological insomnia and other insomnia disorders of the ICSD-2 are mentioned in the text description of the disorder. Objective testing by PSG is not required for diagnosis, but

should be considered if features suggest a concurrent sleep-related breathing disorder. The inclusion of short-term insomnia disorder with similar diagnostic criteria as chronic insomnia disorder applies to insomnia which is less than three months in duration. Excessive time in bed and short sleep are included as isolated symptoms and normal variants, not as specific disorders.

5.2.2. Sleep-Related Breathing Disorders

The sleep-related breathing disorders are characterized by abnormalities of respiration during sleep, which in some can also occur during wakefulness. Sometimes more than one disorder can exist in a patient.

The sleep-related breathing disorders section is organized into four main categories; OSA disorders, CSA syndromes, sleep-related hypoventilation disorders, and sleep-related hypoxemia disorder. The CSA syndromes are divided into eight types; two related to Cheyne-Stokes breathing (CSB), high altitude, substance, three primary CSA disorders of which one is infancy and the other prematurity, and a new entity entitled treatment-emergent CSA. The latter category applies to central apnea that follows CPAP administration.

OSA syndrome maintains the criterion of five or more predominantly obstructive respiratory events per hour of sleep (including respiratory effort related arousals [RERAs]), when studied in a sleep center or by out-of-center sleep studies (OCST), as long as typical symptoms are present, otherwise 15 or more are sufficient to make the diagnosis. The OSA disorders are divided into adult and pediatric types.

In the pediatric criteria, for those less than 18 years of age, one or more obstructive events are required per hour of sleep as long as respiratory symptoms or sleepiness are present, alternatively, obstructive hypoventilation (with a $PaCO_2 > 50$ mmHg) along with symptoms of snoring, labored breathing or sleepiness, and other cognitive effects are required.

CSA-CSB is five or more central apnea or hypopneas per hour of sleep with a pattern that meets criteria for CSB. This diagnosis can occur along with a diagnosis of OSA. CSA without CSB is diagnosed as CSA due to a medical disorder without CSB that occurs as a consequence of a medical or neurological disorder, but not due to a medication or substance. CSA due to high-altitude periodic breathing is central apnea attributable to high-altitude of at least 1500 m, but usually above 2500 m with associated symptoms of sleepiness, difficulty in sleeping, awakening with shortness of breath or headache, or witnessed apneas. PSG is not required. CSA occurs due to a medicine or substance most typically due to an opioid or respiratory depressant, but not associated with CSB. PSG is required. Primary CSA is five or more central apneas or central hypopneas per hour of sleep in the absence of CSB and of unknown etiology. Typical symptoms of sleepiness or abnormal breathing events are present, but may not be present in children. Primary CSA of infancy occurs in an infant with greater than 37 weeks conceptional age with recurrent, prolonged (> 20 s duration) central apneas and periodic breathing for more than 5% of total sleep time. Apnea or cyanosis is seen by an observer or desaturation is detected by monitoring. Primary CSA of prematurity occurs in an infant of less than 37 weeks conceptional age with similar features and respiratory events.

Treatment-emergent CSA is a disorder that has become recognized since the application of positive airway pressure (PAP) despite resolution of obstructive respiratory events. It is diagnosed when five or more obstructive events during a PSG with PAP that shows resolution of obstructive events and presence of central apneas or hypopneas.[6]

Sleep-related hypoventilation disorders consist of seven disorders that meet diagnostic criteria for sleep-related hypoventilation with or without oxygen desaturation. Obesity hypoventilation is hypoventilation during wakefulness ($PaCO_2 > 45$ mmHg) in the presence of obesity (BMI > 30 kg/m^2. Congenital central alveolar hypoventilation syndrome is diagnosed when the sleep-related hypoventilation is associated with the PHOX2B gene. Late-onset central hypoventilation with hypothalamic dysfunction has sleep-related hypoventilation without symptoms in the first few years of life and the PHOX2B gene is not present. Idiopathic central alveolar hypoventilation is sleep-related hypoventilation with the presence of lung or airway disease or any other known cause. Sleep-related hypoventilation due to a medication or substance is sleep-related hypoventilation when a medication or substance is known to be the primary cause. Sleep-related hypoventilation due to a medical disorder is sleep-related hypoventilation due to a lung or airway disease, or other medical cause. Sleep-related hypoxemia disorder is arterial oxygen saturation of <88% in adults or <90% in children for less than 5 min. PSG is not required. Snoring and catathrenia (prolonged expiratory expiration in REM sleep) are regarded as isolated symptoms or normal variants.

5.2.3. Central Disorders of Hypersomnolence

The central disorders of hypersomnolence include disorders in which the primary complaint is daytime sleepiness not caused by disturbed nocturnal sleep or misaligned circadian rhythms. The term hypersomnolence is used to describe the symptom of excessive sleepiness, and hypersomnia refers to specific disorders. The central disorders of hypersomnolence comprise eight disorders. Narcolepsy has undergone a major revision with elimination of the disorder name terms, with or without cataplexy. Type 1 narcolepsy is presumed to be due to hypocretin loss with either measured reduction in CSF hypocretin, or cataplexy with associated electrophysiological findings. Narcolepsy Type 2 is confirmed by electrophysiological studies in the absence of cataplexy, or with a normal CSF hypocretin level. A major change in the narcolepsy criteria is the inclusion of a SOREMP on the nocturnal PSG as one of the two required to meet the MSLT criteria of two SOREMPs for diagnosis. This finding is mainly based upon a study that indicates that the positive predictive value of a SOREMP on the nocturnal PSG for narcolepsy is 92%.[7] Approximately, 50% of patients with narcolepsy will have a SOREMP less than 15 min on the nocturnal PSG.

Idiopathic hypersomnia is now a single entity with elimination of the two ICSD-2 hypersomnia disorders that had specific sleep duration criteria. Idiopathic hypersomnia disorder requires sleepiness for at least three months, an MSLT mean sleep latency of 8 min or less, or a nocturnal sleep duration of at least 660 min. The ICSD-2 category of recurrent hypersomnia has been reduced to a single entry Kleine-Levin syndrome with a subtype of menstrual-related Kleine-Levin syndrome.[8] The sleepiness must persist for two days to five weeks, and at least once every 18 months. There can be only one symptom with the sleepiness of cognitive dysfunction, altered perception, eating disorder, or disinhibited behavior. Normal alertness and cognitive function must be present between episodes. Hypersomnia due to a medical disorder requires association of sleepiness caused by any underlying medical or neurological disorder. Seven subtypes are mentioned: hypersomnia secondary to Parkinson's disease, posttraumatic hypersomnia, genetic disorders associated with primary central nervous system somnolence, hypersomnia secondary to brain tumors, infections, or other central nervous system lesions, hypersomnia secondary to endocrine disorder, hypersomnia secondary to metabolic encephalopathy and residual hypersomnia in patients with adequately treated OSA.

Hypersomnia due to a medication or substance is sleepiness that occurs as a consequence of a current medication or substance, or withdrawal from wake-promoting medication or substance. Hypersomnia associated with a psychiatric disorder is sleepiness in association with a current psychiatric disorder. Insufficient sleep syndrome is the new term for the previous more cumbersome term of behaviorally induced insufficient sleep syndrome. The reduced sleep must be present most days for at least three months. Extension of sleep time must result in resolution of symptoms. Long sleeping is no longer regarded as a disorder, but as a normal variant. There are no diagnostic criteria, but a total sleep time ≥10 h is suggested as being usually accepted.

5.2.4. Circadian Rhythm Sleep–Wake Disorders

CRSWDs are disorders caused by alteration of the circadian time-keeping system, its entrainment mechanisms, or misalignment of the endogenous circadian rhythm, and the external environment. Most CRSWDs are due to substantial misalignment between the internal rhythm and the required timing of the patient's school, work, or social activities. The CRSWDs comprise six specific disorders including delayed sleep–wake phase disorder (DSW-PD), advanced sleep–wake phase disorder (ASWPD), irregular sleep–wake rhythm disorder, non-24 h sleep–wake rhythm disorder, shift-work disorder, and jet lag disorder. These disorders arise when there is a substantial misalignment between the internal circadian rhythm and the desired sleep–wake schedule. Specific general diagnostic criteria are given for CRSWD. A three month duration of symptoms is a requirement for diagnosing all these disorders except for jet lag disorder, which has a requirement of jet travel across at least two time zones.

DSWPD occurs when there is a significant delay in the phase of the major sleep episode in relation to the desired sleep and wake-up time. When allowed to sleep as desired, sleep duration and quality are age appropriate. ASWPD occurs when there is a significant advance in the phase of the major sleep episode in relation to the desired sleep and wake-up time. When allowed to sleep as desired, sleep duration and quality are age appropriate. Irregular sleep–wake rhythm disorder is a recurrent or chronic pattern of irregular sleep and wake episodes throughout the 24 h-period with symptoms of insomnia and daytime sleepiness. Non-24 h sleep–wake rhythm disorder is due to a progressively delayed sleep–wake pattern. Shiftwork disorder is insomnia or excessive sleepiness associated with a recurring work schedule that overlaps with the usual time for sleep. Sleep logs and actigraphy monitoring for at least seven days is recommended for all the above CRSWDs. Jet lag disorder is insomnia or sleepiness accompanied by transmeridian jet travel across at least two time zones.

A circadian sleep–wake disorder not otherwise specified (NOS) is listed for patients who have a CRSWD that meets all the criteria for CRSWD, but not the specific types.

5.2.5. Parasomnias

Parasomnias are undesirable physical events or experiences that occur during entry into sleep, within sleep, or during arousal from sleep. The parasomnias are divided into three groups: the NREM-related parasomnias, REM-related parasomnias, and an other category of parasomnias.

The NREM-related parasomnias comprise general diagnostic criteria for the group heading of disorders of arousal (DA), from NREM sleep. These disorders have similar features of: genetic and familial patterns, pathophysiology or partial arousals from deep sleep, priming by sleep deprivation, and biopsychosocial stressors. They may be triggered by external stimuli and are not secondary to psychiatric disorders, neuropathology and are associated with impaired cognitive functioning and amnesia for the episode. Specific general diagnostic criteria are given for each DA and the detailed text applies to all of the DAs as no text is presented for each of the specific DAs except for diagnostic criteria. Confusional arousals are characterized by mental confusion or confused behavior that occurs while the patient is in bed. Sleep-related abnormal sexual behaviors are listed as a subtype to be classified under confusional arousals. Sleepwalking is arousals associated with ambulation and other complex behaviors out of bed. Sleep terrors are episodes of abrupt terror, typically beginning with alarming vocalizations, such as a frightening scream. The final NREM-related parasomnia, sleep-related eating disorder (SRED) requires an arousal from the main sleep period to distinguish it from night eating syndrome (NES) disorder, which is excessive eating between dinner and bedtime, and SRED requires an adverse health consequence from the disorder.[9] The behavior consists of consumption of peculiar combinations of food or inedible or toxic substances, or injurious behaviors while in the pursuit of food, or adverse effects from recurrent nocturnal eating. There is partial or complete loss of conscious awareness during the episode, with subsequent impaired recall.

The REM-related parasomnias include RBD, recurrent isolated sleep paralysis (RISP) and nightmare disorder. RBD, which is repeated episodes of vocalizations and/or complex motor behaviors, requires the polysomnographic evidence of REM sleep without atonia (RWA).[10] RISP is the recurrent inability to move the trunk and all of the limbs at sleep onset or upon awakening from sleep that causes distress or fear of sleep. Nightmare disorder is repeated occurrences of extended, extremely dysphoric, and well-remembered dreams that usually involve threats to survival, security, or physical integrity that is associated with significant distress or psychosocial, occupational, or other areas of impaired functioning.

The other parasomnia section includes three specific disorders; exploding head syndrome (EHS), sleep-related hallucinations, and sleep enuresis. EHS is a complaint of a sudden noise or sense of explosion in the head either at the wake–sleep transition or upon awakening during the night associated with abrupt arousal. Sleep-related hallucinations are predominantly visual hallucinations that are experienced just prior to sleep onset or upon awakening during the night or in the morning. Sleep enuresis is involuntary voiding during sleep at least twice a week in people older than five years of age. Parasomnias associated with medical disorders, and medication or substance, and non-specific parasomnia comprise the other entries in this category. Sleep talking is a normal variant that can occur in both NREM and REM sleep and can be associated with parasomnias, such as RBD or DAs.

5.2.6. Sleep-Related Movement Disorders

SRMDs are mainly, characterized by relatively simple, usually stereotyped, movements that disturb sleep or its onset, with the exception of restless legs syndrome. The SRMDs comprise seven specific disorders; restless legs syndrome, periodic limb movement disorder (PLMD), sleep-related leg cramps, sleep bruxism, sleep-related rhythmic disorder (RMD), benign sleep myoclonus of infancy (BSMI), and propriospinal myoclonus at sleep onset (PSM). SRMDs are relatively simple, usually stereotyped movements that disturb sleep or its onset.

Restless legs syndrome (also known as Willis-Ekbom disease) is an urge to move the legs, usually accompanied by or thought to be caused by uncomfortable and unpleasant sensations in the legs. The ICSD-3 criteria do not include any frequency or duration criteria as is contained in the DSM-V criteria.

PLMD is defined y the polysomnographic demonstration of periodic limb movements (PLMS) of > 5/h in children and > 15/h in adults that cause significant sleep disturbance or impairment of functioning. Sleep-related leg cramps are painful sensations that occur in the leg or foot with sudden, involuntary muscle hardness, or tightness. Sleep-related bruxism is tooth grinding during sleep that is associated with tooth wear or morning jaw muscle pain or fatigue. RMD is repetitive, stereotyped, and rhythmic motor behaviors involving large muscle groups that are sleep related. BSMI is repetitive myoclonic jerks that involve the limbs, trunk, or whole body that occurs from birth to six months of age during sleep. As PSM mainly occurs during relaxed wakefulness and drowsiness as the patient attempts to sleep, the term "at sleep onset" has been added to the propriospinal myoclonus' name. The three final categories are related to a medical disorder,

medication of substance, and an unspecified parasomnia.

Isolated symptoms and normal variants include excessive fragmentary myoclonus (EFM), hypnagogic foot tremor (HFT) and alternating muscle activation (EMA), and sleep starts (hypnic jerks). EFM is now regarded as a normal variant found on polysomnographic EMG recordings that are characterized by small movements of the corners of the mouth, fingers or toes, or without visible movement. HFT is rhythmic movement of the feet or toes that occurs in the transition between wake and sleep or in light NREM sleep, ALMA is brief activation of the anterior tibialis in one leg with alternation in the other leg. Sleep starts (hypnic jerks) are brief, simultaneous contractions of the body or one or more body segments occurring at sleep onset.

The final category in the ICSD-3 is a general other sleep disorder category for disorders that cannot be classified elsewhere.

5.3. CONCLUSIONS

The new ICSD-3 is a major advance over previous versions, but it is unfortunate that some of the diagnostic criteria differ from that of DSM-V, for example the criteria for narcolepsy. However, the DSM-V serves as an entry-level classification, mainly for psychiatrists, and it is to be hoped that in the future the two classifications will be merged into one that will cause less confusion not only for clinicians, but also for agencies that reimburse for healthcare and provide for treatment options.

Table 5.3. ICD-10-CM sleep disorders.

F51 Sleep disorders not due to a substance or known physiological condition

F51.01 Primary insomnia

F51.02 Adjustment insomnia

F51.03 Paradoxical insomnia

F51.04 Psychophysiologic insomnia

F51.05 Insomnia due to other mental disorder

F51.09 Other insomnia not due to a substance or known physiological condition

F51.1 Hypersomnia not due to a substance or know physiological condition

F51.11 Primary hypersomnia

F51.12 Insufficient sleep syndrome

F51.13 Hypersomnia due to other mental disorder

F51.19 Other hypersomnia not due to a substance or known physiological condition

F51.3 Sleepwalking [somnambulism]

F51.4 Sleep terrors [night terrors]

F51.5 Nightmare disorder

F51.8 Other sleep disorders not due to a substance or known physiological condition

F51.9 Sleep disorder not due to a substance or known physiological condition, unspecified

G47 Organic sleep disorders

G47.0 Insomnia, unspecified

G47.01 Insomnia due to medical condition

G47.09 Other insomnia

G47.1 Hypersomnia, unspecified

G47.11 Idiopathic hypersomnia with long sleep time

G47.12 Idiopathic hypersomnia without long sleep time

G47.13 Recurrent hypersomnia

G47.14 Hypersomnia due to medical condition

G47.19 Other hypersomnia

G47.2 Circadian rhythm sleep disorder, unspecified type

G47.21 Circadian rhythm sleep disorder, delayed sleep phase type

G47.22 Circadian rhythm sleep disorder, advanced sleep phase type

G47.23 Circadian rhythm sleep disorder, irregular sleep–wake type

G47.24 Circadian rhythm sleep disorder, free running type

G47.25 Circadian rhythm sleep disorder, jet lag type

G47.26 Circadian rhythm sleep disorder, shift work type

G47.27 Circadian rhythm sleep disorder in conditions classified elsewhere

G47.29 Other circadian rhythm sleep disorder

G47.3 Sleep apnea, unspecified

G47.31 Primary central sleep apnea

G47.32 High-altitude periodic breathing

G47.33 Obstructive sleep apnea (adult) (pediatric)

G47.34 Idiopathic sleep-related nonobstructive alveolar hypoventilation

G47.35 Congenital central alveolar hypoventilation syndrome

G47.36 Sleep-related hypoventilation in conditions classified elsewhere

G47.37 Central sleep apnea in conditions classified elsewhere

G47.39 Other sleep apnea

G47.4 Narcolepsy and cataplexy

G47.41 Narcolepsy

G47.411 Narcolepsy with cataplexy

G47.419 Narcolepsy without cataplexy , NOS

G47.42Narcolepsy in conditions classified elsewhere

G47.421 Narcolepsy in conditions classified elsewhere with cataplexy

G47.429 Narcolepsy in conditions classified elsewhere without cataplexy

G47.50 Parasomnia, unspecified

G47.51 Confusional arousals

G47.52 REM sleep behavior disorder

G47.53 Recurrent isolated sleep paralysis

G47.54 Parasomnia in conditions classified elsewhere

G47.59 Other parasomnia

G47.6 Sleep-related movement disorders

G47.61 Periodic limb movement disorder

G47.62 Sleep-related leg cramps

G47.63 Sleep-related bruxism

G47.69 Other sleep-related movement disorders

G47.8 Other sleep disorders

G47.9 Sleep disorder, unspecified

Z72.8 Problems related to sleep

Z72.820 Sleep deprivation

Z72.821 Inadequate sleep hygiene

Z73.8 Other problems related to life management difficulty

Z73.810 Behavioral insomnia of childhood, sleep onset association type

Z73.811 Behavioral insomnia of childhood, limit setting type

Z73.812 Behavioral insomnia of childhood, combined type

Z73.819 Behavioral insomnia of childhood, unspecified type

Adapted from: International Classification of Diseases, Tenth Revision, Clinical Modification (ICD-10-CM), National Center for Health Statistics. Centers for Disease Control and Prevention (CDC), December 20, 2010.

Appendix A lists several disorders that are coded in other sections of ICD 10 other than the sleep sections and includes fatal familial insomnia, sleep-related epilepsy, sleep-related headaches, sleep-related laryngospasm, sleep-related gastro-esophageal reflux, sleep-related myocardial ischemia. Appendix B lists the ICD sleep-related substance-induced sleep disorders (Table 5.3).

KEY WORDS

- **classification**
- **diagnosis**
- **sleep**
- **disorders**
- **insomnia**
- **parasomnias**
- **sleep apnea**
- **circadian**

REFERENCES

1. Association of Sleep Disorder Centers. Classification of Sleep and Arousal Disorders. *Sleep.* **1979,** *2 (1),* 1–154.

2. American Psychiatric Association. *Diagnostic and Statistical Manual of Mental Disorders 5th ed. DSM-V.* Washington DC, **2013.**

3. American Academy of Sleep Medicine. *International Classification of Sleep Disorders, 3rd ed.* Darien, IL: American Academy of Sleep Medicine, **2014.**

4. International Classification of Diseases, Tenth Revision, Clinical Modification (ICD-10-CM), National Center for Health Statistics. Centers for Disease Control and Prevention (CDC), December 20, **2010.**

5. Edinger, J. D.; Wyatt, J. K.; Stepanski, E. J., et al. Testing the Reliability and Validity of DSM-IV-TR and ICSD-2 Insomnia Diagnoses: Results of a Multi-Method/Multi-Trait Analysis. *Arch. Gen. Psychiatry.* **2011,** *68,* 992–1002.

6. Westhoff, M.; Arzt, M.; Litterst, P. Prevalence and Treatment of Central Sleep Apnoea Emerging after Initiation of Continuous Positive Airway Pressure in Patients with Obstructive Sleep Apnoea without Evidence of Heart Failure. *Sleep Breath.* **2012,** *16,* 71–78.

7. Andlauer, O.; Moore, H.; Jouhier, L., et al. Nocturnal REM Sleep Latency for Identifying Patients with Narcolepsy/Hypocretin Deficiency. *JAMA Neurol.* **2013,** *6,* 1–12.

8. Arnulf, I.; Lin, L.; Gadoth, N., et al. Kleine-Levin Syndrome: A Systematic Study of 108 Patients. *Ann. Neurol.* **2008,** *63,* 482–493.

9. Brion, A.; Flamand, M.; Oudiette, D.; Voillery, D.; Golmard, J. L.; Arnulf, I. Sleep-Related Eating Disorder Versus Sleepwalking: A Controlled Study. *Sleep Med.* **2012,** *3,* 1094–1101.

10. Schenck, C. H.; Howell, M. J. Spectrum of RBD (Overlap between RBD and Other Parasomnias). *Sleep Biol. Rhythms.* **2013,** *11*(Supplement 1), 27–34.

6

THE CLINICAL APPROACH TO THE PATIENT WITH SLEEP DISORDERS

Ridhwan Y. Baba, Imran M. Ahmed, Shelby Harris, and Michael J. Thorpy

Ridhwan Y. Baba
Department of Neurology, Sleep-Wake Disorders Center, Montefiore Medical Center, Albert Einstein College of Medicine, Bronx, NY, USA

Imran M. Ahmed
Department of Neurology, Sleep-Wake Disorders Center, Montefiore Medical Center, Albert Einstein College of Medicine, Bronx, NY, USA

Shelby Harris
Department of Neurology, Sleep-Wake Disorders Center, Montefiore Medical Center, Albert Einstein College of Medicine, Bronx, NY, USA

Michael J. Thorpy
Department of Neurology, Sleep-Wake Disorders Center, Montefiore Medical Center, Albert Einstein College of Medicine, Bronx, NY, USA

Corresponding author: Michael J. Thorpy MD, Sleep-Wake Disorders Center, Montefiore Medical Center, Albert Einstein College of Medicine, 111 East 210th Street, Bronx, New York, 10467, USA.
Email: michael.thorpy@einstein.yu.edu.
Tel.: +1-718-920-4841, Fax: +1-718-798-4352

ABSTRACT

There have always been sleep disorders, many of which have been recognized for centuries and some of our treatments have even been dated from the decades ago. In the last 20 years, however, there has been a dramatic increase in knowledge about sleep disorders and their treatments. It is now possible to objectively diagnose most sleep disorders and new, specific treatments can be instituted. The recent recognition of chronobiology has led us to be able to explain alterations in the sleep-wake pattern of humans. Genetic causes of sleep disorders, such as advanced sleep phase syndrome have shed new light on disorders that were previously believed to be primarily associated with an individual's behavior. Neurochemical changes have led to a better understanding of the pathophysiology of some disorders such as narcolepsy. Despite advances in our understanding of sleep disorders, however, accurate diagnosis and treatment always requires a detailed understanding of the patient's sleep-wake and medical history. The art of good sleep medicine still lies in the ability of the clinician to take a thorough history, develop a differential diagnosis and formulate a treatment plan. This chapter details the important elements of the clinical evaluation.

A variety of sleep complaints are extremely common in our society. Most of these complaints can be grouped under the categories of insomnia, excessive daytime sleepiness (EDS), or abnormal events occurring during sleep. Approximately, 10% of the general population has a complaint of insomnia that occurs every night for two weeks or more.[1] Yet, only 5–6% of patients will ever seek a physician in order to address their sleep problem.[2] In fact, a majority of patients with insomnia never discuss their complaints with a physician, and usually resort to over-the-counter medications, or self-remedies in order to alleviate their sleep disturbance. Excessive daytime tiredness and sleepiness are also commonly reported complaints, with prevalence in the community estimated to be as high as 10–25%.[3] Further, around 30% of the general population has some sleep disturbance a few nights every month,[4] and nearly everyone has had some type of abnormal intrusion into sleep, such as nightmares, sleepwalking or some other psychological, or physiological intrusion into sleep. Nevertheless, despite this prevalence of sleep disorders in the society most remain under diagnosed and undertreated.

With better understanding of sleep disorders, significant morbidity as a sequela of sleep disruption has been recognized. For example, insomnia can lead to the development of depression[5] or EDS can be a manifestation of obstructive sleep apnea (OSA) that, at its worse, can lead to sudden death during sleep. Even lesser degrees of disturbed sleep and daytime sleepiness can lead to impaired functional ability during the daytime, and a tendency for mood disturbances that might include irritability, anxiety, and depressive feelings. There has also been a growing concern regarding the possibility of sleepiness impairing functional ability often leading to motor vehicle and/ or industrial accidents.[6] Several major catastrophic events that have affected society have been ascribed to disturbances of the sleep-wake cycle in the individuals responsible. The Exxon Valdez ship accident in Alaska that led to a major environmental oil disaster, challenger space shuttle accident, the Chernobyl nuclear power station accident, or more recently, the Metro-North train derailment in New York are some examples of major accidents that were in part caused by human errors associated with an inadequate sleep-wake pattern.

6.1. CLASSIFICATION OF SLEEP DISORDERS

We now recognize that there are many different types of sleep disorders. Classification systems have evolved over the years and in the 1970's a symptom-based classification system brought awareness that some sleep disorders could cause not only disturbed sleep at night but also could cause symptoms during the daytime. In 1990, the international classification of sleep disorders (ICSD) unified the classification criteria for over 80 sleep disorders that greatly enhanced clinical research on patients with sleep disruption.[7] More recently, in 2014, the American Academy of Sleep Medicine introduced the third edition of the ICSD, which updated the earlier classification schema, and is based in part upon both a symptomatic and a pathophysiologic organization.[8] This classification of sleep disorders has led to not only improved clinical research but has also helped improve clinical practice. The third edition of the ICSD classification system divides the sleep disorders into seven major groups. The first group is those sleep disorders that primarily cause disturbed sleep at night – the insomnias. Under the insomnias group, the earlier classification into subtypes, such as primary insomnias like psychophysiological insomnia, idiopathic insomnia, inadequate sleep hygiene, and paradoxical insomnia, as well as secondary insomnia related to other psychiatric or medical disorders has been abandoned for a new diagnostic schema based on the frequency and duration of symptoms: chronic insomnia disorder, short-term insomnia disorder, and other insomnia disorder. The second group is a list of sleep disorders related to breathing disturbance at night, the sleep-related breathing disorders. The prime example in this category is OSA that can occur in both adults and in children. In addition, central sleep apnea, sleep-related hypoventilation disorders, and hypoxemia disorders are listed. The third classification group includes patients, who present with EDS that is not related to a sleep-related breathing disorder - the hypersomnias of central origin. Narcolepsy, idiopathic hypersomnia, Kleine-Levin syndrome, as well as hypersomnias associated with other psychiatric or medical disorders or due to a medication, or substance are in this group. The circadian rhythm sleep-wake disorders are listed as a separate group; the most commonly seen in clinical practice are delayed and advanced sleep phase syndrome, however, Irregular sleep-wake, jet lag and shift work type are not uncommon, and are also described. The fifth group includes the parasomnias disorders that intrude into sleep at night that are not typically associated with EDS or insomnia such as, confusional arousals, sleepwalking, and sleep terrors. The disorders included in this group also include REM sleep behavior disorder (RBD), recurrent isolated sleep paralysis, and nightmare disorder. The sixth group includes those movement disorders that occur in relationship to sleep. In this category are restless legs syndrome (RLS), periodic limb movement disorder (PLMD), and sleep-related leg cramps. The seventh and final clinical group, other sleep disorders, lists disorders that, although not strictly sleep disorders, are commonly seen in the practice of sleep medicine. These include disorders such as sleep-related epilepsy, sleep-related headaches, sleep-related choking, and sleep-related laryngospasm. Each classification category also lists isolated symptoms and normal variants, such as snoring, long sleepers, and excessive fragmentary myoclonus.

6.2. EVALUATION OF SLEEP DISORDERS

6.2.1. Sleep History

A careful history of the primary sleep complaint is essential to formulate a differential diagnosis. As such, when a patient presents to a physician for evaluation of a sleep complaint, a logical progression through symptoms will greatly help in understanding the sleep problem (Tables 6.1 and 6.2).

Table 6.1. Key nocturnal sleep questions.

When do you usually go to bed and wake up on weekdays and weekends?

Do you have any trouble falling asleep, staying asleep, or awaken in the morning prior to your desired time?

Do you snore, gasp, choke, or stop breathing during sleep?

Do you have an unpleasant sensation in your legs at night?

Do you feel refreshed after a full night's sleep?

Do you nap during the daytime?

Would you doze easily or feel sleepy if in a quiet environment or monotonous situations?

Do you take medications to help you sleep?

Table 6.2. Assessment of the patient with sleep disorders.

Structured clinical interview

    Direct patient observation

    Questioning of spouse, family member, and bed partner

    Onset, frequency, detailed description, and effects of sleep disorder

    Complete medical, psychiatric, family, and psychosocial history

Standardised questionnaires

    Epworth sleepiness scale

Comprehensive physical exam

Common sleep studies

    Polysomnography

    Multiple sleep latency test

    Maintenance of wakefulness test

Laboratory investigations

    Blood tests (cell counts, chemistry, thyroid function, etc.)

    Endoscopic upper airway evaluation

    Cephalometric Xrays

    Imaging studies like MRI, etc.

    Nerve conduction velocity studies and EMG

If the primary complaint is related to sleep apnea, that is, episodes of witnessed apnea or gasping and choking in sleep reported by the patient, then a detailed history of other nocturnal behaviors, such as snoring should also be obtained. It is often important to understand how the snoring affects the patient and the bed partner. In many cases, patients who have very loud snoring would have moved out of the bedroom so as to prevent the snoring from disturbing their bed partner. Occasionally, it may the bed partner who moves out of the bedroom. The variation and the quality of snoring are also important, as quiet episodes alternating with loud episodes of snoring may be indicative of apnea in the absence of episodes of gasping and choking. The patient is often unaware of these events, and the bed partner alone may observe the patient's sleep behaviors. Accordingly, interviewing the bed partner about any sleep problems may be of value.

A detailed history of daytime symptoms, especially sleepiness and how it affects the individual is also important, with emphasis on sleepiness while driving. In this context, information regarding sleep behavior during the night can be very helpful in understanding the cause of daytime sleepiness. Relevant questions, such as whether the patient is restless, how frequently they have to get up to go the bathroom at night, whether there is excessive sweating, or whether there is abnormal movement activity, or vocalization during sleep may all be indicative of an underlying sleep disorder. Typically, one would inquire about how the patient feels upon awakening, whether the patient is tired, fatigued or groggy, and whether they have headaches when they awaken. The patient's sleep position may be helpful as supine sleep may often be associated with more severe apnea in patients who have OSA, and varied sleep positions are common in patients who have sleep apnea or insomnia. The patient's bedtime, time to fall asleep, and times of awakening are also very important to rule out insufficient sleep duration. Features regarding narcolepsy, such as cataplexy, sleep paralysis, hypnagogic hal-

lucinations, and automatic behaviors can be helpful in any patient with sleepiness and helps determine if narcolepsy is present.[9] Any history of significant traumatic brain injury should be noted due to its association with narcolepsy and other sleep disorders. In addition, features of RLS and PLMD should be determined. Menstrual-related sleepiness in women, hyperphagia, and hypersexuality are additional symptoms seen in recurrent hypersomnia, and should also be queried.

Since a patient's weight greatly influences the prevalence of sleep apnea, historical information on the height and the weight of the patient and any changes over the previous 2–5 years, can be very helpful. Also, any attempts of weight reduction should be noted.

A detailed medical history should follow the sleep history, and could focus mainly on nasopharyngeal problems and cardiovascular disease especially any evidence of heart failure or ischemic heart disease if a diagnosis of a sleep-related breathing disorder is being entertained. The presence of hypertension and its association with the symptoms of OSA should be determined. The history can also include information about an individual's neurologic or psychiatric problems if other disorders such as insomnia, RBD, RLS, other parasomnias, or movement disorders during sleep are being considered. A complete record of current medications that the patient is taking should be obtained as most medications can have an adverse effect on sleep at night or may cause impaired alertness during the daytime. Any history of cerebrovascular problems should also be recorded.

Patients who have nasopharyngeal problems tend to be mouth breathers, and inquiry as to whether they are an obligatory mouth breather may be helpful. Other complaints such as rhinitis, postnasal drip, or sinus problems may be helpful to determine if there have been any previous visits to an otolaryngologist for treatment of nasal congestion. Whether the patient has had a tonsillectomy and adenoidectomy is also important to note, especially in a patient with OSA.

A childhood history of observed abnormal sleep behaviors or diagnosed sleep disorders might be helpful for some patients, particularly those who have parasomnia activity. A number of sleep disorders including OSA and narcolepsy, as well as RLS have a familial tendency. As such, obtaining a family history of sleep disorders may be helpful.

An assessment of patient's social situation is also important, particularly relationships with other family members, as well as any stressors of financial, personal, or social nature that may be contributing to sleep disturbances. The patient's physical activity should be assessed, as well as a history that focuses on drugs and alcohol.

6.2.2. Physical Examination

A comprehensive physical examination should ideally be obtained for most patients with sleep disorders, with special focus on respiratory, cardiovascular, gastrointestinal, endocrine, and neurological evaluation (Table 6.3). If time does not allow a comprehensive physical or if a physical exam has recently been completed, a focused examination is usually acceptable. In this case, the examination should still include determination of the patient's blood pressure and vital signs.

The patient's body habitus may help determine the etiology of OSA, and as such is important to assess; therefore, the patient's height, weight (that is, body mass Index), and neck circumference should be obtained as well as a determination of the distribution of body fat (e.g. abdominal, neck, android vs gynecoid body habitus,and so forth). An evaluation of the upper airway is important in a patient with possible OSA, particularly to determine if there is a narrow airway and/or the presence of enlarged tonsillar tissue or a large tongue. The size and shape of the soft pal-

---

Table 6.3. Evaluation of a patient with sleep disorders.

Examination of head, ears, nose, eyes, and throat (HEENT)
    Chronic nasal congestion, enlarged turbinates, and deviated nasal septum
    Crowded oropharynx: tonsils, soft palate, uvula, and tongue
    Retrognathia or micrognathia, small maxilla
    Neck circumference, enlarged thyroid
Cardiovascular
    Signs of hypertension, coronary heart disease, and heart failure
    Ankle edema, distal pulses
Neurologic
    Assess for CNS disease (e.g. Parkinson's disease, myotonic dystrophy)
    Assess for peripheral nerve disease (e.g. radiculopathy, peripheral neuropathy)

ate and uvula should be determined. When necessary, the patient should be referred to an otolaryngologist for endoscopic evaluation of the upper airway to determine if more specific obstruction is present particularly in the posterior nasopharynx or oropharynx. In some patients, an evaluation of thyroid size may prove pertinent as a prominent thyroid may also contribute to airway obstruction during sleep. In a patient with RLS, a neurological evaluation of the motor and sensory function of the extremities may suggest an alternative pathology, such as a neuropathy or a radiculopathy. Electromyography and nerve conduction velocity studies may be required for further assessment of such nerve lesions. Additionally, an assessment of the degree of leg edema and palpation of the distal pulses may suggest a vascular etiology of leg symptoms (e.g. vascular claudication) that is occasionally confused with RLS. A neurological examination is also important when evaluating RBD or narcolepsy, as focal abnormalities on exam may suggest an etiology or an associated neurologic disorder.

6.2.3. Questionnaires and Sleep Logs

Sleep questionnaires may assist the physician to more quickly collect extensive information regarding sleep-wake habits. Standardized questionnaires, such as the Epworth sleepiness scale are a valuable instrument for determining the presence of daytime sleepiness over time.[10] The patient scores the likelihood of falling asleep in eight situations on a rating scale of 0–3 leading to a maximum score of 24. Patients with a score of 10 or higher should be considered to have significant daytime sleepiness, and those over 15 have severe daytime sleepiness. Other sleep questionnaires, such as the Pittsburgh sleep quality index, the Stanford sleepiness scale, the Ullanlinna narcolepsy scale, or the Karolinska sleepiness scale may also be useful in some patients depending upon the primary complaint.

Of great importance is the sleep log that documents, over a period of approximately two weeks, the time of daily sleep onset, wake time, and awakenings during the night as well as any daytime naps. Utilizing this form, the physician can readily see if the patient suffers from circadian disruption to their sleep pattern and also can help determine the severity of the sleep difficulty, particularly in a patient with insomnia. A sleep log also allows the patient to record other events, such as parasomnias, so that the frequency of their occurrence on a nightly basis can be determined.

6.2.4. Laboratory Investigations

6.2.4.1. Polysomnography

Polysomnography is usually the main form of investigation of sleep disorders (Table 6.4), and typically involves studying the patient over the duration of their usual major sleep episode with concomitant measurement of sleep, cardiac, respiratory parameters, and any movement activity. The patient has electrodes applied to their head for the recording of the electroencephalogram, electromyogram, and electrooculogram. In addition, measures of ventilation include respiratory effort that involves movements of the chest and abdomen, and airflow through the nose and mouth. Oxygen saturation by means of an infrared oximeter and electrocardiography

Table 6.4. Indications for sleep studies.

Polysomnography	Multiple sleep latency test	Maintenance of wakefulness test
Sleep-related breathing disorder	Narcolepsy	Assessment of treatment response for sleepiness
Narcolepsy REM sleep behavior disorder Sleep-related epilepsy movement activity during sleep CPAP or BPAP titration Evaluation of effectiveness of treatment	Idiopathic hypersomnia assessment of severity of sleepiness	Assessment of alertness for workplace safety e.g. commercial driving, airline pilots

are also recorded. The patient's leg activity is measured by means of electrodes placed on the anterior tibialis muscle of each leg. In addition, often the patient is videotaped throughout the recording for any abnormal behavior that occurs during sleep. On occasions other measures, such as end-tidal carbon dioxide concentration, or gastro esophageal pH, body position or sound of snoring are also determined.

A polysomnogram is typically stored in a digital format on a computer attached to a polysomnograph, and can then be displayed on a computer screen for analysis. A typical recording is performed at a recording speed of 15 mm/s; however, this can be changed to 30 mm/s if abnormal epileptic activity is suspected.[11]

The amount and percentage of each sleep stage recorded as well as the number of breathing events, such as apneas or hypopneas, or the number of periodic limb movements, and whether or not they are associated with arousals are scored. Typically, an apnea/ hypopnea index, which is the number of respiratory events per hour of sleep, is determined.

Although the all-night polysomnogram has its greatest utility in the detection of OSA, it is also very useful for determining abnormal events that occur during specific sleep stages such as, sleep terrors that might arise out of slow wave sleep, or periodic leg movements that typically occur during NREM sleep. Video monitoring can be very helpful in determining the clinical features of an abnormal event that occurs during sleep, such that differentiation can be made between a parasomnia and an epileptic disorder.[12] Electroencephalographic activity is also crucial in making this differentiation.

Relatively more recently, portable home sleep studies are commonly being utilized. These studies record at least four of the above-described information, typically airflow, respiratory effort, heart rate, and oxygen desaturation; however, additional recorded data such as body position is not uncommon. A more detailed discussion of home sleep studies is beyond the scope of this chapter; however, it should be noted that these devices serve as a screen for the diagnosis of a sleep-related breathing disorder. They do not, at present, have a role in the diagnosis of other sleep disorders.

A typical patient with insomnia does not require overnight polysomnography.[13] The etiology of insomnia can usually be determined from the history and treatment is often initiated without the need for a polysomnogram. However, in some patients, particularly in the elderly, OSA may underlie the insomnia and therefore polysomnography may be required.

If OSA is detected, a second night of polysomnography with nasal continuous positive airway pressure (CPAP) may be required to determine if it is an effective treatment modality. Usually a full night of recording with measurement of respiratory variables is taken, and CPAP is added and adjusted as appropriate in order to find the most appropriate pressure setting to relieve the respiratory events.

6.2.4.2. Multiple Sleep Latency Test

A multiple sleep latency test (MSLT) is usually performed to determine a patient's tendency to fall asleep during the main period of wakefulness (Table 6.4).[14] The patient takes four or five brief naps throughout the day with the first nap beginning approximately 2 h after awakening in the morning and subsequent naps are scheduled at 2 h intervals. The patient is placed in a darkened room and asked to lie back, relax, and to sleep if possible. Typically, they are allowed 20 min in order to achieve sleep. Subsequently, the time to falling asleep, that is, sleep latency, and the stage of sleep achieved are determined. Mean sleep latency is determined from the four or five naps; patients with severe sleepiness will fall asleep within a mean of 5 min over

the four or five nap opportunities. Also, any periods of REM sleep during these naps are recorded. One nap episode with REM sleep is not specific to any one disorder; however, two naps with REM sleep are usually indicative of pathology. In the presence of normal amount and percentage of REM sleep at night, two sleep onset REM periods (SOREMP) during the daytime naps are highly suggestive of narcolepsy. One SOREMP occurring within 15 min of sleep onset during the nocturnal polysomnogram can count as one of the two required on the MSLT for making the diagnosis of narcolepsy in the new ICSD-3 diagnostic criteria. For the diagnosis of narcolepsy, a SOREMP on an all night polysomnogram has a 92% positive predictive value for narcolepsy.[15] Other disorders, such as OSA or sleep deprivation, can also cause two or more SOREMPs during the daytime.

The MSLT is especially important in detecting sleepiness in a patient who might otherwise deny sleepiness, such as an older individual who may insist on driving a motor vehicle despite pleas to the contrary by family members who may have witnessed severe daytime sleepiness. In this context, an objective measure of sleepiness may be helpful in illustrating to the patient that significant sleepiness is present.

Another variation of the MSLT called the maintenance of wakefulness test (MWT) can also be utilized (Table 6.4).[16] This test is performed in a similar manner as the MSLT except that the patient is placed in a semi-reclining position and asked to remain awake during four 40 min nap opportunities. The time to the onset of sleep for each nap is recorded. A pathologically sleepy patient typically falls asleep within an average of 8 min over the four nap periods and a healthy alert individual should be able to remain awake for the full 40 min on each of the trials. Sleep latencies between 8 and 40 min are of uncertain significance. Although available normative studies have reported similar mean sleep latency (sleep latency measured to the first epoch of unequivocal sleep), variable lower limits of mean sleep latency have been suggested (8.0–26.1 min).[17–19] This is especially important, as the MWT is often helpful in determining the effects of treatment to relieve daytime sleepiness and assessing workplace safety. In a validity study of 30 male subjects with untreated OSA, a pathologic sleep latency of upto 19 min was better predictive of driving ability (measured by deviation from driving in the center of the road on a driving simulator) and alertness.[20] Consequently, the data available suggests that the mean sleep latency between 8 and 40 min should be used or interpreted in association with clinical history to assess the ability to maintain wakefulness for workplace, occupational, or driving safety.

6.2.4.3. Other Ancillary Tests

Additional investigations may be required for a comprehensive evaluation of sleep disorders (Table 6.1). Many patients with OSA would undergo fiber optic endoscopy of the upper airway to determine the site of obstruction. Occasionally cephalometric radiography may be required for a patient who has microagnathia or retrognathia, to demonstrate the abnormal jaw position.

Blood tests, especially screening blood tests are usually performed by the referring physician. It is important to determine whether the complete blood cell count and blood chemistry is within the normal range in any patient who has impaired daytime alertness or difficulty in sleeping. Thyroid function tests may be indicated if the patient has features suggesting hypothyroidism; however, most patients with daytime sleepiness do not require routine thyroid testing, as the likelihood of a positive return is low.[21] Patients who have features suggestive of RLS should undergo biochemical screening to ensure that there is no evidence of renal impairment or other chemical abnormality. In addition a serum ferritin level should be determined.[22] A serum ferritin level of less than 50 µg/L indicates a need for iron replacement therapy in a patient with RLS; more recently, a higher threshold of 75 µg/L for treatment has been recommended.[23] As mentioned earlier, electromyography and nerve conduction studies may be required for better evaluation of a neuropathy or a radiculopathy that is mimicking RLS symptoms. Urine screening may be required to rule out drug abuse, particularly in young people with unexplained daytime sleepiness or insomnia.

6.3. GENERATING A DIFFERENTIAL DIAGNOSIS

Before making a diagnosis, a careful review of the onset, exacerbation, and remission of patient's complaints is important. Additionally, relevant findings from the physical exam and results of the laboratory evaluation should be considered to formulate a differential diagnosis. The three primary sleep complaints, insomnia, excessive sleepiness, and abnormal events during sleep have been discussed in detail below.

6.4. INSOMNIA

A patient with insomnia will typically complain of either difficulty in falling asleep at night, frequent awakenings during the night, or early morning awakening much before the desired time. A subjective feeling of inadequacy of nighttime sleep is often reported. Complaints of tiredness, fatigue, irritability, and lack of energy constitute some of the most common issues presented to clinicians. These complaints are usually associated with some impairment of daytime function as well. Typically, a patient with insomnia is also unable to sleep during the daytime as a hyperarousal state is common in these individuals; however, occasionally a patient may find that when sitting relaxed in the early evening there is a tendency to fall asleep for a brief period of time. This is a feature that is often seen with the subtype of insomnia, psychophysiologic insomnia.[24]

Approximately 30% of patients with chronic insomnia have either depression or an anxiety disorder. Most patients with insomnia have some depressive or anxiety features even though they do not meet specific criteria for an axis I psychiatric diagnosis.[25] To determine if the patient is clinically depressed, it is important to enquire about symptoms, such as reduced appetite, tendency to feel like crying, depressive affect, and suicidal ideation.

Most patients with chronic insomnia tend to be concerned about their inability to sleep normally, and as nighttime approaches this concern often becomes more intense. As a result, the patient will often delay going to sleep at night until excessively sleepy. In addition, there may be a tendency to stay in bed later the next morning after a bad night of sleep. Similarly, if the patient is fatigued during the daytime, there is a tendency to go to bed early to try to get more sleep. As a result, the sleep-wake pattern often becomes disrupted and spread out over a larger portion of the day. The patient may spend anywhere from, for example, 8:00 p.m. through to 8:00 a.m. in the morning, in bed. Any sleep that occurs, therefore, occurs within a 12 h window and at irregular times. Additionally, patients will often complain about impaired memory, concentration difficulty, and inattention because of their sleep disturbance and they may complain of daytime headaches or an abnormal fuzziness, or grogginess that may occur throughout the day. Difficulty with motor vehicle driving and concern for the potential of accidents is not uncommon.

When a patient presents with a complaint of insomnia, it is important to first identify the duration as well as relevant clinical clues to determine the possible etiology of these symptoms (Fig. 6.1 and Table 6.5). When there are environmental circumstances, such as construction outside the home or a bed partner's snoring that are responsible for the person's insomnia, then an environmental sleep disorder is likely. The timing of the insomnia in association with the person's phase of entrainment to the 24 h sleep-wake pattern may help identify a circadian rhythm disorder, for example, advanced phase or delayed phase type. Once again, a two-week assessment of the patient's bedtime and wake up time is useful to help identify these disorders.

One must always ask about proper sleeping habits. Often poor sleep hygiene is not the primary cause of insomnia, but it frequently maintains a pre-existing in-

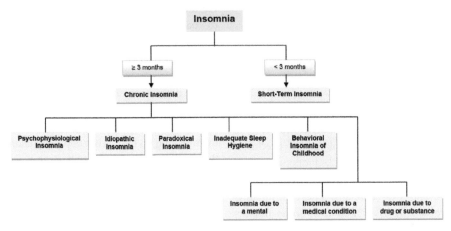

Figure 6.1. Evaluation of insomnia.

Table 6.5. Differential diagnosis of difficulty falling and/ or staying asleep and early morning arousals.

Psychophysiologic insomnia
Inadequate sleep hygiene
Insomnia due to medical or psychiatric disorder
Insomnia due to drug or substance use
Adjustment insomnia
Paradoxical insomnia
Circadian rhythm disorders, e.g. delayed sleep phase type or advance sleep phase type
Other sleep disorders e.g. RLS, OSA
Sleep disruptive environmental circumstances induced insomnnia
Depression and/ or anxiety
Behavioral insomnia of childhood
Idiopathic insomnia

somnia. Sometimes, insomnia may be temporally associated with a stressor, for example, exam or work presentation and resolves, when the stress is relieved. This type of insomnia is known as adjustment insomnia and is identified by inquiring about the duration and potential triggers for the sleep disturbance. Occasionally, a patient may indicate a chronic pattern of little or no sleep with daytime impairment that is significantly less than expected given the extreme level of sleep deprivation reported. In addition, there is a mismatch between objective findings from a polysomnogram or actigraphy and subjective sleep estimates. This has been termed paradoxical insomnia or sleep state misperception. An insomnia complaint present from childhood, with no identifiable precipitant or cause and without any periods of improvement is consistent with a diagnosis of idiopathic insomnia.

A person's medical conditions, psychiatric disorders, drug or substance use might also be the cause of the insomnia and need to be considered. Physicians should emphasize on treating insomnia concurrently with the underlying primary medical or psychiatric disorder. These conditions often include metabolic and endocrine disturbances such as hypothyroidism or renal disease.[26] Depression and anxiety, as mentioned earlier, are common in patients presenting with insomnia. Often other sleep disorders may present with insomnia as well. A complaint of sleep onset insomnia may be due to RLS. A history inquiring about the RLS diagnostic symptoms (urge to move the legs, improvement or relief with movement, symptoms worse when sitting or lying, and circadian pattern to symptoms) would help identify this as the cause of the insomnia. If the patient had trouble maintaining sleep, the possibility of RLS still needs to be considered; however, another likely cause may also be PLMD. The majority of patients with RLS also have periodic limb movements during sleep (in this case, only RLS is diagnosed), but a large portion of patients with periodic limb movements in sleep do not have RLS. A history obtained from a bed partner of kicking or leg movements during sleep would support the diagnosis of PLMD; however, definitive diagnosis requires documentation by a polysomno-

gram. In addition, a sleep-related breathing disorder may present as insomnia in up to 50% of patients.[27] Accordingly, a history of snoring, gasping for air, choking, or witnessed apneas might help identify this as a cause of the insomnia. In short, a thorough history is necessary to diagnose and differentiate the insomnia types; sleep studies may be done in select cases to aid in the diagnostic evaluation.

6.5. SLEEPINESS AND FATIGUE

A complaint of EDS is important to note as it can signal toward an undiagnosed sleep disorder or other potentially treatable condition. Patients, however, may not typically use the word "sleepiness" to describe their symptoms. Very often they will use vague words such as tiredness, fatigue, no energy or other similar terminology. Fatigue is a feeling of tiredness that may be both physical and psychological, and it typically occurs in conditions like depression or multiple sclerosis. Pure fatigue is not associated with significant daytime sleepiness. There is no physiological drive for sleep that occurs in such patients so that when they lie down they do not fall asleep but lie awake and rest. This distinction between fatigue and sleepiness is important because if sleepiness is present the focus will be on sleep at night and specific sleep disorders, whereas if the primary problem is one of fatigue the focus might be on an underlying medical or psychological problem that is not producing any specific sleep disruption.

Some patients will have both fatigue and sleepiness. This is often seen in patients with multiple sclerosis who have a background level of fatigue but also can suffer from sleep disorders such as OSA that can cause them to have significant sleepiness.[28] This also commonly occurs in patients with Parkinson's disease who are often very sleepy during the daytime but even when the sleepiness is relieved there is often a background level of fatigue that is not responsive to usual management of sleep disorders.[29] Sometimes patients with Parkinson's disease are not aware of their sleepiness either because of a lack of awareness of what true alertness means or because they are unaware of the times that they may fall asleep.[30] This is when a history from a bed partner or caretaker can be very helpful for collaborating or refuting the patient's statements on sleepiness.

When a patient presents with excessive daytime sleepiness, a number of diagnoses need to be considered (Fig. 6.2 and Table 6.6). Daytime somnolence would normally result if (1) there were insufficient time allowed for sleep, (2) sleep is occurring at a biologically appropriate time but not

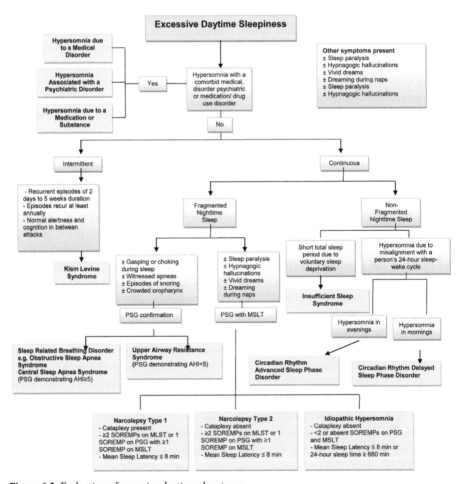

Figure 6.2. Evaluation of excessive daytime sleepiness.

Table 6.6. Differential diagnosis of excessive daytime sleepiness.

Insufficient sleep syndrome

Sleep-related breathing disorders e.g. OSA

Central disorders of hypersomnolence e.g. narcolepsy with or without cataplexy

Circadian rhythm disorders e.g. delayed or advance sleep phase type

Hypersomnia due to medical or psychiatric disorder

Hypersomnia due to drug or substance use

Recurrent hypersomnia e.g. Kleine-Levin syndrome

Idiopathic hypersomnia

a socially appropriate time, (3) nocturnal sleep is disrupted, (4) there is an organic dysfunction of the brain or neurochemicals, (5) a medical or psychiatric disorder is causing the symptom or (6) a drug or other substance is causing the symptom. In many cases, the most likely diagnosis can be identified by the patient's history (as discussed above).

Initially, it is useful to determine whether or not insufficient sleep, a circadian rhythm disorder, a medical disorder, a psychiatric disorder, a drug or a substance is the cause of the patient's hypersomnia. Questions regarding the amount of time allowed for sleep can help identify insufficient sleep syndrome. Circadian rhythm disorders may present with either insomnia or hypersomnia. A person with delayed sleep phase type may complain of hypersomnia in the morning (when awakening prior to completion of his habitual sleep period) or insomnia when going to bed earlier than his habitual sleep period. Similarly, a person with advanced sleep phase type may complain of early morning awakenings (insomnia) or hypersomnia toward the evening hours. A two-week assessment of the patient's bedtime and wake up time is useful to help identify these disorders.

Once these diagnoses are entertained; then sleep disorders like narcolepsy, OSA, and recurrent hypersomnia can be considered.

A history of recurrent episodes of somnolence lasting from two days to four weeks, occurring at least once a year, and normal alertness, and cognition in between attacks is likely to be a recurrent hypersomnia. If a person snores, has episodes of choking or gasping for air, has a crowded oropharynx on exam, is a mouth breather or has witnessed apneas, it is supportive of a diagnosis of a sleep-related breathing disorder. A polysomnogram is required to confirm the diagnosis and assess its severity. Sometimes a patient may give a history of excessive daytime somnolence as well as episodes of muscle weakness that occur in association with emotional extremes or stress (cataplexy). This patient has narcolepsy with cataplexy until proven otherwise. Other features supportive of this diagnosis include sleep paralysis, hypnagogic hallucinations, and automatic behaviors. A polysomnogram documenting the absence of other causes of hypersomnia followed by a MSLT demonstrating two or more SOREMP with a mean sleep latency of less than 8 min also supports the diagnosis. Narcolepsy without cataplexy requires confirmation of the diagnosis with the polysomnogram followed by a MSLT. Usually a history of hypersomnia (without cataplexy) that is present almost daily for at least three months and other causes of hypersomnia are ruled out by polysomnogram suggests a diagnosis of idiopathic hypersomnia. Of course, the daytime hypersomnolence is confirmed with an MSLT. In summary, a history identifies the likely cause of excessive daytime somnolence in most cases and a polysomnogram with or without a MSLT is usually necessary to make or confirm a diagnosis.

6.6. ABNORMAL EVENTS DURING SLEEP

Abnormal events during sleep, or parasomnias (Fig. 6.3 and Table 6.7) are the third most common cause for presentation to a sleep disorders center. These abnormal events may have a variety of features and may be sensations or motor activity. Occasionally, the episodes may be very pronounced with symptoms of violent activity during sleep that may or may not be accompanied with vocalization to the point that screaming or shouting can occur. Abnormal motor activity during sleep should lead the clinician to inquire about symptoms of RBD, sleep terrors, or nocturnal seizures.[31] Again, a history from an observer may be essential in determining the exact nature of these abnormal events.

Periodic limb movements during sleep occur commonly in otherwise asymptomatic individuals, and PLMD is only diagnosed if all other causes of sleep disruption (including RLS) have been ruled out and

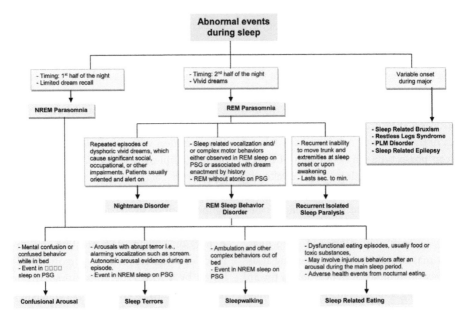

Figure 6.3. Evaluation of abnormal events during sleep.

Table 6.7. Differential diagnosis of abnormal behaviors/ movements during sleep.

NREM parasomnias

Confusional arousals

Sleepwalking

Sleep terrors

Sleep-related eating disorder

REM parasomnias

REM sleep behavioral disorder

Recurrent isolated sleep paralysis

Nightmare disorder

Sleep-related movement disorders e.g. PLMD, RLS

Sleep-related bruxism

Sleep starts (Hypnic jerks)

Parasomnia/ sleep-related movement disorder due to a medical disorder

Parasomnia/ sleep-related movement disorder due to a medication or substance

Sleep-related epilepsy

Other medical disorders e.g. GERD, myocardial ischemia

the limb (typically leg) movements are clearly associated with sleep disruption or daytime symptoms. The limb movements during sleep may not be noticed by the patient and is often only noticed by an observer. Repetitive episodes of limb movements occurring at intervals of 20–40 s throughout sleep, often disappearing during REM sleep, is the typical feature of PLMD.[32]

Parasomnias may occur preferentially during NREM sleep or REM sleep and some may occur at any time during the sleep cycle. The typical NREM parasomnias (confusional arousals, sleepwalking, sleep terrors) usually occur during slow wave sleep (or during transition to and from slow wave sleep) and thus occur during the first half of the night, as there is a greater density of slow wave sleep at this time. The REM parasomnias (e.g. RBD)

usually occur during the second half of the night since there is more REM sleep during this time. Generally, dream recall is vague at best when awoken from a NREM parasomnia and it is relatively vivid and detailed when awoken from a REM parasomnia. A person is typically confused and disoriented when awoken from a NREM parasomnia (occurring in slow wave sleep) and is relatively more alert and oriented when awoken from a REM parasomnia.

Some movement disorders or behaviors may occur at variable times during the night or have no preference for a certain sleep stage (e.g. sleep-related bruxism, sleep-related movement disorder due to a medical disorder, psychiatric disorder, or drug, or substance use). Periodic limb movement during sleep is less likely to occur during REM sleep because of the atonia that accompanies this sleep stage.

Sleep-related epilepsy has a tendency to occur during the transitional sleep stages, but more commonly from transition from wake to sleep; it is least likely to occur during REM sleep. Furthermore, epileptic seizures may occur more than once during the night (and may also occur during the daytime), whereas, the listed NREM parasomnias usually occur once a night.

Many of the sleep-related movement disorders/parasomnias could be identified or diagnosed with a polysomnogram (with additional electroencephalogram and/or electromyogram leads); however, history may be the only aspect of the evaluation that is necessary for some of the diagnoses.

Other abnormal events during sleep include unexpected gastrointestinal, respiratory or cardiac events occurring during sleep. Episodes of chest pain or acute shortness of breath may indicate the need for cardiac or pulmonary evaluation particularly when ischemic heart disease or cardiac arrhythmias are suspected. Such symptoms may also be associated with OSA or even have a psychogenic etiology. Gastroesophageal reflux disorder (GERD) can also result in disrupted sleep. Studies have demonstrated an associated of GERD with OSA. A detailed understanding of the sleep complaint and the patient's medical history may be essential in determining the differential diagnosis.

6.7. SUMMARY

Poor sleep imparts a significant personal and societal burden, one that has been better recognized with widespread knowledge of sleep disorders across all fields of medicine. The ICSD-3 clarifies previously known sleep disorders, with a better understanding of their diagnostic and epidemiological features. It facilitates an accurate diagnosis, improved communication between physicians, and standardization of data for research purposes. Further research to understand the validity of this classification is still necessary. Typically, a comprehensive evaluation of sleep disorders should constitute a detailed clinical history, physical exam, and appropriate investigations, for example, polysomnography. When performed accurately, the physician should be in a good position to identify most sleep disorders and initiate appropriate therapy.

KEYWORDS

- **Evaluation**
- **Polysomnography**
- **History**
- **Investigations**
- **Classification**

KEY TO ABBREVIATIONS

EDS	Excessive Daytime Sleepiness
OSA	Obstructive Sleep Apnea
ICSD	International Classification of Sleep Disorders
RLS	Restless Legs Syndrome
PLMD	Periodic Limb Movement Disorder
RBD	REM Sleep Behavior Disorder
REM	Rapid Eye Movement
NREM	Non-Rapid Eye Movement
GERD	Gastroesophageal Reflux Disease
SOREMP	Sleep Onset REM Period
MSLT	Multiple Sleep Latency Test

MWT   Maintenance of Wakefulness Test

CPAP   Continuous Positive Airway Pressure

REFERENCES

1. Ohayon, M. M. Epidemiology of Insomnia: What We Know and What We Still Need to Learn. *Sleep Med Rev.* **2002,** 6 (2), 97–111.
2. Hatoum, H. T.; Kania, C. M.; Kong, S. X.; Wong, J. M.; Mendelson, W. B. Prevalence of Insomnia: A Survey of the Enrollees at Five Managed Care Organizations. *Am j Manag Care.* **1998,** 4 (1), 79–86.
3. Young, T. B. Epidemiology of Daytime Sleepiness: Definitions, Symptomatology, and Prevalence. *J Clin Psychiatry.* **2004,** 65 Suppl 16, 12–16.
4. Ancoli-Israel, S.; Roth, T. Characteristics of Insomnia in the United States: Results of the 1991 National Sleep Foundation Survey. I. *Sleep.* **1999,** 22 Suppl 2, S347–S353.
5. Breslau, N.; Roth, T.; Rosenthal, L.; Andreski, P. Sleep Disturbance and Psychiatric Disorders: A Longitudinal Epidemiological Study of Young Adults. *Biol Psychiatry.* **1996,** 39 (6), 411–418.
6. Mitler, M. M.; Carskadon, M. A.; Czeisler, C. A.; Dement, W. C.; Dinges, D. F.; Graeber, R. C. Catastrophes, Sleep, and Public Policy: Consensus Report. *Sleep.* **1988,** 11 (1), 100–109.
7. Thorpy, M. J. The International Classification of Sleep Disorders: Diagnostic and Coding Manual. Diagnostic Classification Steering Committee. Rochester, Minnesota: American Sleep Disorders Association, **1990.**
8. American Academy of Sleep Medicine. International Classification of Sleep Disorders, 3rd E. Darien, IL: American Academy of Sleep Medicine, **2014.**
9. Thorpy, M., Current Concepts in the Etiology, Diagnosis and Treatment of Narcolepsy. *Sleep Med.* **2001,** 2 (1), 5–17.
10. Johns, M. W. A New Method for Measuring Daytime Sleepiness: The Epworth Sleepiness Scale. *Sleep.* **1991,** 14 (6), 540–545.
11. The AASM Manual for the Scoring of Sleep and Associated Events. American Academy of Sleep Medicine. Westchester, IL **2007.**
12. Aldrich, M. S.; Jahnke, B., Diagnostic Value of Video-Eeg Polysomnography. *Neurology.* **1991,** 41 (7), 1060–1066.
13. Littner, M.; Hirshkowitz, M.; Kramer, M.; Kapen, S.; Anderson, W. M.; Bailey, D.; Berry, R. B.; Davila, D.; Johnson, S.; Kushida, C.; Loube, D. I.; Wise, M.; Woodson, B. T.; American Academy of Sleep, M.; Standards of Practice, C. Practice Parameters for Using Polysomnography to Evaluate Insomnia: An Update. *Sleep.* **2003,** 26 (6), 754–760.
14. Carskadon, M. A.; Dement, W. C.; Mitler, M. M.; Roth, T.; Westbrook, P. R.; Keenan, S. Guidelines for the Multiple Sleep Latency Test (MSLT): A Standard Measure of Sleepiness. *Sleep.* **1986,** 9 (4), 519–524.
15. Andlauer, O.; Moore, H.; Jouhier, L.; Drake, C.; Peppard, P. E.; Han, F.; Hong, S. C.; Poli, F.; Plazzi, G.; O'Hara, R.; Haffen, E.; Roth, T.; Young, T.; Mignot, E., Nocturnal Rapid Eye Movement Sleep Latency for Identifying Patients with Narcolepsy/Hypocretin Deficiency. *JAMA Neurol.* **2013,** 70 (7), 891–902.
16. Mitler, M. M.; Gujavarty, K. S.; Browman, C. P. Maintenance of Wakefulness Test: A Polysomnographic Technique for Evaluation Treatment Efficacy in Patients with Excessive Somnolence. *Electroencephalogr Clin Neurophysiol.* **1982,** 53 (6), 658–661.
17. Doghramji, K.; Mitler, M. M.; Sangal, R. B.; Shapiro, C.; Taylor, S.; Walsleben, J.; Belisle, C.; Erman, M. K.; Hayduk, R.; Hosn, R.; O'Malley, E. B.; Sangal, J. M.; Schutte, S. L.; Youakim, J. M. A Normative Study of the Maintenance of Wakefulness Test (MWT). *Electroencephalogr Clin Neurophysiol.* **1997,** 103 (5), 554–562.
18. Banks, S.; Barnes, M.; Tarquinio, N.; Pierce, R. J.; Lack, L. C.; McEvoy, R. D. The Maintenance of Wakefulness Test in Normal Healthy Subjects. *Sleep.* **2004,** 27 (4), 799–802.
19. Littner, M. R.; Kushida, C.; Wise, M.; Davila, D. G.; Morgenthaler, T.; Lee-Chiong, T.; Hirshkowitz, M.; Daniel, L. L.; Bailey, D.; Berry, R. B.; Kapen, S.; Kramer, M.; Standards of Practice Committee of the American Academy of Sleep, M. Practice Parameters for Clinical Use of the Multiple Sleep Latency Test and the Maintenance of Wakefulness Test. *Sleep.* **2005,** 28 (1), 113–121.
20. Sagaspe, P.; Taillard, J.; Chaumet, G.; Guilleminault, C.; Coste, O.; Moore, N.; Bioulac, B.; Philip, P. Maintenance of Wakefulness Test as

a Predictor of Driving Performance in Patients with Untreated Obstructive Sleep Apnea. *Sleep.* **2007,** *30* (3), 327–330.

21. Mickelson, S. A.; Lian, T.; Rosenthal, L. Thyroid Testing and Thyroid Hormone Replacement in Patients with Sleep Disordered Breathing. *Ear Nose Throat J.* **1999,** *78* (10), 768–771, 774–775.

22. Kryger, M. H.; Otake, K.; Foerster, J. Low Body Stores of Iron and Restless Legs Syndrome: A Correctable Cause of Insomnia in Adolescents and Teenagers. *Sleep Med.* **2002,** *3* (2), 127–132.

23. Wang, J.; O'Reilly, B.; Venkataraman, R.; Mysliwiec, V.; Mysliwiec, A. Efficacy of Oral Iron in Patients with Restless Legs Syndrome and a Low-Normal Ferritin: A Randomized, Double-Blind, Placebo-Controlled Study. *Sleep Med.* **2009,** *10* (9), 973–975.

24. Spielman, A. J.; Saskin, P.; Thorpy, M. J. Treatment of Chronic Insomnia by Restriction of Time in Bed. *Sleep.* **1987,** *10* (1), 45–56.

25. Ford, D. E.; Kamerow, D. B. Epidemiologic Study of Sleep Disturbances and Psychiatric Disorders. An Opportunity for Prevention? *JAMA.* **1989,** *262* (11), 1479–1484.

26. Katz, D. A.; McHorney, C. A. Clinical Correlates of Insomnia in Patients with Chronic Illness. *Arch Intern* Med. **1998,** *158* (10), 1099–1107.

27. Luyster, F. S.; Buysse, D. J.; Strollo, P. J., Jr. Comorbid Insomnia and Obstructive Sleep Apnea: Challenges for Clinical Practice and Research. *J Clin Sleep Med.* **2010,** *6* (2), 196–204.

28. Stradling, J. R.; Davies, R. J. Sleep. 1: Obstructive Sleep Apnoea/Hypopnoea Syndrome: Definitions, Epidemiology, and Natural History. *Thorax.* **2004,** *59* (1), 73–78.

29. Chaudhuri, K. R., Nocturnal Symptom Complex In PD and its Management. *Neurology.* **2003,** *61* (6 Suppl 3), S17–S23.

30. Merino-Andreu, M.; Arnulf, I.; Konofal, E.; Derenne, J. P.; Agid, Y. Unawareness of Naps in Parkinson's Disease and in Disorders with Excessive Daytime Sleepiness. *Neurology.* **2003,** *60* (9), 1553–1554.

31. Mahowald, M. W.; Schenck, C. H. Diagnosis and Management of Parasomnias. *Clin Cornerstone.* **2000,** *2* (5), 48–57.

32. Hening, W.; Allen, R.; Earley, C.; Kushida, C.; Picchietti, D.; Silber, M. The Treatment of Restless Legs Syndrome and Periodic Limb Movement Disorder. An American Academy of Sleep Medicine Review. *Sleep.* **1999,** *22* (7), 970–999.

7

MELATONIN SIGNALING AS A LINK BETWEEN SLEEP AND CIRCADIAN BIOLOGY: PRACTICAL IMPLICATIONS

Daniel P. Cardinali and Gregory M. Brown

ABSTRACT

Normal circadian rhythms are synchronized to a regular 24 h environmental light-dark cycle. Both the suprachiasmatic nucleus (SCN) and melatonin are essential for this adaptation. Melatonin exerts its chronophysiological action in part by acting through specific membrane receptors (MT_1, MT_2), which have been identified in SCN cells as well as in several neural and non-neural tissues. Both receptors have

Daniel P. Cardinali
UCA-BIOMED-CONICET, Faculty of Medical Sciences, Pontificia Universidad Católica Argentina, 1107 Buenos Aires, Argentina

Gregory M. Brown
Centre for Addiction and Mental Health, University of Toronto, 100 Stokes St., Toronto, ON M6J 1H4, Canada

Corresponding author: Dr. Daniel P. Cardinali, UCA-BIOMED-CONICET, Faculty of Medical Sciences, Pontificia Universidad Católica Argentina, 1107 Buenos Aires, Argentina. E-mail: danielcardinali@uca.edu.ar, danielcardinali@fibertel.com.ar. Phone: +54-911-44743547

been cloned and share general features with other G protein linked receptors. Melatonin also exerts direct effects on intracellular proteins, such as calmodulin or tubulin, has strong free radical scavenger properties, which are non-receptor mediated, is an effective mitochondrial protector and may interact with proteasome to affect intracellular physiology. Within the SCN, melatonin reduces neuronal activity in a time-dependent manner. The disruption of these circadian mechanisms causes a number of sleep disorders known as circadian rhythm sleep disorders (CRSDs). CRSDs include delayed or advanced sleep phase syndromes; non-24 h sleep-wake rhythm disorder, time zone change syndrome ("jet lag") and shift work sleep disorder. Disturbances in the circadian phase position of plasma melatonin levels have been found in all these disorders. In addition, comorbidity of severe circadian alterations with neurodegenerative diseases like Alzheimer's disease (AD) has been documented. Currently there is sufficient evidence to implicate endogenous melatonin as an important mediator in CRSD pathophysiology. The documented efficacy of melatonin to reduce chronic benzodiazepine/Z drug use in insomnia patients is also discussed.

7.1. INTRODUCTION

The objectives of this chapter are to discuss the manner, in which the circadian system regulates melatonin and the sleep–wake cycle, the linkages between melatonin, the circadian system and the sleep–wake cycle and studies on the use of melatonin to treat Circadian Rhythm Sleep disorders (CSRDs), the circadian alterations seen in Alzheimer's disease (AD), and benzodiazepine (BZD)/Z drug abuse.

CRSDs have become a major focus of attention in recent years.[1,2] Major industrial, air, and train accidents have been generally attributed to inefficient handling of situations by individuals suffering from fatigue due to a malfunctioning circadian time keeping system. Also contributing to in-job accidents is the scheduling of the work itself.[3] There is evidence that work performance is negatively impacted by night shift work, especially when the hours of work include the period when the hormone melatonin is normally at its peak of production (the "circadian trough"). The resulting decrements in alertness and performance are further exacerbated by poor quality sleep, another condition, which often afflicts night shift workers. Similarly, affected are long distance truck drivers and others who must do extended highway driving. It has been reported that sleep-related motor vehicle accidents are about 20 times greater at 0600 h than at 1000 h.[3]

Synchronization of the sleep-wake rhythm and the rest-activity cycles with the light-dark (LD) cycle of the external environment is essential for maintaining man's normal mental and physical health. Circulating melatonin, which is produced mainly in the pineal gland, is essential for this physiological adaptation. This is particularly apparent in pathological conditions, such as CRSDs, some of which are known to result from disturbances in the rhythm of melatonin secretion.[1,2,4,5] The melatonin secretion cycle represents a convenient means for observing the body's circadian time keeping system. Since a disruption in the rhythm of melatonin secretion is a central feature of CRSDs, an increasing amount of evidence now shows that the strategic application exogenous melatonin itself can be of benefit in resynchronizing the altered circadian pattern.

CRSDs include delayed sleep phase syndrome (DSPS), advanced sleep phase syndrome (ASPS), non-24 h sleep-wake rhythm disorder, time zone change syndrome ("jet lag"), and shift work sleep disorder.[2,6] Disturbances in the circadian phase position of plasma melatonin levels have been found in all these disorders. CRSDs respond better to chronobiological manipulations involving, for example, the use of phototherapy or melatonin, rather than to conventional hypnotic therapy.[2,6]

In AD patients melatonin secretion decreases and exogenous melatonin administration improves sleep efficiency, sundowning and, to some extent, cognitive function (see recent review).[7] This effect can be particularly important in mild cognitive impairment (MCI), an etiologically heterogeneous syndrome characterized by cognitive impairment preceding dementia. Approximately, 12% of MCI patients convert to AD or other dementia disorders every year. Recent studies indicate that melatonin can be a useful add-on drug for treating MCI in a clinical setting.[7]

The ultimate goal of anti-insomnia therapy is symptomatic and functional recovery that helps a return to everyday life. However, a large proportion of patients under BZD/Z drug treatment (the most common antiinsomnia drugs prescribed) fail to achieve a complete and sustained recovery and are left with residual symptoms produced by the treatment itself, like tolerance or dependency, that make relapse or recurrence more likely, and poorer quality of life a reality. Thus BZD/Z drug abuse has become a public health issue and has led to multiple campaigns to reduce both prescription and consumption of BZD/Z-drugs. Melatonin has been promoted as a drug to improve sleep in patients with insomnia mainly because it does not cause hangover or show any addictive potential.[8]

7.2. MOLECULAR AND NEURAL BASES OF THE MAMMALIAN CIRCADIAN TIMING SYSTEM

Circadian timing provides temporal organization of most biochemical, physiological, and neurobehavioral events in a manner beneficial to the organism. This is the basis of a predictive homeostasis that allows the organism to anticipate events for an optimal adaptation. For example, every day prior to waking, plasma cortisol, sympathetic tone, and body temperature rise, anticipatory to increased activity, and postural change.

The circadian timing system comprises a hierarchy of pacemakers with the hypothalamic suprachiasmatic nucleus (SCN) as the master pacemaker. The SCN includes a small neuronal group that co-ordinates timing of the sleep–wake cycle as well as coordinating it with circadian rhythms in other parts of the brain and peripheral tissues.[9] The SCN consists of a set of individual oscillators that are coupled to form a pacemaker. Anatomically, the mammalian SCN comprises two major subdivisions, a core and a shell. The core lies adjacent to the optic chiasm, predominantly comprises neurons producing vasoactive intestinal polypeptide or gastrin-releasing peptide co-localized with γ-aminobutyric acid (GABA) and receives dense visual inputs from the retino-hypothalamic tract (RHT), and geniculo-thalamic tract (GHT) as well as midbrain raphe afferents. It contains a population of non-rhythmic cells that are responsive to light.[9,10] In contrast the shell surrounds the core, contains a large population of arginine vasopressin-producing neurons in its dorsomedial portion, and a smaller population of calretinin-producing neurons dorsally and laterally, co-localized with GABA, and largely receives input from non-visual hypothalamic, brainstem, and medial forebrain regions. There is overlap in cell populations and functions between these anatomical regions.[9–11]

At a molecular level, circadian clocks are based on clock genes, some of which encode proteins able to feedback their own transcription (Fig. 7.1).[12–14] The mammalian circadian oscillator is composed of two interlocking transcription/translation feedback loops: that is, core and auxiliary loops. The positive drive to the daily clock is constituted by helix-loop-helix, PAS-domain containing transcription factor genes, called *Bmal1* and *Clock* (or its paralog *Npas2*). The protein products of

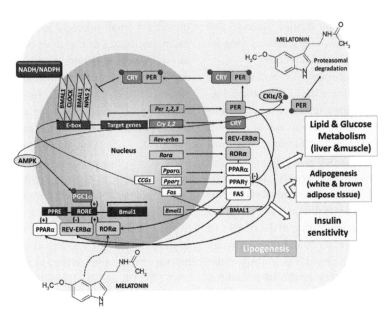

Figure 7.1. A network of transcription-translation feedback loops constitutes the mammalian circadian clock. The transcription factor BMAL1 forms heterodimers with CLOCK. These heterodimers drive the transcription of three mammalian period genes *Per1*, *Per2*, and *Per3*, two Cryptochrome genes *Cry1* and *Cry2*, and *Rev-Erb-a* genes by binding to their respective CACGTG E-Box elements present in the promoters. Rhythmic output of the clock is achieved through E-box elements in clock-controlled genes (CCGs), which can impact a range of cell processes and physiology. In the figure 7.1 a number of genes involved in lipid and glucose metabolism are shown. The BMAL1–CLOCK heterodimer can also inhibit BMAL1 transcription. After transcription and translation, the Rev–Erb-a protein enters the nucleus to suppress the transcription of *Bmal1* and *Cry* genes. A second cluster of genes are driven by RORA (RAR-Related Orphan Receptor-A), which are sensitive to negative regulation by Rev–Erb-a and so are expressed in phase with BMAL1. Rev-Erb-α/RORA is involved in gene expression during circadian night, which is in phase with BMAL1 and in antiphase to Per2 oscillations. As the Per proteins, such as PER2, accumulate in the cytoplasm, they become phosphorylated by CSNK-I-Epsilon/Delta (Casein Kinase-I-Epsilon/Delta). The phosphorylated forms of PER are unstable and are degraded by ubiquitylation and proteasomal degradation. Late in the subjective day, however, CRY accumulates in the cytoplasm, promoting the formation of stable CSNK-I-Epsilon/Per/Cry complexes, which enter the nucleus at the beginning of a subjective night. Once in the nucleus, Cry1 disrupts the CLOCK/BMAL1-associated transcriptional complex, resulting in the inhibition of *Cry*, *Per*, and Rev–Erb-a transcription and derepression of BMAL1 transcription. The interacting positive and negative feedback loops of circadian genes ensure low levels of Per and Cry, and a high level of BMAL1 at the beginning of a new circadian day. In the SCN neurons, the intracellular levels of CLOCK remain steady throughout the 24 h period, whereas BMAL1 expression levels are high at the beginning of a subjective day and low at the beginning of a subjective night. In addition, a number of signaling molecules derived from cellular metabolism fine-tune these molecular clock loops (e.g. AMPK, an AMP-dependent kinase; NADP/NADPH via sirtuins). The effect of melatonin on SCN clock genes is seen after a lag period of about 24 h suggesting that clock gene transcription was not the immediate target in situations in melatonin-induced phase advances of circadian rhythms mediated by MT$_1$ and MT$_2$ receptors. Thus the involvement of nuclear RORα receptors and the inhibition of the ubiquitin–proteasome system deserve to be considered.

these genes form heterodimeric complexes that control the transcription of other clock genes, notably three period (*Per1/Per2/Per3*) genes and two cryptochrome (*Cry1/Cry2*) genes, which in turn provide the negative feedback signal that shuts down the *Clock/Bmal1* drive to complete the circadian cycle. Accumulated PER and CRY proteins intensively repress E-box-mediated transcription until their levels sufficiently decrease once again. Additionally, CLOCK and BMAL1 also control the transcription of the nuclear receptors RORα and REV-ERBα, which modulate the BMAL1 mRNA levels by competitive actions on the rev responsive element (RRE) residing in the *Bmal1* promoter. Collectively, the cycling of clock components also determines the levels of clock-controlled genes by transcription via the E-box and/or RRE, thus achieving an oscillating pattern and generating rhythmic physiological outputs. In addition, a number of signaling molecules, including kinases, fine-tune these molecular clock loops. For example, Casein kinase Ie (CKIe) and CKId form a complex with PERs and CRYs, phosphorylate PERs and then promote proteasome-dependent degradation of these negative regulators. AMPK, an AMP-dependent kinase, phosphorylates CRYs, and promotes their degradation so as to terminate the suppressive effects of CRYs on the CLOCK/BMAL1 heterodimer. Thus clock gene expression oscillates because of the delay in the feedback loops, regulated in part by phosphorylation of the clock proteins that control their stability, nuclear re-entry, and transcription complex formation.[12–14]

In murine SCN the transcription of *Per* and *Cry* occurs during the light phase of daily photoperiod, while translation into their respective proteins occurs several hours later.[15] Another important observation was that *Bmal1* transcription occurs during most of the scotophase with a rhythm about 180 deg out of phase with that of *Per* and *Cry*.[15–17] Circadian expression of over 300 SCN transcripts was documented.[18] In mice at least 120 SCN proteins are expressed in a circadian manner, 23 of them being up-regulated or down regulated by light.[19]

In the absence of periodic environmental synchronizers, the circadian pacemaker is free running with a period very near to 24 h in mammals. In humans, the interindividual differences are small. However, a large scale epidemiologic study showed that differences in sleep–wake times follow a near Gaussian distribution with extreme cases at each end; extreme early cases woke as extreme late ones fell asleep.[20] The rhythm is adjusted to 24 h by the action of light, the main (but not the unique) Zeitgeber in humans. Brief exposures to light are sufficient to entrain the SCN clockwork to solar time, adjusting the oscillator to a precise 24 h cycle.

Genetic screening has shown that polymorphisms in human clock genes are correlated with alterations in sleep or diurnal preferences.[21] For example, in DSPS a correlation with certain polymorphisms in the clock gene hPer3 has been demonstrated[22] while a mutation in the hPer2 gene is associated with familiar ASPS.[23]

7.3. SYNTHESIS AND MECHANISM OF ACTION OF MELATONIN

Melatonin is the primary hormonal output of the circadian system. Its synthesis as well as other circadian rhythms is controlled by a circadian signal from the SCN. The photoperiod is the major entraining influence on the SCN with inputs arriving from the retina via the RHT and the GHT.

Partially originating from a subset of directly photosensitive retinal ganglion cells that contain melanopsin as a photopigment, the neurotransmitters participating in the RHT are glutamate and the pituitary adenylyl cyclase-activating peptide, both released at the SCN level.[24] The action spectrum for melatonin suppression in man is in keeping with a shortwave non-rod, non-cone photopigment.[25] Moreover, it has been shown that alerting effects of light are most pronounced at very short (420–460 nm) wavelengths.[26] Retinal projections to the intergeniculate leaflet of the lateral geniculate complex subsequently project to the SCN via the GHT.[9] Light acts via these tracts to entrain the rhythm in the SCN as well as acting downstream of the SCN clock to block the activity of the pineal gland.

The SCN acts on the pineal gland via a complex multisynaptic pathway. It projects to the autonomic subdivision of the hypothalamic paraventricular nucleus, which in turn projects directly to the upper thoracic intermediolateral cell column. Preganglionic sympathetic noradrenergic fibers project to the superior cervical ganglion, which sends postganglionic sympathetic noradrenergic fibers to the pineal gland to stimulate synthesis of melatonin.[27] Norepinephrine released from the sympathetic fibers acts on the pineal via a dual receptor mechanism. It activates adenylyl cyclase via β_1-adrenergic receptors[28] and protein kinase C activity via α_{1B} –adrenergic receptors,[29] which potentiate β_1-adrenergic

receptor activation of adenylyl cyclase. There is therefore a very rapid, large increase in cyclic 3', 5'-adenosine monophosphate, which leads to phosphorylation of the enzyme arylalkylamine N-acetyltransferase (AANAT). When phosphorylated, AANAT is activated by formation of a reversible regulatory complex with 14-3-3 proteins.[30]

AANAT, the enzyme that converts serotonin to N-acetylserotonin, has a pivotal role in the timing of melatonin synthesis. It increases speedily with a doubling time of about 15 min in response to darkness onset and in response to light it shows an even more rapid half-life of degradation of 3.5 min.[30] Since melatonin itself has a half-life in the circulation of about 30 min in man, its levels change rapidly in response to circadian signals and light.[31] Hydroxyindole-O-methyltransferase (HIOMT), the enzyme that catalyzes production of melatonin from N-acetylserotonin, is responsible for the amplitude of the nocturnal peak of melatonin.[32,33] By using a combination of molecular approaches together with a sensitive in-vivo measurement of pineal indoles it was shown that N-acetylserotonin is present in vast excess during the night allowing the conclusion that although AANAT is the rhythm-generating enzyme it is not rate limiting for nocturnal production.[34]

Once formed, melatonin is not stored in the pineal gland but is immediately secreted into the bloodstream. In plasma melatonin binds mostly to albumin.[35] It then passes through the choroid plexus to the cerebrospinal fluid.[36] In a recent study, melatonin concentration was measured in CSF sampled during neurosurgery in both lateral and third ventricles in patients and compared with their plasma levels.[37] A significant difference in melatonin concentration was observed between lateral and third ventricles, with the highest levels in the third ventricle. Melatonin levels were significantly higher in third ventricle than in the plasma, suggesting that melatonin may enter directly the CSF through the pineal recess in humans.[37]

Endogenous melatonin, whether measured in saliva or in urine, is often referred to as a "hormonal finger print", having a profile, which is both unique and yet consistently predictable (on a daily and weekly basis) within the individual. This differs from the high interindividual variability of circulating melatonin levels, presumably of a genetic origin.[38] In humans, plasma melatonin begins to increase steadily after 1900–2300 h and reaches its peak value between 0200 and 0400 h. The levels then decline, reaching their lowest values during daytime hours. The rhythm is well preserved from childhood to adulthood but after approximately the age of 55 the nocturnal peak of melatonin production begins to decline, a possible contributing factor to insomnia, which is often seen among the elderly.

The first attempts to identify brain melatonin receptors employed ^{3}H-melatonin as a radioligand.[39,40] This was followed by the discovery of the first functional melatonin receptor in a neuronal mammalian tissue, the rabbit retina.[41] In 1984, Vakkuri et al.[42] introduced the radioligand 2-^{125}I-iodomelatonin as a tracer for use in melatonin radioimmunoassay. This molecule turned out to be the silver bullet of melatonin receptor research as its selectivity and high specific activity allowed the field to move forward. By using this ligand, binding sites for melatonin were identified in a wide variety of central and peripheral tissues.[43] Molecular cloning of the first high affinity melatonin receptor (MT_1) was accomplished using a cDNA library constructed from a dermal cell line of melanophores,[44] the first tissue in which melatonin's action had been demonstrated. This initial finding led to the discovery of another G$_i$-protein coupled melatonin receptor in humans (MT_2),[45] which is 60% identical in amino acid sequence to the MT_1 receptor. Yet a third receptor, now called GPR50, shares 45% of the amino acid sequence with MT_1 and MT_2 but does not bind melatonin.[46,47] It is unusual in that it lacks N-linked glycosylation sites and that it has a C-terminal that is over 300 amino acids long. Many G protein-coupled receptors (GPCR), including the MT_1 and MT_2 receptors, exist in living cells as dimers. The relative propensity of the MT_1 homodimer and MT_1/MT_2 heterodimer formation are similar, whereas that of the MT_2 homodimer is 3-4 fold lower.[48] It is of considerable interest that the GPR50 receptor, though lacking the ability to bind melatonin, abolishes high affinity binding of the MT_1 receptor through heterodimerization.[49,50] Thus the GPR50 receptor may have a role in melatonin function by altering binding to the MT_1 receptor.[51]

A fourth 2-^{125}I-iodomelatonin binding site was also identified in mammals $(MT_3$, initially called ML-2).[52] Unlike the picomolar membrane receptors it is a nanomolar binding site with a specific pharmacologic profile and fast kinetics of association/ dissociation.[53] It has now been purified from hamster kidney and characterized as the analog of quinone reductase type 2.[54]

A combination of reagents derived from the molecular clones and pharmacologic tools have revealed a considerable amount of information about the MT_1 and MT_2 receptors.[55] For example, it has been shown that the MT_1 receptor inhibits firing acutely in SCN slices, and that both MT_1 and MT_2 may contribute to phase shifting in these slices. MT_1 and MT_2 have also been shown to differentially regulate $GABA_A$ receptor function in SCN.[56]

Mapping of the MT_1 and MT_2 receptors in brain though not yet complete has revealed much information. As expected, MT_1 and MT_2 receptors are present in the SCN.[57] They are also found in several other brain areas and in the periphery. The MT_1 receptor is extremely widely distributed in the hypothalamus of particular note, it is co-localized with corticotrophin in the PVN and with oxytocin and vasopressin in the PVN and supraoptic nucleus.[57] MT_1 receptors are found in the cerebellum>= occipital cortex>= parietal cortex> temporal cortex> thalamus> frontal cortex >= hippocampus.[58] MT_2 receptors have been identified in the hippocampus,[59] MT_1 plus MT_2 in the occipital cortex[60] and MT_1 in the dopaminergic system.[61]

Melatonin also binds to transcription factors belonging to the retinoic acid receptor superfamily, in particular, splice variants of RORα and the product of another gene, RZRβ (for ref. see).[62] Although these nuclear binding proteins have for quite some time been a matter of debate and although their affinity to melatonin is lower, compared to MT membranes receptors, their classification as nuclear receptors now seems to be justified. RORα subforms are ubiquitously expressed in all mammalian tissues tested to date.[63] RORα expression levels frequently depend on the differentiation state of cells or may vary within the cell cycle.[63–65]

Melatonin is the prototype of a class of drugs that influence the circadian apparatus and are referred to as chronobiotics.[66–69] The response to melatonin follows a phase response curve (PRC), so that morning administration causes a delay, while evening administration causes an advance in circadian rhythms.[70] This PRC is about 12 h out of phase with the PRC to light which causes a phase advance in the morning and a phase delay in the evening.[71] A detailed PRC for melatonin (3 mg) established that the maximum advance portion peaked about 5 h before DMLO in the afternoon, the maximum delay portion was about 11 h after DMLO shortly after habitual awakening and a dead zone was in the first half of usual sleep.[72] Maximum advance and delay shifts were 1.8 and 1.5 h, respectively.

In a study of one hour sleep schedule advance with early morning light it was shown that the addition of afternoon melatonin treatment (0.5 or 3 mg) caused a significantly greater phase advance of 2.5 h, again with minimal side effects.[73] There was no difference between the two melatonin doses. Thus effects of morning light and evening melatonin are additive and can be used to cause pre-adaptation prior to eastward flight.

Current recommendations for light treatment require the patients to sit in front of a bright light box for at least 1 h daily, thus limiting their willingness to comply. Light applied through closed eyelids during sleep might not only be efficacious for changing circadian phase but also lead to better compliance because patients would receive light treatment while sleeping. Two studies investigated the impact of a train of 480 nm (blue) light pulses presented to the retina through closed eyelids on melatonin suppression and on delaying circadian phase.[74] Both studies employed a sleep mask that provided narrow-band blue light pulses of 2 sec duration every 30 sec. The results demonstrated that blue light pulses given early in the sleep episode significantly delayed circadian phase in older adults therefore indicating the efficacy and practicality of light treatment by a sleep mask aimed at adjusting circadian phase in a home setting.[74] In a same vein, the effects of filtering short wavelengths (< 480 nm) during night shifts on sleep and performance in shift-work nurses were assessed indicating that this manipulation can be an approach to reduce sleep disruption and improve performance in rotating-shift workers.[75]

7.4. MELATONIN AND THE REGULATION OF THE SLEEP–WAKE CYCLE

Melatonin phase-shifts circadian rhythms in the SCN by acting on MT_1 and MT_2 melatonin receptors expressed by SCN neurons, thus creating a reciprocal interaction between the SCN and the pineal gland. Melatonin's phase-altering effect is caused by its direct influence on the electrical and metabolic activity of the SCN.

The effect of melatonin on SCN clock genes is seen after a lag period of about 24 h suggesting that clock gene transcription was not the immediate target in situations in melatonin-induced phase advances of circadian rhythms mediated by MT_1 and MT_2 receptors.[76] Thus the involvement of nuclear RORα receptors should be consid-

ered. This may result in the inhibition of the ubiquitin–proteasome system recently attributed to melatonin.[77] Melatonin inhibition of the proteasome would tend to stabilize clock proteins in the SCN (such as BMAL1) transcribed during the scotophase. Additionally, melatonin-mediated phosphorylation of proteins through classical receptor-mediated pathways could indirectly influence the susceptibility of proteins to degradation by the proteasome.[77]

The circadian rhythm in the secretion of melatonin has been shown to be responsible for the timing of the sleep rhythm in both normal and blind subjects (i.e. in the absence of the synchronizing effect of light).[78] The daily sleep-wake cycle is influenced by two separate processes: (1) the endogenous biological clock that drives the circadian rhythm of sleep-wake cycle (process C, for "circadian") and (2) a homeostatic component (process S, for "sleep") that influences "sleep propensity", a state which is determined by the immediate history of sleep and wakefulness and additionally the duration of previous sleep episodes.[79,80] These two processes, which interact continuously, determine the consolidated bout of sleep at night and the consolidated bout of wakefulness during daytime. Observations of subjects whose circadian rhythms had been experimentally desynchronized have supported the inference that homeostatic processes drive slow sleep while rapid eye movement (REM) sleep is driven by the circadian component.[81]

The role of SCN in the regulation of sleep-wake cycle is relevant, interacting with both sleep regulatory mechanisms,[82] as first studied in the squirrel monkey. In this primate species lesions made in the

SCN caused either loss of sleep or prolonged sleep.[83] It is suggested that circadian signals emanating from the SCN promote wakefulness during the day. Additionally, SCN activity may facilitate sleep during the subjective night. It is interesting to note that the neural pathways from the SCN that promote wakefulness are complemented by those that are involved in the promotion of sleep.[82]

Studies in humans under constant routine conditions have defined the so-called "biological night" that corresponds to the period during, which melatonin is produced and secreted into the bloodstream. The beginning of the biological night is characterized by onset of the melatonin surge, an accompanying increase in sleep propensity as well as a decrease in core body temperature; the opposite occurs as the biological night and sleep end.[84] Due to its neuronal suppressive actions on MT_1 receptors[85] and as a result of its phase shifting activity, mediated via its actions on MT_1 and MT_2 receptors,[86] melatonin modulates the electrical activity of the SCN. The firing rate of the SCN decreases during the transition from non-REM to REM sleep.[82]

Crucial for optimal treatment of CRSD, melatonin and other treatments should be administered at a time related to individual circadian timing (typically assessed using the dim light melatonin onset (DLMO).[87] If not administered according to the individual patient's circadian timing, melatonin and other treatments may not only be ineffective, they may even result in contrary effects. It has been proposed that knowing the patient's individual circadian timing by assessing DLMO improves diagnosis and treatment of CRSD with melatonin as well as other therapies such as light or chronotherapy.[88]

7.5. MELATONIN AND DELAYED SLEEP PHASE SYNDROME

DSPS is mainly encountered in young individuals. A common sleep-wake disorder that accounts for 10% of insomniacs who are diagnosed in sleep laboratory, DSPS is due to altered physiological timing in the biological clock.[6] In this condition, the timing of sleep onset and wake time are delayed. The onset of sleep is delayed in some cases to 0200 – 0600 h. Neither sleep architecture nor the maintenance of sleep is affected.[6] However, persons suffering from this disorder experience chronic sleep onset insomnia so that forced early awakening results in daytime sleepiness. It has been shown that the peak melatonin secretion occurs between 0800 h and 1500 h in some DSPS patients demonstrating the abnormal phase position of melatonin in this sleep disorder.[89,90]

DSPS is the most frequently occurring CRSD. Dagan and Eisenstein[91] found that 83.5% of 322 CRSD patients were of the DSPS type. The prevalence of DSPS in adolescence is more than 7% .[89] Among those with DSPS onset of symptoms, which occurred in early childhood was reported by 64.3% of the sample, in the beginning of puberty by 25.3%, and during adulthood by 10.4%.[91]

Even a minor brain injury or head trauma can act as a trigger for the development of DSPS.[92] DSPS can also follow whiplash injury.[93] Frequently occurring jet lag or frequently occurring shift-work are also risk factors for the development of DSPS.[94] A great proportion of patients have a prior history of depression.[95] DSPS persists even after remission of the depression thus suggesting that DSPS may be a cause rather than a consequence of depression.[96]

Dahlitz and co-workers were the first to report a placebo-controlled study that demonstrated the efficacy of melatonin in the treatment of DSPS patients.[97,98] A 5 mg dose of melatonin was administered orally at 2200 h to patients suffering from DSPS for a period of four weeks. In those studies, it was noted that melatonin significantly advanced the sleep onset time by an average of 82 min, with a range of 19 to 124 min. The mean wakefulness time also was advanced by 117 min.[97,98] Though the total duration of sleep remained unaltered (mean about 8 h) after melatonin treatment, there was a significant improvement in sleep quality. Several other confirming reports followed.[99–102] For example, Wilhelmsen-Langeland et al.[103] reported that a combined bright light and melatonin treatment improved subjective daytime sleepiness, fatigue, and cognitive function in a three month study of DSPS patients. Although an alternative strategy, gradual advancement of rise times produced positive effects on subjective sleepiness, fatigue, and cognitive performance during short-term treatment of patients, the benefits from gradually advanced rise times wear off, indicating that the continuation of bright light and melatonin treatment is needed to maintain positive effects over time.[103]

Remarkably, a recent consensus on the roles of melatonin in children and on treatment guidelines concluded that the best evidence for melatonin efficacy in children is in DSPS.[104] Melatonin is most effective when administered 3–5 h before DLMO. It was also concluded that there was no evidence that extended-release melatonin confers advantage over immediate release preparations.[104] For maximum treatment effectiveness the timing of administration is just as critical for melatonin administration as it is for the application of bright light therapy.[105]

No adverse effects of melatonin were noted. Melatonin was also shown in placebo-controlled studies to be effective for treating children with idiopathic chronic sleep onset insomnia, which is related to child onset DSPS.[106,107]

7.6. MELATONIN AND ADVANCED SLEEP PHASE SYNDROME

The changes in sleep patterns, which occur with advancing age can, in part, be attributed to changes in the functioning of the circadian oscillator.[108] The characteristic pattern of ASPS includes complaints of persistent early evening sleep onset and early morning awakenings.[6] Typically in ASPS, sleep onset occurs at around 2000 h and wakefulness occurs at around 0300 h. The quality of sleep is progressively impaired by increased awakenings occurring during the night. It has been suggested that this impairment is due to an attenuation of amplitude of the rhythm in melatonin secretion, which in turn may disrupt the phase relationships of the sleep-wake cycle as well as other circadian rhythms.

Leger et al.[109] in studies undertaken in 517 human subjects aged 55 years and above noted a significant decline in the secretion of 6-sulfatoxymelatonin, the principal urinary melatonin metabolite, in subjects suffering from insomnia. Melatonin replacement therapy was administered in dosages of 2 mg of controlled release tablets and was found to improve significantly the sleep quality of patients in the insomnia group [109]. Also affected were measures of alertness and behavioral integrity, which also showed improvements in these subjects. These findings were interpreted to support the conclusion that decrements in sleep quality, which are often seen among the elderly, are largely attributable to a decline in the production of melatonin, which also occurs with advancing age.[8] The evidence was also taken to support the conclusion that melatonin promotes sleep possibly through circadian entraining effects as well as by a sleep-regulating effect.

Genetic testing has provided support for the conclusion that familial ASPS is an inherited sleep–wake rhythm disorder with an autosomal dominant mode of inheritance.[21] Indeed, alteration in the function of clock genes has been documented as one of the major causes of CRSDs.[110–112]

7.7. MELATONIN AND TIME ZONE CHANGE SYNDROME (JET LAG)

Rapid transmeridian flight across several time zones results in a temporary mismatch between the endogenous circadian rhythms and the new environmental LD cycle.[113,114] As a result endogenous rhythms shift in the direction of the flight; an eastbound flight will result in a phase advance of rhythms while a westbound flight will produce a phase delay. The re-establishment of normal phase relationships differs from one rhythm to another. Because of this phenomenon a transient desynchronization of circadian rhythms occurs, giving rise to a cluster of symptoms. These symptoms typically include transiently altered sleep patterns (e.g. disturbed night time sleep, impaired daytime alertness, and performance), mood and cognitive performance (e.g. irritability and distress),

appetite (e.g. anorexia), along with other physical symptoms such as disorientation, fatigue, gastrointestinal disturbances and light-headedness that are collectively referred to as "jet lag".

Several studies have reported an improved quality and duration of sleep or accelerated resynchronization of rhythms by giving melatonin (for ref. see).[113,114] Doses used in these studies ranged from 0.5 to 10 mg, with more soporific effect with doses of 5 or 10 mg and little difference in phase response between doses. Melatonin at bedtime after eastbound flight provides benefits from both the soporific and phase resetting effects. Studies using melatonin on westbound flights have been few and have revealed less benefit than after eastbound flight. An exception is a study[115] using a combination of 3 mg bed time melatonin with exercise and light exposure in 2–3 h time blocks (08:00 to 11:00 and 13:00 to 16:00) for six days at the destination after 12 h westward flight in athletes. Although this study lacked a placebo control, resynchronization was reported after 2.13 days, much more rapidly than would be expected. Use of light in this manner is theorized to cause blunting or masking of the endogenous circadian signal and sensitization to Zeitgebers.[116] In a follow-up study that again used 3 mg melatonin at bedtime in combination with only 30 min or more of outdoor exercise in the same two time blocks it was reported that after westward travel of 11 h resynchronization occurred in 2.54 days while after 13 h eastward travel it occurred in 2.27 days.[68] Although there was no placebo control, adaptation to the new environment occurred much more rapidly than would be expected.

Recommendations for using melatonin and light therapy in jet lag from the American Academy of Sleep Medicine indicate that for an individual flying eastward, in whom advancing the sleep phase is desired, 0.5–5 mg melatonin administered following flight at the desired time of bed along with early morning bright light exposure will assist with circadian shift .[106] In contrast, for westward travel, which is considered easier to adapt to because of endogenous circadian periodicity of 24 h, only bright light in the evening is recommended.

7.8. MELATONIN AND SHIFT WORK SLEEP DISORDER

It has been estimated that in our modern 24/7 society at least one fifth of total work force is engaged in rotating shift work[117]. These individuals are forced to forego their nocturnal sleep while they are on a nightshift, and sleep during the day.[118] This inversion of the sleep-wake rhythm with work at night at the low phase of the circadian temperature rhythm and sleep at the time of peak body temperature has given rise to insomnia-like sleep disturbances. Sleep loss impairs the individual's alertness and performance that affects not only work productivity but also has been found to be a major cause for industrial and sleep-related motor vehicle crashes.[117] Sleep-related crashes occur most commonly in the early morning hours (0200–0600 h).[3] Sleep deprivation and the associated desynchronization of circadian rhythms are common in shift-work sleep disorder.[119–121]

Many treatment procedures have been advocated. Czeisler and co-workers administered bright light for improving the physiological adaptation of the circadian rhythms of night-shift workers to their inverted sleep-wake schedules.[122] In their study bright light was found effective for resynchronizing alertness, cognition, performance, and body temperature to the new work schedules. Following the successful application of bright light, melatonin has been used in shift workers to accelerate adaptation of their circadian rhythms and sleep-wake rhythms to the new work schedules (see, e.g.).[84,123] A phase delay in plasma melatonin was noted in shift workers, when melatonin was administered at the morning bedtime following the night shift. The shift in melatonin secretion has been associated with increase in work performance as well.[124] Correctly timed administration of melatonin is advocated for hastening adaptation of circadian rhythms in shift-workers, inasmuch as melatonin administration in the evening (1600 h) does not affect daytime sleepiness and mood.[125] Melatonin (1.5 mg at 1600 h) was found to advance the timing of both endogenous melatonin and cortisol rhythms without causing any deleterious effects on endocrine function or daytime mood and sleepiness. There is thus evidence that combination of both bright light and melatonin can be an effective and reliable strategy for treating shift work disorder.

Recently, a Cochrane meta-analysis to evaluate the effects of pharmacological interventions to reduce sleepiness or to improve alertness at work and decrease sleep disturbances whilst off work was published.[126] Primary outcomes were sleep length and sleep quality while off work, alertness, and sleepiness, or fatigue at work of 15 randomized placebo-controlled trials including 718 participants. Nine trials evaluated the effect of melatonin and two the effect of hypnotics for improving sleep problems. One trial assessed the effect of modafinil, two of armodafinil and one examined caffeine plus naps to decrease sleepiness or to increase alertness. Melatonin (1–10 mg) after the night shift increased sleep length during daytime sleep as compared to placebo. Both modafinil and armodafinil increased alertness and reduce sleepiness to some extent in employees who suffer from shift work sleep disorder but they were associated with adverse events. Caffeine plus naps reduced sleepiness during the night shift, but the quality of evidence was assessed as low.[126]

7.9. MELATONIN AND NON 24 HOUR SLEEP-WAKE DISORDER

As mentioned above, the neuroendocrine effects of light are mediated primarily via specialized non-rod, non-cone ganglion cell photoreceptors. Consequently, most blind people with no perception of light have either non-entrained or abnormally phased circadian rhythms due to this inability to detect light. Conversely, most visually impaired individuals with some degree of light perception should exhibit normal entrainment, emphasizing the functional separation of visual and "non-visual" photoreception. This was re-inforced in a recent study on the prevalence of circadian disorders in blind women.[127] The circadian type of 127 individuals was determined from the timing and time course of the melatonin rhythm as measured by urinary 6-sulfatoxymelatonin. The results indicated that 37% of participants with no perception of light were classified as normally entrained (bilateral enucleation,

retinopathy of prematurity) while 69% of participants with some degree of light perception (retinitis pigmentosa. age-related macular degeneration) were normally entrained.[127]

Patients with non-24 h sleep-wake disorder suffer from recurrent insomnia and daytime sleepiness. The circadian rhythm of sleepiness has shifted out of phase with the desired time for sleeping. Melatonin has been employed to correct right abnormal sleep-wake rhythms in blind human subjects[128] and the Food & Drug Administration approved the melatonin analog tasimelteon to be used with this aim in 2013.

In a recent observational study of 21 patients with non-24 h sleep-wake disorder who kept a sleep-wake schedule of their choosing the circadian phase was determined using the melatonin onset from plasma or saliva samples and actigraphy.[129] A total of 469 assessments were conducted over 5,536 days of study. Although subjects lacked entrainment such that circadian phase drifted an average of 0.39 h later per day there was notable intersubject and intrasubject variability in the rate of drift including relative coordination and periods of transient entrainment, during which there was little to no drift in the circadian phase. Hence the relative coordination and transient entrainment may result in misdiagnosis and responsiveness to environmental time cues and influence treatment success with oral melatonin.[129] The authors recommended the use of actigraphy to adjust the timing of melatonin in this situation.

The prevalence of non-24 h sleep-wake rhythm disorder among sighted patients is unknown. Most of the reported cases belong to the Japanese population and very few cases have been documented outside of Japan and these have been predominantly male and associated with avoidant or schizoid personalities (see for ref.).[2,6] Thus, non-24 h sleep-wake disorder is a rare sleep disorder among sighted patients in Western populations. In the Japanese population, it has been estimated that non-24 h sleep-wake rhythm disorder comprises 23% of all CRSDs.[130] The prevalence of circadian rhythm disorders in this population (0.13–0.4%) is consistent with that observed in other populations.[131]

It is likely that this sleep disorder is rare in Western populations because it is underdiagnosed. Diagnosis is complicated by the fact that at times non-24 h sleep-wake rhythm disorder can resemble both ASPS and DSPS and in fact exhibit the same polysomnographic features.

Under certain circumstances, irregular CRSD may arise. For example, treatment with psychotropic drugs such as haloperidol[132] can trigger a CRSD of an irregular type. In addition, prolactin secreting microadenomas,[133] or occupational inadequate exposure to bright light can be related to the development of irregular CRSD.[134] Minor head trauma can cause irregular CRSD as well as DSPS.[92] Most of reports of this type of CRSD refer it to the influence of environmental and medical conditions.

7.10. MELATONIN AND CIRCADIAN RHYTHM ABNORMALITIES IN ALZHEIMER'S DISEASE

AD is an age-associated neurodegenerative disease that is characterized by a progressive loss of cognitive function, loss of memory, and several neurobehavioral manifestations.[135] Concomitantly, melatonin levels are lower in AD patients compared to age-matched control subjects. Decreased CSF melatonin levels observed in AD patients reflect a decrease in pineal melatonin production rather than a diluting effect of CSF. CSF melatonin levels decrease even in preclinical stages when the patients do not manifest any cognitive impairment (at Braak stages I–II), suggesting that the reduction in CSF melatonin may be an early marker for the first stages of AD.[136]

AD patients with disturbed sleep–wake rhythms did not only exhibit reduced amounts of melatonin secreted, but also a higher degree of irregularities in the melatonin pattern, such as variations in phasing of the peak.[137] Therefore, the melatonin rhythm has not only lost signal strength in clock resetting, but also reliability as an internal synchronizing time cue. Loss or damage of neurons in the hypothalamic SCN and other parts of the circadian timing system may account for the circadian rhythm abnormalities seen in demented patients, especially as the number of neurons in the SCN of AD patients is reduced.

In both elderly subjects and in patients suffering from dementias the administration of exogenous melatonin has been found to not only to enhance sleep quality but also to improve the sleep-wake rhythm in clinical setting studies.[138–142] In a double-blind study, the major findings with regard to melatonin effects on sleep–wake rhythmicity, cognitive and non-cognitive functions were confirmed.[143] In a multicenter, randomized, placebo-controlled clinical trial[144] the increase in nocturnal total sleep time and decreased wake after sleep onset, as determined on an actigraphic basis, were only apparent as trends in the melatonin-treated groups. On subjective measures, however, caregiver ratings of sleep quality showed significant improvement in the 2.5 mg sustained-release melatonin group relative to placebo. Large interindividual differences between patients suffering from a neurodegenerative disease are not uncommon.

To test whether the addition of melatonin to bright-light therapy enhances the efficacy in treating rest-activity (circadian) disruption in institutionalized patients with AD 50 subjects were examined in a randomized, controlled trial.[145] Light treatment alone did not improve nighttime sleep, daytime wake, or rest-activity rhythm while light treatment plus melatonin (5 mg/day) increased daytime wake time and activity levels and strengthened the rest-activity rhythm. In another randomized controlled trial on the effect of combined bright light and melatonin on cognitive and non-cognitive function in 189 elderly residents of group care facilities bright light had a benefit in improving cognitive and non-cognitive symptoms of dementia, which was amplified by the conjoint administration of melatonin.[146] Melatonin alone had a slight adverse effect on mood. It must be noted that in another double-blind randomized placebo-controlled trial of 24 institutionalized patients with AD melatonin failed to improve sleep or agitation.[147]

Since the circadian oscillator system is obviously affected in AD patients showing severe sleep disturbances, the efficacy of melatonin should be expected to depend heavily on disease progression. One cannot expect a profound inhibition of disease progression once a patient is already in an advanced demented state, notwithstand-

ing a very few case reports with anecdotal evidence of slight mental improvements. Therefore, the use of melatonin in the early stages of AD should be important, like MCI.

MCI is an etiologically heterogeneous syndrome characterized by cognitive impairment shown by objective measures adjusted for age and education in advance of dementia. Approximately, 12% of MCI convert to AD or other dementia disorders every year.[148] Since MCI may represent prodromal AD it should be adequately diagnosed and treated. The degenerative process in AD brain starts 20–30 years before the clinical onset of the disease. During this phase, plaques and tangles loads increase and at a certain threshold the first symptom appears. As already mentioned, CSF melatonin levels decrease even in preclinical stages when the patients do not manifest any cognitive impairment (at Braak stages I–II),[136] suggesting that the reduction in CSF melatonin may be an early trigger and marker for AD.

In a recent study, 30 patients with MCI and 28 age-matched controls underwent psychiatric, medical, and neuropsychological assessment, followed by overnight polysomnography and dim light melatonin onset assessment.[149] Participants also performed an episodic memory task while in the laboratory. Patients with MCI had advanced timing of their melatonin secretion onset relative to controls, but the levels of melatonin secreted did not differ between groups. The MCI group also had greater wake after sleep onset and increased rapid eye movement sleep latency. There were differential associations between dim light melatonin onset and cognition between the two groups, with earlier dim light melatonin onset being associated with poorer memory performance in MCI patients. The authors concluded that circadian misalignment and sleep disruption is evident in patients with MCI, and is consistent with changes observed in Alzheimer's disease.[149]

We previously reported a retrospective analysis, in which daily 3–9 mg of a fast-release melatonin preparation p.o. at bedtime for up to three years significantly improved cognitive and emotional performance and daily sleep-wake cycle in 25 MCI patients.[150]

Another series of 96 MCI outpatients, 61 of who had received daily 3–24 mg of a fast-release melatonin preparation p.o. at bedtime for 15–60 months in comparison to a similar group of 35 MCI patients who did not receive it, was published.[151] Patients treated with melatonin exhibited significantly better performance in mini–mental state examination and the cognitive subscale of the AD assessment scale. After application of a neuropsychological battery comprising a Mattis' test, digit-symbol test, Trail A and B tasks and the Rey's verbal test, better performance was found in melatonin-treated patients for every parameter tested. Abnormally, high Beck Depression Inventory scores decreased in melatonin-treated patients, concomitantly with the improvement in the quality of sleep and wakefulness.[151] These results further supported that melatonin is a useful add-on drug for treating MCI in a clinic environment.

Thus, an early initiation of melatonin treatment can be decisive for therapeutic success. In recent evaluation of the available data in the literature,[7] six double blind, randomized placebo-controlled trials and two open-label retrospective studies (N = 782), consistently showed that the administration of daily evening melatonin improves sleep quality and cognitive performance in MCI patients indicating that melatonin treatment could be effective at early stages of the neurodegenerative disease.

There are two reasons why it is convenient to use melatonin in MCI patients. In the course of the neurodegenerative process, the age-related deterioration in circadian organization becomes significantly exacerbated and is responsible for behavioral problems like sundowning. Age-related cognitive decline in healthy older adults can be predicted by the fragmentation of the circadian rhythm in locomotor behavior. Hence, replacement of the low melatonin levels occurring in brain can be highly convenient in MCI patients. Moreover, the bulk of information on the neuroprotective properties of melatonin derived from experimental studies (see for ref.)[7] indicates that it may be highly desirable to employ pharmacological doses in MCI patients with the aim of arresting or slowing disease's progression.

The mechanisms accounting for the therapeutic effect of melatonin in MCI patients remain to be defined. Melatonin treatment mainly promotes slow wave sleep in the elderly[152] and can therefore be beneficial in MCI by augmenting the restorative phases of sleep, including the augmentation of the secretion of GH, and neurotrophins. Additionally, the antioxidant, mitochondrial, and antiamyloidogenic effects of melatonin can be seen as potentially interfering with the onset of the disease.[7] Therefore, when melatonin treatment begins can be decisive for final response.

One important aspect to be considered is the melatonin dose employed, which may be unnecessarily low when one takes into consideration the binding affinities, half-life and relative potencies of the different melatonin agonists on the market. In addition to being generally more potent than the native molecule, melatonin analogs are employed in considerably higher amounts.[153] Indeed, melatonin has a high safety profile, it is usually remarkably well tolerated and, in some studies, it has been administered to patients at very large doses. Moreover, some of the melatonin agonists would not be expected to have antioxidant effects.

7.11. MELATONIN AND BZD/Z DRUG ABUSE

In aging individuals a combination of altered sleep and sleep pathologies increases the risk of drug-induced insomnia or excessive diurnal somnolence.[154–156] Many old adult patients are treated for longer periods or with higher dosages of hypnotic drugs than are recommended, generally with a lack of individual dosage titration. An ideal hypnotic drug should not produce undesired side effects, such as impairment of memory, cognition, next psychomotor retardation, and day hangover effects, or potentiality of abuse. Melatonin as a chronobiotic fulfills many of these requirements[8] with recent meta-analysis supporting the efficacy of melatonin in primary sleep disorders.[157]

The chronic and extensive use of BZD/Z drugs has become a public health issue and has led to multiple campaigns to reduce both their prescription and consumption. Since melatonin and BZD shared some neurochemical (i.e. interaction with GABA-mediated mechanisms in brain)[158] and behavioral properties (e.g. a similar day-dependent anxiolytic activity),[159] melatonin therapy was postulated as an effective tool to decrease the dose of BZD needed in patients. Two early observations pointed to the possible beneficial effect of melatonin in this respect. One of us reported in an open label study that eight out of 13 insomnia patients either discontinued or reduced BZD use by 50–75% after receiving a 3 mg dose of fast release melatonin.[160] Dagan et al. published a case report on the efficacy of 1 mg of controlled release melatonin to completely cease any BZD use in a 43 year old woman who had suffered from insomnia for the past 11 years.[161]

A double-blind, placebo controlled, study followed by a single blind period enlisted 34 primary insomnia outpatients aged 40–90 years who took BZD and had low urinary 6-sulphatoxy melatonin levels, 14 out of 18 subjects who had received prolonged-release melatonin (2 mg), but only four out of 16 in the placebo group, discontinued BZD therapy.[162] An open label study further supported the efficacy of fast release melatonin to decrease BZD use, that is, 13 out of 20 insomnia patients taking BZD together with melatonin (3 mg) could stop BZD use while another four patients decreased BZD dose to 25–66% of initial doses.[163] In the retrospective analysis of 96 MCI outpatients mentioned above the comparison of the medication profile indicated that 9.8% in the melatonin group received BZD verses 62.8% in the non-melatonin group thus supporting administration of fast release melatonin to decrease BZD use.[151]

A retrospective analysis of a German prescription database identified 512 patients who had initiated treatment with prolonged release melatonin (2 mg) over a 10 month period.[164] From 112 patients in this group who had previously used BZD, 31% discontinued treatment with BZD three months after beginning melatonin treatment. Therefore, melatonin helps to facilitate BZD discontinuation in older insomniacs.

In a study aimed to analyze and evaluate the impact of anti-BZD/Z-drugs campaigns and the availability of alternative pharmacotherapy (melatonin) on the consumption of BZD and Z-drugs in several European countries it was found that campaigns failed when they were not associated with the availability of melatonin in the market.[165] In this pharmacoepidemiological study, the reimbursement of melatonin supports better penetration rates and a higher reduction in sales for BZD/Z-drugs.

A post marketing surveillance study of prolonged release melatonin (2 mg) was performed in Germany. It examined the effect of three weeks of treatment on sleep in 597 patients. Most of the patients (77%) who used traditional hypnotics before melatonin treatment had stopped using them and only 5.6% of naïve patients started such drugs after melatonin discontinuation.[166] A major advantage of melatonin use as a chronobiotic is that it has a very safe profile, it is usually remarkably well tolerated and, in some studies, it has been administered to patients at very large doses

and for long periods of time without any potentiality of abuse.

7.12. CONCLUSIONS

The evidence at present suggests that dysfunctionality in melatonin secretion may play a central role in the pathophysiology of CRSDs. Melatonin has been found useful in treating the disturbed sleep-wake rhythms seen in DSPS, non-24 h sleep-wake rhythm disorder and irregular type of CRSD, as well as in shift-work sleep disorder and jet lag.

A number of melatonin analogs are now in the market that shares the potential activity to treat CRSDs.[167] Ramelteon (Rozerem™) is a MT_1/MT_2 melatonin receptor agonist, synthesized by Takeda Chemical Industries with a half-life much longer (1–2 h) than that of melatonin. Ramelteon acts on both MT_1 and MT_2 melatonergic receptors present in the SCN (for ref. see).[168]

Agomelatine (Valdoxan™, Servier) is a novel antidepressant drug, which acts as both a melatonin MT_1 and MT_2 receptor agonist and as a $5\text{-}HT_{2c}$ antagonist (for ref see).[169] Animal studies indicate that agomelatine accelerates re-entrainment of wheel running activity and a study performed in healthy older men indicated that agomelatine phase-shifts 24 h rhythms of hormonal release and body temperature.[170]

The melatonin MT_1/MT_2 agonist tasimelteon (Hetlioz™, Vanda) was recently introduced in the market to treat non 24 h sleep-wake disorders in blind people. A randomized controlled trial of for transient insomnia after sleep time shift was published.[171] After an abrupt advance in sleep time, tasimelteon improved sleep initiation, and maintenance concurrently with a shift in endogenous circadian rhythms indicating that it may have therapeutic potential for transient insomnia in CRSDs.[171]

Several studies underline the efficacy of melatonin to treat circadian and cognitive symptoms in AD. In this case, however, the neuroprotective activity of melatonin seems not to be mediated by melatonin membrane receptors but rather by the strong antioxidative activity that melatonin and its endogenous metabolites have. So far none of melatonin analogs has been shown to display such a neuroprotective activity.

ACKNOWLEDGMENTS

DPC is a Research Career Awardee from the Argentine Research Council (CONICET). Studies in DPC's Laboratory were supported by a grant from the Agencia Nacional de Promoción Científica y Tecnológica, Argentina (PICT 2012, 0984).

KEY WORDS

- melatonin
- circadian rhythm sleep disorders
- suprachiasmatic nucleus
- Alzheimer's disease
- Benzodiazepine/Z drug abuse

REFERENCES

1. Barion, A.; Zee, P. C. A Clinical Approach to Circadian Rhythm Sleep Disorders. *Sleep Med.* **2007**, *8*, 566–577.

2. Zee, P. C.; Attarian, H.; Videnovic, A. Circadian Rhythm Abnormalities. *Continuum (Minneap Minn).* **2013**, *19*, 132–147.

3. Wheaton, A. G.; Shults, R. A.; Chapman, D. P.; Ford, E. S.; Croft, J. B. Drowsy Driving and Risk Behaviors – 10 States and Puerto Rico, 2011–2012. *MMWR (Morb Mortal Wkly Rep).* **2014**, *63*, 557–562.

4. Lockley, S. W.; Skene, D. J.; Arendt, J.; Tabandeh, H.; Bird, A. C; Defrance, R. Relationship between Melatonin Rhythms and Visual Loss in the Blind. *J Clin Endocrinol Metab.* **1997**, *82*, 3763–3770.

5. Pandi-Perumal, S. R.; Trakht, I.; Spence, D. W.; Srinivasan, V.; Dagan, Y.; Cardinali, D. P. The Roles of Melatonin and Light in the Pathophysiology and Treatment of Circadian Rhythm Sleep Disorders. *Nat Clin Pract Neurol.* **2008**, *4* (8), 436–447.

6. Zhu, L.; Zee, P. C. Circadian Rhythm Sleep Disorders. *Neurol Clin.* **2012**, *30*, 1167–1191.

7. Cardinali, D. P.; Vigo, D, E.; Olivar, N.; Vidal, M. F.; Brusco, L. I. Melatonin Therapy in Patients with Alzheimer's Disease. *Antioxidants.* **2014**, *3*, 245–277.

8. Wilson, S. J.; Nutt, D. J.; Alford, C.; Argyropoulos, S.V.; Baldwin, D. S.; Bateson, A. N.; Britton, T. C.; Crowe, C.; Dijk, D.J.; Espie, C. A.; Gringras, P.; Hajak, G.; Idzikowski, C.; Krystal, A. D.; Nash, J. R.; Selsick, H.; Sharpley, A. L.; Wade, A. G. British Association for Psychopharmacology Consensus Statement on Evidence-Based Treatment of Insomnia, Parasomnias and Circadian Rhythm Disorders. *J Psychopharmacol.* **2010**, *24*, 1577–1601.

9. Moore, R. Y. The Suprachiasmatic Nucleus and the Circadian Timing System. *Prog Mol Biol Transl Sci.* **2013**, *119*, 1–28.

10. Lee, H. S.; Nelms, J. L.; Nguyen, M.; Silver, R.; Lehman, M. N. The Eye is Necessary for a Circadian Rhythm in the Suprachiasmatic Nucleus. *Nat Neurosci.* **2003**, *6*, 111–112.

11. Morin, L. P.; Allen, C. N. The Circadian Visual System, 2005. *Brain Res Rev.* **2006**, *51*, 1–60.

12. Dibner, C.; Schibler, U.; Albrecht, U. The Mammalian Circadian Timing System: Organization and Coordination of Central and Peripheral Clocks. *Annu Rev Physiol.* **2010**, *72*, 517–549.

13. Partch, C. L., Green, C. B.; Takahashi, J. S. Molecular Architecture of the Mammalian Circadian Clock. *Trends Cell Biol.* **2014**, *24*, 90–99.

14. Buhr, E. D.; Takahashi, J. S. Molecular Components of the Mammalian Circadian Clock. *Handb Exp Pharmacol.* **2013**, *217*, 3–27.

15. Reppert, S. M.; Weaver, D. R. Molecular Analysis of Mammalian Circadian Rhythms. *Annu Rev Physiol.* **2001**, *63*, 647–676.

16. Tournier, B. B.; Menet, J. S.; Dardente, H.; Poirel, V. J.; Malan, A.; Masson-Pevet, M.; Pevet, P.; Vuillez, P. Photoperiod Differentially Regulates Clock Genes' Expression in the Suprachiasmatic Nucleus of Syrian Hamster. *Neuroscience.* **2003**, *118*, 317–322.

17. Lincoln, G.; Messager, S.; Andersson, H.; Hazlerigg, D. Temporal Expression of Seven Clock Genes in the Suprachiasmatic Nucleus and the Pars Tuberalis of the Sheep: Evidence for an Internal Coincidence Timer. Proc Natl Acad Sci U S A. **2002**, *99*, 13890–13895.

18. Panda, S.; Antoch, M. P.; Miller, B. H.; Su, A. I.; Schook, A. B.; Straume, M.; Schultz, P. G.; Kay, S. A.; Takahashi, J. S.; Hogenesch, J. B. Coordinated Transcription of Key Pathways in the Mouse by the Circadian Clock. *Cell.* **2002**, *109*, 307–320.

19. Tian, R.; Alvarez-Saavedra, M.; Cheng, H. Y.; Figeys, D. Uncovering the Proteome Response of the Master Circadian Clock to Light Using an AutoProteome System. *Mol Cell Proteomics.* **2011**, *10*, M110.

20. Roenneberg, T.; Merrow, M. Entrainment of the Human Circadian Clock. *Cold Spring Harb Symp Quant Biol.* **2007**, *72*, 293–299.

21. Ebisawa, T. Circadian Rhythms in the CNS and Peripheral Clock Disorders: Human Sleep Disorders and Clock Genes. *J Pharmacol Sci.* **2007**, *103*, 150–154.

22. Archer, S. N.; Robilliard, D. L.; Skene, D. J.; Smits, M.; Williams, A.; Arendt, J.; von Schantz, M. A Length Polymorphism in the Circadian Clock Gene Per3 is Linked to Delayed Sleep Phase Syndrome and Extreme Diurnal Preference. *Sleep.* **2003**, *26*, 413–415.

23. Toh, K. L.; Jones, C. R.; He, Y.; Eide, E. J.; Hinz, W. A.; Virshup, D. M.; Ptacek, L. J.; Fu, Y. H. An hPER2 Phosphorylation Site Mutation in Familial Advanced Sleep Phase Syndrome. *Science.* **2001**, *291*, 1040–1043.

24. Hannibal, J. Roles of PACAP-Containing Retinal Ganglion Cells in Circadian Timing. *Int Rev Cytol.* **2006**, *251*, 1–39.

25. Brainard, G. C.; Hanifin, J. P.; Greeson, J. M.; Byrne, B.; Glickman, G.; Gerner, E.; Rollag, M. D. Action Spectrum for Melatonin Regulation in Humans: Evidence for a Novel Circadian Photoreceptor. *J Neurosci.* **2001**, *21*, 6405–6412.

26. Cajochen, C.; Munch, M.; Kobialka, S.; Krauchi, K.; Steiner, R.; Oelhafen, P.; Orgul, S.; Wirz-Justice, A. High Sensitivity of Human Melatonin, Alertness, Thermoregulation, and Heart Rate to Short Wavelength Light. *J Clin Endocrinol Metab.* **2005**, *90*, 1311–1316.

27. Maronde, E.; Stehle, J. H. The Mammalian Pineal Gland: Known Facts, Unknown Facets. *Trends Endocrinol Metab.* **2007**, *18*, 142–149.

28. Sugden, D; Klein, D. C. Beta-Adrenergic Receptor Control of Rat Pineal Hydroxyindole-O-Methyltransferase. *Endocrinology.* **1983**, *113*, 348–353.

29. Chik, C. L.; Ho, A. K.; Klein, D. C. Alpha 1-Adrenergic Potentiation of Vasoactive Intestinal Peptide Stimulation of Rat Pinealocyte Adenosine 3',5'-Monophosphate and Guanosine 3',5'-Monophosphate: Evidence for a Role of Calcium and Protein Kinase-C. *Endocrinology.* **1988**, *122*, 702–708.

30. Klein, D. C. Arylalkylamine N-acetyltransferase: "the Timezyme". *J Biol Chem.* **2007**, *282*, 4233–4237.

31. Claustrat, B.; Brun, J.; Chazot, G. The Basic Physiology and Pathophysiology of Melatonin. *Sleep Med Rev.* **2005**, *9*, 11–24.

32. Ribelayga, C.; Pevet, P.; Simonneaux, V. HIOMT Drives the Photoperiodic Changes in the Amplitude of the Melatonin Peak of the Siberian Hamster. *Am J Physiol Regul Integr Comp Physiol.* **2000**, *278*, R1339–R1345.

33. Ceinos, R. M.; Chansard, M.; Revel, F.; Calgari, C.; Miguez, J. M.; Simonneaux, V. Analysis of Adrenergic Regulation of Melatonin Synthesis in Siberian Hamster Pineal Emphasizes the Role of HIOMT. *Neurosignals.* **2004**, *13*, 308–317.

34. Liu, T.; Borjigin, J. N-Acetyltransferase is not the Rate-Limiting Enzyme of Melatonin Synthesis at Night. *J Pineal Res.* **2005**, *39*, 91–96.

35. Cardinali, D. P.; Lynch, H. J.; Wurtman, R. J. Binding of Melatonin to Human and Rat Plasma Proteins. *Endocrinology.* **1972**, *91*, 1213–1218.

36. Tricoire, H.; Moller, M.; Chemineau, P.; Malpaux, B. Origin of Cerebrospinal Fluid Melatonin and Possible Function in the Integration of Photoperiod. *Reprod Suppl.* **2003**, *61*, 311–321.

37. Leston, J.; Harthe, C.; Mottolese, C.; Mertens, P.; Sindou, M.; Claustrat, B. Is Pineal Melatonin Released in the Third Ventricle in Humans? A Study in Movement Disorders. *Neurochirurgie.* **2014**, *61*, 85–89.

38. Griefahn, B.; Brode, P.; Remer, T.; Blaszkewicz, M. Excretion of 6-Hydroxymelatonin Sulfate (6-OHMS) in Siblings during Childhood and Adolescence. *Neuroendocrinology.* **2003**, *78*, 241–243.

39. Cardinali, D. P.; Faigon, M. R.; Scacchi, P.; Moguilevsky, J. Failure of Melatonin to Increase Serum Prolactin Levels in Ovariectomized Rats Subjected to Superior Cervical Ganglionectomy or Pinelaectomy. *J Endocrinol.* **1979**, *82*, 315–319.

40. Niles, L. P.; Wong, Y. W.; Mishra, R. K.; Brown, G. M. Melatonin Receptors in Brain. *Eur J Pharmacol.* **1979**, *55*, 219–220.

41. Dubocovich, M. L. Melatonin is a Potent Modulator of Dopamine Release in the Retina. *Nature.* **1983**, *306*, 782–784.

42. Vakkuri, O.; Leppaluoto, J.; Vuolteenaho, O. Development and Validation of a Melatonin Radioimmunoassay using Radioiodinated Melatonin as Tracer. *Acta Endocrinol (Copenh).* **1984**, *106*, 152–157.

43. Morgan, P. J.; Barrett, P.; Howell, H. E.; Helliwell, R. Melatonin Receptors: Localization, Molecular Pharmacology and Physiological Significance. *Neurochem Int.* **1994**, *24*, 101–146.

44. Reppert, S. M.; Weaver, D. R.; Ebisawa, T. Cloning and Characterization of a Mammalian Melatonin Receptor that Mediates Reproductive and Circadian Responses. *Neuron.* **1994**, *13*, 1177–1185.

45. Reppert, S.M.; Godson, C.; Mahle, C.D.; Weaver, D. R.; Slaugenhaupt, S. A.; Gusella, J. F. Molecular Characterization of a Second Melatonin Receptor Expressed in Human Retina and Brain: The Mel1b Melatonin Receptor. *Proc Natl Acad Sci U S A.* **1995**, *92*, 8734–8738.

46. Batailler, M.; Mullier, A.; Sidibe, A.; Delagrange, P.; Prevot, V.; Jockers, R.; Migaud, M. Neuroanatomical Distribution of the Orphan GPR50 Receptor in Adult Sheep and Rodent Brains. *J Neuroendocrinol.* **2012**, *24*, 798–808.

47. Dufourny, L.; Levasseur, A.; Migaud, M.; Callebaut, I.; Pontarotti, P.; Malpaux, B.; Monget, P. GPR50 is the Mammalian Ortholog of Mel1c: Evidence of Rapid Evolution in Mammals. *BMC Evol Biol.* **1999**, *8*, 105.

48. Daulat, A. M.; Maurice, P.; Froment, C.; Guillau, J. L.; Broussard, C.; Monsarrat, B.; Delagrange, P.; Jockers, R. Purification and Identification of G Protein-Coupled Receptor Protein Complexes under Native Conditions. *Mol Cell Proteomics.* **2007**, *6*, 835–844.

49. Levoye, A.; Dam, J.; Ayoub, M. A.; Guillaume, J. L.; Couturier, C.; Delagrange, P.; Jockers, R. The Orphan Gpr50 Receptor Specifically Inhibits Mt1 Melatonin Receptor Function through Heterodimerization. *EMBO J.* **2006**, *25*, 3012–3023.

50. Levoye, A.; Jockers, R.; Ayoub, M. A.; Delagrange, P.; Savaskan, E.; Guillaume, J. L. Are G Protein-Coupled Receptor Heterodimers of Physiological Relevance?--Focus on Melatonin Receptors. *Chronobiol Int.* **2006**, *23*, 419–426.

51. Bechtold, D. A.; Sidibe, A.; Saer, B. R.; Li, J.; Hand, L. E.; Ivanova, E. A.; Darras, V. M.; Dam, J.; Jockers, R.; Luckman, S. M.; Loudon, A. S. A Role for the Melatonin-Related Receptor GPR50 in Leptin Signaling, Adaptive Thermogenesis, and Torpor. *Curr Biol.* **2012**, *22*, 70–77.

52. Pickering, D. S.; Niles, L. P. Pharmacological Characterization of Melatonin Binding Sites in Syrian Hamster Hypothalamus. *Eur J Pharmacol.* **1990**, *175*, 71–77.

53. Dubocovich, M. L.; Delagrange, P.; Krause, D. N.; Sugden, D.; Cardinali, D. P.; Olcese, J. International Union of Basic and Clinical Pharmacology. LXXV. Nomenclature, Classification, and Pharmacology of G Protein-Coupled Melatonin Receptors. *Pharmacol Rev.* **2010**, *62*, 343–380.

54. Nosjean, O.; Ferro, M.; Coge, F.; Beauverger, P.; Henlin, J. M.; Lefoulon, F.; Fauchere, J. L.; Delagrange, P.; Canet, E.; Boutin, J. A. Identification of the Melatonin-Binding Site Mt3 as the Quinone Reductase 2. *J Biol Chem.* **2000**, *275*, 31311–31317.

55. Audinot ,V.; Bonnaud, A.; Grandcolas, L.; Rodriguez, M.; Nagel, N.; Galizzi, J. P.; Balik, A.; Messager, S.; Hazlerigg, D. G.; Barrett, P.; Delagrange, P.; Boutin, J. A. Molecular Cloning and Pharmacological Characterization of Rat Melatonin Mt1 and Mt2 Receptors. *Biochem Pharmacol.* **2008**, *75*, 2007–2019.

56. Wan, Q.; Man, H. Y.; Liu, F.; Braunton, J.; Niznik, H. B.; Pang, S. F.; Brown, G. M.; Wang, Y. T. Differential Modulation of GABAA Receptor Function by Mel1a and Mel1b Receptors. *Nat Neurosci.* **1999**, *2*, 401–403.

57. Wu, Y. H.; Zhou, J. N.; Balesar, R.; Unmehopa, U.; Bao, A.; Jockers, R.; van, H. J.; Swaab, D. F. Distribution of Mt1 Melatonin Receptor Immunoreactivity in the Human Hypothalamus and Pituitary Gland: Colocalization of Mt1 With

Vasopressin, Oxytocin, and Corticotropin-Releasing Hormone. *J Comp Neurol.* **2006**, *499*, 897–910.

58. Mazzucchelli, C.; Pannacci, M.; Nonno, R.; Lucini, V.; Fraschini, F.; Stankov, B. M. The Melatonin Receptor in the Human Brain: Cloning Experiments and Distribution Studies. *Brain Res Mol Brain Res.* **1996**, *39*, 117–126.

59. Savaskan, E.; Ayoub, M. A.; Ravid, R.; Angeloni, D.; Fraschini, F.; Meier, F.; Eckert, A.; Muller-Spahn, F.; Jockers, R. Reduced Hippocampal MT2 Melatonin Receptor Expression in Alzheimer's Disease. *J Pineal Res.* **2005**, *38*, 10–16.

60. Brunner, P.; Sozer-Topcular, N.; Jockers, R.; Ravid, R.; Angeloni, D.; Fraschini, F.; Eckert, A.; Muller-Spahn, F.; Savaskan, E. Pineal and Cortical Melatonin Receptors MT1 and MT2 are Decreased in Alzheimer's Disease. *Eur J Histochem.* **2006**, *50*, 311–316.

61. Uz, T.; Arslan, A. D.; Kurtuncu, M.; Imbesi, M.; Akhisaroglu, M.; Dwivedi, Y.; Pandey, G. N.; Manev, H. The Regional and Cellular Expression Profile of the Melatonin Receptor MT1 in the Central Dopaminergic System. *Brain Res Mol Brain Res.* **2005**, *136*, 45–53.

62. Hardeland, R.; Cardinali, D. P.; Srinivasan, V.; Spence, D. W.; Brown, G. M.; Pandi-Perumal, S. R. Melatonin--A Pleiotropic, Orchestrating Regulator Molecule. *Prog Neurobiol.* **2011**, *93*, 350–384.

63. Carlberg, C. Gene Regulation by Melatonin. *Ann N Y Acad Sci.* **2000**, *917*, 387–396.

64. Steinhilber, D.; Brungs, M.; Werz, O.; Wiesenberg, I.; Danielsson, C.; Kahlen, J. P.; Nayeri, S.; Schrader, M.; Carlberg, C. The Nuclear Receptor for Melatonin Represses 5-Lipoxygenase Gene Expression in Human B Lymphocytes. *J Biol Chem.* **1995**, *270*, 7037–7040.

65. Kobayashi, H.; Kromminga, A.; Dunlop, T. W.; Tychsen, B.; Conrad, F.; Suzuki, N.; Memezawa, A.; Bettermann, A.; Aiba, S.; Carlberg, C.; Paus, R. A Role of Melatonin in Neuroectodermal-Mesodermal Interactions: The Hair Follicle Synthesizes Melatonin and Expresses Functional Melatonin Receptors. *FASEB J.* **2005**, *19*, 1710–1712.

66. Dawson, D.; Armstrong, S. M. Chronobiotics-Drugs that Shift Rhythms. *Pharmacol Ther.* **1996**, *69*, 15–36.

67. Arendt, J.; Skene, D. J. Melatonin as a Chronobiotic. *Sleep Med Rev.* **2005**, *9*, 25–39.

68. Cardinali, D. P.; Furio, A. M.; Reyes, M. P.; Brusco, L. I. The Use of Chronobiotics in the Resynchronization of the Sleep-Wake Cycle. *Cancer Causes Control.* **2006**, *17*, 601–609.

69. Touitou, Y.; Bogdan, A. Promoting Adjustment of the Sleep-Wake Cycle by Chronobiotics. *Physiol Behav.* **2007**, *90*, 294–300.

70. Lewy, A. J.; Ahmed, S.; Jackson, J. M.; Sack, R. L. Melatonin Shifts Human Circadian Rhythms According to a Phase-Response Curve. *Chronobiol Int.* **1992**, *9*, 380–392.

71. Lewy, A. J.; Sack, R. L. The Role of Melatonin and Light in the Human Circadian System. *Prog Brain Res.* **1996**, *111*, 205–216.

72. Burgess, H. J.; Revell, V. L.; Eastman, C. I. A Three Pulse Phase Response Curve to Three Milligrams of Melatonin in Humans. *J Physiol.* **2008**, *586*, 639–647.

73. Revell, V. L.; Burgess, H. J.; Gazda, C. J.; Smith, M. R.; Fogg, L. F.; Eastman, C. I. Advancing Human Circadian Rhythms with Afternoon Melatonin and Morning Intermittent Bright Light. *J Clin Endocrinol Metab.* **2006**, *91*, 54–59.

74. Figueiro, M. G.; Plitnick, B.; Rea, M. S. Pulsing Blue Light through Closed Eyelids: Effects on Acute Melatonin Suppression and Phase Shifting of Dim Light Melatonin Onset. *Nat Sci Sleep.* **2014**, *6*, 149–156.

75. Rahman, S. A.; Shapiro, C. M.; Wang, F.; Ainlay, H.; Kazmi, S.; Brown, T. J.; Casper, R. F. Effects of Filtering Visual Short Wavelengths during Nocturnal Shiftwork on Sleep and Performance. *Chronobiol Int.* **2013**, *30*, 951–962.

76. Poirel, V. J.; Boggio, V.; Dardente, H.; Pevet, P.; Masson-Pevet, M.; Gauer, F. Contrary to other Non-Photic Cues, Acute Melatonin Injection Does not Induce Immediate Changes of Clock Gene mRNA Expression in the Rat Suprachiasmatic Nuclei. *Neuroscience.* **2003**, *120*, 745–755.

77. Vriend, J.; Reiter, R. J. Melatonin Feedback on Clock Genes: A Theory Involving the Proteasome. *J Pineal Res.* **2015**, *58*, 1–11.

78. Lewy, A. J. Melatonin and Human Chronobiology. *Cold Spring Harb Symp Quant Biol.* **2007**, *72*, 623–636.

79. Achermann, P.; Borbely, A. A. Mathematical Models of Sleep Regulation. *Front Biosci.* **2003**, *8*, s683–s693.

80. Dijk, D. J.; von, S. M. Timing and Consolidation of Human Sleep, Wakefulness, and Performance

by a Symphony of Oscillators. *J Biol Rhythms.* **2005**, *20*, 279–290.

81. Scheer, F. A.; Wright, K. P. Jr.; Kronauer, R. E.; Czeisler, C. A. Plasticity of the Intrinsic Period of the Human Circadian Timing System. *PLoS One.* **2007**, *2*, e721.

82. Saper, C. B. The Neurobiology of Sleep. *Continuum (Minneap Minn).* **2013**, *19*, 19–31.

83. Edgar, D. M.; Dement, W. C.; Fuller, C. A. Effect of SCN Lesions on Sleep in Squirrel Monkeys: Evidence for Opponent Processes in Sleep-Wake Regulation. *J Neurosci.* **1993**, *13*, 1065–1079.

84. Lewy, A. J.; Emens, J.; Jackman, A.; Yuhas, K. Circadian Uses of Melatonin in Humans. *Chronobiol Int.* **2006**, *23*, 403–412.

85. Liu, C.; Weaver, D. R.; Jin, X.; Shearman, L. P.; Pieschl, R. L.; Gribkoff, V. K.; Reppert, S. M. Molecular Dissection of two Distinct Actions of Melatonin on the Suprachiasmatic Circadian Clock. *Neuron.* **1997**, *19*, 91–102.

86. Mason, R.; Brooks, A. The Electrophysiological Effects of Melatonin and a Putative Melatonin Antagonist (N-Acetyltryptamine) on Rat Suprachiasmatic Neurones in Vitro. *Neurosci Lett.* **1988**, *95*, 296–301.

87. Pandi-Perumal, S. R.; Smits, M.; Spence, W.; Srinivasan, V.; Cardinali, D. P.; Lowe, A. D.; Kayumov, L. Dim Light Melatonin Onset (DLMO): A Tool for the Analysis of Circadian Phase in Human Sleep and Chronobiological Disorders. *Prog Neuropsychopharmacol Biol Psychiatry.* **2007**, *31*, 1–11.

88. Keijzer, H.; Smits, M. G.; Duffy, J. F.; Curfs, L. M. Why the Dim Light Melatonin Onset (DLMO) should be Measured before Treatment of Patients with Circadian Rhythm Sleep Disorders. *Sleep Med Rev.* **2014**, *18*, 333–339.

89. Gradisar, M. Crowley, S. J. Delayed Sleep Phase Disorder in Youth. *Curr Opin Psychiatry.* **2013**, *26*, 580–585.

90. Micic, G.; de, B. A.; Lovato, N.; Wright, H.; Gradisar, M.; Ferguson, S.; Burgess, H. J.; Lack, L. The Endogenous Circadian Temperature Period Length (Tau) in Delayed Sleep Phase Disorder Compared to Good Sleepers. *J Sleep Res.* **2013**, *22*, 617–624.

91. Dagan, Y.; Eisenstein, M.; Circadian Rhythm Sleep Disorders: Toward a More Precise Definition and Diagnosis. *Chronobiol Int.* **1999**, *16*, 213–222.

92. Ayalon, L.; Borodkin, K.; Dishon, L.; Kanety, H.; Dagan, Y. Circadian Rhythm Sleep Disorders Following Mild Traumatic Brain Injury. *Neurology.* **2007**, *68*, 1136–1140.

93. Nagtegaal, J. E.; Kerkhof, G. A.; Smits, M. G.; Swart, A. C.; van der Meer, Y. G. Traumatic Brain Injury-Associated Delayed Sleep Phase Syndrome. *Funct Neurol.* **1997**, *12*, 345–348.

94. Bjorvatn, B.; Pallesen, S. A Practical Approach to Circadian Rhythm Sleep Disorders. *Sleep Med Rev.* **2009**, *13*, 47–60.

95. Regestein, Q. R.; Pavlova, M. Treatment of Delayed Sleep Phase Syndrome. *Gen Hosp Psychiatry.* **1995**, *17*, 335–345.

96. Campos, C. I.; Nogueira, C. H.; Fernandes, L. Aging, Circadian Rhythms and Depressive Disorders: A Review. *Am J Neurodegener Dis.* **2013**, *2*, 228–246.

97. Dahlitz, M.; Alvarez, B.; Vignau, J.; English, J.; Arendt, J.; Parkes, J. D. Delayed Sleep Phase Syndrome Response to Melatonin. *Lancet.* **1991**, *337*, 1121–1124.

98. Alvarez, B.; Dahlitz, M. J.; Vignau, J.; Parkes, J. D. The Delayed Sleep Phase Syndrome: Clinical and Investigative Findings in 14 Subjects. *J Neurol Neurosurg Psychiatry.* **1992**, *55*, 665–670.

99. Bartlett, D. J.; Biggs, S. N.; Armstrong, S. M. Circadian Rhythm Disorders among Adolescents: Assessment and Treatment Options. *Med J Aust.* **2013**, *199*, S16–S20.

100. Braam, W.; Smits, M. G.; Didden, R.; Korzilius, H.; van Geijlswijk, I. M.; Curfs, L. M. Exogenous Melatonin for Sleep Problems in Individuals with Intellectual Disability: A Meta-Analysis. *Dev Med Child Neurol.* **2009**, *51*, 340–349.

101. Eckerberg, B.; Lowden, A.; Nagai, R.; Akerstedt, T. Melatonin Treatment Effects on Adolescent Students' Sleep Timing and Sleepiness in a Placebo-Controlled Crossover Study. *Chronobiol Int.* **2012**, *29*, 1239–1248.

102. Rahman, S. A.; Kayumov, L.; Shapiro, C. M. Antidepressant Action of Melatonin in the Treatment of Delayed Sleep Phase Syndrome. *Sleep Med.* **2010**, *11*, 131–136.

103. Wilhelmsen-Langeland, A.; Saxvig, I. W.; Pallesen, S.; Nordhus, I. H.; Vedaa, O.; Lundervold, A. J.; Bjorvatn, B. A Randomized Controlled Trial with Bright Light and Melatonin for the Treatment of Delayed Sleep Phase Disorder: Effects on Subjective and Objective Sleepiness and Cognitive Function. *J Biol Rhythms.* **2013**, *28*, 306–321.

104. Bruni, O.; Alonso-Alconada, D.; Besag, F.; Biran, V.; Braam, W.; Cortese, S.; Moavero, R.; Parisi, P.; Smits, M.; Van der Heijden, K.; Curatolo, P. Current Role of Melatonin in Pediatric Neurology: Clinical Recommendations. *Eur J Paediatr Neurol.* **2015**, *19*, 122–133.

105. Lewy, A. J.; Sack, R. L.; Miller, L. S.; Hoban, T. M.; Singer, C. M.; Samples, J. R.; Krauss, G. L. The Use of Plasma Melatonin Levels and Light in the Assessment and Treatment of Chronobiologic Sleep and Mood Disorders. *J Neural Transm Suppl.* **1986**, *21*, 311–322.

106. Morgenthaler, T. I.; Lee-Chiong, T.; Alessi, C.; Friedman, L.; Aurora, R. N.; Boehlecke, B.; Brown, T.; Chesson, A. L. Jr.; Kapur, V.; Maganti, R.; Owens, J.; Pancer, J.; Swick, T. J.; Zak, R. Practice Parameters for the Clinical Evaluation and Treatment of Circadian Rhythm Sleep Disorders. An American Academy of Sleep Medicine Report. *Sleep.* **2007**, *30*, 1445–1459.

107. van Geijlswijk, I. M.; Korzilius, H. P.; Smits, M. G. The Use of Exogenous Melatonin in Delayed Sleep Phase Disorder: A Meta-Analysis. *Sleep.* **2010**, *33*, 1605–1614.

108. Singletary, K. G.; Naidoo, N. Disease and Degeneration of Aging Neural Systems that Integrate Sleep Drive and Circadian Oscillations. *Front Neurol.* **2011**, *2*, 66.

109. Leger, D.; Laudon, M.; Zisapel, N. Nocturnal 6-Sulfatoxymelatonin Excretion in Insomnia and its Relation to the Response to Melatonin Replacement Therapy. *Am J Med.* **2004**, *116*, 91–95.

110. Hida, A.; Kusanagi, H.; Satoh, K.; Kato, T.; Matsumoto, Y.; Echizenya, M.; Shimizu, T.; Higuchi, S.; Mishima, K. Expression Profiles of PERIOD1, 2, and 3 in Peripheral Blood Mononuclear Cells from Older Subjects. *Life Sci.* **2009**. *84*, 33–37.

111. Xu, Y.; Padiath, Q. S.; Shapiro, R. E.; Jones, C. R.; Wu, S. C.; Saigoh, N.; Saigoh, K.; Ptacek, L. J.; Fu, Y. H. Functional Consequences of a Ckidelta Mutation Causing Familial Advanced Sleep Phase Syndrome. *Nature.* **2005**, *434*, 640–644.

112. Archer, S. N.; Carpen, J. D.; Gibson, M.; Lim, G. H.; Johnston, J. D.; Skene, D. J.; Von, S. M. Polymorphism in the PER3 Promoter Associates with Diurnal Preference and Delayed Sleep Phase Disorder. *Sleep.* **2010**, *33*, 695–701.

113. Srinivasan, V.; Singh, J.; Pandi-Perumal, S. R.; Brown, G. M.; Spence, D. W.; Cardinali, D. P.

Jet Lag, Circadian Rhythm Sleep Disturbances, and Depression: The Role of Melatonin and its Analogs. *Adv Ther.* **2010**, *27*, 796–813.

114. Brown, G. M.; Pandi-Perumal, S. R.; Trakht, I.; Cardinali, D. P. Melatonin and its Relevance to Jet Lag. *Travel Med Infect Dis.* **2009**, *7*, 69–81.

115. Cardinali, D. P.; Bortman, G. P.; Liotta, G.; Perez, L. S.; Albornoz, L. E.; Cutrera, R. A.; Batista, J.; Ortega, G. P. A Multifactorial Approach Employing Melatonin to Accelerate Resynchronization of Sleep-Wake Cycle After a 12 Time-Zone Westerly Transmeridian Flight in Elite Soccer Athletes. *J Pineal Res.* **2002**, *32*, 41–46.

116. Czeisler, C. A.; Kronauer, R. E.; Allan, J. S.; Duffy, J. F.; Jewett, M. E.; Brown, E. N.; Ronda, J. M. Bright Light Induction of Strong (Type 0) Resetting of the Human Circadian Pacemaker. *Science.* **1989**, *244*, 1328–1333.

117. Wright, K. P., Jr.; Bogan, R. K.; Wyatt, J. K. Shift Work and the Assessment and Management of Shift Work Disorder (SWD). *Sleep Med Rev.* **2013**, *17*, 41–54.

118. Morrissette, D. A. Twisting the Night Away: A Review of the Neurobiology, Genetics, Diagnosis, and Treatment of Shift Work Disorder. *CNS Spectr.* **2013**, *18* Suppl 1, 45–53.

119. Gumenyuk, V.; Howard, R.; Roth, T.; Korzyukov, O.; Drake, C. L. Sleep Loss, Circadian Mismatch, and Abnormalities in Reorienting of Attention in Night Workers with Shift Work Disorder. *Sleep.* **2014**, *37*, 545–556.

120. Rajaratnam, S. M.; Howard, M. E.; Grunstein, R. R. Sleep Loss and Circadian Disruption in Shift Work: Health Burden and Management. *Med J Aust.* **2013**, *199*, S11–S15.

121. Herichova, I. Changes of Physiological Functions Induced by Shift Work. *Endocr Regul.* **2013**, *47*, 159–170.

122. Czeisler, C. A.; Johnson, M. P.; Duffy, J. F.; Brown, E. N.; Ronda, J. M.; Kronauer, R. E. Exposure to Bright Light and Darkness to Treat Physiologic Maladaptation to Night Work. *N Engl J Med.* **1990**, *322*, 1253–1259.

123. Folkard, S.; Arendt, J.; Clark, M. Can Melatonin Improve Shift Workers' Tolerance of the Night Shift? Some Preliminary Findings. *Chronobiol Int.* **1993**, *10*, 315–320.

124. Quera-Salva, M. A.; Guilleminault, C.; Claustrat, B.; Defrance, R.; Gajdos, P.; McCann, C. C.; De, L. J. Rapid Shift in Peak Melatonin Secretion Associated with Improved Performance

in Short Shift Work Schedule. *Sleep.* **1997**, *20,* 1145–1150.

125. Rajaratnam, S. M.; Dijk, D. J. Middleton, B.; Stone, B. M.; Arendt, J. Melatonin Phase-Shifts Human Circadian Rhythms with no Evidence of Changes in the Duration of Endogenous Melatonin Secretion or the 24-Hour Production of Reproductive Hormones. *J Clin Endocrinol Metab.* **2003**, *88,* 4303–4309.

126. Liira, J.; Verbeek, J. H.; Costa, G.; Driscoll, T. R.; Sallinen, M.; Isotalo, L. K.; Ruotsalainen, J. H. Pharmacological Interventions for Sleepiness and Sleep Disturbances Caused by Shift Work. *Cochrane Database Syst Rev.* **2014**, *8,* CD009776.

127. Flynn-Evans, E. E.; Tabandeh, H.; Skene, D. J.; Lockley, S. W. Circadian Rhythm Disorders and Melatonin Production in 127 Blind Women with and without Light Perception. *J Biol Rhythms.* **2014**, *29,* 215–224.

128. Skene, D. J.; Arendt, J. Circadian Rhythm Sleep Disorders in the Blind and their Treatment with Melatonin. *Sleep Med.* **2007**, *8,* 651–655.

129. Emens, J. S.; Laurie, A. L.; Songer, J. B.; Lewy, A. J. Non-24-Hour Disorder in Blind Individuals Revisited: Variability and the Influence of Environmental Time Cues. *Sleep.* **2013**, *36,* 1091–1100.

130. Yazaki, M.; Shirakawa, S.; Okawa, M.; Takahashi, K. Demography of Sleep Disturbances Associated with Circadian Rhythm Disorders in Japan. *Psychiatry Clin Neurosci.* **1999**, *53,* 267–268.

131. Schrader, H.; Bovim, G.; Sand, T. The Prevalence of Delayed and Advanced Sleep Phase Syndromes. *J Sleep Res.* **1993**, *2,* 51–55.

132. Ayalon, L.; Hermesh, H.; Dagan, Y. Case Study of Circadian Rhythm Sleep Disorder Following Haloperidol Treatment: Reversal by Risperidone and Melatonin. *Chronobiol Int.* **2002**, *19,* 947–959.

133. Borodkin, K.; Ayalon, L.; Kanety, H.; Dagan, Y. Dysregulation of Circadian Rhythms Following Prolactin-Secreting Pituitary Microadenoma. *Chronobiol Int.* **2005**, *22,* 145–156.

134. Doljansky, J. T.; Kannety, H.; Dagan, Y. Working under Daylight Intensity Lamp: An Occupational Risk for Developing Circadian Rhythm Sleep Disorder? *Chronobiol Int.* **2005**, *22,* 597–605.

135. Slats, D.; Claassen, J. A.; Verbeek, M. M.; Overeem, S. Reciprocal Interactions between Sleep, Circadian Rhythms and Alzheimer's Disease: Focus on the Role of Hypocretin and Melatonin. *Ageing Res Rev.* **2013**, *12,* 188–200.

136. Wu, Y. H.; Feenstra, M. G.; Zhou, J. N.; Liu, R. Y.; Torano, J. S.; Van Kan, H. J.; Fischer, D. F.; Ravid, R.; Swaab, D. F. Molecular Changes Underlying Reduced Pineal Melatonin Levels in Alzheimer Disease: Alterations in Preclinical and Clinical Stages. *J Clin Endocrinol Metab.* **2003**, *88,* 5898–5906.

137. Meeks, T. W.; Ropacki, S. A.; Jeste, D. V. The Neurobiology of Neuropsychiatric Syndromes in Dementia. *Curr Opin Psychiatry.* **2006**, *19,* 581–586.

138. Brusco, L. I.; Marquez, M.; Cardinali, D. P. Monozygotic Twins with Alzheimer's Disease Treated with Melatonin: Case Report. *J Pineal Res.* **1998**, *25,* 260–264.

139. Brusco, L. I.; Marquez, M.; Cardinali, D. P. Melatonin Treatment Stabilizes Chronobiologic and Cognitive Symptoms in Alzheimer's Disease. *Neuro Endocrinol Lett.* **1998**, *19,* 111–115.

140. Cohen-Mansfield, J.; Garfinkel, D.; Lipson, S. Melatonin for Treatment of Sundowning in Elderly Persons with Dementia - A Preliminary Study. *Arch Gerontol Geriatr.* **2000**, *31,* 65–76.

141. Cardinali, D. P.; Brusco, L. I.; Liberczuk, C.; Furio, A. M. The Use of Melatonin in Alzheimer's Disease. *Neuro Endocrinol Lett.* **2002**, *23* Suppl 1, 20–23.

142. Mahlberg, R.; Kunz, D.; Sutej, I.; Kuhl, K. P.; Hellweg, R. Melatonin Treatment of Day-Night Rhythm Disturbances and Sundowning in Alzheimer Disease: An Open-Label Pilot Study Using Actigraphy. *J Clin Psychopharmacol.* **2004**, *24,* 456–459.

143. Asayama, K.; Yamadera, H.; Ito, T.; Suzuki, H.; Kudo, Y.; Endo, S. Double Blind Study of Melatonin Effects on the Sleep-Wake Rhythm, Cognitive and Non-Cognitive Functions in Alzheimer Type Dementia. *J Nippon Med Sch.* **2003**, *70,* 334–341.

144. Singer, C.; Tractenberg, R. E.; Kaye, J.; Schafer, K.; Gamst, A.; Grundman, M.; Thomas, R.; Thal, L. J. A Multicenter, Placebo-Controlled Trial of Melatonin for Sleep Disturbance in Alzheimer's Disease. *Sleep.* **2003**, *26,* 893–901.

145. Dowling, G. A.; Burr, R. L.; Van Someren, E. J.; Hubbard, E. M.; Luxenberg, J. S.; Mastick, J.; Cooper, B. A. Melatonin and Bright-Light Treatment for Rest-Activity Disruption in Insti-

tutionalized Patients with Alzheimer's Disease. *J Am Geriatr Soc.* **2008**, *56*, 239–246.

146. Riemersma-van der Lek, R. F.; Swaab, D. F.; Twisk, J.; Hol, E. M.; Hoogendijk, W. J.; van Someren, E. J. Effect of Bright Light and Melatonin on Cognitive and Noncognitive Function in Elderly Residents of Group Care Facilities: A Randomized Controlled Trial. *JAMA.* **2008**, *299*, 2642–2655.

147. Gehrman. P. R.; Connor, D. J.; Martin, J. L.; Shochat, T.; Corey-Bloom, J.; Ancoli-Israel, S. Melatonin Fails to Improve Sleep or Agitation in Double-Blind Randomized Placebo-Controlled Trial of Institutionalized Patients with Alzheimer Disease. *Am J Geriatr Psychiatry.* **2009**, *17*, 166–169.

148. Gauthier, S.; Reisberg, B.; Zaudig, M.; Petersen, R. C.; Ritchie, K.; Broich, K.; Belleville, S. Brodaty, H. Bennett, D. Chertkow, H. Cummings, J. L.; de Leon, M.; Feldman, H.; Ganguli, M.; Hampel, H.; Scheltens, P.; Tierney, M. C.; Whitehouse, P.; Winblad, B. Mild Cognitive Impairment. *Lancet.* **2006**, *367*, 1262–1270.

149. Naismith, S. L.; Hickie, I. B.; Terpening, Z.; Rajaratnam, S. M.; Hodges, J. R.; Bolitho, S.; Rogers, N. L.; Lewis, S. J. Circadian Misalignment and Sleep Disruption in Mild Cognitive Impairment. *J Alzheimers Dis.* **2014**, *38*, 857–866.

150. Furio, A. M.; Brusco, L. I.; Cardinali, D. P. Possible Therapeutic Value of Melatonin in Mild Cognitive Impairment. A Retrospective Study. *Journal of Pineal Research.* **2007**, *43* (4), 404–409.

151. Cardinali, D. P.; Vigo, D. E.; Olivar, N.; Vidal, M. F.; Furio, A. M.; Brusco, L. I. Therapeutic Application of Melatonin in Mild Cognitive Impairment. *Am J Neurodegener Dis.* **2012**, *1*, 280 291.

152. Monti, J. M.; Alvarino, F.; Cardinali, D. P.; Savio, I.; Pintos, A. Polysomnographic Study of the Effect of Melatonin on Sleep in Elderly Patients with Chronic Primary Insomnia. *Arch Gerontol Geriatr.* **1999**, *28*, 85–98.

153. Cardinali, D. P.; Srinivasan, V.; Brzezinski, A.; Brown, G. M. Melatonin and its Analogs in Insomnia and Depression. *J Pineal Res.* **2012**, *52*, 365–375.

154. Neikrug, A. B.; Ancoli-Israel, S. Sleep Disorders in the Older Adult - A Mini-Review. *Gerontology.* **2010**, *56*, 181–189.

155. Wolkove, N.; Elkholy, O.; Baltzan, M.; Palayew, M. Sleep and Aging: 1. Sleep Disorders Commonly Found in Older People. *CMAJ.* **2007**, *176*, 1299–1304.

156. Wolkove, N.; Elkholy, O.; Baltzan, M.; Palayew, M. Sleep and Aging: 2. Management of Sleep Disorders in Older People. *CMAJ.* **2007**, *176*, 1449–1454.

157. Ferracioli-Oda, E.; Qawasmi, A.; Bloch, M. H. Meta-Analysis: Melatonin for the Treatment of Primary Sleep Disorders. *PLoS One.* **2013**, *8*, e63773.

158. Cardinali, D. P.; Pandi-Perumal, S. R.; Niles, L. P. Melatonin and its Receptors: Biological Function in Circadian Sleep-Wake Regulation; in Monti, J. M.; Pandi-Perumal, S. R.; Sinton, C. M., (Eds): Neurochemistry of Sleep and Wakefulness. Cambridge UK, Cambridge University Press, **2008**, pp 283–314.

159. Golombek, D. A.; Pevet, P.; Cardinali, D. P. Melatonin Effects on Behavior: Possible Mediation by the Central Gabaergic System. *Neurosci Biobehav Rev.* **1996**, *20*, 403–412.

160. Fainstein, I.; Bonetto, A.; Brusco, L. I.; Cardinali, D. P. Effects of Melatonin in Elderly Patients with Sleep Disturbance. A Pilot Study. *Curr Ther Res.* **1997**, *58*, 990–1000.

161. Dagan, Y.; Zisapel, N.; Nof, D.; Laudon, M.; Atsmon, J. Rapid Reversal of Tolerance to Benzodiazepine Hypnotics by Treatment with Oral Melatonin: A Case Report. *Eur Neuropsychopharmacol.* **1997**, *7*, 157–160.

162. Garfinkel, D.; Zisapel, N.; Wainstein, J.; Laudon, M.: Facilitation of Benzodiazepine Discontinuation by Melatonin: A New Clinical Approach. *Arch Intern Med.* **1999**, *159*, 2456–2460.

163. Siegrist, C.; Benedetti, C.; Orlando, A.; Beltran, J. M.; Tuchscherr, L.; Noseda, C. M.; Brusco, L. I.; Cardinali, D. P. Lack of Changes in Serum Prolactin, FSH, TSH, and Estradiol after Melatonin Treatment in Doses that Improve Sleep and Reduce Benzodiazepine Consumption in Sleep-Disturbed, Middle-Aged, and Elderly Patients. *J Pineal Res.* **2001**, *30*, 34–42.

164. Kunz, D.; Bineau, S.; Maman, K.; Milea, D.; Toumi, M. Benzodiazepine Discontinuation with Prolonged-Release Melatonin: Hints from a German Longitudinal Prescription Database. *Expert Opin Pharmacother.* **2012**, *13*, 9–16.

165. Clay, E.; Falissard, B.; Moore, N.; Toumi, M. Contribution of Prolonged-Release Melatonin and Anti-Benzodiazepine Campaigns to the Reduction of Benzodiazepine and Z-Drugs

Consumption in Nine European Countries. *Eur J Clin Pharmacol.* **2013**, *69*, 1–10.

166. Hajak, G.; Lemme, K.; Zisapel, N. Lasting Treatment Effects in a Postmarketing Surveillance Study of Prolonged-Release Melatonin. *Int Clin Psychopharmacol.* **2015**, *30*, 36–42.

167. Cardinali, D. P.; Pandi-Perumal, S. R.; Srinivasan, V.; Spence, D. W.; Trakht I: Therapeutic Potential of Melatonin Agonists. *Expert Rev Endocrinol Metab.* **2008**, *3* (2), 269–279.

168. Pandi-Perumal, S. R.; Srinivasan, V.; Spence, D. W.; Moscovitch, A.; Hardeland, R.; Brown, G. M.; Cardinali, D. P. Ramelteon: A Review of its Therapeutic Potential in Sleep Disorders. *Adv Ther.* **2009**, *26*, 613–626.

169. Cardinali, D. P.; Vidal, M. F.; Vigo, D. E. Agomelatine: Its Role in the Management of Major Depressive Disorder. *Clinical Medicine Insights: Psychiatry.* **2012**, *4*, 1–23.

170. Leproult, R.; Van, O. A.; L'Hermite-Baleriaux, M.; Van, C. E.; Copinschi, G. Phase-Shifts of 24-H Rhythms of Hormonal Release and Body Temperature Following Early Evening Administration of the Melatonin Agonist Agomelatine in Healthy Older Men. *Clin Endocrinol (Oxf).* **2005**, *63*, 298–304.

171. Rajaratnam, S. M.; Polymeropoulos, M. H.; Fisher, D. M.; Roth, T.; Scott, C.; Birznieks, G.; Klerman, E. B. Melatonin Agonist Tasimelteon (VEC-162) For Transient Insomnia After Sleep-Time Shift: Two Randomised Controlled Multicentre Trials. *Lancet.* **2009**, *373*, 482–491.

8

CIRCADIAN RHYTHM SLEEP DISORDERS

Lenise Jihe Kim, Sergio Tufik, and Monica Levy Andersen

8.1. DEFINITION AND CLASSIFICATION OF BIOLOGICAL RHYTHMS

Rhythm can be defined as the occurrence and recurrence of events in a regular interval of time (Palmer, 1976). When these events involve a biological system, we consider as a biological rhythm (Aschoff, 1981). One of the first reports about the existence of a biological rhythmicity was performed by the French astronomer De Mairan in 1729. De Mairan observed the regular pattern of opening and closing of the leaf from the *Mimosa pudica* during the daytime. To verify whether this pattern was influenced by the sun lighting, De Mairan placed the plant in a closet, maintaining a condition of constant darkness. Independent of the absence of light stimulus, the movements of opening and closing the leafs remained with the same rhythmicity. This observation indicated that the biological rhythms could be modulated by endogenous mechanisms even with the lack of environmental cues.

As described by Aschoff (1981), some factors distinguish the types of biological rhythms, comprising:

- the involved biological system,
- the endogenous process responsible for the rhythm induction,
- the specific function of the rhythm.

In addition, another important factor in the classification of the biological rhythms is the period of the rhythm. The period of a biological rhythm is defined as the duration to complete one cycle of the endogenous event (Minors and Waterhouse, 2013). According to the period, the biological rhythms are classified as circadian, infradian, and ultradian (Halberg, 1969) as demonstrated in the Table 8.1.

Lenise Jihe Kim
Departamento de Psicobiologia, Universidade Federal de São Paulo, Rua Napoleão Barros, Vila Clementino, São Paulo, Brazil

Sergio Tufik
Departamento de Psicobiologia, Universidade Federal de São Paulo, Rua Napoleão Barros, Vila Clementino, São Paulo, Brazil

Monica Levy Andersen
Departamento de Psicobiologia, Universidade Federal de São Paulo, Rua Napoleão Barros, Vila Clementino, São Paulo, Brazil

Corresponding author: Monica Levy Andersen, E-mail: ml.andersen12@gmail.com

Table 8.1. Classification of biological rhythms according to the endogenous period duration.

Rhythm	Definition	Examples
Circadian	Rhythms with a near 24-h endogenous period (20–28 h)	Sleep–wake cycle, core body temperature
Infradian	Rhythms with an endogenous period longer than 28 h	Menstrual cycle, lifetime of erythrocytes
Ultradian	Rhythms with an endogenous period shorter than 20 h	Heart rate, sleep stages cycle

8.2. CIRCADIAN RHYTHMS: SLEEP–WAKE CYCLE

The sleep–wake cycle is a circadian rhythm which frequency is synchronized according to the 24 h light-dark cycle. The rhythmicity of the sleep–wake cycle is regulated by both intrinsic and extrinsic factors, constituting a circadian temporization system (Fig. 8.1). Via an afferent pathway, the environmental cues, especially the light stimulus, are processed in the retina, and are send to the suprachiasmatic nucleus by the retinohypothalamic tract (Lima and Vargas, 2014). The suprachiasmatic nucleus is the central pacemaker of the sleep–wake cycle (Brown et al., 2012). The rhythmicity of the sleep–wake cycle is regulated by the clock genes present in the cells of suprachiasmatic nucleus. The expression

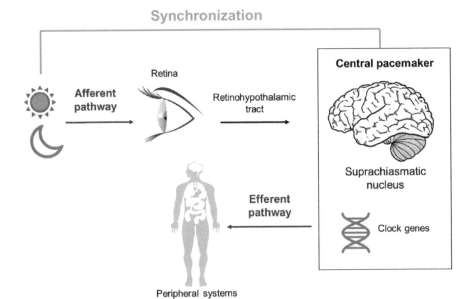

Figure 8.1. Schematic illustration about the circadian temporization system of the sleep–wake cycle.

of the clock genes' proteins occurs in a near 24 h period. This rhythmicity of the central pacemaker is send to the peripheral systems by hormonal and neural efferent pathways, leading to the synchronization between the biological and the environmental rhythms (Lima and Vargas, 2014).

Thus, to guarantee the expression of a sleep–wake cycle according to the circadian preference of the individuals, a correct synchronization between the biological and the environmental rhythms is necessary. In other words, the sleep time should match the timing of the circadian rhythm of sleep and wake propensity (American Academy of Sleep Medicine, 2014). Therefore, alterations in the sleep and wake schedules may lead to an impairment in the process of synchronization between the biological clock and the environmental times. Consequently, we observe an imbalance or a misalignment between the biological and the social circadian times that characterize the Circadian Rhythm Sleep–Wake Disorders (CRSWD).

8.3. DEFINITION OF CIRCADIAN RHYTHM SLEEP–WAKE DISORDERS

According to the American Academy of Sleep Medicine (2014), the CRSWD can be defined as:

"A sleep disorder caused by alterations of the circadian time-keeping system, its entrainment mechanisms, or a misalignment of the endogenous circadian rhythm and the external environment"

In this sense, the diagnosis of CRSWD occurs when the patients met the following criteria of the American Academy of Sleep Medicine (2014):

A. A chronic or recurrent pattern of sleep–wake rhythm disruption primarily due to alteration of the endogenous circadian timing system or misalignment between the endogenous circadian rhythm and the sleep–wake schedule desired or required by an individual's physical environment or social/work schedules.

B. The circadian rhythm disruption leads to insomnia symptoms, excessive sleepiness, or both.

C. The sleep and wake disturbances cause clinically significant distress or impairment in mental, physical, social, occupational, educational, or other important areas of functioning.

Considering these criteria, the CRSWD are subdivided in:

- delayed sleep-wake phase disorder,
- advanced sleep-wake phase disorder,
- irregular sleep-wake rhythm disorder,
- non-24 h sleep-wake rhythm disorder,
- shift work disorder,
- jet lag disorder,
- circadian sleep-wake disorder not otherwise specified (NOS).

In this chapter, we will focus on the definition and the diagnostic criteria of the delayed sleep–wake phase, advanced sleep–wake phase, shift work, and jet lag disorders.

8.4. METHODS FOR ASSESSING CIRCADIAN TIMING AND PREFERENCE

8.4.1. Habitual Circadian Timing

Since CRSWD are characterized by a misalignment between the biological and the environmental rhythms, the evaluation of the circadian timing may improve the accuracy of the diagnosis. This comprises the assessment of the habitual social activities schedule, such as the work and school times. Several methods may be used to evaluate the sleep–wake disruption due to social factors, including the patients' self-report during the anamneses and the application of questionnaires. However, the most precise and recommended method by the American Academy of Sleep Medicine (2014) is the actigraphy. The actigraphy should be conducted for at least seven days, preferably for 14 days, including both weekdays and weekend. In addition, the collection of collateral information is essential (e.g. data from sleep diaries) (Ancoli-Israel et al., 2003). Compared with the polysomnography (PSG), the gold standard method for the diagnosis of sleep disorders, the actigraphy shows some advantages, which makes the PSG not mandatory for circadian rhythms disorders diagnosis (Table 8.2) (Ancoli-Israel et al., 2003).

Nevertheless, we also have to consider that actigraphy shows some disadvantages that limit the use of this technique in some cases (Table 8.3) (Ancoli-Israel et al., 2003).

Table 8.2. Advantages of actigraphy use compared with PSG in the diagnosis of CRSWD.

Advantages of actigraphy
1. Actigraphy allows the assessment of the habitual wake up and bedtime of patients during working and non-working days
2. Actigraphy is an important tool for the evaluation of the effect night-to-night variations in sleep
3. The costs for conducting a actigraphy is much lower than a PSG

Table 8.3. Disadvantages of actigraphy use in the diagnosis of CRSWD.

Disadvantages of actigraphy
1. Actigraphy is a method that allows only an estimation of the total sleep time since it detects body movements
2. Actigraphy does not distinguish sleep stages and the presence of other sleep disorders
3. Actigraphy is not sensitive to the evaluation of some sleep parameters (e.g. sleep onset latency)

8.4.2. CIRCADIAN PREFERENCES

Another factor that should be evaluated in the diagnosis of CRSWD is the individual circadian preference. Circadian preference may be defined as the pattern of synchronization between the biological and the environmental rhythms, with individual clocks entraining differently over the 24 h light-dark cycle. Circadian preferences are classified as chronotypes, including morning-type, evening-type, and intermediate (Horne and Ostberg, 1976). Morning-type individuals are characterized by phase advancing, preferring to wake up and sleep earlier than the average population. On the other hand, evening-type individuals show a phase delay, preferring to wake up and sleep later than the average population (Horne and Ostberg, 1976). In the general population, approximately, 52% of the individuals are morning-type, followed by 40% of intermediate and only 8% of evening-type (Kim et al., 2015). The assessment of chronotype can be performed by questionnaires and physiological measures of endogenous circadian timing (e.g. salivary dim light melatonin onset) (American Academy of Sleep Medicine, 2014).

8.5. DELAYED SLEEP–WAKE PHASE DISORDER

According to the American Academy of Sleep Medicine (2014), the Delayed Sleep–Wake Phase Disorder (DSWPD) occurs when the habitual sleep–wake time is delayed, often more than 2 h, compared

with the socially acceptable timing (Fig. 8.2). The patients frequently show a sleep duration constraint and experience difficult in to arise at morning (Wittmann et al., 2006). For the diagnosis of DSWPD, the following criteria should be present:

A. There is a significant delay in the phase of the major sleep episode in relation to the desired or required sleep time and wake-up time, as evidenced by a chronic or recurrent complaint by the patient or a caregiver of inability to fall asleep and difficulty awakening at a desired or required clock time.

B. The symptoms are present for at least 3 months.

C. When patients are allowed to choose their *ad libitum* schedule, they will exhibit improved sleep quality and duration for age and maintain a delayed phase of the 24 h sleep–wake pattern.

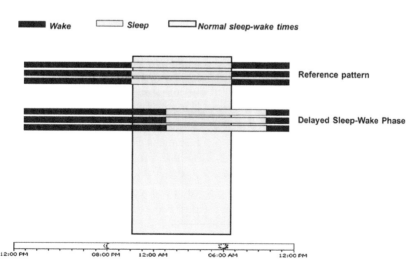

Figure 8.2. Schematic actogram of a patient with DSWPD and the comparison with a normal sleep–wake schedule.

D. Sleep log and, whenever possible, actigraphy monitoring for at least seven days (preferably 14 days) demonstrate a delay in the timing of the habitual sleep period. Both work/school days and free days must be included within this monitoring.

E. The sleep disturbance is not better explained by another current sleep disorder, medical or neurological disorder, mental disorder, medication use, or substance use disorder.

8.6. ADVANCED SLEEP–WAKE PHASE DISORDER

The Advanced Sleep–Wake Phase Disorder (ASWPD) is characterized by a phase advance, leading to habitual sleep onset and offset occurrence 2, or more hours earlier than the required or desired times (American Academy of Sleep Medicine, 2014) (Fig. 8.3). For the diagnosis, the following criteria should be present:

A. There is an advance (early timing) in the phase of the major sleep episode in relation to the desired or required sleep time and wake-up time, as evidenced by a chronic or recurrent complaint of difficulty staying awake until the required or desired conventional bedtime, together with an inability to remain asleep until the required or desired time for awakening.

B. Symptoms are present for at least 3 months.

C. When patients are allowed to sleep in accordance with their internal biological clock, sleep quality, and duration are improved with a consistent but advanced timing of the major sleep episode.

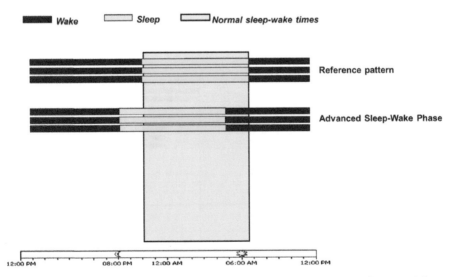

Figure 8.3. Schematic actogram of a patient with ASWPD and the comparison with a normal sleep–wake schedule.

D. Sleep log and, whenever possible, actigraphy monitoring for at least seven days (preferably 14 days) demonstrate a stable advance in the timing of the habitual sleep period. Both work/school days and free days must be included within this monitoring.

E. The sleep disturbance is not better explained by another current sleep disorder, medical or neurological disorder, mental disorder, medication use, or substance use disorder.

8.7. SHIFT WORK DISORDER

The shift work disorder (SWD) can be defined as occurrence of work hours during the usual sleep episode (Fig. 8.4), leading to several sleep complaints, such as excessive daytime somnolence (American Academy of Sleep Medicine, 2014). The total sleep time in shift workers is often reduced and the patients experience a poor quality of sleep. For the diagnosis, the following criteria should be present:

A. There is a report of insomnia and/ or excessive sleepiness, accompanied by a reduction of total sleep time, which is associated with a recurring work schedule that overlaps the usual time for sleep.

B. The symptoms have been present and associated with the shift work schedule for at least 3 months.

C. Sleep log and actigraphy monitoring (whenever possible and preferably with concurrent light exposure

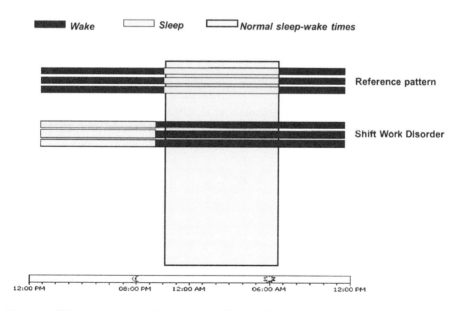

Figure 8.4. Schematic actogram of a patient with SWD and the comparison with a normal sleep–wake schedule.

measurement) for at least 14 days (work and free days) demonstrate a disturbed sleep and wake pattern.

D. The sleep and/or wake disturbance are not better explained by another current sleep disorder, medical or neurological disorder, mental disorder, medication use, poor sleep hygiene, or substance use disorder.

8.8. JET LAG DISORDER

Jet lag disorder can be defined as a temporary alteration in the sleep–wake cycle pattern required by a change in time zone. The duration and the severity of the jet lag disorder depends on several conditions, including the number of time zones traveled and the tolerance to circadian misalignment when awake during the biological night. In addition, the direction of the travel influence the jet lag disorder course (American Academy of Sleep Medicine, 2014). In this sense, eastward travel is usually more difficult to adjust to than westward travel. For diagnosis, the following criteria should be present:

A. There is a complaint of insomnia or excessive daytime sleepiness, accompanied by a reduction of total sleep time, associated with transmeridian jet travel across at least 2 time zones.

B. There is associated impairment of daytime function, general malaise, or somatic symptoms (e.g. gastrointestinal disturbance) within one to 2 days after travel.

C. The sleep disturbance is not better explained by another current sleep disorder, medical or neurological disorder, mental disorder, medication use, or substance use disorder.

8.9. CONSEQUENCES OF CRSWD

In general, excessive daytime somnolence is the most frequent symptom among the patients with CRSWD (American Academy of Sleep Medicine, 2014). Considering the imbalance or misalignment between the biological and the social circadian times, the patients often have to readjust their sleep–wake schedule according to the social demands (e.g. work, school, travels). Consequently, these individuals become chronically sleep deprived, leading to several health impairments (Fig. 8.5). Some of these injuries comprise an increase in the mortality rate (Gu et al., 2015), impairments in academic performance (Sivertsen et al., 2013), and higher susceptibility for work accidents (Asaoka et al., 2013) due to the excessive daytime somnolence.

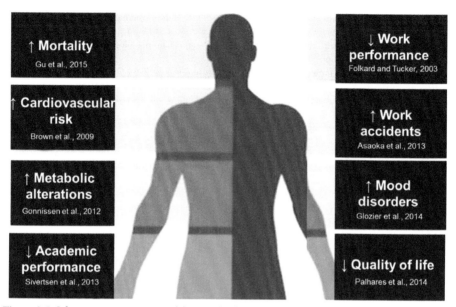

Figure 8.5. Schematic representation of the main health impairments in patients with CRSWD.

8.10. CONCLUSIONS AND PERSPECTIVES

CRSWD are a group of sleep disorders characterized by a misalignment between the endogenous circadian rhythm and the environmental schedules. As a consequence of this imbalance, we observed several health and social problems, including an increased risk of mortality, cardiovascular outcomes, and academic/work performance impairments. Therefore, patients with circadian rhythms disorders seem to be more likely to suffer with daily socioeconomic pressures. Although the diagnostic criteria are already well established in the clinical practice, there is still a lack of preventive measures aiming to improve the quality of sleep of these patients. In this sense, the evaluation of the social demands (work, school, travels) and the reorganiza-

tion of these activities according to the individual circadian preferences and endogenous clock could be an important strategy to guarantee the correct synchronization between the biological and social rhythms.

KEY WORDS

- circadian rhythm
- sleep-wake cycle
- sleep disorders

REFERENCES

American Academy of Sleep Medicine. International classification of sleep disorders, 3rd.; Westchester, IL: American Academy of Sleep Medicine, **2014**.

Ancoli-Israel, S.; Cole, R.; Alessi, C.; Chambers, M.; Moorcroft, W.; Pollak, C.P. The Role of Actigraphy in the Study of Sleep and Circadian Rhythms. *Sleep.* **2003**, *26*, 342–392.

Asaoka, S.; Aritake, S.; Komada, Y.; Ozaki, A.; Odagiri, Y.; Inoue, S.; Shimomitsu, T.; Inoue, Y. Factors Associated with Shift Work Disorder in Nurses Working with Rapid-Rotation Schedules in Japan: The Nurses' Sleep Health Project. *Chronobiol Int.* **2013**, *30*, 628–636.

Aschoff, J. Biological Rhythms, 1st ed.; Plenum Press: New York, **1981**.

Brown, D. L.; Feskanich, D.; Sánchez, B. N.; Rexrode, K. M.; Schernhammer, E. S.; Lisabeth, L. D. Rotating Night Shift Work and the Risk of Ischemic Stroke. *Am J Epidemiol.* **2009**, *169*, 1370–1377.

Brown, R. E.; Basheer, R.; McKenna, J. T.; Strecker, R. E.; McCarley, R. W. Control of Sleep and Wakefulness. *Physiol Rev.* **2012**, *92*, 1087–1187.

De Mairan, J. Observation Botanique. Histoire de l'Academie Royale des Sciences, **1729**.

Folkard, S.; Tucker, P. Shift Work, Safety and Productivity. *Occup Med (Lond).* **2003**, *53*, 95–101.

Glozier, N.; O'Dea, B.; McGorry, P. D.; Pantelis, C.; Amminger, G. P.; Hermens, D. F.; Purcell, R.; Scott, E.; Hickie, I. B. Delayed Sleep Onset in Depressed Young People. *BMC Psychiatry.* **2014**, *14*, 33.

Gonnissen, H. K.; Rutters, F.; Mazuy, C.; Martens, E. A.; Adam, T. C.; Westerterp-Plantenga, M. S. Effect of a Phase Advance and Phase Delay of the 24-h Cycle on Energy Metabolism, Appetite, and Related Hormones. *Am J Clin Nutr.* **2012**, *96*, 689–697.

Gu, F.; Han, J.; Laden, F.; Pan, A.; Caporaso, N. E.; Stampfer, M. J.; Kawachi, I.; Rexrode, K. M.; Willett, W. C.; Hankinson, S. E.; Speizer, F. E.; Schernhammer, E. S. Total and Cause-Specific Mortality of U.S. Nurses Working Rotating Night Shifts. *Am J Prev Med.* **2015**, *48*, 241–252.

Halberg, F. Chronobiology. *Annu Rev Physiol.* **1969**, *31*, 675–725.

Horne, J. A.; Ostberg, O. A Self-Assessment Questionnaire to Determine Morningness-Eveningness in Human Circadian Rhythms. *Int J Chronobiol.* **1976**, *4*, 97–110.

Kim, L. J.; Coelho, F. M.; Hirotsu, C.; Bittencourt, L.; Tufik, S.; Andersen, M. L. Is the Chronotype Associated with Obstructive Sleep Apnea? *Sleep Breath.* **2015**, *19*, 645–651.

Lima, L. E. B.; Vargas, N. N. G. O Relógio Biológico e Os Ritmos Circadianos De Mamíferos: Uma Contextualização Histórica. *Revista da Biologia.* **2014**, *12*, 1–7.

Minors, D. S.; Waterhouse, J. M. Circadian Rhythms and the Human, 2nd Ed.; Butterworth-Heinemann, **2013**.

Palhares, V. C., Corrente, J. E.; Matsubara, B. B. Associação between Sleep Quality and Quality of Life in Nursing Professionals Working Rotating Shifts. *Rev Saude Publica.* **2014**, *48*, 594–601.

Palmer, J. An Introduction to Biological Rhythms, 1st Ed.; Academic Press: New York, **1976**.

Sivertsen, B.; Pallesen, S.; Stormark, K. M.; Bøe, T.; Lundervold, A. J.; Hysing, M. Delayed Sleep Phase Syndrome in Adolescents: Prevalence and Correlates in a Large Population Based Study. *BMC Public Health.* **2013**, *13*, 1163.

Wittmann, M.; Dinich, J.; Merrow, M.; Roenneberg, T. Social Jetlag: Misalignment of Biological and Social Time. *Chronobiol Int.* **2006**, *23*, 497–509.

9

PHARMACOLOGICAL MANAGEMENT OF INSOMNIA

David N. Neubauer, Kenneth Buttoo,
and Seithikurippu R. Pandi-Perumal

ABSTRACT

Insomnia is not only the most common sleep disorder in the general population, it is a frequent complaint in the primary care setting as well as specialists setting alike. The prevalence of Insomnia is high among people with comorbid conditions. While the pharmacological treatment of insomnia is widely practiced, the compounds used for treating insomnia vary quite widely in terms of its pharmacokinetic profile and pharmacodynamic actions, which ul-

timately influence the efficacy and safety profile. The medications approved with indications for the treatment of insomnia are in four distinct pharmacodynamic categories: benzodiazepine receptor agonists (BZRAs), melatonin receptor agonists, histamine receptor antagonists, and orexin receptor antagonists. All of the OTC products specifically sold as sleep aids in the United States are first-generation antihistamines – almost entirely diphenhydramine and to a lesser extent doxylamine. The unregulated substances taken to treat insomnia typically are classified as dietary supplements. Melatonin has been shown to have therapeutic efficacy in selected circadian rhythm disorders that may involve insomnia symptoms. While alcohol may be sedating and facilitate sleep onset, the subsequent sleep often is of poor quality and may be associated with an exacerbation of sleep disorders, such as sleep apnea. Many medications recommended and prescribed for insomnia do not have formal approvals for this indication, but do have pharmacologic properties that may result in sedating effects or sleep enhancement by other

David N. Neubauer
Department of Psychiatry and Behavioral Sciences, Johns Hopkins Bayview Medical Center, Baltimore, MD , USA. Phone: +1 (410) 550-0066; Fax: +1 410-550-1407

Kenneth Buttoo
601 Harwood Avenue South, Ajax, ON L1S 2J5, Canada

Seithikurippu R. Pandi-Perumal
Somnogen Canada Inc., College Street, Toronto, ON, Canada

Corresponding author: David N. Neubauer, E-mail: neubauer@jhmi.edu; Tel.: +1 410 550 0066; Fax: +1 410 550 1407.

mechanisms. It should be emphasized here that pharmacotherapeutic strategies for addressing insomnia are most effective when incorporated into a broader treatment context. Optimal effectiveness most likely will be achieved with the concurrent treatment of comorbid conditions and in the context of healthy sleep habits and in combination with cognitive-behavioral strategies.

Insomnia is among the most prevalent health concerns in the general population and it is present especially at high rates among people with comorbid health problems. Approximately one-third of adults have occasional difficulty with insomnia and about 10% have persistent problems achieving sufficient sleep.[1] While some have argued that insomnia is a disorder resulting from the stressors of modern life, descriptions of frustrating sleep difficulty date back thousands of years.[2] Although the word insomnia may be used colloquially to represent poor sleep, specific diagnostic criteria for insomnia disorders are detailed in the newly updated Diagnostic and Statistical Manual, 5th Edition (DSM-5) and International Classification of Sleep Disorders, 3rd Edition (ICSD-3).[3,4] Essentially, insomnia disorder is persistent difficulty falling asleep or remaining asleep, or awakening earlier than desired in the context of an adequate opportunity for sleep, and is in association with daytime consequences (for example, fatigue, poor concentration, irritability, and impaired performance). People often try a variety of remedies in the attempt to improve their sleep. These may include sleep hygiene measures, behavioral, and psychotherapeutic strategies, and an extraordinarily diverse assortment of substances and medications – with or without actual data supporting their use

to improve sleep. However, there is abundant evidence demonstrating the efficacy of cognitive behavioral therapy and the use of selected pharmaceutical agents in the treatment of insomnia disorder. The major focus of this chapter will be medications with formal indications for the treatment of insomnia, although issues related to other commonly used medications and substances also will be discussed.

It should be emphasized here that pharmacotherapeutic strategies for addressing insomnia are most effective when incorporated into a broader treatment context. Any treatment plan for insomnia patients should evolve from a comprehensive evaluation that has included a detailed review of the sleep history, comorbid conditions (for example, medical, psychiatric, substance use, and other sleep disorders) and associated treatments, a review of systems, mental and physical examinations, and a family history.[5] The assessment should consider sleep hygiene issues, modifiable behaviors, daytime and evening routines, any alcohol and caffeine use, and work or school schedules. The results of past treatment efforts should be reviewed. Patients should be asked about any benefits and adverse effects associated with previous medication trials. When treating insomnia, behavioral changes typically are necessary – rarely is a pill the sole solution.

Specific medication selections should be made based on the pharmacodynamic and pharmacokinetic properties of a drug. Medication recommendations may be influenced by an individual's pattern of sleep disturbance, co-administered medications, substance use history, age, sex, and reproductive status, in addition to patient preference, cost, insurance coverage, and availability.[5] It is important to monitor patients over time for efficacy, possible side effects, and the need for continued treatment. Generally, the medication for older adults and people with debilitating physical conditions is started at the lower available doses. In all cases the formal prescribing guidelines should be reviewed regarding dosing recommendations and warnings for specific populations.

The broad domain of compounds that people take in the attempt to improve their sleep is conveniently divided into four categories: 1) medications specifically approved with an indication for the treatment of insomnia; 2) prescription medications without an insomnia indication; 3) over-the-counter (OTC) drugs approved and marketed as sleep aids; and 4) unregulated substances, some of which are marketed as sleep aids.

9.1. UNREGULATED SUBSTANCES

The unregulated substances taken to treat insomnia typically are classified as dietary supplements. Often they are promoted as homeopathic, natural, or "drug-free." While some include a single active compound, most are combinations of assorted plant extracts or other preparations. The ingredients may include various vitamins, minerals (for example, magnesium), and synthesized compounds (for example, melatonin). Most common are the plant-based ingredients, such as valerian root, kava-kava, hops, chamomile, skullcap, lavender, and passionflower. While minimal evidence supports the efficacy of these products in enhancing sleep or treating insomnia, these products are generally regarded as safe. One exception is kava-kava,

for which the US Food and Drug Administration (FDA) has issued a warning about potential severe liver injury.[6] Melatonin has been shown to have therapeutic efficacy in selected circadian rhythm disorders that may involve insomnia symptoms.[7] However, the bedtime dosing for an insomnia disorder appears to have limited benefits for most patients.[8] While melatonin is unregulated in the US, it is available only by prescription in the European Union (EU) and several other countries. Although not promoted as an insomnia treatment, many people use alcohol with the hope of improving their sleep. While alcohol may be sedating and facilitate sleep onset, the subsequent sleep often is of poor quality and may be associated with an exacerbation of sleep disorders, such as sleep apnea.[9]

9.2. OVER-THE-COUNTER MEDICATIONS

The manufacture and marketing of OTC medications are regulated. All of the OTC products specifically sold as sleep aids in the United States are first-generation antihistamines – almost entirely diphenhydramine and to a lesser extent doxylamine. The OTC sleep aids are available as single active ingredients or combined with other medications, such as analgesics – typically as a "PM" or bedtime formulation. Histamine is a potent wake-promoting hypothalamic neurotransmitter, so there is an excellent pharmacodynamic rationale for these histamine antagonists in promoting sleep.[10] While possibly beneficial for occasional use, the efficacy of OTC medications in the treatment of insomnia patients has not been established. Several potential problems with the use of the OTC antihistamine sleep aids have been identified. The elimination half-lives are relatively long, so may cause next-morning grogginess following bedtime use. In addition to the antihistaminic effect, there is anticholinergic activity that in some individuals may lead to serious problems including urinary retention, confusion, and delirium.[11,12] Older adults and people concomitantly taking other anticholinergic medications are most vulnerable to these anticholinergic side effects of OTC sleep aids.[13] Finally, people may become tolerant to the sedating action of the OTC sleep aids and find them ineffective which lead them to take rather high doses.[14]

9.3. ALTERNATE PRESCRIPTION MEDICATIONS

Many medications recommended and prescribed for insomnia do not have formal approvals for this indication, but do have pharmacologic properties that may result in sedating effects or sleep enhancement by other mechanisms (for example, increased slow-wave activity).[15] The sedation associated with the off-label use of most of these psychotropic agents likely is due to varying degrees of postsynaptic antagonism at assorted serotonin, histamine, muscarinic, and alpha-adrenergic receptors.[16] Antidepressants are by far the most commonly prescribed medications in this category; however, atypical antipsychotics, anticonvulsants, and antihypertensives (for example, clonidine) are sometimes recommended. The use of all of these medications typically is without evidence-based assessments for efficacy and safety in populations of insomnia patients, and the risk–benefit ratio may not be as favorable

for insomnia as for the approved indications.

The antidepressants most often prescribed for insomnia have included amitriptyline, doxepin, mirtazapine, and trazodone. These may offer benefits for sleep onset and sleep maintenance for some patients. The relatively long half-lives may cause sedation the following morning or later in the daytime. Dose-dependent orthostatic hypotension may occur with several of these medications. Weight gain may be an issue with amitriptyline and mirtazapine. In recent years, the use of trazodone for insomnia has been quite common.[17,18]

9.4. APPROVED INSOMNIA MEDICATIONS

The medications approved with indications for the treatment of insomnia are in four distinct pharmacodynamic categories: 1) benzodiazepine receptor agonists (BZRAs); 2) melatonin receptor agonists, 3) histamine receptor antagonists, and 4) orexin receptor antagonists. These medications all have been evaluated for efficacy and safety in randomized, controlled clinical trials with populations of insomnia patients. In the United States, there are numerous compounds and an assortment of formulations among the BZRAs, but only single drugs represented in the remaining three insomnia pharmacotherapy groups. While all are approved for the treatment of insomnia, selected medications have indications specifying efficacy for sleep onset and/or sleep maintenance. Some of the medications have indications for short-term treatment while others have no implied limitations on their duration of use.

Table 9.1 lists all of the US FDA-approved insomnia medication with the generic and brand names, available doses, approximate elimination half-lives, most common side effects, indications, US Drug Enforcement

Table 9.1. FDA-Approved Medications Indicated for Treating Insomnia Disorder[45–59]

Generic (brand) name	Doses (mg)	Half-life (h)	Indications	Most common side effects	DEA class	Pregnancy category
Benzodiazepine immediate release						
Flurazepam (Dalmane)	15, 30	48–120	"treatment of insomnia characterized by difficulty in falling asleep, frequent nocturnal awakenings, and/or early morning awakening"	Dizziness, drowsiness, lightheadedness, staggering, loss of coordination, falling	IV	X
Temazepam (Restoril)	7.5, 15, 22.5, 30	8–20	"short-term treatment of insomnia"	Drowsiness, dizziness, lightheadedness, difficulty with coordination	IV	X
Triazolam (Halcion)	0.125, 0.25	2–4	"short-term treatment of insomnia"	Drowsiness, headache, dizziness, lightheadedness, "pins and needles" feelings on your skin, difficulty with coordination	IV	X
Quazepam (Doral)	7.5, 15	48–120	"treatment of insomnia characterized by difficulty in falling asleep, frequent nocturnal awakenings, and/or early morning awakenings"	Drowsiness, headache	IV	X
Estazolam (ProSom)	1, 2	8–24	"short-term management of insomnia characterized by difficulty in falling asleep, frequent nocturnal awakenings, and/or early morning awakenings ….administered at bedtime improved sleep induction and sleep maintenance"	Somnolence, hypokinesia, dizziness, abnormal coordination	IV	X

Table 9.1. (*Continued*)

Nonbenzodiazepine immediate release

Zolpidem (Ambien)	5, 10	1.5–2.4	"short-term treatment of insomnia characterized by difficulties with sleep initiation"	Drowsiness, dizziness, diarrhea, drugged feelings	IV	C
Zaleplon (Sonata)	5, 10	1	"short-term treatment of insomnia....shown to decrease the time to sleep onset"	Drowsiness, lightheadedness, dizziness, "pins and needles" feeling on your skin, difficulty with coordination	IV	C
Eszopiclone (Lunesta)	1, 2, 3	5–7	"treatment of insomnia....administered at bedtime decreased sleep latency and improved sleep maintenance"	Unpleasant taste in mouth, dry mouth , drowsiness, dizziness, headache , symptoms of the common cold	IV	C

Nonbenzodiazepine extended release

Zolpidem ER (Ambien CR)	6.25, 12.5	2.8–2.9	"treatment of insomnia characterized by difficulties with sleep onset and/or sleep maintenance (as measured by wake time after sleep onset)"	Headache, sleepiness, dizziness	IV	C

Nonbenzodiazepine alternate delivery

Zolpidem Oral spray (ZolpiMist)	5, 10	~2.5	"short-term treatment of insomnia characterized by difficulties with sleep initiation"	Drowsiness, dizziness, diarrhea, drugged feelings	IV	C
Zolpidem Sublingual (Edluar)	5, 10	~2.5	"short-term treatment of insomnia characterized by difficulties with sleep initiation"	drowsiness, dizziness, diarrhea, drugged feelings	IV	C
Zolpidem Sublingual (intermezzo)	1.75, 3.5	~2.5	"for use as needed for the treatment of insomnia when a middle-of-the-night awakening is followed by difficulty returning to sleep. Not indicated ... when the patient has fewer than 4 h of bedtime remaining before the planned time of waking"	Headache, nausea, fatigue	IV	C

Table 9.1. (*Continued*)

Selective melatonin receptor agonist						
Ramelteon (Rozerem)	8	1–2.6	"treatment of insomnia characterized by difficulty with sleep onset"	Drowsiness, tiredness, dizziness	None	C
Selective histamine H$_1$ receptor antagonist						
Doxepin (Silenor)	3, 6	15.3	"treatment of insomnia characterized by difficulties with sleep maintenance"	Somnolence/sedation, nausea, upper respiratory tract infection	None	C
Dual orexin receptor antagonist						
Suvorexant (Belsomra)	5, 10, 15, 20	12	"treatment of insomnia characterized by difficulty with sleep onset or sleep maintenance, or both"	Somnolence	IV	C

Agency (DEA) classification, and the FDA Pregnancy Category.

While these insomnia medications have specific safety concerns and warnings associated with their pharmacologic characteristics, the FDA also has issued broad warnings for all hypnotic medications in relation to rare but severe anaphylactic or anaphylactoid reactions and for abnormal thinking or behavioral changes. Behavioral changes may include complex behaviors with amnesia, such as driving, eating, having telephone conversations, or pursuing sexual activities when not fully awake.[19] Broad prescribing guidelines for hypnotics suggest using the lowest effective dose for the shortest possible duration.[5]

9.4.1. Benzodiazepine Receptor Agonists

The BZRA hypnotics are positive allosteric modulators of gamma-aminobutyric acid (GABA) responses at the GABA-A receptor complex.[20] GABA is the primary central nervous system (CNS) inhibitory neurotransmitter and GABA-A receptors are widely distributed in the CNS, including hypothalamic regions (for example, ventrolateral preoptic nucleus) central to the regulation of sleep and waking. The BZRA hypnotics are positive allosteric modulators because they enhance the normal inhibitory GABA effect through action at a separate benzodiazepine recognition site on this receptor complex. Normally the pentameric transmembrane structure controls the entry of negative chloride ions through a central ion channel. The most common configuration is two alpha, two beta, and one gamma subunit. When GABA attaches at the GABA site (alpha–beta interface) on the receptor complex, chloride ions enter the neuron thereby decreasing the charge across the cell membrane. When a BZRA compound is present at the benzodiazepine recognition site (alpha–gamma interface), a greater number of the chloride ions are able enter the neuron in response to GABA's action with a resulting hyperpolarization enhancing the inhibitory effect. While certain BZRA compounds have anxiolytic, anticonvulsant, or muscle relaxant properties, the BZRA hypnotics have pharmacokinetic properties that facilitate sleep onset and/or maintenance while minimizing the potential for impairment during the typical waking period.[21]

BZRA hypnotics were introduced in 1970, within a decade of the release of the first BZRA compounds, chlordiazepoxide and diazepam. The use of BZRAs to enhance sleep was welcomed at that time due to the safety problems with barbiturates and related drugs commonly used then. The BZRA hypnotics comprise two types of compounds: 1) those with the characteristic benzodiazepine structure (benzene and diazepine rings) and 2) nonbenzodiazepines drugs with alternate structures but similar pharmacodynamic properties. The first five BZRA hypnotics released in the United States were structural benzodiazepines (flurazepam, temazepam, triazolam, quazepam, and estazolam) while all of the subsequent products have been nonbenzodiazepines (zolpidem, zaleplon, and eszopiclone). The licensed BZRA hypnotics vary somewhat internationally. Benzodiazepine hypnotics available in certain countries outside of the United States include nitrazepam, flunitrazepam, and lormetazepam, whereas a widely available nonbenzodiazepine compound is zopiclone.

The BZRA hypnotics all are relatively rapidly absorbed and so should be beneficial for sleep onset. The pharmacokinetic properties influence the duration of action and efficacy for sleep maintenance, as well as the potential for undesired next-day sedation or impairment.[22] All of the BZRA

hypnotics are in the DEA schedule IV category due to their identified but low potential for abuse. The benzodiazepine hypnotics all are Pregnancy Category X (contraindicated in pregnancy) and the nonbenzodiazepines all are classified as C (risk during pregnancy not ruled out).

The benzodiazepine BZRA hypnotics were released in the United States between 1970 and 1990. Except for triazolam, these benzodiazepines have moderate to long elimination half-lives, so may benefit sleep maintenance, but have elevated risks for undesired daytime effects. A meta-analysis of studies assessing the safety of benzodiazepine hypnotics in the treatment of insomnia showed that headaches, dizziness, nausea, and fatigue were the most commonly reported side effects.[23]

The nonbenzodiazepine hypnotics became available with zopiclone in the late 1980s and in the United States, beginning in the early 1990s. Zolpidem, zaleplon, and eszopiclone are the three marketed in the United States. These compounds range from short to intermediate half-lives relative to the benzodiazepines, so in this regard are less likely to cause undesired next-day symptoms. It also has been speculated that some degree of GABA-A alpha subunit differential affinity might offer unique safety and efficacy advantages for these compounds.[24] Headaches, dizziness, nausea, and somnolence were the more common reported side effects noted in the above-mentioned meta-analysis.[23]

Among the nonbenzodiazepine BZRA hypnotics, zaleplon has the shortest and eszopiclone has the longest half-life, with zolpidem in between these two. Although estimated half-lives are provided, these are considerable variations in the metabolism of these compounds among these drugs with corresponding effects on the duration of positive and possible negative actions. In the United States, the FDA has revised the prescribing guidelines toward lower initial doses for zolpidem and eszopiclone. In the case of zolpidem, the use of lower doses is especially important for woman, as generally they metabolize the medication slower than men.[25,26]

Eszopiclone and zaleplon so far have been available in the United States only as immediate-release tablets or capsules. However, zolpidem is manufactured as both immediate-release and extended-release tablets, and in a variety of alternate delivery formulations that include a dissolvable tablet and liquid spray for bedtime use and a low-dose dissolvable tablet for as-needed middle-of-the-night use. The middle-of-the-night dose should be taken only when patients have at least 4 h still available to remain in bed.

9.4.2. Melatonin Receptor Agonists

Melatonin plays a central role in the regulation of sleep and waking. The production and release of melatonin is driven by the anterior hypothalamic suprachiasmatic nucleus (SCN), which is the master timekeeper of the circadian cycle. Normally melatonin levels are very low throughout the daytime, rise in the evening as bedtime approaches, plateau at a higher level during the typical sleeping hours, and then decline in the morning. The circadian system supports wakefulness in the evening with a stimulating signal that declines with the rise of melatonin as a result of the action of the SCN melatonin receptors.[27] Sleep onset is facilitated by melatonin agonist activity at the MT1 receptor. Agonist effects at the MT2 receptor are thought to reinforce the circadian rhythmicity. In response to photoperiod exposure via the retino hypothalamic tract, the SCN is able to optimize sleep during the nighttime rather than occurring at random times.[28]

The use of bedtime exogenous melatonin as a sleep aid, while commonly used, has not been shown to have robust beneficial effects in controlled clinical trials.[8] However, the efficacy of strategically timed exogenous melatonin has been demonstrated in the treatment of selected circadian rhythm sleep–wake disorder, especially the delayed sleep phase type.[29] A prescription extended-release melatonin product is available in selected countries outside of the United States.

Ramelteon is the only melatonin receptor agonist approved in the United States for treating insomnia. It is selective for the MT1 and MT2 receptors.[30] The specific indication is for the treatment of insomnia characterized by difficulty with sleep onset. Ramelteon is available only in an 8 mg dose, which is recommended for both adult and elderly individuals. The prescribing guidelines suggest that the medication be taken about 30 min prior to bedtime. Though generally well-tolerated, the more common side effects are drowsiness, tiredness, and dizziness. Studies have shown that ramelteon is safe for patients with any severity of obstructive sleep apnea and for those with mild to moderate degrees of hepatic impairment or chronic obstructive pulmonary disease.[31–33] While generally there is a low risk of drug–drug interaction with ramelteon, it should not be combined with strong CYP1A2 inhibitors (for example, fluvoxamine) due to a possible marked elevation of the ramelteon blood level. The DEA considers ramelteon a non-scheduled medication due to an absence of abuse potential.[34] The Pregnancy Category is C.

Tasimelteon also is a selective MT1 and MT2 melatonin receptor agonist approved by the FDA (2014). The indication for tasimelteon is not insomnia disorder, but rather the non-24 h type of circadian rhythm sleep–wake disorder that occurs most commonly among totally blind individuals. Patients with non-24 h disorder follow a free-running circadian pattern that periodically causes nighttime sleep disturbance and daytime sleepiness alternating with phases of normal sleep–wake cycles.[35–37]

9.4.3. Histamine Receptor Antagonists

Histamine is a potent wake-promoting neurotransmitter that is produced in the CNS in the hypothalamic tuberomammillary bodies.[38] Antagonists for the histamine H1 receptor have sedating effects.[39] This is the basis of the active ingredients in the OTC sleep aids and the use of various prescription psychotropic medications recommended for insomnia symptoms.

The administration of a low dose of a highly selective H1 antagonist with a sufficiently long half-life would be expected to offer benefits for sleep maintenance.[10] Among the antidepressants, the tricyclic doxepin is the most selective for the H1 receptor. While doxepin sometimes has been prescribed at various higher doses to promote sedation, clinical trials have demonstrated the efficacy and safety of 3 and 6 mg doses in the treatment of insomnia disorder in adult and elderly patients.[40,41]

It has been argued that the doxepin sleep-promoting effects may be optimum at the lower dosage range since other receptor stimulating effects could compete with the antihistaminic sedation when administered at higher doses.[10] The doxepin 3 and 6 mg doses were approved in 2010 by the FDA for the treatment of insomnia with a specific indication for insomnia associated with sleep maintenance. The prescribing guidelines suggest bedtime doses of 6 mg for adults and 3 mg for elderly patients. There is no limitation in the duration of use. The contraindications are severe urinary retention, untreated narrow angle glaucoma, and concomitant use with monoamine oxidase inhibitors. The most common side effects are somnolence/sedation, nausea, and upper respiratory tract infection. Low-dose doxepin has no abuse potential so is classified as a non-scheduled medication. The Pregnancy Category is C.

9.4.4. Orexin Receptor Antagonists

The orexin (also called hypocretin) system has a key function in promoting and stabilizing wakefulness. Orexin-producing neurons originate in the lateral hypothalamus and have widespread projections in the CNS, including other wake-promoting centers. Two orexin neurotransmitters (orexin A and orexin B) are formed from a precursor prepro-orexin peptide. The orexin neurotransmitters bind with two similar g-protein-coupled receptors (OX1R and OX2R). Orexin activity is high during the waking state and low during sleep.[42] People with narcolepsy, which is associated with excessive sleepiness, have deficits in orexin functioning. Soon after the 1998 discovery of the orexin system, it was speculated that a pharmacologic approach to dampening orexin activity could be a novel therapeutic approach for insomnia patients who may have central hyperarousal in relation to the underlying pathology of their disorder.[43]

While several types of orexin antagonists have been evaluated with insomnia patients, at present suvorexant is the only compound with this pharmacodynamic activity approved for insomnia disorder. The specific FDA indication is for the treatment of insomnia characterized by difficulty with sleep onset or sleep maintenance, or both. Suvorexant functions as a potent and reversible dual orexin receptor antagonist.[44] It is available as immediate-release tablets in doses from 5 to 20 mg. The prescribing guidelines recommend that suvorexant be taken no more than once at night within 30 min of bedtime and with at least 7 h remaining before the planned wakeup time. The suggested starting dose for adults is 10 mg. If the medication is well-tolerated and insufficiently effective, it may be increased to a maximum of 20 mg. Patients concomitantly taking a moderate CYP3A inhibitor should begin with the 5 mg dose. Somnolence is the most commonly reported side effect. Patients prescribed suvorexant should be educated about the potential for excessive sleepiness or impairment the morning following bedtime dosing. There is no limitation on the duration of use, and discontinuation is not associated with withdrawal effects. The medication is contraindicated in people with narcolepsy. The DEA classifies suvorexant as a Schedule IV controlled substance due to likeability ratings in reactional drug abusers, although physiologic dependence has not been observed. Suvorexant is Pregnancy Category C.

9.5. CONCLUSIONS

As illustrated above, the compounds used for treating insomnia vary quite widely in their pharmacodynamic actions and pharmacokinetic properties, both of which may influence the efficacy and safety profile for patients. Even among the medications with formal indications for insomnia, there are abundant choices allowing customization for individual patients moving the field further into the realm of personalized medicine. For most patients, the lowest effective dose for the shortest possible time is an appropriate strategy, although monitored chronic hypnotic dosing may be the best approach for selected individuals. Optimal effectiveness most likely will be achieved with the concurrent treatment of comorbid conditions and in the context of healthy sleep habits and in combination with cognitive-behavioral strategies.

KEY WORDS

- BZD
- BZRA
- CBT
- GABA
- OTC
- SCN
- antihistamine
- benzodiazepines
- hypnotics
- insomnia
- melatonin
- orexin
- pharmacology
- polysomnography
- suprachiasmatic nucleus

REFERENCES

1. Ohayon, M. M. Prevalence and Comorbidity of Sleep Disorders in General Population. *Rev. Prat.* **2007**, *57*, 1521–1528.
2. Maharg, L. *The Savvy Insomniac*; Fine Fettle Books: Ann Arbor, MI, **2013**, p 320.
3. American Psychiatric Association, In *Diagnostic and Statistical Manual of Mental Disorders, Fifth Edition*; American Psychiatric Association: Arlington, VA, 2013.
4. American Academy of Sleep Medicine, *International Classification of Sleep Disorders, 3rd Edition*; American Academy of Sleep Medicine: Darien, IL, **2014**, p 383.
5. Schutte-Rodin, S.; Broch, L.; Buysse, D.; Dorsey, C.; Sateia, M. Clinical Guideline for the Evaluation and Management of Chronic Insomnia in Adults. *J. Clin. Sleep Med.* **2008**, *4*, 487–504.
6. U. S. Food and Drug Administration Center for Food Safety and Applied Nutrition Kava-containing Dietary Supplements May Be Associated with Severe Liver Injury. http://www.cfsan.fda.gov/~dms/addskava.html.
7. Lewy, A. J. Clinical Applications of Melatonin in Circadian Disorders. *Dialogues Clin. Neurosci.* **2003**, *5*, 399–413.
8. Buscemi, N.; Vandermeer, B.; Hooton, N.; Pandya, R.; Tjosvold, L.; Hartling, L.; Baker, G.; Klassen, T. P.; Vohra, S. The Efficacy and Safety of Exogenous Melatonin for Primary Sleep Disorders. A Meta-Analysis. *J. Gen. Intern. Med.* **2005**, *20*, 1151–1158.
9. Roehrs, T.; Roth, T. Sleep, Sleepiness, Sleep Disorders and Alcohol Use and Abuse. *Sleep Med. Rev.* **2001**, *5*, 287–297.
10. Krystal, A. D.; Richelson, E.; Roth, T. Review of the Histamine System and the Clinical Effects of H1 Antagonists: Basis for a New Model for Understanding the Effects of Insomnia Medications. *Sleep Med. Rev.* **2013**, *17*, 263–272.
11. Thomas, A.; Nallur, D. G.; Jones, N.; Deslandes, P. N. Diphenhydramine Abuse and Detoxification: A Brief Review and Case Report. *J. Psychopharmacol.* **2009**, *23*, 101–105.
12. Gerretsen, P.; Pollock, B. G. Drugs with Anticholinergic Properties: A Current Perspective on Use and Safety. *Expert Opin. Drug Saf.* **2011**, *10*, 751–765.
13. Rudolph, J. L.; Salow, M. J.; Angelini, M. C.; McGlinchey, R. E. The Anticholinergic Risk Scale and Anticholinergic Adverse Effects in Older Persons. *Arch. Intern. Med.* **2008**, *168*, 508–513.
14. Richardson, G. S.; Roehrs, T. A.; Rosenthal, L.; Koshorek, G.; Roth, T. Tolerance to Daytime Sedative Effects of H1 Antihistamines. *J. Clin. Psychopharmacol.* **2002**, *22*, 511–515.
15. Krystal, A. D. Antidepressant and Antipsychotic Drugs. *Sleep Med. Clin.* **2010**, *5*, 571–589.
16. Mayers, A. G.; Baldwin, D. S. Antidepressants and Their Effect on Sleep. *Hum. Psychopharmacol.* **2005**, *20*, 533–559.
17. Mendelson, W. B. A Review of the Evidence for the Efficacy and Safety of Trazodone in Insomnia. *J. Clin. Psychiatry.* **2005**, *66*, 469–476.
18. Morlock, R. J.; Tan, M.; Mitchell, D. Y. Patient Characteristics and Patterns of Drug Use for Sleep Complaints in the United States: Analysis of National Ambulatory Medical Survey data, 1997–2002. *Clin. Ther.* **2006**, *28*, 1044–1053.
19. U. S. Food and Drug Administration FDA Requests Label Change for All Sleep Disorder Drug Products. http://www.fda.gov/NewsEvents/Newsroom/PressAnnouncements/2007/ucm108868.htm.
20. Gottesmann, C. GABA Mechanisms and Sleep. *Neuroscience.* **2002**, *111*, 231–239.
21. Mohler, H.; Fritschy, J. M.; Rudolph, U. A New Benzodiazepine Pharmacology. *J. Pharmacol. Exp. Ther.* **2002**, *300*, 2–8.
22. Krystal, A. D. A Compendium of Placebo-Controlled Trials of the Risks/Benefits of Pharmacological Treatments for Insomnia: The empirical Basis for U.S. Clinical Practice. *Sleep Med. Rev.* **2009**, *13*, 265–274.
23. Buscemi, N.; Vandermeer, B.; Friesen, C.; Bialy, L.; Tubman, M.; Ospina, M.; Klassen, T. P.; Witmans, M. The Efficacy and Safety of Drug Treatments for Chronic Insomnia in Adults: A Meta-Analysis of RCTs. *J. Gen. Intern. Med.* **2007**, *22*, 1335–1350.
24. Rudolph, U.; Mohler, H. GABA-Based Therapeutic Approaches: GABAA Receptor Subtype Functions. *Curr. Opin. Pharmacol.* **2006**, *6*, 18–23.
25. U.S. Food and Drug Administration Zolpidem Containing Products: Drug Safety Communication - FDA Requires Lower Recommended Doses. http://www.fda.gov/Safety/MedWatch/SafetyInformation/SafetyAlertsforHumanMedicalProducts/ucm334738.htm.
26. U. S. Food and Drug Administration FDA warns of next-day impairment with sleep aid Lunesta

(eszopiclone) and lowers recommended dose. http://www.fda.gov/Drugs/DrugSafety/ucm397260.htm.

27. Dijk, D. J.; von Schantz, M. Timing and Consolidation of Human Sleep, Wakefulness, and Performance by a Symphony of Oscillators. *J. Biol. Rhythms.* **2005**, *20*, 279–290.

28. Zee, P. C.; Manthena, P. The Brain's Master Circadian Clock: Implications and Opportunities for Therapy of Sleep Disorders. *Sleep Med. Rev.* **2006**, *11*, 59–70.

29. Barion, A.; Zee, P. C. A Clinical Approach to Circadian Rhythm Sleep Disorders. *Sleep Med.* **2007**, *8*, 566–577.

30. Kato, K.; Hirai, K.; Nishiyama, K.; Uchikawa, O.; Fukatsu, K.; Ohkawa, S.; Kawamata, Y.; Hinuma, S.; Miyamoto, M. Neurochemical Properties of Ramelteon (TAK-375), A Selective MT1/MT2 Receptor Agonist. *Neuropharmacology.* **2005**, *48*, 301–310.

31. Karim, A.; Tolbert, D.; Zhao, Z. Single and Multiple Dose Pharmacokinetic Evaluation of Ramelteon (TAK-375) in Subjects with and Without Hepatic Impairment. *J . Clin. Pharmacol.* **2004**, *44*, 1210.

32. Sainati, S.; Tsymbalov, S.; Demissie, S.; Roth, T. Double-Blind, Placebo-Controlled, Two-Way Crossover Study of Ramelteon in Subjects with Mild to Moderate Chronic Obstructive Pulmonary Disease. *Sleep.* **2005**, *28* (*Abstract Supplement*), A162.

33. Kryger, M.; Wang-Weigand, S.; Roth, T. Safety of Ramelteon in Individuals with Mild to Moderate Obstructive Sleep Apnea. *Sleep Breath.* **2007**, *11*, 159–164.

34. Griffiths, R. R.; Johnson, M. W. Relative Abuse Liability of Hypnotic Drugs: A Conceptual Framework and Algorithm for Differentiating Among Compounds. *J. Clin. Psychiatry.* **2005**, *66 Suppl9*, 31–41.

35. Lockley, S. W.; Dressman, M. A.; Xiao, C.; Fisher, D. M.; Torres, R.; Lavedan, C.; Licamele, L.; Polymeropoulos, M. H. Tasimelteon Treatment Entrains the Circadian Clock and Demonstrates a Clinically Meaningful Benefit in Totally Blind Individuals with Non-24-Hour Circadian Rhythms. In 95th Annual Meeting of the Endocrine Society, San Francisco, CA, June 15–18, 2013; Abstract SUN-134.

36. Lockley, S. W.; Dressman, M. A.; Xiao, C.; Licamele, L.; Polymeropoulos, M. H. RESET Study Demonstrates that Tasimelteon Maintains Entrainment of Melatonin and Cortisol in Totally Blind Individuals with Non-24-Hour Circadian Rhythms. In 95th Annual Meeting of the Endocrine Society, San Francisco, CA, June 15–18, 2013; Abstract SUN-137.

37. Zlotos, D. P.; Jockers, R.; Cecon, E.; Rivara, S.; Witt-Enderby, P. A. MT1 and MT2 Melatonin Receptors: Ligands, Models, Oligomers, and Therapeutic Potential. *J. Med. Chem.* **2014**, *57*, 3161–3185.

38. Tashiro, M.; Yanai, K. Molecular Imaging of Histamine Receptors in the Human Brain. *Brain Nerve.* **2007**, *59*, 221–231.

39. Montoro, J.; Sastre, J.; Bartra, J.; del Cuvillo, A.; Davila, I.; Jauregui, I.; Mullol, J.; Valero, A. L. Effect of H1 Antihistamines upon the Central Nervous System. *J. Investig. Allergol. Clin. Immunol.* **2006**, *16 Suppl1*, 24–28.

40. Roth, T.; Rogowski, R.; Hull, S.; Schwartz, H.; Koshorek, G.; Corser, B.; Seiden, D.; Lankford, A. Efficacy and Safety of Doxepin 1 mg, 3 mg, and 6 mg in Adults with Primary Insomnia. *Sleep.* **2007**, *30*, 1555–1561.

41. Scharf, M.; Rogowski, R.; Hull, S.; Cohn, M.; Mayleben, D.; Feldman, N.; Ereshefsky, L.; Lankford, A.; Roth, T. Efficacy and Safety of Doxepin 1 mg, 3 mg, and 6 mg in Elderly Patients with Primary Insomnia: A Randomized, Bouble-Blind, Placebo-Controlled Crossover Study. *J. Clin. Psychiatry.* **2008**, *69*, 1557–1564.

42. Nishino, S. The Hypocretin/Orexin Eeceptor: Therapeutic Prospective in Sleep Disorders. *Expert. Opin. Investig. Drugs.* **2007**, *16*, 1785–1797.

43. Mignot, E. A Commentary on the Neurobiology of the Hypocretin/Orexin System. *Neuropsychopharmacology.* **2001**, *25*, S5–13.

44. Winrow, C. J.; Gotter, A. L.; Cox, C. D.; Doran, S. M.; Tannenbaum, P. L.; Breslin, M. J.; Garson, S. L.; Fox, S. V.; Harrell, C. M.; Stevens, J.; Reiss, D. R.; Cui, D.; Coleman, P. J.; Renger, J. J. Promotion of Sleep by Suvorexant-A Novel Dual Orexin Receptor Antagonist. *J. Neurogenet.* **2011**, *25*, 52–61.

45. Abbott Laboratories ProSom Prescribing Information. **2006**.

46. ECR Pharmaceuticals ZolpiMist Prescribing Information. **2010**.

47. King Pharmaceuticals Inc. Sonata Prescribing Information. **2006**.

48. Meda Pharmaceuticals Edluar Prescribing Information. **2013**.

49. Merck Sharp & Dohme Corp Belsomra Prescribing Information. **2014.**

50. Mylan Pharmaceuticals Inc. Restoril Prescribing Information. **2010.**

51. Neuro Pharma, I. Doral Prescribing Information. **2013.**

52. Purdue Pharma LP Intermezzo Prescribing Information. **2012.**

53. Roxane Laboratories Inc. Halcion Prescribing Information. **2012.**

54. Sanofi-Aventis Ambien CR Prescribing Information. **2013.**

55. Sanofi-Aventis Ambien Prescribing Information. **2013.**

56. Somaxon Pharmaceuticals, I. Silenor Prescribing Information. **2010.**

57. Sunovion Pharmaceuticals, I. Lunesta Prescribing Information. **2012.**

58. Takeda Pharmaceuticals North America Rozerem Prescribing Information. **2010.**

59. West-Ward Pharmaceutical Corp. Dalmane Prescribing Information. **2010.**

10

NON-PHARMACOLOGICAL MANAGEMENT OF INSOMNIA

Matthew Ebben and Forrest Armstrong

ABSTRACT

Although many clinicians may be inclined to first give patients medications to treat chronic insomnia, the National Institutes of health has advocated for the use of non-pharmacological interventions as first line treatments for insomnia. This recommendation is based on evidence that non-pharmacological interventions are equally as effective as hypnotic medications and have longer lasting treatment effects, while also equipping patients with skill sets to help cope with future bouts of insomnia. The two most important models of insomnia are Spielman's 3-p model and Bootzin's operant conditioning model. The 3-p model by Spielman and colleagues views chronic insomnia through a diathesis-stress based lens and argues that constitutional diatheses are aggravated by acute stressors, resulting in the initial episode of insomnia, which is then maintained by compensatory behaviors such as spending excess time in bed. On the other hand, Bootzin's model posits that insomnia is a problem of operant learning, and the failure of the bedroom to serve as a discriminate cue for sleep onset. Thus, the intervention formulated by Spielman, Sleep Restriction Therapy, relies on restricting time in bed to achieve its aim, while Bootzin's intervention, stimulus control therapy, removes people from the bedroom if they are not falling asleep or sleepy. These two techniques have been combined with relaxation techniques, cognitive therapy, and sleep hygiene to create a multicomponent intervention called Cognitive Behavioral Therapy for Insomnia. A separate, emerging technique, known as Intensive Sleep Retraining also views insomnia through the operant model, and attempts to re-associate the bed with sleep by giving patients a great number of learning trials in rapid succession.

Matthew Ebben
Center for Sleep Medicine, Weill Medical College of Cornell University, NY 10065, USA

Forrest Armstrong
Center for Sleep Medicine, Weill Medical College of Cornell University, NY 10065, USA

Corresponding author: Matthew Ebben, mae2001@med.cornell.edu

10.1. INTRODUCTION

10.1.1. Importance of Non-pharmacological Management of Insomnia

While many clinicians may be inclined to employ hypnotic medications as a first-line treatment for insomnia, there are a myriad of reasons for which practitioners ought to consider making use of behavioral treatments before employing pharmacological agents. Foremost, a substantial number of publications have demonstrated that cognitive and behavioral interventions for insomnia produce a reduction of insomnia symptoms comparable to hypnotics,[1] and that treatment effects of non-pharmacologic interventions are maintained for longer periods of time,[2] as patients are equipped with a skill set which allows them to effectively cope with factors underlying the insomnia.[3]

In contrast, hypnotics, while potentially effective, likely fail to address underlying cognitive and behavioral factors responsible for the course of insomnia. As a result, insomnia symptoms are likely to rebound in cases where patients do not have immediate access to medication, develop a tolerance to the drug, or if the substance loses its effectiveness over time. Hypnotic use also carries the potential for abuse and adverse events,[4,5] the range of which depends on the substance used. Moreover, a few recent studies have suggested a relationship between hypnotic use and increased risk of mortality.[6,7]

10.1.2. 3P Model of Insomnia

In order to facilitate an understanding of the mechanisms of non-pharmacological treatments for insomnia, it is important to consider how insomnia manifests and persists. The 3P model of insomnia, first introduced by Spielman et al.,[8] proposes that three factors – predisposing traits, precipitating events, and perpetuating behaviors account for the development and course of insomnia. Precipitating factors are diatheses that make individuals more vulnerable to insomnia, such as constitutionally high anxiety or physiological arousal. Precipitating factors are stressful events (for example, personal loss or financial difficulties), that result in transient difficulty sleeping. Perpetuating factors are maladaptive behaviors and cognitions that may extend transient episodes of sleep difficulties into a chronic insomnia. Here, cognitive processes, such as catastrophizing the impact that a lack of sleep might have on daytime performance, may lead to excess worry and anxiety about the amount of sleep obtained, that, in turn, interferes with sleep. Further, the anxiety associated with sleep onset may result in conditioned arousal associated with the bedroom, making sleep even more difficult to initiate. In order to make up for lost sleep, individuals may also engage in maladaptive compensatory behaviors that perpetuate insomnia. Such behaviors include watching the clock in order to determine how much time remains to sleep, spending excess time in bed attempting to accrue more sleep which may enhance the learned association between the bedroom and sleeplessness, or napping during the day which may decrease sleep drive for the nighttime sleep period.

10.3. COGNITIVE BEHAVIORAL THERAPY AND ITS COMPONENTS

Cognitive behavioral therapy for insomnia (CBT-I) is a highly efficacious [2,9,10] multimodal treatment targeting cognitive and behavioral phenomena that perpetuate insomnia. The five subcomponents of CBT-I are sleep restriction, stimulus control instructions, cognitive therapy, sleep hygiene, and relaxation techniques. Before application of CBT-I or any of its component treatments, it is critical to ascertain a patient's sleep schedule by means of a sleep diary for at least two weeks. Although a consensus sleep diary, developed by the American Academy of Sleep Medicine, now exists,[11] this diary has been constructed mainly for research purposes. Clinicians may find a graphic diary to be of greater utility as such diary formats typically afford clinicians more nuanced information about a patient's sleep habits.

10.3.1. Sleep Restriction Therapy

10.3.1.1. Rationale

A technique introduced by Spielman et al.,[8] sleep restriction therapy (SRT) focuses on reducing time in bed. For individuals suffering with insomnia, much of the time in bed is spent in pained, futile attempts to initiate sleep, or worrying about the impact of the lack of sleep on daytime performance or health, all of which exacerbate insomnia. Sleep restriction therapy (SRT) acts through creating a sleep deficit by reducing the amount of time spent in bed. This restriction produces mild to moderate sleep deprivation (increased sleep drive), that facilitates sleep onset, and the maintenance and depth of sleep, while simultaneously reducing time in bed that might otherwise be given to rumination.

10.3.1.2. Description

Using two or more weeks of sleep diary data, an average sleep time is calculated, and the new prescribed amount of time in bed is set equal to this amount, with a minimum of at least 5 h in bed (for example, an individual who 6½ h on average, as demonstrated by sleep diaries, will be limited to 6½ h in bed). The patient is also required to get out of bed at the same time every morning, at a time that coincides with the time that the patient wakes up for work. During this time, patients are asked to fill out sleep diaries each morning. At the end of each week, the clinician uses the diary to determine the patient's sleep efficiency, which is the proportion of time asleep to the total amount of time spent in bed (time asleep in minutes/total time in bed in minutes × 100). After each week of treatment, if sleep efficiency is above 90%, and the patient remains sleepy during the day, he or she is given an additional quarter-hour to half-hour in bed. Similarly, if the patient's sleep efficiency is less than 85%, the amount of time in bed is decreased by 15 min. If sleep efficiency falls between 85% and 90%, there is no change to the amount of time in bed. Of course, changes to sleep and wake times should be made in a manner that is compatible with the individual's schedule.

10.3.1.3. Notes

Due to the sleep deprivation that occurs during the course of sleep restriction, patients may experience a number of undesired symptoms, such as fatigue, impaired cognitive performance, and excessive daytime sleepiness. It is important to recognize sleepiness is a desired outcome of sleep restriction, although the degree and timing of this sleepiness may impact treatment adherence. Nonetheless, patients may find these side effects difficult to tolerate, particularly during the first few weeks of treatment.[12] Patients should be informed of these potential side effects and instructed to refrain from driving or operating heavy machinery, should they feel impaired.

10.3.1.4. Variant: Sleep Compression

During sleep compression, the amount of time in bed is slowly scaled back, rather than radically reduced. For example, for an individual who sleeps an average of 6 h a night but remains in bed an additional 2 h a night, either bedtime or wake time would be reduced by 20 min each week over the course of six weeks.

Caveats: Either a fixed bed or rise time must be adhered to throughout the protocol. In addition, the reduction of time in bed should not be too rapid, as faster reductions may result in an unstable circadian rhythm.

10.3.2. Stimulus Control Therapy

10.3.2.1. Rationale

Stimulus control Therapy (SCT), a technique developed by Bootzin [13] , is derived from learning theory. From this perspective, insomnia can be viewed as a learned association between sleeplessness and the bed, resulting from repeated learning trials wherein the stimulus (bedroom or bed-

time) is consistently followed by difficulty sleeping caused by arousal. Thus, stimulus control therapy (SCT) is an intervention that attempts to eliminate conditioned arousal related to the bedroom and to re-establish the bedroom as a discriminant cue for sleep onset.

10.3.2.2. Description

Stimulus control has five general instructions to patients:

a) Use the bed only for sleep and sex. This instruction aims to eliminate arousing stimuli, which may come to be associated with the bedroom. For instance, watching stimulating television programs in bed may lead to conditioned arousal.

b) Get into bed only when sleepy. Attempting to fall asleep when sleep drive is high, of course, increases the likelihood of rapid sleep onset. In addition, this instruction aims to create learning trials in which the bed is strongly associated with swift sleep onset. One caveat to this instruction is that patients may need to learn to distinguish sleepiness and fatigue. Although a patient maybe fatigued and feel a need to rest, he or she may have low homeostatic sleep drive (sleepiness) and should therefore stay out of the bed.

c) If unable to sleep after 10–20 min, get out of bed, move to another room and engage in a low-stimulation activity with as little light as possible. Return to bed only when feeling sleepy. Repeat this as many times during the night as necessary.

Awakenings are often followed by long, pained attempts to return to sleep that are arousing, resulting in further arousal and increased difficulty returning to sleep; thus, getting out of bed after a short time attempts to quash this cycle. Quiet activities are meant to prevent arousal that may arise from more stimulating activities that may leave the patient unable to fall asleep as easily. Of course, as in the first rule, patients should only return to the bed if they feel sleepy.

d) Set a standard wake time. That is, rise at the same time every day. This instruction aims to prevent patients from spending excess time in bed, thereby consolidating sleep in addition to helping entrain the circadian rhythm.

e) Do not nap during the day. Napping has the potential to decrease the homeostatic sleep drive, thereby making sleep onset and maintenance more difficult.

10.3.2.3. Notes

Stimulus control may be contraindicated for individuals who have difficulty getting out of bed unassisted, or who are prone to falls. While SCT and SRT have been found to be generally comparable in efficacy,[14] SCT is more easily implemented than SRT and may produce fewer adverse effects.

10.3.3. Relaxation Therapy

From the perspective of the 3P model, there are a number of constitutional factors that predispose individuals to insomnia,

such as, increased heart rate, cortisol levels, and other metabolic properties.[17] Unlike sleep restriction and stimulus control, relaxation techniques most often attempt to dampen the physiological and psychological hyperarousal that predispose individuals to an episode of insomnia, rather than eliminating perpetuating behaviors. Potential relaxation techniques include progressive muscle relaxation, mindfulness-based stress reduction, diaphragmatic breathing, and guided imagery.

10.3.3.1. Progressive Muscle Relaxation

Of these techniques, progressive muscle relaxation (PMR) has been the most studied. PMR attempts to reduce physiological arousal by means of focusing the patient's awareness on the relief of muscle tension. At first, during the day, the seated patient is asked to tighten a muscle from 10–15 s, and then to relax that muscle an equal amount of time, while noting the change in sensation. This process is then repeated for a number of different muscles, moving from a proximal region of the body upward. For example, a patient may begin with the calves, and from this point move to the quadriceps, then abdominals, and so on. Interestingly, while PMR has repeatedly been demonstrated to an effective intervention for insomnia, it is not an effective technique for ameliorating psychological tension. Indeed, one study found that PMR had a negative effect in insomniacs without muscle tension.[18]

10.3.3.2. Breathing Techniques

Diaphragmatic breathing techniques have their roots in Yogic practices in India dat-ing from circa 500 BCE.[19] Here, the patient is instructed inhale deeply through the nose such that there is a marked expansion of the belly caused by flexion of the diaphragm. The breath is held for approximately 10 s, and then slowly exhaled over a similar duration for as many cycles as the patient feels comfortable with. Again, this technique is to be used during the day at first, and after sufficient practice may be employed during the night before bed or during awakenings.

10.3.3.3. Mindfulness-Based Stress Reduction

A number of studies have demonstrated that the mindfulness-based stress reduction (MBSR) technique introduced by Kabat-Zinn may be somewhat effective in reducing anxiety in a number of different illnesses.[20] Recently, a few studies have demonstrated the efficacy CBT-I incorporating mindfulness training.[21,22] MBSR's underlying tenants are based on acceptance, non-judgment, and non-reactivity. In its application to the treatment of insomnia, MBSR aims to promote non-judgment and acceptance of sleeplessness and its potential daytime consequences, rather than anxiety or catastrophization.

10.3.3.4. Guided Imagery

Guided imagery is a relaxation technique that aims to replace anxious thoughts with peaceful ones. During guided imagery, patients are asked to close their eyes and recall an experience or location in which they were alone and felt highly relaxed. The practitioner then guides the patient by having the patient recall particular details of the experience, especially sensory experiences such as scents, images, sounds, and so forth. A few rules are generally helpful to ensure the patients' story is conducive to sleep. These include: 1) excluding all other people from the story; 2) make sure the story has a beginning, middle, and end; and 3) make the visualization as detailed as possible. Once the patient is proficient with this technique, it is typically used at bedtime, but can also be used during nighttime awakenings, if racing thoughts are a problem. Guided imagery is often paired with "thought stopping" and using a "worry list" in order to control unwanted cognitions from intruding on the patients' story.

10.3.4. Cognitive Therapy

Insomnia is frequently marked by a number of dysfunctional beliefs and attitudes pertaining to sleep loss and its consequences. For example, some individuals may believe that they will suffer daytime deficits disproportionate to the amount of sleep loss, or that their insomnia will have severe long-term health impacts. In general, these maladaptive cognitions can be grouped into three categories: 1) mistaking passing difficulties sleeping for a chronic problem; 2) attributing poor daytime performance to insomnia; 3) overestimation of sleep need; 4) sleep-related anxiety; and 5) Other common cognitive errors, such as over generalization or magnification.[23] These cognitive errors aggravate nighttime anxiety, and may begin a cycle in which beliefs about sleeplessness' perceived negative outcomes produce anxiety that prevents falling back to sleep. During cognitive therapy, the clinician challenges the patient's erroneous beliefs surrounding sleep loss. While one study [24] has shown cognitive therapy to be an effective stand-alone treatment for insomnia, cognitive therapy is typically viewed as a component of a multimodal treatment. In particular, some of the components of CBT-I such as sleep restriction may trigger anxiety or catastrophization; therefore, cognitive therapy may prove integral to successful implementation of CBT-I's other components.

10.3.5. Sleep Hygiene

10.3.5.1. Rationale

Sleep may be negatively impacted by an individual's lifestyle and habits; thus, poor "sleep hygiene" may exacerbate some cases of insomnia. Sleep hygiene instructions guide patients in altering those lifestyle factors that prove deleterious to sleep. In spite of the role that poor sleep hygiene may play in insomnia, sleep hygiene has not been shown as an effective standalone treatment for insomnia, but is a frequent addition to behavioral treatments.

10.3.5.1. Description

In presenting the rules of sleep hygiene, it is critical that the clinician explain that the listed factors that may contribute to disturbed sleep are not necessarily causal. Further, the clinician should offer both a rationale for each rule as well as discuss the relevance to the patient.

Although the rules of sleep hygiene are quite variable within the literature, they can generally be reduced to the following:[25]

a) Do not take daytime naps.
Taking naps during the day decreases homeostatic sleep drive.

b) Do not use caffeine within 6 h of getting into bed.
Caffeine remains in the body and has a long chemical life. Taking caffeine later in the day may prevent the patient from falling asleep easily, and may decrease sleep quality as well.

c) Avoid using alcohol to sleep.
While alcohol can bring a swift sleep onset, it often fragments sleep.

d) Avoid using nicotine.
Nicotine is an activating substance that may make falling asleep more difficult.

e) Use earplugs or a white noise machine.
Noises may lead to awakenings, Use earplugs or a white noise machine to mask other noises.

f) Avoid sleeping with pets.
Pets may make disturbing movements or noises during the night, resulting in sleep disruption.

g) Make the bedroom a comfortable place to sleep.
An uncomfortable bed will likely contribute to sleep disruption. Additionally, the bedroom should be kept cool, dry, and dark for optimal sleeping conditions.

h) Keep a regular schedule.
A regular sleep schedule strengthens circadian entrainment.

10.3.6. INTENSIVE SLEEP RETRAINING

10.3.6.1 Rationale

Intensive sleep retraining (ISR), Is a treatment developed by Harris et al.[15] Like

SCT, ISR rests on the principles of learning theory and aims to create a learned association between the bed and a swift sleep onset. In contrast to SCT, whose course of treatment runs a number of weeks, ISR consists of a great number of learning trials that occur in rapid succession. Further, treatment is conducted in the laboratory, rather than at home.

10.3.6.2. Description

ISR comprises three distinct phases:

a) Preparation for Treatment

Daily sleep diary data should be collected over the two weeks preceding treatment. In order to facilitate sleep onset during treatment, patients are restricted to 5 h in bed the night before treatment and are awoken by an alarm. Naps and caffeine are forbidden in the 24 h before treatment.

b) Treatment

Patients are first outfitted with polysomnographic recording equipment and asked to complete a subjective sleepiness self-report measure. Afterward, every 30 min the patient undergoes a sleep trial, in which the lights are turned out, and the patient is given a 20 min opportunity to sleep. Technicians monitoring the patient's sleep, wake the patient after 3 min of any stage of sleep. After the trial, patients are asked whether feel they had slept, and are then given objective feedback. After this awakening, or 20 min without sleep, patients are asked to get out of bed and perform quiet activities, such as reading or watching television until the next trial. Trials continue in this fashion over the course of 25 h. Thus, during treatment, there are a total of 50 sleep opportunities.

c) Post-Treatment

Given that, certain factors that contribute to insomnia may persist when patients return to their home; a six-week regimen of SCT should be initiated the day after treatment and sleep diaries should be completed during this time. Indeed, this treatment sequence has been found to produce greater response to treatment and longer lasting gains than ISR alone.16 The night following treatment, the patient should limit their time in bed to 8 h to ensure stability of the circadian rhythm.

10.3.6.3.Notes

ISR is contraindicated in cases where acute sleep deprivation may aggravate pre-existing conditions such as seizures or bipolar disorder.

While IRS has proven to be an effective treatment for sleep onset insomnia, there is little evidence supporting its effectiveness for other insomnia variants.

10.4. SUMMARY

Because the nature of insomnia is complex and generally multifactorial, CBT-I includes a number of components, each with different therapeutic targets. Further, patients may have any number idiosyncratic difficulties or responses to a given component over time; therefore, each component must be tailored to the needs of the patient, and patients' progress must be well-monitored. For example, in cases where individuals have worries or fears about curtailing sleep time, particular components may be difficult for individuals to adhere to. This may require that cognitive therapy and stimulus control be introduced before sleep restriction. Given this need for constant refinement and modification, implementation of CBT-I requires a fair deal of clinical skill.

The clinical gains of CBT-I are both large and well maintained. In light of the potential adverse effects of medications, patients' preference for behavioral interventions and the equal efficacy of behavioral and pharmacological treatments, first-line treatments consist of behavioral, rather than pharmacological interventions. Indeed, this guideline has found support from both the American Academy of Sleep Medicine and National Institutes of Health.[26]

KEY WORDS

- **behavioral**
- **cognitive**
- **insomnia**
- **non-pharmacological**
- **CBT-I**

REFERENCES

1. Smith, M. T.; Perlis, M. L.; Park, A.; Smith, M. S.; Pennington, J.; Giles, D. E.; Buysse, D. J. Comparative Meta-Analysis of Pharmacotherapy and Behavior Therapy for Persistent Insomnia. *Am. J. Psychiatry.* **2002**, *159*, 5–11.

2. Morin, C. M.; Colecchi, C.; Stone, J.; Sood, R.; Brink, D. *JAMA.* **1999**, *281*, 991–999.

3. Morin, C. M. *Sleep Med.* **2006**, *7 Suppl 1*, S15–S19.

4. O'brien, C. P. *J. Clin. Psychiatry.* **2005**, *66 Suppl 2*, 28–33.

5. Hajak, G.; Müller, W. E.; Wittchen, H. U.; Pittrow, D.; Kirch, W. Abuse and Dependence Potential for the Non-Benzodiazepine Hypnotics Zolpidem and Zopiclone: A Review of Case Reports and Epidemiological Data. *Addiction.* **2003**, *98*, 1371–1378.

6. Mallon, L.; Broman, J. E.; Hetta, J. *Sleep Med.* **2009**, *10*, 279–286.

7. Kripke, D. F.; Langer, R. D.; Kline, L. E. Hypnotics' Association with Mortality or Cancer: A Matched Cohort Study. *BMJ Open.* **2012**, *2*, e000850–e000850.

8. Spielman, A. J.; Saskin, P.; Thorpy, M. J. *Sleep.* **1987**, *10*, 45–56.

9. Morin, C. M.; Culbert, J. P.; Schwartz, S. M. *Am. J. Psychiatry.* **1994**, *151*, 1172–1180.

10. Edinger, J. D.; Wohlgemuth, W. K.; Radtke, R. A.; Marsh, G. R.; Quillian, R. E. *JAMA.* **2011**, *285*, 1856–1864.

11. Carney, C. E.; Buysse, D. J.; Ancoli-Israel, S.; Edinger, J. D.; Krystal, A. D.; Lichstein, K. L.; Morin, C. M. The Consensus Sleep Diary: Standardizing Prospective Sleep Self-Monitoring. *Sleep.* **2012**, *35*, 287–302.

12. Kyle, S. D.; Morgan, K.; Spiegelhalder, K.; Espie, C. A. *Sleep Med.* **2011**, *12*, 735–747.

13. Bootzin, R. R. Stimulus Control Treatment for Insomnia. Proceedings of the 80th Annual Convention of the American Psychological Association, Honolulu, Hawaii, September 2–8, 1973; Vol 7, 395–396

14. Epstein, D. R.; Sidani, S.; Bootzin, R. R.; Belyea, M. J. Dismantling Multicomponent Behavioral Treatment for Insomnia in Older Adults: A Randomized Controlled Trial. *Sleep.* **2012**, *35*, 797–805.

15. Harris, J.; Lack, L.; Wright, H.; Gradisar, M.; Brooks, A. *J. Sleep Res.* **2007**, *16*, 276–284.

16. Harris, J.; Lack, L.; Kemp, K.; Wright, H.; Bootzin, R. A Randomized Controlled Trial of Intensive Sleep Retraining (ISR): A Brief Conditioning Treatment for Chronic Insomnia. *Sleep.* **2012**, *35*, 49–60.

17. Bonnet, M. H.; Arand, D. L. *Sleep Med. Rev.* **2010**, *14*, 9–15.

18. Hauri, P. J., Percy, L., Hellekson, C., Hartmann, E., & Russ, D. *Biofeedback Self. Regul.* **1982**, *7*, 223–234.

19. Samuel, G. *The Origins of Yoga and Tantra: Indic Religions to the Thirteenth Century*; Cambridge University Press: Cambridge, UK, 2008, p. 9.

20. Bohlmeijer, E.; Prenger, R.; Taal, E.; Cuijpers, P. *J. Psychosom. Res.* **2010**, *68*, 539–544.

21. Gross, C. R.; Kreitzer, M. J.; Reilly-Spong, M.; Wall, M.; Winbush, N. Y.; Patterson, R.; Mahowald, M.; Cramer-Bornemann, M. *Explore (NY).* **2011**, *7*, 76–87.

22. Ong, J. C.; Manber, R. *Behav. Ther.* **2011**, *39*, 171–182.

23. Morin, C. M.; Stone, J.; Trinkle, D.; Mercer, J.; Remsberg, S. *Psychol. Aging.* **1993**, *8*, 463–467.

24. Harvey, A. G.; Sharpley, A. L.; Ree, M. J.; Stinson, K.; Clark, D. M. *Behav. Res. Ther.* **2007**, *45*, 2491–2501.

25. Posner, D.; Gehrman, P. R. In *Behavioral Treatments for Sleep Disorders*; Perlis, M.; Aloia, M.; Kuhn, B., Eds.; Academic Press: Massachusetts, MA, 2011; pp 31–43.

26. National Institutes of Health. National Institutes of Health State of the Science Conference statement on Manifestations and Management of Chronic Insomnia in Adults, June 13-15, 2005. *Sleep.* **2005**, *28*, 1049–1057.

11

SLEEP AND ITS DISORDERS IN WOMEN

There are clear sex based physiological differences in both normal sleep and the pathogenesis of sleep disorders. The hormonal changes that occur during a woman's life span significantly impact both normal sleep patterns and risk and presentation of sleep disorders. The prevalence of sleep disorders and its symptomatology exhibit clear sex based differences. For instance, chronic insomnia and sleep disordered breathing (SDB) both sex specific incidences and variable presentations. The impact of restless legs syndrome (RLS) on pregnancy health and fetal outcomes has only been recently explored. Recent data has also shown increased risk of endometrial cancer in shift working women. The incidence of comorbidities and the therapeutic responses to various pharmacological and non-pharmacological interventions also show gender and sex differences. The

Hrayr Attarian
Department of Neurology, Circadian Rhythms and Sleep Research Lab, Northwestern University, Chicago, IL, USA

Corresponding author: Hrayr Attarian, E-mail: hattaria@nm.org

doses of some sleep aids such as zolpidem are different for women and men as is the prevalence of depression and its association with CPAP noncompliance.

A fair amount of these sex differences remain unknown and under-recognized.

Complex reproductive endocrine changes that a woman's body goes through, from puberty to menopause and with pregnancy, affect her sleep and the presentation of sleep disorders significantly. Some of this uniqueness is due purely to sex hormones but there are also some other physiological elements that make the specificity of a woman's sleep patterns and the pathophysiology of some of the disorders multifactorial.

This chapter will explore some of the normal sleep differences in women compared to men and will also discuss individual disorders and how they impact women at different points in their life span.

11.1. NORMAL SLEEP PHYSIOLOGY

Gender-specific sleep differences start early on in life. Female infants from ages 6–36 months appear to sleep better, particularly they fall back to sleep faster after middle of the night awakenings. Male infants have roughly twice the trouble falling back to sleep than females (1). Girls, starting at age three, have higher N3 sleep and higher sleep efficiency, findings that remain consistent throughout their lifespan. As they get older, girls also have higher percentage of N1 sleep (2).

With puberty and adolescence these sex-based differences become more pronounced. Although cultural differences exist, in general teenage girls spend more time in bed but get less sleep and have longer sleep latencies. The normal circadian delay of adolescence starts and ends earlier in females than males (14–19 versus 16–21) and is less pronounced (3).

The majority of studies documenting differences in sleep parameters between the genders are done in adults. Sleep architecture studies demonstrate a significantly higher N3 percentage among women that remains relatively unchanged from mid 30s through past age 70, while men experience a gradual decrement in N3 or slow wave sleep as they age (4).

Expanding the age to 22–88 years demonstrates the same decline in stage N3 sleep in men and a corresponding increase in stage N1 that does not occur in women. Sleep efficiency (SE); percentage of time in bed spent sleeping, in the same cohort shows that women have higher SE most of their lives that remains higher than that of men despite an age-related decline. Only in very early and very late adulthood do women experience less SE than men (5)

During the menstrual cycle itself, there are consistent sleep architecture changes. Percentage of stage R declines from 25 to 28% in the follicular phase (FP) to about

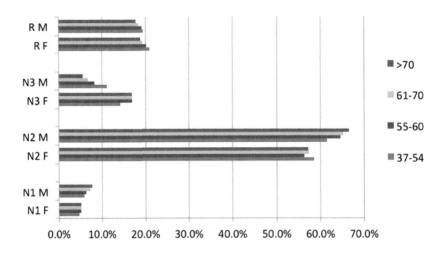

Adapted from Redline S, Kirchner HL, Quan SF, Gottlieb DJ, Kapur V, Newman A. The effects of age, sex, ethnicity, and sleep-disordered breathing on sleep architecture. Archives of internal medicine. 2004;164(4):406-18.

Figure 11.1. Demonstrates the differences in different sleep stages between the genders and in different age group.

22–25% in the luteal phase (LP). Stage N3 sleep also decreases from around 20% in FP to 15% in LP. In addition sleep onset latency (SL) decreases from a maximum of 25 min in FP to a nadir of 10 min in LP and SE increases from a minimum of 88% in FP to a maximum 94% in LP. These changes correlate with an increase in progesterone level and gradual decline in estrogen levels. Minimum core body temperature is lower by 0.4 °C during LP than in FP. There are no significant changes in melatonin except for a sharp drop during ovulation coinciding with a sharp rise in luteal hormone (6).

When women take hormonal contraception, their body temperature remains

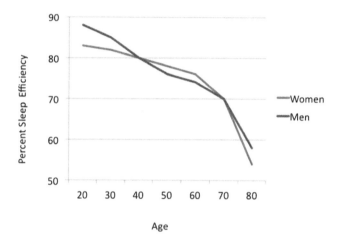

Adapted from Bixler EO, Papaliaga MN, Vgontzas AN, et al. Women sleep objectively better than men and the sleep of young women is more resilient to external stressors: effects of age and menopause. Journal of Sleep Research 2009;18(2):221-8.

Figure 11.2. Differences in percent sleep efficiency between the genders and in different age group.

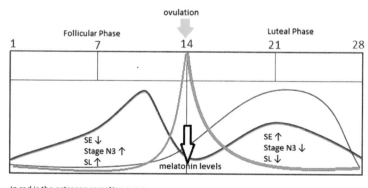

In red is the estrogen secretion curve.
In blue is the progesterone secretion curve.

Figure 11.3. Normal sleep architecture changes and hormonal changes during a woman's menstrual cycle.

constant throughout their cycle, higher than usual, and with less diurnal changes. There is also an increase in melatonin levels without a change in dim light melatonin onset or DLMO (7).

Women with premenstrual dysphoric disorder (PMDD), on the other hand, have lower overall melatonin levels (a difference of 14.25 pg/ml in FP and 24.18 pg/ml in LP), higher prolactin levels, and increased subjective complaints with only minor changes in sleep architecture (6).

11.2. OBSTRUCTIVE SLEEP APNEA (OSA)

For decades, OSA was thought to be a masculine illness as it is much more prevalent in men. OSA defined as an apnea–hypopnea index (AHI) of 5/h or more occurs in 9% of adult women versus 24% of adult men. If the definition is changed to include complaints of subjective daytime sleepiness, then the estimate becomes 14% in men and 5% in women and if only moderate to severe OSA is taken into consideration (AHI > 15/h) then the prevalence becomes 13% for men and 6% for women. As in men, increase in body mass index (BMI) proportionately increases the severity of OSA as does age. There is, however, a more dramatic increase (about 5 fold)

in prevalence of OSA among women after menopause (8). Table 11.1 summarizes the prevalence of OSA (AHI $\geq$ 5/h) based on BMI and age [adapted from Reference (8)].

Women with OSA go undiagnosed more often and for longer periods of time. Almost 93% of women compared to 80% of men with OSA remains undiagnosed, and for men it takes an average of 7.8 years to get diagnosed with OSA from onset of symptoms while for women it takes approximately twice as long. Men are eight times more likely to be diagnosed with OSA than women despite the M/F ratio being around 2.6/1 (9).

Some of this delay and under-diagnosis in women is due to gender bias but some is due to the fact that women do not present with the same symptomatology as men do. The establishment of the classic symptoms of OSA was done mostly in male cohorts hence another source of gender bias was introduced there (10). Regardless, women are 2–3 times less likely to report excessive daytime sleepiness, snoring, and observed apneas (male bed partners also tend to be less observant which may influence the reporting of the last two symptoms in heterosexual women). Most common symptoms of OSA in women are insomnia, fatigue, depressed mood, and morning

Table 11.1. Prevalence of OSA in women by age and weight.

Weight	Premenopausal	Postmenopausal
Normal weight (BMI < 25 kg/m^2)	1.44%	9.3%
Overweight (BMI 26–29.9 kg/m^2)	4.2%	20.2%
Obese (BMI 30–39.9 kg/m^2)	13.5%	41.1%
Severely obese (BMI > 40 kg/m^2)	43%	67.9%

headache; hence they are often evaluated for hypothyroidism, depression, and anxiety. Insomnia is present in 20% of women with OSA and has lead to the recommendation of screening all menopausal women with insomnia for OSA (11). Sleep disturbances and neurobehavioral impact of these disruptions is greater in women and the response to treatment and compliance with it is the same as it is with men (12).

Hormonal factors play an important role in development of OSA in women. OSA defined as daytime sleepiness and an AHI > 10/h occurred in 2.7% of menopausal women not on hormone replacement therapy (HRT) versus 0.5% among menopausal women on HRT. Combined HRT also reduced the severity of OSA by 50% while estrogen alone reduced it by 25% (13). Despite these studies, given the potential morbidities associated with HRT, it is not recommended as treatment of OSA in menopausal women.

Polycystic ovarian syndrome (PCOS), a condition characterized by increased testosterone levels also increases the risk of OSA in women independent of weight. The prevalence of OSA in PCOS is about 44.4–47% compared to 5.5–15% age and weight matched controls. Testosterone levels and central adiposity correlate with increased risk of OSA and its severity (14).

Women with OSA tend to be more obese than their male counterparts even after controlling for severity of OSA. The contribution of obesity, however, as a risk factor for OSA in and of itself is not different between sexes (15). Few of the contributing factors to this may be the differential size of men's upper airway diameter versus women's and the difference in ventilator

drive in response to O_2 and CO_2 between males and females.

OSA tends to be more rapid eye movement (REM)-dependent in women than in men. About 78% of patients with REM-dependent OSA tend to be women and 62% of women with OSA have predominantly REM-related OSA (16). Figure 11.4 illustrates a hypnogram of a woman with almost exclusively REM-dependent OSA.

The known complications of OSA tend to manifest in women slightly differently than men. For instance, the effect of OSA on risk of stroke seems to be highest in young women. In large cohort of about 30,000 women with OSA (AHI $\geq$ 5/h), the impact of OSA on stroke risk decreased with age (adjusted HR 4.90, 95% CI 1.93–12.4 for subgroup aged 20–35 years; adjusted HR 1.64, 95% CI 1.01–2.65 for subgroup aged 36–50 years; adjusted HR 1.38, 95% CI 1.01–1.89 for subgroup aged 51–65 years) (17). OSA also increases the risk of cardiovascular disease in women with an HR of 1.6 (CI, 0.52–4.90) to 3.5 (CI, 1.23–9.98) depending on severity of OSA. *Continuous positive airway pressure* (CPAP) treatment reduces the risk to 0.19–0.55 after a median usage of 72 months (18).

Glucose tolerance is another complication of OSA that tends to be more prominent in premenopausal women. It is also more responsive to weight loss and improvement in nocturnal hypoxemia (19).

Lastly cognitive and functional decline are more prominent in older women with OSA. After adjusting for age, obesity, minimental state examination score, depressive symptoms, history of hypertension, sleep quality and duration and chronic obstructive pulmonary disease, the adjusted OR

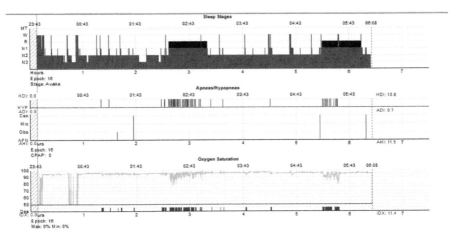

Figure 11.4. Hypnogram showing REM dependent OSA in a woman

for a decline in activities of daily living was 2.22–3.07 (depending on measures used) in women aged > 75 years with moderate-severe OSA (AHI > 15/h) (20).

Older women were also at an OR of 1.7–2.04 of developing cognitive impairment and dementia with moderate and severe OSA after controlling for all the above variables (21). This risk almost doubled in a subgroup of women with the APOE epsilon4 allele (22).

11.2.1. Pregnancy and OSA

Pregnancy is a state of significant albeit temporary changes in respiratory physiology. Some of these alterations are conducive to the development of OSA while others are protective against it. First there are the obvious anatomical alterations due to a gravid uterus. Shifting of intra-abdominal organs, elevation of diaphragm and an increase in intra-abdominal pressure affect breathing during sleep. Second there is the impact of hormonal changes on sleep and breathing as both estrogen and progesterone change sleep microstructure and basic respiratory physiology.

Functional residual capacity (FRC) gradually decreases by about 20% in the latter months of pregnancy primarily due to the anatomical changes. This leads to increase in minute ventilation and tidal volume, which results in higher PaO_2 pressures, and lower $PaCO_2$ pressures. The purpose of this PaO_2 gradient is to shunt oxygen to the fetus and allow its excretion of CO_2 but it can lead to maternal hypoxemia that can worsen OSA (23). This normal decline in nocturnal oxygen saturation in late pregnancy is exacerbated by sleeping in the supine and in increased likelihood of upper airway collapse. These same anatomical changes however make it hard for women to lie on their back hence, in a way, becoming a protective mechanism by inducing more side sleep (24). The accelerated increase in BMI in pregnancy although not an independent risk factor for gestational OSA, certainly can worsen pre-existing OSA in women with preconception baseline obesity (25).

Upper airway mucosal congestion and hyperemia that raise estrogen levels, particularly in the third trimester, directly lead to snoring and upper airway resistance. Nasal congestion is reported in up to 30% of women in their third trimester and the Mallampati score (a measure of oropharyngeal crowding) increases in the third trimester compared to the first (26). Estrogen, however, decreases REM sleep and progesterone increases non-rapid eye movement (NREM) sleep. OSA in young women is often worse in REM sleep therefore reduction in time spent in that vulnerable state may be another protective mechanism. Lastly progesterone induced increase in minute volume is in itself against protective airway occlusion.

Habitual snoring, often considered a marker for OSA increases from 4% in non-pregnant women to 25% in women during gestation (27). The association between habitual snoring and adverse fetal outcomes has been a point of much debate with two trials showing intrauterine growth retardation (IUGR) and two not showing any adverse fetal outcomes (28). None of these studies utilized poly somnography (PSGs) hence the prevalence of OSA could not be ascertained in any of these cohorts. An increase in inflammatory markers in umbilical cord blood has been demonstrated among habitual snorers although this does not always correlate with adverse fetal outcomes (29).

Snoring, however, may not necessarily mean that there is underlying OSA. The prevalence of PSG-confirmed OSA among pregnant habitual snorers has been reported to be only 11.4 % (30). Neither standardized questionnaires nor overnight oximetry measurements have high predictive value for OSA in pregnancy (31). A novel four-variable model that includes pregestational hypertension (assigned a value of 15 points), habitual snoring (assigned another 15 points), obesity and age, is so far, the screening algorithm with the highest positive and negative predictive values. A score of > 75 (15 + 15 + BMI + Age) conferred a high likelihood of OSA in pregnant women (32).When OSA is confirmed, the one fetal adverse outcome that is strongly associated with it is *intrauterine growth restriction* (IUGR) in the form of both small for gestational age (SGA) infants and low birth weight (LBW) (33). Pregnant women with OSA who use CPAP give birth to infants with higher average birth weights (516 gm or more) (33). The most plausible pathophysiology of IUGR in OSA is placental ischemia because of maternal hypoxemia. In addition, smaller infants born to mothers with confirmed OSA tend to have lower Apgar scores and are more likely to be born preterm and by cesarean section (33).

OSA is also a major risk factor for pregnancy-induced hypertension, pre-eclampsia, and eclampsia. Habitual snoring during pregnancy especially if its onset was after conception is significantly associated with gestational hypertension and pre-eclampsia. Odds ratios (OR) for hypertension are 2.03 for any habitual snoring during pregnancy, 2.36 for pregnancy onset snoring (27), and 1.59 for pre-eclampsia (27). When OSA is confirmed by PSG, then it emerges as an independent and significant risk factor for pre-eclampsia and gestational hypertension with ORs ranging from 1.6 to 7.5 (34,35).

There is threefold increase in odds of gestational diabetes mellitus (GDM) with OSA in pregnancy (34,35). Short sleep duration, less than or equal to 7 h per night, and frequent snoring; more than three nights a week are also independent risk factors for GDM (33). The risk ratio (RR) of GDM with short sleep (≥ 4 h/night) alone is 5.56 while with habitual snoring alone it is 1.86. When short sleep, weight and snoring are combined the GDM risk is increased by 3.23 folds in lean women (pre-pregnancy BMI of $< 25\text{kg/m}^2$) and by almost 10 fold in overweight/obese women (pre-pregnancy BMI of $> 25\text{kg/m}^2$). Finally when considering weight and habitual snoring without sleep duration, the RR for GDM among overweight and obese women is 6.9 (34). The potential mechanisms for GDM in OSA are oxidative stress and lowering of adiponectin levels, both of which lead to insulin resistance (36,37). Table 11.2 summarizes the major complications of OSA in pregnancy.

Table 11.2. Potential complications of OSA in pregnancy.

Complication	Likelihood
Pregnancy induced hypertension	Definite
Gestational diabetes	Definite
IUGR	Possible
C-Section	Possible
Preterm birth	Possible

The treatment for OSA in pregnancy is primarily CPAP and in an auto mode if possible. Given that the OSA gets progressively worse as the pregnancy nears, its completion in auto CPAP has the flexibility to treat the dynamic nature of the OSA in pregnancy. CPAP tends to both alleviate OSA and incrementally improve BP control and may even improve fetal health (36).

11.3. RESTLESS LEGS SYNDROME (RLS)

RLS is a neurological disorder characterized by an urge to move the limbs (legs more than arms) often accompanied by uncomfortable sensations. These become worse in the evening or at night and when sedentary for a prolonged period of time. The diagnostic criteria for RLS are the following 1) an urge to move the limbs (legs>arms) usually caused or accompanied by uncomfortable or painful sensations in the limbs, 2) periods of rest or inactivity, such as lying or sitting worsen the urge to move or unpleasant sensations, 3) movements such as walking or stretching partially or totally relieve the urge and 4) the urge to move or unpleasant sensations are worse in the evening or at night or occur exclusively in the evening or at night.

A meta-analysis of all epidemiological studies from across the globe has shown varying degrees of prevalence among different ethnicities and geographical locations (37). One thing, however, that has remained constant throughout is the female to male ratio of 1.5–2/1. See Figure 11.5: Prevalence of RLS Around the World.

The blank areas on the map have no data on RLS.

Primary reason for this increased female preponderance is the significantly increased risk of RLS with pregnancy. In

Figure 11.5. Prevalence of RLS in various countries.

fact RLS is one of the most prevalent sleep disorders in pregnancy. RLS occurs in 13–25% of women in the late second and third trimester of pregnancy. This is a 2–3-fold increase in prevalence as compared to the general population. There is a precipitous improvement in RLS a month after delivery with up to 87% of those who had gestational RLS (gRLS) reporting complete resolution of symptoms and the remainder have 50% or more improvement in severity scores (38). Two studies of gRLS report almost half of subjects having a severe case of it based on IRLSG guidelines (38) while one study reported 14% in the severe range and 75% in the moderate range (39)

Risk factors for gRLS include personal history of RLS (OR of 12.9), family history of RLS (OR 8.43), gRLS in previous pregnancies (OR 53.7), smoking and obesity (40). Other variables with conflicting results include estrogen levels, anemia, fer-

ritin levels, folic acid levels and fetal growth induced stretching and compression of nerves.

The pathophysiology of gRLS has not been clearly elucidated. Although iron stores and its surrogate marker ferritin levels have been implicated this does not explain the entire phenomenon of gRLS. Iron supplementation improves gRLS only in women whose ferritin levels are < 75 ng/ml. In addition the precipitous improvement within a month of delivery does not correlate with replenishment of iron stores, which may take up to 3 months to return to prepregnancy levels. Estrogen levels have correlated with periodic limb movement sleep indices (PLMS index) but not with gRLS. PLMS is a flawed surrogate for RLS. Estrogen replacement therapy in menopausal women was associated with an OR of 2.5 for RLS but hormonal contraceptives containing estrogen do not increase the risk of RLS in younger, premenopausal women.

Progesterone levels do not correlate with RLS and although prolactin levels are associated with PLMS and it is a dopamine antagonist, the data otherwise does not support its association with RLS symptoms. For one the diurnal levels of prolactin do not correlate with worsening of RLS symptoms in both pregnant and non-pregnant women. Another piece of evidence against a prolactin-gRLS association is that prolactin remains high in women who breastfeed but the postpartum rapid improvement in RLS is no different in this cohort versus those who do not breastfeed.

Treatment:

Given that most approved pharmacological measures of RLS are pregnancy category C or higher the treatment of gRLS is problematic. Table 11.3 summarizes the treatment recommendations of gRLS based on clinical consensus (adapted from Picchietti et al.) (41).

First step is ascertaining the accuracy of diagnosis by doing a thorough face-to-face interview and making sure the four diagnostic features all exist. It is also recommended ruling out mimics of RLS such as lower extremity edema, leg cramps, positional discomfort, legs "going to sleep", compression neuropathies, leg pain and tendon/ligament sprain. In mild cases of gRLS reassurance may be sufficient.

Second step is to identify and avoid (if possible) triggering factors. These include assessing iron status and supplementing orally until ferritin is >75 ng/ml or if oral supplementation fails then considering IV iron treatment. Other triggers to avoid include serotoninergic agents, periods of prolonged immobility, antihistamines and antiemetics, caffeine, tobacco and alcohol and lastly evaluating and treating OSA.

Third line therapies are the non-pharmacological measures. These include yoga, moderate exercise, pneumatic compression devices, massage, and hypnosis.

If all these fail and pharmacotherapy is needed then low dose carbidopa/levodopa 25/100–50/200 mg or clonazepam 0.25–1 mg at bedtime is recommended during pregnancy. In refractory, very severe cases low dose oxycodone is recommended. During lactation gabapentin 300–900 mg in the evening or clonazepam 0.251 mg

Table 11.3. gRLS treatment recommendations.

First line	Accurate diagnosis	Ruling out mimics
Second line	Reassurance in mild cases	Avoiding precipitating factors (meds, immobility, OSA)
Third line	Iron replacement to Ferritin >75 ng/ml	Massage, moderate exercise or pneumatic devices
Fourth line: pregnancy	Low dose clonazepam or carbidopa levodopa	If refractory and severe: low dose oxycodone
Fourth line: lactation	Low dose clonazepam or gabapentin	If refractory and severe: low dose tramadol

may be used. In severe refractory cases low dose tramadol may be a good substitute (41).

11.4. INSOMNIA

Chronic insomnia disorder, as defined by the *International Classification of Sleep Disorders* (ICSD) 3, is characterized by difficulty falling and/or staying asleep associated with daytime sequelae attributable to poor sleep occurring at least three times a night and for three months or more. Chronic insomnia disorder that affects around 9–11% of the general population is slightly more common in women than it is in men with a ratio of 1.2–1.5/1 (42,43).

Figure 11.6 Prevalence of Insomnia in Women (adapted from Leger et al.) (43,44)

The cause of this female preponderance is multifactorial and includes the strong, bidirectional association between insomnia and depression. Depressive symptoms are more common in women and chronic insomnia itself can increase the risk of a major depressive disorder by four folds. Another variable is the high prevalence of insomnia at particular points in a woman's life cycle; primarily during pregnancy and menopause.

11.4.1. Insomnia During Pregnancy

Difficulty falling asleep, staying asleep and overall both subjectively and objectively reduced sleep are common pregnancy related complaints. These generally are worse in the third trimester with over half of women complaining of poor sleep on a nightly basis in the last eight weeks of their pregnancy.

There is a strong association between depression before and during gestation and pregnancy- induced insomnia as well as between habitual smoking and 3rd trimester insomnia (45,46).

Manifestations of insomnia in late pregnancy include two to three nightly awakenings that result in an average wake after sleep onset (WASO) of 30–45 min and reduced sleep time to less than 7 h and sometimes as little as 4 h a night. Most women also report a commensurate increase in daytime napping. A total of sleep time of 6 h or less per night occurs more commonly in Nulliparous women, and those older than 30 years of age, as compared to those who are younger or have been pregnant before (47). The secretion of oxytocin, a stimulating hormone, makes insomnia even worse as labor approaches. Insomnia tends to persist up to three

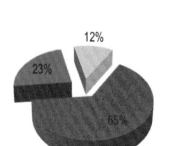
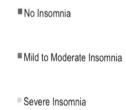

12%

23%

65%

■ No Insomnia

■ Mild to Moderate Insomnia

▪ Severe Insomnia

Figure 11.6. Prevalence of insomnia among women.

months postpartum especially among women who have undergone C-sections. Also women who bottle-feed sleep on average, 45 min less per night than those who breastfeed. Other factors impacting maternal sleep during the post-partum period including co-bedding and infant temperament as a third of infants can be colicky the first 2–3 months of life (48).

of poor gestational sleep is post partum depression (PPD). Poor sleep quality and curtailed sleep during pregnancy increases the risk of PPD with an OR of 7.7 especially if maternal sleep was reduced to 4 h or less between midnight and 6 AM. Napping less than one hour a day during pregnancy is also an independent risk factor for PPD as is perceived poor sleep quality (48).

11.4.2. Complication of Pregnancy Related Insomnia

Curtailed and poor quality sleep reduces a woman's pain tolerance during labor and makes it harder for her to cope with it. In addition women who sleep 6 h or less on average per night tend to have longer labors and higher chance of C-sections than those who sleep seven or more hours per night (49). The most important complication

11.4.2.1. Treatment of Pregnancy-Related Insomnia

As most sedative hypnotics are potentially teratogenic the treatment of pregnancy-related insomnia must be primarily based on cognitive and behavioral approaches (CBT).

Figure 11.7 summarizes the CBT approaches to insomnia treatment (Sánchez-Ortuño and Edinger) (50).

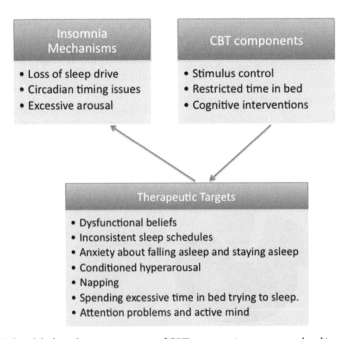

Figure 11.7. Simplified graphic representation of CBT measures in pregnancy related insomnia.

11.4.3. Insomnia During Menopause

Chronic insomnia is common in the setting of menopause. A wide range of prevalence from 11.8 to 56.6% is reported in the medical literature (11). The heterogeneity of assessment tools and regional and cultural differences account for to this large variation. Vasomotor symptoms (VMS); hot flushes and night sweats, psychosocial stress, obesity, and age are all commonly associated factors with insomnia in menopause. Insomnia in menopause is strongly associated with poor health related quality of life, decreased work productivity, and increased healthcare utilization. In addition it has been correlated with anxiety and depression, aortic disease, coronary and cardiovascular disease (51).

11.4.3.1. Treatment of Chronic Insomnia in Menopause

Other sleep disorders such as RLS and OSA have a high prevalence in women with menopausal insomnia. Evaluating for these conditions and treating them when present should be first line when addressing chronic insomnia in menopausal women.

Although there is very little written on CBT in this population, given that CBT is safe and extremely effective in other forms of insomnia, it should be second line.

In those in whom both RLS and OSA have been ruled out or ruled in and treated and they have failed CBT, Table 11.4 summarizes the efficacy of various pharmacological and non pharmacological agents and their efficacy (adapted from Attarian et al.) (11).

Table 11.4. Interventions for insomnia in menopause.

Treatment	Recommendation	Concerns
Low dose conjugated estrogen	Suggested	Potential risk of vascular and oncogenic adverse events.
Isoflavones	Suggested	Issues with purity
Valerian	Suggested	Issues with purity
Escitalopram	Suggested	Common SSRI related side effects
Gabapentin	Suggested	Weight gain, next day sedation
Zolpidem	Suggested	Next day sedation, automatic behaviors, falls
Eszopiclone	Suggested	Next day sedation, automatic behaviors
Exercise	Suggested	None
Hypnosis	Suggested	None
Phytofemale complex	May be considered	Issues with purity extract
Mirtazapine+melatonin	May be considered	Weight gain/issues with purity of melatonin
Citalopram	May be considered	Common SSRI related side effects
Ramelteon	May be considered	Sedation, nightmares
Yoga	May be considered	None
Massage	May be considered	None

HRT has the best available data supporting its efficacy. Caution should be exercised, however, when recommending HRT because of the potentially high risk of adverse events associated with its use.

Other interventions that have been found reasonably effective include, gabapentin, eszopiclone, escitalopram and zolpidem. The dose of the latter should not exceed 5 mg a night in women per the revised FDA guidelines. Valerian root, isoflavones, hypnosis, and high intensity exercise also have been found reasonably effective and are much safer than the prescription medications. Interventions of questionable efficacy include pycogenol and phytofemale complex (proprietary blend of black cohosh, dong quai, milk thistle, red clover, ginseng, and chaste tree berry fruit extract), yoga, massage, citalopram, mirtazapine + long acting melatonin, and ramelteon.

11.5. PARASOMNIAS

The NREM parasomnias that fall on the continuum of disorders of arousal (confusional arousals, sleep terror, and sleep walking) do not exhibit any significant sex and gender predilection in their prevalence or incidence. The only NREM parasomnia that is clearly female predominant is Sleep Related Eating Disorder (SRED). SRED is characterized by repeated episodes of eating arising out of the major sleep periods. There is complete or partial loss of awareness during these episodes and the eating of unusual food combinations and even toxic non-food substance may lead to injury or harm. Female to male ratio of prevalence ranges from 1.5/1 to 4/1 and there is a strong association with depres-

sion and daytime eating disorders. Workup includes ruling out other sleep disorders such as OSA, RLS, irregular sleep wake disorder and exposure to zolpidem or other benzodiazepine receptor agonist sedative hypnotics. Treatment, when an underlying condition is not identified, is usually low dose topiramate or low dose pramipexole (52). This and other parasomnias are discussed in detail elsewhere in the textbook.

Of the remaining parasomnias the only one with a slight but clear sex predilection in prevalence is REM sleep behavior disorder or RBD. Initially thought to be a predominantly male disease, later studies have demonstrated that the M/F ratio actually is about 1.62/1. The initial impression of it being primarily a man's disease had to do with the differential content of dreams that men and women experience with RBD. Women tend to have less violence themed dreams and their dream enactment tends to be less physical (53).

11.6. FEW WORDS ON SHIFT WORK DISORDER

Shift work disorder in women has been associated with increased risk of endometrial and breast cancer but not that of other malignancies such as ovarian cancer or melanoma.

The odds ratio for endometrial cancer is 1.47 in night shift workers and 2.09 among obese night shift workers (54). The OR for breast cancer with shift work is 1.34 but increased to 3.69 for women who were shift workers and slept ≥ 9 h per day.

Suppression of endogenous melatonin has been postulated to play a role in increased risk of these malignancies (55). Other illnesses for which shift work increases a woman's risk include metabolic syndrome, OR of 2.29, diabetes (OR 1.17–1.42 depending on years of shift work) (56), cardiovascular disease (57) and increased risk of miscarriages and infertility (58).

11.7. CONCLUSION

Certain sleep disorders are more common in women while others have a unique, sex-specific presentation. It is important to recognize the symptoms and signs and the treatments available for women in specific periods of their lifecycle. The impact of untreated sleep disorders can lead to significant morbidity and mortality while therapeutic interventions, that are relatively simple and safe, can make a big difference in their overall wellbeing and health (Table 11.5).

Table 11.5. Key Questions when evaluating a woman for OSA.

Do you feel fatigued during the day?	Unlike men, women with OSA are less likely to be sleepy so they will not report high scores on standardized questionnaires like the Epworth Sleepiness Scale. So eliciting fatigue is very important
Do you snore or have been told you stop breathing in your sleep? Do you wake up with a sore throat or wake up in the middle of the night gasping?	Although this is an important question to ask not all women will have astute enough bed partners to give an account of these symptoms. The second set of questions may be more telling
Do you have non-refreshing sleep and difficulty concentrating?	These are often surrogate symptoms for daytime sleepiness
Do you have difficulty staying asleep?	Often women with OSA present with difficulty staying asleep characterized by multiple short awakenings
Is your mood depressed? If so did the mood problem precede or follow the fatigue?	Although most women with OSA may endorse depression it often is in the setting of the profound fatigue and related to the disability that the fatigue causes
History of lack of response to sedative hypnotics and/ or antidepressants.	Many women present to their primary physicians with these complaints and are often treated with sedative hypnotics or misdiagnosed with depression
In premenopausal women with a BMI < 30 kg/m2 with a high clinical suspicion of OSA it helps eliciting a history of PCOS	History of PCOS is elicited by inquiring about hirsutism, regularity of menses, infertility etc. Often these symptoms may go unexplored in the primary care setting

Table 11.6. Questions to ask when evaluating a woman with insomnia.

Do you have difficulty falling asleep or staying asleep?	Unlike with OSA most women with primary insomnia complain of both falling asleep and staying asleep
Are you able to nap?	Another key feature of differentiating primary insomnia from OSA or other common comorbid sleep disorders is their inability to nap as they are hyposomnolent day and night. A woman with OSA disrupting their sleep may be able to fall asleep during the day for short naps.
Do you feel sleepy during the day?	Women with primary insomnia clearly state that they are not sleepy during the day but feel the lack of sleep is making them function poorly at home and at work. They often score very low on the Epworth Sleepiness Scale with a score of 0–2
Were you a light sleeper before these symptoms?	Although not necessarily the case with everyone with primary insomnia a majority of women with it report not being "great sleepers" most of their lives even before these symptoms started
Did the symptoms start with menopause?	If the symptoms started with menopause it is very important to explore for climacteric symptoms of hot flashes and night sweats, symptoms of OSA and RLS
Do you feel anxious about sleep?	Most of women with primary insomnia complain that their anxiety is particularly around sleep while those where the insomnia is due to a generalized anxiety disorder have an overwhelming anxiety permeating all aspects of their life

KEY WORDS

- **sex differences**
- **RLS**
- **SBD**
- **OSA**
- **parasomnias**
- **circadian rhythms**
- **sleep architecture**
- **shift work disorder**
- **circadian rhythm disorder**
- **pregnancy**
- **insomnia**

REFERENCES

1. Weinraub, M.; Bender, R. H.; Friedman. S. L.: Susman, E. J., Knoke, B.; Bradley, R.; et al. Patterns of Developmental Change in Infants' Nighttime Sleep Awakenings from 6 Through 36 Months of Age. *Dev. Psychol.* **2012**, *48*(6), 1511–1528.

2. Montgomery-Downs, H. E.; O'Brien, L. M.; Gulliver, T. E.; Gozal, D. Polysomnographic Characteristics in Normal Preschool and Early School-Aged Children. *Pediatrics* **2006**, *117*(3), 741–753.

3. Hysing, M.; Pallesen, S.; Stormark, K. M.; Lundervold, A. J.; Sivertsen, B. Sleep Patterns and Insomnia Among Adolescents: A Population-Based Study. *J. Sleep. Res.* **2013**, *22*(5), 549–556.

4. Redline, S.; Kirchner, H. L.; Quan, S. F.; Gottlieb, D. J.; Kapur, V.; Newman, A. The Effects of Age, Sex, Ethnicity, and Sleep-Disordered Breathing on Sleep Architecture. *Adv. Intern. Med.* **2004**, *164(4)*, 406–418.

5. Bixler, E. O.; Papaliaga, M. N.; Vgontzas, A. N.; Lin, H. M.; Pejovic, S.; Karataraki, M.; et al. Women Sleep Objectively Better Than Men and the Sleep of Young Women is More Resilient to External Stressors: Effects of Age and Menopause. *J. Sleep. Res.* **2009**, *18(2)*, 221–228.

6. Shechter, A.; Lesperance, P.;, Ng Ying Kin, N. M.; Boivin, D. B. Nocturnal Polysomnographic Sleep Across the Menstrual Cycle in Premenstrual Dysphoric Disorder. *Sleep Med.* **2012**, *13(8)*, 1071–1078.

7. Baker, F. C.; Driver, H. S. Circadian Rhythms, Sleep, and the Menstrual Cycle. *Sleep Med.* **2007**, *8(6)*, 613–622.

8. Peppard, P. E.; Young, T.; Barnet, J. H.; Palta, M.; Hagen, E. W.; Hla, K. M. Increased Prevalence of Sleep-Disordered Breathing in Adults. *Am. J. Epidemiol.* **2013**, *177(9)*, 1006–1014.

9. Fuhrman, C.; Fleury, B.; Nguyen, X. L.; Delmas, M. C. Symptoms of Sleep Apnea Syndrome: High Prevalence and Underdiagnosis in the French Population. *Sleep Med.* **2012**, *13(7)*, 852–858.

10. Quintana-Gallego, E.; Carmona-Bernal, C.;, Capote, F.; Sanchez-Armengol, A.; Botebol-Benhamou, G.; Polo-Padillo, J.; et al. Gender Differences in Obstructive Sleep Apnea Syndrome: A Clinical Study of 1166 Patients. *Respir. Med.* **2004**, *98(10)*, 984–989.

11. Attarian, H.; Hachul, H.; Guttuso, T.; Phillips, B. Treatment of Chronic Insomnia Disorder in Menopause: Evaluation of Literature. *Menopause* **2014**. 22(6):674–684.

12. Ye, L.; Pien, G. W.; Ratcliffe, S. J.; Weaver, T. E. Gender Differences in Obstructive Sleep Apnea and Treatment Response to Continuous Positive Airway Pressure. *J. Clin. Sleep. Med.* **2009**, *5(6)*, 512–518.

13. Wesstrom, J.; Ulfberg, J.; Nilsson, S. Sleep Apnea and Hormone Replacement Therapy: A Pilot Study and a Literature Review. *Acta. Obstet. Gynecol. Scand.* **2005**, *84(1)*, 54–57.

14. Mokhlesi, B.; Scoccia, B.; Mazzone, T.; Sam, S. Risk of Obstructive Sleep Apnea in Obese and Nonobese Women with Polycystic Ovary Syndrome and Healthy Reproductively Normal Women. *Fertil. Steril.* **2012**, *97(3)*, 786–791.

15. Phillips, B. A.; Collop, N. A.; Drake, C.; Consens, F.; Vgontzas, A. N.; Weaver, T. E. Sleep Disorders and Medical Conditions in Women. Proceedings of the Women & Sleep Workshop, National Sleep Foundation, Washington, DC, March 5-6, 2007. *J. Womens Health* **2008**, *17(7)*, 1191–1199.

16. Conwell, W.; Patel, B.; Doeing, D.; Pamidi, S.; Knutson, K. L.; Ghods, F.; et al. Prevalence, Clinical Features, and CPAP Adherence in REM-Related Sleep-Disordered Breathing: A Ccross-Sectional Analysis of a Large Clinical Population. *Sleep. Breath.* **2012**, *16(2)*, 519–526.

17. Chang, C. C.; Chuang, H. C.; Lin, C. L.; Sung, F. C.; Chang, Y. J.; Hsu, C. Y.; et al. High Incidence of Stroke in Young Women with Sleep Apnea Syndrome. *Sleep Med.* **2014**, *15(4)*, 410–414.

18. Campos-Rodriguez, F.; Martinez-Garcia, M. A.; de la Cruz-Moron, I.; Almeida-Gonzalez, C.; Catalan-Serra, P.; Montserrat, J. M. Cardiovascular Mortality in Women with Obstructive Sleep Apnea with or without Continuous Positive Airway Pressure Treatment: A Cohort Study. *Ann. Intern. Med.* **2012**, *156(2)*, 115–122.

19. Gilardini, L.; Lombardi, C.; Redaelli, G.; Vallone, L.; Faini, A.; Mattaliano, P.; et al. Glucose Tolerance and Weight Loss in Obese Women with Obstructive Sleep Apnea. *PloS one* **2013**, *8(4)*, e61382.

20. Spira, A. P.; Stone, K. L.; Rebok, G. W.; Punjabi, N. M.; Redline, S.;, Ancoli-Israel, S.; et al. Sleep-Disordered Breathing and Functional Decline in Older Women. *J. Am. Geriatr. Soc.* **2014**, *62(11)*,2040–2046.

21. Yaffe, K.; Laffan, A. M.; Harrison, S. L.; Redline, S.; Spira, A. P.; Ensrud, K. E.; et al. Sleep-Disordered Breathing, Hypoxia, and Risk of Mild Cognitive Impairment and Dementia in Older Women. *Jama.* **2011**, *306(6)*, 613–619.

22. Spira, A. P.; Blackwell, T.; Stone, K. L.; Redline, S.; Cauley, J. A.; Ancoli-Israel, S.; et al. Sleep-Disordered Breathing and Cognition in Older Women. *J. Am. Geriatr. Soc.* **2008**, *56(1)*, 45–50.

23. Bobrowski, R. A. Pulmonary Physiology in Pregnancy. *Clin. Obstet. Gynecol.* **2010**, *53*(2), 285–300.

24. Morong, S.; Hermsen, B.; de Vries, N. Sleep-Disordered Breathing in Pregnancy: A Review of the Physiology and Potential Role for Positional Therapy. *Sleep. Breath.* **2014**, *18*(1), 31–37.

25. Davies, G. A.; Maxwell, C.; McLeod, L.; Gagnon, R.; Basso, M.; Bos, H.; et al. Obesity in Pregnancy. *J. Obstet. Gynaecol. Can.* **2010**, *32*(2), 165–173.

26. Izci, B.; Vennelle, M.; Liston, W. A.; Dundas, K. C.; Calder, A. A.; Douglas, N. J. Sleep-Disordered Breathing and Upper Airway Size in Pregnancy and Post-Partum. *Eur. Respir. J.* **2006**, *27*(2), 321–327.

27. O'Brien, L. M.; Bullough, A. S.; Owusu, J. T.; Tremblay, K. A.; Brincat, C. A.; Chames, M. C.; et al. Pregnancy-Onset Habitual Snoring, Gestational Hypertension, and Preeclampsia: Prospective Cohort Study. *Am. J. Obstet. Gynecol.* **2012**, *207*(6), 487 e1–9.

28. Micheli, K.; Komninos, I.; Bagkeris, E.; Roumeliotaki, T.; Koutis, A.; Kogevinas, M.; et al. Sleep Patterns in Late Pregnancy and Risk of Preterm Birth and Fetal Growth Restriction. *Epidemiology* **2011**, *22*(5),738–744.

29. Tauman, R.; Many, A.; Deutsch, V.; Arvas, S.; Ascher-Landsberg, J.; Greenfeld, M.; et al. Maternal Snoring During Pregnancy is Associated with Enhanced Fetal Erythropoiesis--A Preliminary Atudy. *Sleep Med.* **2011**, *12*(5), 518–522.

30. Guilleminault, C.; Querra-Salva, M.; Chowdhuri, S.; Poyares, D. Normal Pregnancy, Daytime Sleeping, Snoring and Blood Pressure. *Sleep Med.* **2000**, *1*(4), 289–297.

31. Tantrakul, V.; Sirijanchune, P.; Panburana, P.; Pengjam, J.; Suwansathit, W.; Boonsarngsuk, V.; et al. Screening of Obstructive Sleep Apnea during Pregnancy: Differences in Predictive Values of Questionnaires across Trimesters. *J. Clin. Sleep. Med.* **2015**, *11*(2), 157–163.

32. Facco, F. L.; Ouyang, D. W.; Zee, P. C.; Grobman, W. A. Development of a Pregnancy-Specific Screening Tool for Sleep Apnea. *J. Clin. Sleep. Med.* **2012**, *8*(4), 389–394.

33. Fung, A. M.; Wilson, D. L.; Lappas, M.; Howard, M.; Barnes, M.; O'Donoghue, F.; et al. Effects of Maternal Obstructive Sleep Apnoea on Fetal Growth: A Prospective Cohort Study. *PloS one* **2013**, *8*(7), e68057.

34. Facco, F. L.; Ouyang, D. W.; Zee, P. C.; Grobman, W. A. Sleep Disordered Breathing in a High-Risk Cohort Prevalence and Severity Across Pregnancy. *Am. J. Perinatol.* **2014**, *31*(10), 899–904.

35. O'Brien, L. M.; Bullough, A. S.; Chames, M. C.; Shelgikar, A. V.; Armitage, R.; Guilleminualt, C.; et al. Hypertension, Snoring, and Obstructive Sleep Apnoea During Pregnancy: A Cohort Study. *BJOG-Int. J. Obstet. Gy.* **2014**, *121*(13), 1685–1693.

36. Blyton, D. M.; Skilton, M. R.; Edwards, N.; Hennessy, A.; Celermajer, D. S.; Sullivan, C. E. Treatment of Sleep Disordered Breathing Reverses Low Fetal Activity Levels in Preeclampsia. *Sleep* **2013**, *36*(1), 15–21.

37. Ohayon, M. M.; O'Hara, R.; Vitiello, M. V. Epidemiology of Restless Legs Syndrome: A Synthesis of the Literature. *Sleep. Med. Rev.* **2012**, *16*(4), 283–295.

38. Hubner, A.; Krafft, A.; Gadient, S.; Werth, E.; Zimmermann, R.; Bassetti, C. L. Characteristics and Determinants of Restless Legs Syndrome in Pregnancy: A Prospective Study. *Neurology* **2013**, *80*(8), 738–742.

39. Vahdat, M.; Sariri, E.; Miri, S.; Rohani, M.; Kashanian, M.; Sabet, A.; et al. Prevalence and Associated Features of Restless Legs Syndrome in a Population of Iranian Women During Pregnancy. *Int. J. Gynecol. Obstetrics.* 2013, *123*, 46e9.

40. Sikandar, R.; Khealani, B. A.; Wasay, M. Predictors of Restless Legs Syndrome in Pregnancy: A Hospital Based Cross Sectional Survey from Pakistan. *Sleep Med.* **2009,** *10*(6), 676–678.

41. Picchietti, D. L.; Hensley, J. G.; Bainbridge, J. L.; Lee, K. A.; Manconi, M. McGregor, J. A.; et al. Consensus Clinical Practice Guidelines for the Diagnosis and Treatment of Restless Legs Syndrome/Willis-Ekbom Disease During Pregnancy and Lactation. *Sleep. Med. Rev.* **2014**, *18*, 1–14.

42. Chung, K. F.; Yeung, W. F.; Ho, F. Y.; Yung, K. P.; Yu. Y. M.; Kwok, C. W. Cross-Cultural and Comparative Epidemiology of Insomnia: The Diagnostic and Statistical Manual (DSM), International Classification of Diseases (ICD) and International Classifica-

tion of Sleep Disorders (ICSD). *Sleep Med.* **2015**;16(4):477–482.

43. Leger, D.; Partinen, M.; Hirshkowitz, M.; Chokroverty, S.; Hedner, J.; Investigators, E. S. Characteristics of Insomnia in a Primary Care Setting: EQUINOX Survey of 5293 Insomniacs from 10 Countries. *Sleep Med.* **2010**, *11*(*10*), 987–998.

44. Leger, D.; Guilleminault, C.; Dreyfus, J. P.; Delahaye, C.; Paillard, M. Prevalence of Insomnia in a Survey of 12,778 Adults in France. *J. Sleep. Res.* **2000**, *9*(*1*), 35–42.

45. Okun, M. L.; Kiewra, K.; Luther. J. F.; Wisniewski, S. R.; Wisner, K. L. Sleep Disturbances in Depressed and Nondepressed Pregnant Women. *Depress. Anxiety.* **2011**, *28*(*8*), 676–685.

46. Fernandez-Alonso, A. M.; Trabalon-Pastor, M.; Chedraui, P.; Perez-Lopez, F. R. Factors Related to Insomnia and Sleepiness in the

Late Third Trimester of Pregnancy. *Arch. Gynecol. Obstet.* **2012**, *286*(*1*), 55–61.

47. Tsai, S.Y.; Lin, J. W.; Kuo, L. T.; Thomas, K. A. Daily Sleep and Fatigue Characteristics in Nulliparous Women During the Third Trimester of Pregnancy. *Sleep* **2012**, *35*(*2*), 257–262.

48. Abbott, S. M.; Attarian, H.; Zee, P. C. Sleep

Disorders in Perinatal Women. *Best. Pract. Res. Clin. Obstet. Gynaecol.* **2014**, *28*(*1*), 159–168.

49. Chang, J. J.; Pien, G. W.; Duntley, S. P.; Macones, G. A. Sleep Deprivation During Pregnancy and Maternal and Fetal Outcomes: Is There a Relationship? *Sleep. Med. Rev.* **2010**, *14*(*2*),107–114.

50. Sanchez-Ortuno, M. M.; Edinger, J. D. Cognitive-Behavioral Therapy for the Manage-

ment of Insomnia Comorbid with Mental Disorders. *Curr. Psychiatry. Rep.* **2012**, *14*(*5*), 519–528.

51. Bolge, S. C.; Balkrishnan, R.; Kannan, H.; Seal, B.; Drake, C. L. Burden Associated with Chronic Sleep Maintenance Insomnia Characterized by Nighttime Awakenings among Women with Menopausal Symptoms. *Menopause* **2010**, *17*(*1*), 80–86.

52. Inoue, Y. Sleep-Related Eating Disorder and its Associated Conditions. *Psychiatry. Clin. Neurosci.* 2014;69(6):309–320.

53. Bjornara, K. A.; Dietrichs, E.; Toft, M. REM Sleep Behavior Disorder in Parkinson's Disease--Is There a Gender Difference? *Parkinsonism Relat. Disord.* **2013**, *19*(*1*), 120–122.

54. Viswanathan, A. N.; Hankinson, S. E.; Schernhammer, E. S. Night Shift Work and the Risk of Endometrial Cancer. *Cancer Res.* **2007**, *67*(*21*), 10618–10622.

55. Wang, P.; Ren, F. M.; Lin, Y.; Su, F. X.; Jia, W. H.; Su, X. F.; et al. Night-Shift Work, Sleep Duration, Daytime Napping, and Breast Cancer Risk. *Sleep Med.* **2015**.

56. Vimalananda, V. G.; Palmer, J. R.; Gerlovin, H.; Wise, L. A.; Rosenzweig, J. L.; Rosenberg, L.; et al. Night-Shift Work and Incident Diabetes Among African-American Women. *Diabetologia* **2015**, *58*(*4*), 699–706.

57. Gu, F.; Han, J.; Laden, F.; Pan, A.; Caporaso, N. E.; Stampfer, M. J.; et al. Total and Cause-Specific Mortality of u.s. Nurses Working Rotating Night Shifts. *Am. J. Prev. Med.* **2015**, *48*(*3*), 241–252.

58. Stocker, L. J.; Macklon, N. S.; Cheong, Y.C.; Bewley, S. J. Influence of Shift Work on Early Reproductive Outcomes: A Systematic Review and Meta-Analysis. *Obstet. Gynecol.* **2014**, *124*(*1*), 99–110.

12

SLEEP AND ANXIETY DISORDERS

Malcolm Lader

ABSTRACT

Sleep disturbances, in particular insomnia, are common in anxiety disorders. Consequently, complaints such as insomnia or nightmares have been incorporated in some definitions of anxiety disorder, such as generalized anxiety disorder and post-traumatic stress disorder. In this chapter, the various anxiety disorders are defined and described according to widely-used schemata. The include Generalized Anxiety Disorder, Panic Disorder, Post-traumatic Stress Disorder, ,Panic Disorder, and Social Anxiety Disorder. The epidemiology, treatment and sleep aspects are discussed. The physiology of Anxiety Disorders and the relationship to arousal, stress, and sleep-wake regulation are outlined.. Polysomnographic studies document limited alteration of sleep in anxiety disorders. There is some indication for alteration in sleep maintenance in generalized anxiety disorder and for both sleep initiation and maintenance in panic disorder; no clear picture emerges for obsessive-compulsive disorder or posttraumatic stress disorder

Malcolm Lader
Institute of Psychiatry, Neurology and Neuroscience, King's College, London, UK

Corresponding author: Malcolm Lader, E-mail: malcolm.lader@kcl.ac.uk

12.1 INTRODUCTION

Anxiety disorders are the most frequent type of psychiatric disorder with a lifetime prevalence of 29% in the general population (Kessler et al., 2005). Anxious individuals have substantial disability and functional impairment (Roy-Byrne et al., 1999). Several studies have examined anxiety symptoms in people with *sleep problems*. In a parallel fashion, others have evaluated sleep problems in *anxious individuals*. Thus, among people complaining of sleep problems, 35–50% is comorbidly anxious. One study reported a 17 fold overrepresentation of clinically significant anxiety among those with sleep difficulties (Taylor et al., 2005).

The temporal relationship between anxiety and sleep disorders is unclear. Some studies report that chronic insomnia predates the anxiety symptoms; other studies suggest the opposite with anxiety leading on to insomnia. It is likely that both patterns can occur.

Clinically, sleep disorders predominantly comprise complaints of insomnia.

But other disorders such as nightmares and the group of hypersomnias are encountered but the relationship with anxiety has been far less extensively studied. It is generally considered that a close relationship exists between sleep and anxiety disorders with stress postulated as the common mediating factor.

Puri and Treasden pages 844 onwards Types of AD – DSM - 5.

Insomnias DSM – 5 Diagnostic point in rating scales.

In this chapter the relationships between anxiety and sleep disorders and *vice versa* are examined. Clinically, the disorders are reviewed under the diagnostic categories set out in the latest widely used diagnostic schema, the Diagnostic and statistical manual version 5 (DSM 5) of the American Psychiatric Association.

12.2 Generalized Anxiety Disorder (GAD)

12.2.1 Introduction

The DSM 5 criteria for GAD include:
- Excessive anxiety and worry, occurring more days than not for at least 6 months, concerning a number of events;
- The individual finds it difficult to control the worry;
- The anxiety and worry are associated with at least three of the following six symptoms (only one item required in children);
- Restlessness, feeling keyed up or on edge.
- Being easily fatigued;
- Difficulty concentrating;
- Irritability;
- Muscle tension;
- Sleep disturbance;

- The anxiety, worry or physical symptoms cause clinically significant distress or impairment in important areas of functioning;
- The disturbance is not due to the physiological effects of a substance or medical condition;
- The disturbance is not better explained by another medical disorder;

Further discussion and explanation of the terms used has been provided by Andrews et al. (2010). Sleep disturbance is one of the prominent features but is not necessary to make the diagnosis. The sleep symptoms include difficulty falling or staying asleep or restless and unsatisfying sleep. The sufferer feels tired the next day.

12.2.2 Epidemiology

About three-quarters of primary care patients with anxiety disorders report a concomitant sleep disorder (Marcks et al., 2010). Of those meeting criteria for GAD the prevalence of insomnia was over twice that of non-anxious people. Conversely, anxiety disorders of one type or another are the most common diagnoses in patients who present with insomnia. Thus GAD and sleep disturbance have a high co-morbidity. This makes clinical sense as the onset of worry before going to bed can obviously cause sleep disturbance (Monti and Monti, 2000). This relationship has been explained in terms of emotional dysregulation (Tsypes et al., 2013).

12.2.3 Subjective

GAD, as the sufferers will readily attest, is typified by excessive, pervasive, and largely uncontrollable worries which are inchoate and ineffable in character. Compared with healthy controls, GAD patients take longer to fall asleep than normal and wake for longer during the night. A questionnaire study compared 59 GAD patients with 66 non-anxious controls. The GAD patients scored significantly higher with respect to several sleep variables including daytime dysfunction, habitual sleep efficiency, sleep latency, and frequency of nightmares (Tsypes et al., 2013). Comorbid depression and secondary anxiety diagnoses did not affect this relationship.

12.2.4 Objective Polysomnography (PSG)

Objectively latency to sleep is prolonged. There are more frequent and longer waking episodes and sleep efficiency is impaired.

12.2.4.1 Effects of Insomnia

Sleep disorder in anxiety disorders have been shown to be associated with daytime dysfunctioning (Ramsawh et al., 2009).

Treatment outline

The sleep difficulties outlined above are difficult to treat and may linger after the anxiety itself has resolved. Thus, a meta-analysis of Cognitive Behavioral Therapy found only modest effects on sleep symptoms (Belleville et al., 2010).

12.3 POST-TRAUMATIC STRESS DISORDER (PTSD)

The anxiety disorder with the most likelihood of being associated with disturbed sleep is PTSD. This typically can develop following a traumatic event that is highly threatening and in which helplessness is marked. The DSM 5 criteria are complex and the following is a résumé.

The affected person was exposed to death, threatened death, actual or threatened serious injury, or actual or threatened sexual violence. This can involve direct exposure with witnessing in person; indirectly, by learning that a close relative or close friend was exposed to trauma; repeated or extreme indirect exposure to aversive details of the event(s), usually in the course of professional duties.

Traumatic events that can lead to PTSD include war, rape, natural disasters, car or plane crashes, kidnapping, assault, terrorist attacks, sexual or physical abuse, sudden death of a loved one, and childhood neglect. Most people associate PTSD with battle-scarred soldiers: military combat is usually the most common cause in men. But any catastrophic life experience can trigger PTSD, especially if the event feels unpredictable and uncontrollable.

The development of PTSD progresses differently from person to person. Most commonly the symptoms of PTSD develop in the hours or days following the traumatic event, but it can sometimes take much longer before they appear.

Among the numerous symptoms are recurrent, involuntary, and intrusive memories, traumatic nightmares, dissociative reactions (e.g., flashbacks), and intense or prolonged distress after exposure to traumatic reminders; persistent effortful avoidance of distressing trauma-related stimuli after the event; negative alterations in cognitions and mood that began or worsened after the traumatic event (such as Inability to recall key features of the traumatic event); persistent (and often distorted) negative beliefs and expectations about oneself; persistent distorted blame of self or others for causing the traumatic event or for resulting consequences; persistent negative trauma-related emotions (e.g., fear, horror, anger, guilt, or shame); markedly diminished interest; feeling alienated from others; persistent inability to experience positive emotions. Other features are irritable or aggressive behavior, self-destructive or reckless behavior; hyper vigilance and exaggerated startle response.

12.3.1 Epidemiology

Maher et al. (2006) reviewed the topic in some detail. Subjective reports of sleep disturbance estimate that 70–91% of patients with PTSD have difficulty falling or staying asleep (e.g., Ohayon and Shapiro, 2000). Nightmares are reported by 19–71% of patients, depending on the severity of their PTSD and their exposure to physical aggression. However, objective measures of sleep disturbance vary in their findings, some note that the PSG is indicative of poor sleep whereas others find no differences compared with non-PTSD controls. Sleep disordered breathing (SDB) and sleep movement disorders are more common in patients with PTSD than in the general population, and these disorders may contribute to the brief awakenings, insomnia, and daytime fatigue in patients with PTSD. In turn, sleep problems impact on the development and symptom severity of PTSD and on the quality of life and functioning of patients.

12.3.2 Treatment Outline

Selective serotonin reuptake inhibitors (SSRIs) are commonly used to treat PTSD, and evidence suggests that they have a small but significant positive effect on sleep disruption. Trazodone may induce significant reductions in insomnia and nightmares. In small studies of patients with PTSD, prazosin, a centrally acting alpha1-adrenoceptor antagonist, has been associated with large reductions in nightmares and insomnia. Augmentation of SSRIs with olanzapine, an atypical antipsychotic, may be effective for treatment-resistant nightmares and insomnia, although adverse effects can be dose limiting. Additional medications, including zolpidem, buspirone, gabapentin and mirtazapine, have been found to improve sleep in patients with PTSD. In contrast, benzodiazepines, tricyclic antidepressants (TCAs) and monoamine oxidase inhibitors (MAOIs) are not useful for the treatment of PTSD-related sleep disorders, and have significant adverse effects.

Cognitive behavioral interventions for sleep disruption in patients with PTSD include strategies targeting insomnia and imagery rehearsal therapy (IRT) for nightmares. One large randomized controlled trial of group IRT demonstrated significant reductions in nightmares and insomnia. Similarly, uncontrolled studies combining IRT and insomnia strategies have demonstrated good outcomes. Sleep disturbance predicts the course of PTSD over the ensuing five years: those with sleep disturbance being less likely to remit (Marcks et al., 2010). By year five only a third of those with sleep problems remitted compared with 56% of those without sleep disturbance.

12.4 PANIC DISORDER (PD)

The DSM 5 criteria for PD include diagnoses of PD with Agoraphobia and PD without Agoraphobia. They are:

A. Recurrent unexpected panic attacks.

B. At least one of the attacks has been followed by one month (or more) of one or both of the following:

1. Persistent concern or worry about additional panic attacks or their consequences (e.g., losing control, having a heart attack, and going crazy).

2. Significant maladaptive change in behavior related to the attacks (e.g., behaviors designed to avoid having panic attacks, such as avoidance of exercise or unfamiliar situations).

C. The Panic Attacks are not restricted to the direct physiological effects of a substance (e.g., a drug of abuse, and a medication) or a general medical condition (e.g., hyperthyroidism, and cardiopulmonary disorders).

D. The Panic Attacks are not restricted to the symptoms of another mental disorder, such as Social Phobia (e.g., in response to feared social situations), Specific Phobia (e.g., in response to a circumscribed phobic object or situation), Obsessive-Compulsive Disorder (e.g., in response to dirt in someone with an obsession about contamination), Posttraumatic Stress Disorder (e.g., in response to stimuli associated with a traumatic event), or Separation Anxiety Disorder (e.g., in response to being away from home or close relatives).

12.4.1 Sleep Studies

Data on sleep abnormalities in PD are limited. In one earlier small study, sleep PSGs were recorded in 13 patients with PD, six of whom experienced panic during sleep, and compared with those from seven controls (Mellman and Uhde 1989). Sleep was disturbed in the patients, as shown by increased sleep latency, decreased total sleep time, and decreased sleep efficiency. Rapid eye movement (REM) latencies were not reduced in the patient group. All six of the panic awakenings were preceded by non-REM sleep, which appeared to be a transition from stage II toward delta sleep. The overall degree of sleep disturbance (i.e., sleep latency, sleep efficiency) did not appear to be influenced by the occurrence of sleep panic. There was also an association of increased REM latency with nights of sleep panic.

A later more extensive study assessed the prevalence of sleep complaints in PD patients, compared them with sleep complaints in a normal population, and investigated the role of comorbid depression and nocturnal panic attacks in sleep complaints in the PD patients (Overbeek et al., 2005). Seventy patients with P D and 70 healthy controls were questioned about their subjective sleep experiences using the Sleep-Wake Experience List, which assesses sleep/arousal complaints over a 24 h period. Two-third of the PD patients reported sleep problems, compared with a fifth of the controls. A total of 76% of PD patients with comorbid depression and 59% of the non-depressed had sleep difficulties; 77% of the PD patients with nocturnal panic attacks reported sleep complaints, versus 53% of the PD patients without nocturnal panic. It was concluded that PD patients demonstrate a higher prevalence of sleep complaints than normal controls. However, this can only partly be explained by comorbid depression, and the presence of nocturnal panic attacks.

Another study has evaluated the role of depression in sleep complaints among patients with PD (Stein et al., 1993a, 1993b).The Pittsburgh Sleep Quality Index (PSQI) was completed by 34 untreated patients with PD and 34 age-matched healthy controls. The patients reported significantly more impaired sleep than the controls with indicated higher global index and on four of seven of its subscales. Sleep was worst among those PD patients with a prior history of major depression. Two-third of patients with PD reported moderately or severely impaired sleep compared to only 15% of controls. A total of 26% of PD patients – but none of the controls – complained of frequent awakenings in the preceding month because of respiratory discomfort. The one-month prevalence of sleep panic in the patients was 18%; lifetime prevalence was 68%.

Patients with PD frequently report being woken from sleep by a panic attack. Most nocturnal panic attacks are characterized by difficulty breathing, hyperventilation, and a rapid or irregular heartbeat. These two symptoms together can cause arm and leg twitching or numbness and tingling in the hands and feet. Other people might feel tightness or pain in the chest. They might make a gagging sound or flail with their arms and legs. They may jump out of bed and rush to the nearest window to get some air. Some people who have nocturnal panic attacks believe they are having a stroke or heart attack, and the fear of dy-

ing can compound the symptoms. These sleep panic attacks occur in up to 70% of people with PD. A vicious cycle may be set up with secondary anxieties and reluctance to go to bed. Nocturnal panic attacks differ from night terrors because patients usually remember the attack in the morning. Panic attacks tend to happen in the middle stages of sleep, usually as a person is moving from stage two to stage three sleep.

12.5 SOCIAL ANXIETY DISORDER (SAD)

The DSM 5 Definition of SAD (otherwise known as Social Phobia) is:

A. A persistent fear of one or more social or performance situations in which the person is exposed to unfamiliar people or to possible scrutiny by others. The individual fears that he or she will act in a way (or show anxiety symptoms) that will be embarrassing and humiliating.

B. Exposure to the feared situation almost invariably provokes anxiety, which may take the form of a situationally bound or situationally pre-disposed panic attack.

C. The person recognizes that this fear is unreasonable or excessive.

D. The feared situations are avoided or else are endured with intense anxiety and distress.

E. The avoidance, anxious anticipation, or distress in the feared social or performance situation(s) interferes significantly with the person's normal routine, occupational (academic) functioning, or social activities or relationships, or there is marked distress about having the phobia.

F. The fear, anxiety, or avoidance is persistent, typically lasting six or more months.

G. The fear or avoidance is not due to direct physiological effects of a substance (e.g., drugs, medications) or a general medical condition not better accounted for by another mental disorder.

It can be seen that anxiety or more specifically fear is a cardinal part of the disorder. In clinical terms the patient is fearful of social situations in which embarrassment is believed to be a distinct possibility. This apprehension can precede the exposure to the phobic situation; it may be overwhelming during the actual exposure; and it may dominate the thoughts and affect of the sufferer when she or he mulls over the last exposure, reactions to it, and behavioral consequences.

12.5.1 Epidemiology

SAD is a common psychiatric disorder with a prevalence rate of around 10% depending on the criteria for "caseness" and the epidemiological methods used to detect such cases (Furmar., 2002). As with other anxiety disorders, comorbidity is high, other anxiety problems or depression are being commonly diagnosed.

12.5.2 Subjective and PSG

The sleep data in patients with SAD are limited and rather inconsistent. In one study (Brown et al., 1994), no differences in sleep architecture were found between SAD patients and normal controls. Another study using the PSQI showed that most patients with SAD reported impaired sleep quality, longer latency to sleep onset, more nocturnal disturbance and more daytime dysfunction than controls (Stein et al.,1993b). Other studies are inconsistent.

Questionnaire data suggest that subjective insomnia is associated with SAD even after the influences of depression severity, number of comorbid diagnoses and medication status was controlled for (Kushnir et al., 2014). Thus, depression is unlikely to be the mediating link between SAD and insomnia.

12.5.2.1. Effects of insomnia

12.5.2.1.1 Treatment outline

Cognitive behavioral therapy (CBT) has been advocated to treat SAD, but usually no emphasis is placed on the specific targeting of sleep problems. A retrospective study included 63 patients with SAD receiving CBT (Kushnir et al., 2014). The Liebowitz Social Anxiety Scale, Beck Depression Inventory, and PSQI among others were completed, together with the Sheehan Disabilities Scale. These are all well established and widely used instruments. The study results indicated that subjective insomnia is associated with the severity level of SAD, after correcting for depression comorbidity. Although CBT led to a reduction in both SAD and depression symptoms, no effect was seen on the sleep difficulties. Specific sleep remedies were advocated to address the latter issues.

12.6 OTHER ANXIETY DISORDERS

Separation anxiety may include nightmares about the theme of the separation.

Specific phobias may occasionally be accompanied by sleep problems, for example a subject with fear of heights (acrophobia) that has checked into a hotel room on an upper floor.

12.6.1 Obsessive-Compulsive Disorder (OCD)

The American Psychiatric Association (APA), the publisher of the DSM 5, changed classification for obsessive-compulsive disorder so that it and related disorders now have their own chapter. Accordingly, they are no longer considered to be "anxiety disorders". This takes account of accumulating research evidence demonstrating that common threads run through a number of OCD related disorders – obsessive thoughts and/or repetitive behaviors. Although accompanied in certain circumstances by anxiety, this affect is not pivotal as it is with the anxiety disorders.

OCD is the main disorder together with body dysmorphic disorder and trichotillomania (hair-pulling disorder). Two new disorders are now included, namely hoarding disorder and excoriation (skin-picking) disorder.

In any case, sleep disorders are not clearly characterized in the group of obsessive disorders (Hohagen et al., 1994; Robinson et al., 1998).

Epidemiology

Nightmares are commonly reported by anxious patients and overall are more prevalent in general psychiatric populations (Ohayon et al., 1997).
Subjective
PSG
Effects of insomnia

Treatment outline
Constructive worry pre-sleep (Carney and Waters, 2006). Mindfulness-based practice (Harvey and Farrell, 2003).
Other disorders — Psychiatry (Gelder p. 930)
Alcohol relationship to Anxiety and insomnia
Other disorders — Neurology and medicine (Gelder p. 931)
Caffeine Puri 846.

12.7 PHYSIOLOGY

The physiological, biochemical and endocrine mechanisms involved in sleep processes have been researched in detail in the past decades. The elucidation of these mechanisms does not help much with analyzing the underlying relationships between sleep and various anxiety disorders. Recourse must be made to higher-level concepts including stress and arousal (or alertness). These have been exhaustively discussed by Staner (2003), among others. The following is a résumé of their consensus.

12.7.1 Anxiety

Anxiety in some degree is a universal emotion. Indeed, it is a necessary constituent of the response of the person to a stress. As such the stress thereby comprises a challenge to the point of stress to the psychological or the physiological integrity or usually both of an individual. Not to experience anxiety and to act appropriately would be maladaptive.

Anxiety constitutes a state and a trait. It may supervene at some time point, usually young adulthood as an emotion or affect and attain proportions that meet criteria for one of the recognized anxiety disorders. This is state anxiety of pathological proportions. By contrast, the trait of anxiety is a persistent and pervading disorder of personality, present throughout life. It represents a lifelong maladaptive response to stress, internal or external, probably reflecting genetic background, developmental influences, parental influences, and early life experiences. In clinical the nature of the stress may not be immediately identifiable.   Psychotherapy of various types may help to uncover these covert influences. It is differentiated from the experience of fear, in that the putative stress is identified by the person experiencing it, although the apparent relationship may not always be valid. Whether normal or pathological, the constituent features of anxiety always incorporate indices of increased arousal or alertness that, could lead to sleep disturbances. Accordingly, anxiety could be an intrinsic part of the arousal response to stress, whether the stress is a real understandable one, identified rightly or wrongly by the person, or exaggerated subconsciously to the point of a pathological reaction.

12.7.2 Arousal and Stress

Response to stress implicates two systems that both play a key role in physiological responses to stressful situation by promoting arousal. These classic mechanisms comprise the corticotropin-releasing hormone (CRH) system and the locus ceruleus autonomic nervous (LC-AN) system. The CRH system is located in the paraventricular nucleus (PVN), and the central nucleus of the amygdala (NA). The autonomic nervous (AN) system has the locus ceruleus (LC) as its chief centre. The arousal response to stress comprises three components (hormonal, autonomic, and behavioral) in which CRH and autonomic systems have been implicated. Different stressors activate different components of the stress system, e.g., the autonomic system will be more implicated in the response to physiological stressors such as extreme exercise, while the CRH system will be recruited for more complex environmental dangers such as emotional stress that in PTSD. The two systems interact in a complex way.

12.7.3 Stress and Sleep-Wake Regulation

Both acute and chronic stresses have pronounced effects on sleep that are mediated through the activation of the hypothalamic-pituitary-adrenal (HPA) axis and the sympathetic system. In humans, there is a close temporal relationship between HPA activity and sleep structure. The HPA axis is inhibited markedly during the early phase of nocturnal sleep, during which *slow wave sleep* (SWS) predominates. In contrast, during late sleep, when REM sleep predominates, HPA activity increases to reach a diurnal maximum shortly after morning awakening. During SWS sympathetic activity is reduced and there is a positive correlation among the amount of REM sleep and activities of the HPA axis and the sympathetic system. Most relevantly, major interactions occur between adrenocorticotropic, autonomic, and Electroencephalography (EEG) indices of arousal during the sleep-wake cycle. For example, cortisol produces prolonged sleep onset, reduced SWS, and increased sleep fragmentation. Accordingly, patients with complaints of insomnia show electrophysiological and psychomotor evidence of increased daytime arousal, as well as indications of increased HPA activity and increased sympathetic tone.

12.7.4 Arousal, Anxiety Disorders and Sleep Disturbances

Thus close relationships exist between stress responses, general arousal mechanisms, anxiety disorders, and insomnia. Sleep would appear to be on the same continuum as anxiety. Strong confirmation comes from pharmacological data. Most of the medications used to treat sleep complaints such as the benzodiazepines and the z-drugs will induce anxiety-relieving actions at lower doses through the day. Conversely, most general anxiolytics will induce sleep in higher doses, especially when given at night. Only those drugs believed to have more specific action are exceptions. The common denominator would appear to be actions on the gamma-aminobutyric acid (GABA) system which induces non-specific reductions in the level of arousal. An example is melatonin which helps improve sleep especially in the elderly but has only weak sedative actions.

12.8 FEEDBACK MECHANISMS

These are probably quite important in mediating the relationship between anxiety and sleep. It has been proposed that sleep disturbance in individuals may be maintained, at least in part, by irrational or exaggerated beliefs such as worry about the consequences of poor sleep, effect on performance the next day, misconceptions about the causes of insomnia, and unreal expectations about duration of sleep (Morin et al., 2007).

Thus, insomnia will act as a stress which can be quite marked. In turn, this will increase anxiety which could be associated with heightened arousal and further insomnia.

12.9 CONCLUSIONS

The interrelationships between sleep, anxiety, and insomnia are complex and intimate. The common thread of stress and arousal runs through these themes. Rather little research has been carried out evaluating recent advances in our understanding of sleep and their relevance to anxiety. More research will elucidate these connections and should give us new and profitable insights into both mechanisms and new treatment approaches.

KEY WORDS

- **sleep**
- **insomnia**
- **anxiety disorders**
- **sleep–wake regulation**
- **insomnia**
- **sleep–wake regulation**

REFERENCES

Andrews, G.; Hobbs, M. J.; Borkovec, T. D. Generalized Worry Disorder: A Review of DSM-IV Generalized Anxiety Disorder and Options for DSM-V. *Depress. Anxiety.* **2010**, 2, 134–147.

Belleville, G.; Cousineau, H.; Levrier, K.; et al. Meta-Analytic Review of the Impact of Cognitive-Behavior Therapy for Insomnia on Concomitant Anxiety. *Clin. Psychol. Rev.* **2011**, 31, 638–652.

Brown,T.; Black, B.; Uhde, T. W. The Sleep Architecture of Social Phobia. *Biol. Psychiatry.* **1994**, *35*, 420–421.

Carney, C. E.; Waters, W. F. Effects of a Structured Problem-Solving Procedure on Pre-Sleep Cognitive Arousal in College Students with Insomnia. *Behav. Sleep. Med.* **2006**, *4*, 13–28.

Furmar, T. Social Phobias: Overview of Community Surveys. *Acta. Psychiatrica. Scandinavica.* **2002**, *105*, 84–93.

Harvey, A. G.; Farrell, C. A Transdiagnostic Approach to Treating Sleep Disturbances in Psychiatric Disorders. *Cogn. Behav. Ther.* **2003**, *38*, 35–42.

Hohagen, F.; Lis, S.; Krieger, S.; et al. Sleep EEG of Patients with Obsessive-Compulsive Disorder. *Eur. Arch. Psychiatry. Clin. Neurosci.* **1994**, *243*, 273–278.

Kessler, R. C.; Bergland, P.; Denler, O. Lifetime Prevalence and Age of Onset of DSM-IV Disorders in the National Comorbidity Survey Replication. *Arch. Gen. Psychiatry.* **2005**, *62*, 593–768.

Kushnir, J.; Marom, S.; Mazar, M.; et al. The link Between Social Snxiety Disorder, Treatment Outcome, and Sleep Difficulties Among Patients Receiving Cognitive Behavioural Group Therapy. *Sleep. Med.* **2014**, *15*, 515–521.

Maher, M. J.; Rego, S. A., Asnis, G. M. Sleep Disturbances in Patients with Post-Traumatic Stress Disorder: Epidemiology, Impact and Approaches to Management. *CNS Drugs.* **2006**, *20*, 567–590.

Marcks, B. A.; Weisberg, R. B.; Edelen, M. O.; Keller, M. B. The Relationship Between Sleep Disturbance and the Course of Anxiety Disorders in Primary Care Patients. *Psychiatry. Res.* **2010**, *178*, 487–492.

Mellman, T. A.; Uhde, T. Electroencephalographic Sleep in Panic Disorder: : A Focus on Sleep-Related Panic Attacks. *Arch. Gen. Psychiatry.* **1989**, *46*, 178–184.

Monti, J.M.; Monti, D. Sleep Disturbance in Generalized Anxiety Disorder and its Treatment. *Sleep. Med. Rev.* **2000**, *4*, 263–276.

Morin, C. M.; Vallières, A.; Ivers, H. Dysfunctional Beliefs and Attitudes About Sleep (DBAS). Validation of a Brief Version (DBAS-16). *Sleep.* **2007**, *30*, 1547–1554.

Ohayon, M. M.; Morselli, P.L.; Guilleminault, C. Prevalence of Nightmares and Their Relationship to Psychopathology and Daytime Functioning in Insomnia Dysregulation Subjects. *Sleep.* **1997**, *20*, 340–348.

Ohayon, M. M.; Shapiro, C. M. Sleep Disturbances and Psychiatric Disorders Associated with Post-traumatic Disorder in the General Population. *Compr. Psychiatry.* **2000**, *41*, 469–478.

Overbeek, T.; van Diest, R.; Schruers, K.; et al. Sleep Complaints in Panic Disorder Patients. *J. Nerv. Ment. Dis.* **2005**, *193*, 488–493.

Ramsawh, H. J.; Stein, M. B.; Belik, S.; et al. Relationship of Anxiety Disorders, Sleep Quality and Functional Impairment in a Community Sample. *J. Psychiatr. Res.* **2009**, *43*, 926–933.

Robinson, D.; Walseben, J.; Pollack, S.; Lerner, G. Nocturnal Polysomnography in Obsessive-Compulsive Disorder. *Psychiatry. Res.* **1998**, *80*, 257–263.

Roy-Byrne, P. P.; Stein, M. B.; Russo, J.; et al. Panic Disorder in the Primary Care Setting: Comorbidity, Disability Service Utilization, and Treatment. *J. Clin. Psychiatry.* **1999**, *60*, 492–499.

Staner, L. Sleep and Anxiety Disorders. *Dialogues. Clin. Neurosci.* **2003**, *5*, 249–258.

Stein, M. B.; Chartier, M.; Walker, J. R. Sleep In Nondepressed Patients with Panic Disorder: I. Systematic Assessment of Subjective Sleep Quality and Sleep Disturbance. *Sleep.* **1993a**, *16*, 724–726.

Stein, M. B.; Kroft, C. D.; Walker, J. R. Sleep Impairment in Patients with Social Phobia. *Psychiat. Res.* **1993b**, *49*, 251–256.

Taylor, D. J.; Lichstein, K. L.; Durrence, H. H.; et al. Epidemiology of Insomnia, Depression, and Anxiety. *Sleep.* **2005**, *28*, 1457–1464.

Tsypes, A.; Aldao, A.; Mennin, D.S.;et al. Emotion Dysregulation and Sleep Difficulties in Generalized Anxiety Disorder. *J. Anxiety. Disord.* **2013**, *27*, 197–203.

13

SLEEP RELATED BREATHING DISORDERS IN ADULTS

Shaden O. Qasrawi, Ruwaida N. Al Ismaili,
Seithikurippu R. Pandi-Perumal, and Ahmed S. BaHammam

ABSTRACT

"Sleep related breathing disorders" is a term used to describe a wide spectrum of abnormalities of respiration during sleep, including obstructive sleep apnea (OSA), central sleep apnea disorders (CSAS) and sleep related hypoventilation disorders. In OSA, there is a cessation or a significant reduction in airflow in the presence of breathing effort due to the repetitive collapse of the upper airway during sleep compared to CSAS, in which the cessation of air flow is due to the absence of respiratory effort. Sleep related hypoventilation disorders are characterized by insufficient sleep related ventilation leading to abnormally elevated P_aCO_2 during sleep. This includes obesity hypoventilation syndrome (OHS), which is defined as $P_aCO_2 > 45$ mm Hg during wakefulness in patients with BMI > 30 kg/m^2 in the absence of other causes for elevated P_aCO_2. Obtaining a detailed history and performing a thorough clinical examination are essential in diagnosing these conditions. This should be followed by the necessary investigations, including polysomnography. Management of such patients involves patient education about the risk factors, natural history and consequences of these disorders as well as treating the underlying medical problems. The treatment is mainly

Shaden O. Qasrawi
University Sleep Disorders Center, College of Medicine, King Saud University, Riyadh, Saudi Arabia; The Strategic Technologies Program of the National Plan for Sciences and Technology and Innovation in the Kingdom of Saudi Arabia

Ruwaida N. Al Ismaili
University Sleep Disorders Center, College of Medicine, King Saud University, Riyadh, Saudi Arabia; The Strategic Technologies Program of the National Plan for Sciences and Technology and Innovation in the Kingdom of Saudi Arabia

Seithikurippu R. Pandi-Perumal
Somnogen Canada Inc, College Street, Toronto, Canada

Ahmed S. BaHammam
University Sleep Disorders Center, College of Medicine, King Saud University, Riyadh, Saudi Arabia; The Strategic Technologies Program of the National Plan for Sciences and Technology and Innovation in the Kingdom of Saudi Arabia

Corresponding author: Ahmed S. BaHammam, University Sleep Disorders Center, College of Medicine, King Saud University, Box 225503, Riyadh 11324, Saudi Arabia. E-mail:ashammam2@gmail.com, ashammam@ksu.edu.sa. Tel.: +966-1-467-1521, Fax: +966-1-467-2558 directed toward improving ventilation during sleep with positive airway pressure (PAP) therapy, which provides pneumatic splinting of the upper airway with or without oxygen. There are other alternative therapeutic options, which should be tailored to patients' conditions, including the use of oral appliances (OA), surgery, hypoglossal nerve stimulation and pharmacological treatment.

13.1 INTRODUCTION

"Sleep related breathing disorders" is a term used to describe a wide spectrum of abnormalities of respiration during sleep including abnormal respiratory pattern (e.g., apneas, hypopneas, or respiratory effort related arousals) or abnormal reduction in gas exchange (e.g., hypoventilation) during sleep. The International Classification of Sleep Disorders – Third Edition (ICSD-3)[1] has defined four major categories of sleep related breathing disorders: 1) obstructive sleep apnea (OSA) disorders, 2) central sleep apnea disorders, 3) sleep related hypoventilation disorders, and 4) sleep related hypoxemia disorder in addition to isolated symptoms and normal variants like snoring and catathrenia.

13.2 DEFINITIONS[2]

Respiratory Events are defined as breathing abnormalities during sleep.

Apnea cessation of airflow (90% decrease in apnea sensor excursions compared to baseline of a minimum duration of 10 s in adults. Apneas are classified as obstructive, mixed, or central based on the pattern of respiratory effort.

Obstructive apnea occurs when airflow is absent or nearly absent, in the presence of respiratory effort.[2]

Central apnea occurs when both airflow and ventilatory effort are absent.

Mixed apnea occurs when there is an interval during which there is no respiratory effort (i.e. central apnea pattern) followed by an interval during which there is obstructed respiratory efforts.

Hypopnea is a reduction in airflow with the minimum amplitude and duration as specified in the hypopnea rules for adults. The reduction in airflow must be accompanied by a 3% desaturation or an arousal or a 4% desaturation.

Respiratory effort related arousals are sequences of breaths characterized by increasing respiratory effort (esophageal manometry); inspiratory flattening in the nasal pressure or positive airway pressure (PAP) device flow channel; or an increase in end-tidal partial pressure of carbon dioxide (PCO2) (children) leading to an arousal from sleep. Respiratory effort related arousals do not meet criteria for hypopnea and have a minimum duration of 10 s in adults or the duration of at least two breaths in children. Respiratory effort-related arousals (RERAs) (> 5 events/h) associated with daytime sleepiness were previously called upper airway resistance syndrome (UARS), which was considered as a subtype of OSA. These patients have abnormal sleep and cardio respiratory changes typically found in OSA.

Hypoventilation is a specified period of increased $PaCO_2$ of > 50 mmHg in children or > 55 mmHg in adults, or a rise of $PaCO_2$ during sleep of 10 mmHg that exceeds 50 mmHg for a specified period of time in adults.

Cheyne-Stokes breathing (CSB) is a breathing rhythm with a specified crescendo and decrescendo change in breathing amplitude separating central apneas or hypopneas.

Apnea index (AI) is the total number of apneas per hour of sleep.

Apnea hypopnea index (AHI) is the total number of apneas and hypopneas per hour of sleep.

Respiratory disturbance index (RDI) is the total number of events (e.g., apneas, hypopneas, and RERAs) per hour of sleep. The RDI is generally higher than the AHI, because the RDI includes the frequency of RERAs, while the AHI does not.

Oxygen desaturation is a frequent consequence of apnea and hypopnea. Several measures are used to quantify the severity of desaturation.

Arousal index (ArI) is the total number of arousals per hour of sleep.

13.3 OBSTRUCTIVE SLEEP APNEA (OSA)

OSA, also referred to as obstructive sleep apnea-hypopnea, is a sleep disorder characterized by cessation or significant decrease in airflow in the presence of breathing effort caused by repetitive collapse of the upper airway during sleep. It is the most common type of sleep-disordered breathing. Recent data demonstrated that the estimated prevalence of moderate to severe OSA ranges from 10 to 17% in middle-aged and elderly men.[3]

13.3.1 Diagnosis of OSA

13.3.1.1 History and Physical Examination

13.3.1.1.1 History

While obtaining history, it is usually helpful to have the patient's bed partner or a family member present during the interview because they may be able to give extra information about patient condition when asleep. Most patients with OSA present with complaint of daytime sleepiness, snoring, gasping or interruptions in breathing while asleep. These symptoms may be reported during the evaluation of another complaint, detected during health maintenance screening, reported during preoperative screening or as a part of the comprehensive evaluation of patients at high risk for OSA. Table 13.1 shows patients' groups at a high risk for OSA, who needs lower threshold for evaluation. Sleep history for patients suspected to have OSA should include evaluation for excessive day time sleepiness not explained by other factors, including assessment of sleepiness severity by the Epworth Sleepiness Scale (ESS).[4] Snoring is the other common feature of OSA, it is associated with a sensitivity of 80–90% for the diagnosis of OSA, but its specificity is below 50%.[5] History should also cover other features of OSA including witnessed apneas, gasping/choking episodes, total sleep amount, nocturia, morning headaches, sleep fragmentation/sleep maintenance insomnia, and decreased concentration and memory.

Information should be obtained about history of secondary conditions that may occur as a result of OSA, including hypertension, stroke, ischemic heart disease, cor pulmonale, decreased daytime alertness and motor vehicle accidents.

Table 13.1. Patients at high risk for OSA who should be evaluated for OSA symptoms.

Obesity (BMI > 35)
Congestive heart failure
Ischemic heart disease
Atrial fibrillation
Pulmonary hypertension
Treatment refractory hypertension
Type 2 diabetes
Stroke
Operators of heavy machinery
Preoperative for bariatric surgery

13.3.1.1.2 Physical examination

Apart from obesity (body mass index > 30 kg/m^2) and a crowded oropharyngeal airway; the physical exam is frequently normal in OSA patients. Blood pressure should always be checked.

Physical findings that are common among patients with OSA include the following:

- Narrow airway: numerous conditions can lead to narrowing of the upper airway such as retrognathia, micrognathia, lateral peritonsillar narrowing, macroglossia, tonsillar hypertrophy, an elongated or enlarged uvula, a high arched or narrow palate, nasal septal deviation, and nasal polyps. The modified Mallampati classification is commonly used to quantify upper airway narrowing, with classes 3 and 4 considered positive for airway narrowing. Both the Mallampati classification and Friedman tongue position have been shown to correlate with OSA severity.[6]
- Large neck and/or waist circumference: OSA has a strong correlation with increased neck size or waist circumference than general obesity, and is prominent among men who have a collar size greater than 17 inches and women who have a collar size greater than 16 inches.
- Patients should be also evaluated for complications of OSA including elevated blood pressure, signs of pulmonary hypertension or corpulmonale and nocturnal cardiac dysrhythmias.

13.3.1.2 Diagnostic Testing

Objective diagnostic testing is necessary to accurately diagnose OSA.[5] Patients who are suspected to have OSA, are usually investigated by full night (i.e., diagnostic only) attended, in-laboratory polysomnography (PSG) which is the gold standard diagnostic test for OSA. Patients who are found to have OSA and who are planned to have PAP therapy are then brought back for another study in which their PAP device is titrated. Split-night attended, in-laboratory PSG is another approved diagnostic testing, and it is diagnostic in the first part and therapeutic in the second part if PAP therapy is indicated.[7] Figure.13.1 shows a hypnogram of a patient with OSA who underwent a split-night sleep study.

Unattended, out-of-center sleep testing (OCST) can be considered in patients without comorbidities and in whom there is a high likelihood of moderate or severe OSA.[8] Most OCST devices record airflow, respiratory effort, and blood oxygenation. Some OCST devices measure peripheral arterial tonometry.[8] OCST should not be used if another sleep disorder is suspected or if the patient has comorbid medical conditions that predispose to non-OSA sleep related breathing disorders (e.g., heart failure predisposes patients to central sleep apnea).[8]

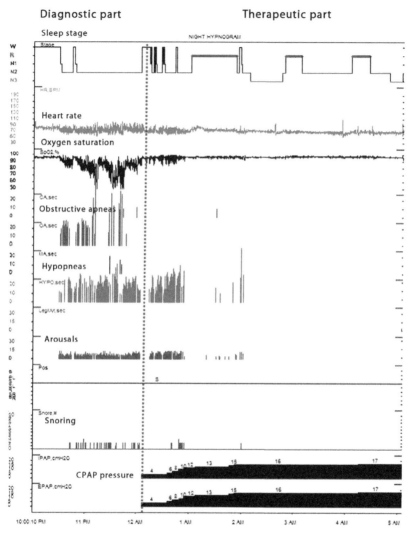

Figure 13.1. A hypnogram of a patient with obstructive sleep apnea. A split-night sleep study was performed. The first part of the study is the diagnostic part. It shows high number of apneas and hypopneas associated with significant desaturation and arousals. The patient spent his sleep in stages N1 and N2 only. The second part of the study is the therapeutic part on CPAP. When the optimal CPAP pressure was reached, apneas and hypopneas were eliminated and oxygen saturation remained >90%. The patients' sleep progressed into stages N3 and REM sleep.

13.3.1.3 Diagnostic Criteria

The diagnosis of OSA is based on the presence or absence of related symptoms, as well as the frequency of respiratory events during sleep (i.e., apneas, hypopneas, and RERAs as measured by polysomnography or OCST). Table 13.2 presents the diagnostic criteria for OSA in adults. Patients who meet criteria for the diagnosis of OSA are classified into mild, moderate or severe disease on the basis of symptoms and the apnea hypopnea index (AHI):[9]

Mild: Patients with an AHI between 5 and 14.

Moderate: Patients with an AHI between 15 and 29.

Severe: Patients with an AHI greater than 30.

Table 13.2. Diagnostic criteria for obstructive sleep apnea (OSA).1

(A and B) or C satisfy the criteria for the diagnosis of OSA

A. The presence of one or more of the following:

    1. The patient complains of sleepiness, nonrestorative sleep, fatigue, or insomnia symptoms.

    2. The patient wakes with breath holding, gasping, or choking.

    3. The bed partner or other observer reports habitual snoring, breathing interruptions, or both during the patient's sleep.

    4. The patient has been diagnosed with hypertension, a mood disorder, cognitive dysfunction, coronary artery disease, stroke, congestive heart failure, atrial fibrillation, or type 2 diabetes mellitus.

B. Polysomnography (PSG) or OCST demonstrates:

Five or more predominantly obstructive respiratory events (obstructive and mixed apneas, hypopneas, or respiratory effort related arousals (RERAs) per hour of sleep during a PSG or per hour of monitoring (OCST).

OR

C. PSG or OCST demonstrates:

Fifteen or more predominantly obstructive respiratory events (apneas, hypopneas, or RERAs) per hour of sleep during a PSG or per hour of monitoring (OCST)

13.3.2 Management of OSA

OSA is a chronic disease that requires long-term, multidisciplinary care. Management starts with patient education. The management of OSA is directed mainly at improving sleep quality, relieving symptoms, normalizing the AHI and oxygen saturation levels. Untreated OSA has several potential consequences and complications, such as excessive daytime sleepiness, impaired daytime function, metabolic dysfunction and an increased risk of cardiovascular diseases[10] and mortality.[11] The patient should be actively involved in making decisions about treatment options. PAP therapy is the treatment of choice for mild, moderate, and severe OSA and should be offered to all patients. Other alternative therapies may be considered based on the severity of the OSA and the patient's condition and preferences. There has been many published clinical practice guidelines for the management of OSA in adults, we will focus in this chapter on the American Academy of Sleep Medicine (AASM) guidelines.[12]

13.3.2.1 Patient Education and Behavioral Strategies

Once the diagnosis of OSA is confirmed and its severity has been determined, it is important to educate the patients about the risk factors, natural history and consequences of OSA,[12] including increased risk of motor vehicle accidents associated with untreated OSA and the risks associated with operating other dangerous equipment while sleepy.[13] In certain patients; behavior modification could improve OSA. For example, weight loss and exercise should be recommended to all patients with OSA who are overweight or obese.[12] Being overweight is a major risk factor for OSA; weight reduction is associated with improvement in breathing pattern, quality of sleep and daytime sleepiness.[14] Because of its low success rate; weight loss programs should be combined with a primary treatment for OSA.[12,15] After significant weight loss is achieved (i.e., 10% or more of body weight), a follow-up PSG should be performed to ascertain if PAP therapy is still needed or whether adjustments in PAP level are necessary.[7] The effect of weight loss achieved with bariatric surgery on OSA appears to be similar to weight loss achieved through diet and exercise, with reductions in AHI proportional to weight loss but with few complete remissions.[16]

Patients with positional OSA should be encouraged to avoid sleep positions that worsen OSA. Usually supine sleep position tends to affect airway size and patency with a decrease in the area of the upper airway.[17] Positional therapy uses methods and devices that keep patients in a non-supine position (e.g. alarm, pillow, backpack, and tennis ball).[12,15] Positional therapy can lead to moderate reductions in (AHI) especially in younger patients or those with a low AHI and less obese group, but it is inferior to PAP, and therefore cannot be recommended except in carefully selected

patients.[14] Since not all patients will normalize their AHI when in non-supine sleep position, positional therapy should also be tested with PSG before initiating this form of treatment as a primary therapy.[12]

All patients with untreated OSA should be advised to avoid alcohol, even during the day, because it can suppress the central nervous system, worsen OSA and sleepiness and increase weight.[18] Other sedatives like Benzodiazepines should also be avoided.[19]

13.3.2.2 POSITIVE AIRWAY PRESSURE THERAPY (PAP)

PAP therapy provides pneumatic splinting of the upper airway; as a result, respiratory events due to upper airway collapse (e.g., apneas and hypopneas) are prevented. PAP therapy has become the standard therapy for OSA.[12] PAP may be delivered in continuous (CPAP), bi-level (BPAP), or auto-titrating positive airway pressure (APAP) modes. PAP is applied through a nasal, oral, or oro-nasal interface during sleep.[12]

Different health bodies advocate different thresholds for the initiation of PAP therapy in OSA. The AASM recommends offering positive airway pressure therapy to all patients who have been diagnosed with OSA.[12]

The available modes of positive airway pressure therapy are summarized below:[12]

- CPAP provides positive airway pressure at a constant level throughout the respiratory cycle. It is often the initial therapy because it is the simplest and best-studied method.
- BPAP provides inspiratory positive airway pressure (IPAP) and expira-

tory positive airway pressure (EPAP). The tidal volume is related to the difference between the IPAP and EPAP (pressure support). There is no proved advantage in using BPAP over CPAP for the routine management of OSA.

- APAP increases or decreases the level of positive airway pressure in relation to signs indicating upper airway resistance changes such as airflow changes, circuit pressure changes, or vibratory snore.
- Adaptive servo-ventilation (ASV) gives a varying amount of inspiratory pressure superimposed on a low level of CPAP with a backup respiratory rate and is mainly used to treat central sleep apneas.[20]

Positive airway pressure has proved that it can improve symptoms of OSA, normalize the risk of traffic and workplace accidents, reduce risk for cardiovascular morbidities particularly arterial hypertension, and recently, it has been shown that CPAP reduces mortality in patients with severe OSA.[21,22] CPAP is a safe mode of treatment with generally minor side effects such as dryness of the upper airways, sneezing and nasal drip, local skin irritation, nasal congestion, eye irritation, aerophagia, sinusitis, and epistaxis. Serious complications like corneal abrasions, bacterial meningitis, massive epistaxis, and pneumocephalus are rare. Poor compliance is likely to lessen the potential benefits of CPAP therapy and it is a major problem with CPAP since large proportion of patients do not use their device and many do not use it long enough.[23] Previous studies showed that OSA patients on CPAP therapy use their machines on average from 4.5 to 5.5 hours per night with compliance rate ranging from 30 to 85%.[24]

Patient education and regular follow up especially during the first few weeks of therapy may improve compliance and help to identify device side effects and problems.[25]

Nasal EPAP is a relatively new treatment for snoring and OSA. Nasal EPAP utilizes the power of the user's own breathing to create positive airway pressure in order to keep the airways patent.[26] Nasal EPAP has been shown to provide statistically significant reductions in AHI, oxygen desaturation index (ODI) and sleepiness, with high patient acceptance and compliance.[23,27]

13.3.2.3 Alternative Therapies

13.3.2.3.1 Oral appliances (OA)

OAs are an alternative treatment for the group of patients with mild or moderate OSA who decline, do not respond or fail to adhere to PAP.[12,28] OAs may improve the patency of the upper airway during sleep by enlarging the upper airway and/or by decreasing its collapsibility.[12] There are several types of oral appliances, including, mandibular advancement devices (MADs) that cover the upper and lower teeth and hold the mandible in an advanced position with respect to the resting position, and tongue retaining devices (TRD), which hold the tongue in a forward position with respect to the resting position, without mandibular repositioning.[12] MADs improve sleep apneas, subjective daytime sleepiness, and quality of life. There is emerging evidence that MAD might have beneficial cardiovascular effects. However, tongue retaining devices (TRDs) are not recommended.[14] Side effects of MADs are generally minor. Excessive salivation and discomfort are the most common side effects. Persistent temporomandibular joint pain and changes in occlusive alignment may necessitate cessation of treatment.

13.3.2.3.2 Upper airway surgery

Upper airway surgical treatments aim to modify airway anatomy through a variety of upper airway reconstructive or bypass procedures, which have to be tailored according to the specific airway obstruction needs of individual patients. Surgery is not considered as a first line treatment for OSA but could be the primary treatment in patients with mild OSA who have severe obstructing anatomy, which is surgically correctible (e.g., tonsillar hypertrophy obstructing the pharyngeal airway). Nevertheless, surgery is reserved as a secondary approach for patients who fail to adequately respond or to comply with positive airway pressure or oral appliances.[12,14]

There are a number of different surgical procedures that may be performed including:[14,29]

- Nasal surgery, including turbinectomy or straightening of the nasal septum.
- Tonsillectomy and/or adenoidectomy.
- Uvulopalatopharyngoplasty (UPPP) or laser-assisted uvulopalatoplasty (LAUP). Modern variants of this procedure sometimes use radio frequency waves to heat and remove tissue.
- Reduction of the tongue base, either with laser excision or radio frequency ablation.
- Genioglossus advancement, in which a small portion of the lower jaw that attaches to the tongue is moved forward, to pull the tongue away from the back of the airway.
- Hyoid suspension, in which the hyoid bone in the neck, another attachment point for tongue muscles, is pulled forward in front of the larynx.
- Maxillo-mandibular advancement, which involves forward fixing of the maxilla and mandible.

13.3.2.3.3 Hypoglossal nerve stimulation or upper airway stimulation

Hypoglossal nerve stimulation using an implantable neurostimulator device is a novel strategy in treating patients with moderate to severe OSA. Based on data that have shown significant reductions in AHI and oxygen saturation index as well as improvement in subjective measures of sleepiness after device implantation in selected patients,[30,31,32] the Inspire device (Upper Airway Stimulation device) was approved by the US Food and Drug Administration in April 2014.

13.3.2.3.4 Pharmacologic treatment

Many pharmacologic agents have been investigated as therapeutic agents for the management of OSA. This includes drugs that might act by stimulating respiratory drive directly (e.g., theophylline) or indirectly (e.g., acetazolamide). However, none of these agents has proven to be sufficiently effective to replace conventional therapies.[14,33]

In patients who continue to have excessive daytime sleepiness despite adequate conventional therapy and are severe enough to warrant treatment, they may benefit from adjunctive pharmacologic therapy with agents like Modafinil or Ar-modafinil.[12]

13.4 CENTRAL SLEEP APNEA SYNDROMES

Central sleep apnea syndromes (CSAS) are a group of disorders, which are characterized by cessation of airflow due to absence of respiratory effort compared to patients with OSA, in whom the respiratory event occurs despite the presence of respiratory effort. CSAS comprise several disorders. According to the ICSD-3,[1] central sleep apnea syndromes (CSAS) are classified into:

- Central sleep apnea (CSA) with Cheyne-Stokes breathing (CSB)
- Central apnea due to a medical disorder without CSB
- Central sleep apnea due to high altitude periodic breathing
- Central sleep apnea due to a medication or substance
- Primary central sleep apnea
- Primary central sleep apnea of infancy
- Primary central sleep apnea of prematurity
- Treatment-emergent central sleep apnea

CSAS can be primary or secondary. Categorizing CSAS into hyperventilation related and hypoventilation related central apnea helps in planning the management. Hyperventilation related CSAS includes primary CSA, CSA associated with CSB, CSA due to a medical condition, and CSA due to high altitude periodic breathing. Hypoventilation related central apnea includes CSA due to central nervous system diseases, central nervous system suppressing drugs or substances, neuromuscular diseases, or severe abnormalities in pulmo-

nary mechanics. Treatment is indicated in symptomatic patients and the first step in the management usually focuses on resolving or ameliorating the underlying cause or the precipitating factor. CSB is the most frequently encountered CSA seen in daily practice. Therefore, we will focus our discussion on CSB.

13.4.1 Cheyne-Stokes Breathing

The AASM defines CSB as a breathing disorder in which there are cyclical fluctuations in breathing, with periods of central apneas or hypopneas that alternate with periods of hyperpnea in a gradual waxing and waning fashion. CSB is seen mostly in patients with heart failure (HF) but has also been described in patients recovering from acute pulmonary edema, advanced renal failure, and central nervous system lesions.[34] Numerous clinical and epidemiological studies have shown that a large proportion of patients with left ventricular systolic dysfunction suffer from CSB or another sleep-related breathing disorders.[34]

The presence of CSB in patients with HF is associated with increased morbidity and mortality and impaired quality of life.[35] Therefore, clinicians who manage HF patients should actively look for symptoms and signs suggestive of sleep-related breathing disorders, in order to implement appropriate interventions. Subjects suffering from advanced HF, with an ejection fraction as low as 35–40%, developed pulmonary congestion and elevated pulmonary venous pressure, which results in the stimulation of pulmonary "J" receptors, and eventually leading to hyperventilation at night and during the daytime.[34] When the $PaCO_2$ level falls below the apnea threshold, central apneas are triggered. And this cycle of hyperventilation-apnea goes on during sleep.

13.4.1.1 History, Physical Examination and Diagnosis

It is usually difficult to differentiate symptoms of HF from symptoms of sleep related breathing disorders as both problems may have overlapping symptoms such as poor sleep quality, daytime somnolence or insomnia, paroxysmal nocturnal dyspnea, and easy fatigability. Therefore, a high index of suspicion is required to prevent unjustified delays in diagnosis. Patients with HF and CSB tend usually to have lower body mass index than patients with concomitant OSA.[36] Moreover CSB has been shown to be more common among males and patients with atrial fibrillation.[34] Table 13.3 shows the diagnostic criteria of CSB employed by the AASM.[1]

An important step in the diagnosis of CSB is the measurement of arterial blood gases (ABG) while the patient is awake. Patients with CSB usually have low or low-normal $PaCO_2$ and increased ventilatory response to CO_2.[37] Previous studies have demonstrated that the severity of underlying HF does not always correlate with the severity of CSB, and that other factors must be considered.[34] It is not uncommon to find that patients with HF have co-existing OSA and CSB.[34]

Overnight PSG typically shows cyclical fluctuations in ventilation, with periods of central apnea or hypopnea that alternate with periods of hyperpnea that predominates during stages N1 and N2 of non-rapid eye movement (NREM) sleep.

CSB among patients with heart failure has been linked to poor prognosis. Studies that assessed the systemic consequences of CSB in patients with heart failure found that laboratory markers indicative of en-

Table 13.3. Diagnostic criteria for Cheyne-Stokes breathing.[1]

(A or B) + C + D satisfy the criteria:

A. The presence of one or more of the following:

1. Sleepiness

2. Difficulty initiating or maintaining sleep, frequent awakenings, or nonrestorative sleep

3. Awakening short of breath

4. Snoring

5. Witnessed apneas

B. The presence of atrial fibrillation/flutter, congestive heart failure, or a neurological disorder.

C. PSG (during diagnostic or positive airway pressure titration) shows all of the following:

1. Five or more central apneas and/or central hypopneas per hour of sleep

2. The total number of central apneas and/or central hypopneas is > 50% of the total number of apneas and hypopneas[2]

3. The pattern of ventilation meets criteria for Cheyne-Stokes breathing (CSB)

4. The disorder is not better explained by another current sleep disorder, medication use (e.g., opioids), or substance use disorder

hanced sympathetic nervous system activity (e.g. plasma and urinary catecholamine levels) are elevated in such patients.[34,38]

13.4.1.2 Management of CSB

There is no consensus yet regarding the best management for CSB in patients with heart failure. Management should aim initially to optimize cardiac function. Medical management of HF is not merely needed to improve sleep related breathing disorders; it is the cornerstone for the management of a dysfunctional heart and thereby increases patient survival.[34]

13.4.1.2 Pharmacological Therapy

Several drugs have been used to treat patients with HF and CSB. Although these drugs frequently reduce sleep related breathing disorders, they may cause undesirable side effects and/or interactions with other drugs. The most commonly administered drugs are methylxanthines (theophyllines), acetazolmide, and benzodiazepines. Unfortunately, all studies that assessed the effects of these medications were short-term and used small sample sizes.[34] Thus, there is limited knowledge of the long-term efficacy and safety profile of these medications. In particular, theophylline appears to be associated with arrhythmia, which could have deleterious effects on a poorly functioning left ventricle. Currently, none of these drugs are recommended as a first line treatment to manage the CSB of patients with HF.

13.4.1.3 Oxygen

The theory behind using oxygen supplementation is that oxygen will offset the hypoxic ventilatory drive and hence suppress periodic breathing. Several studies that have examined the role of home oxygen therapy reported conflicting results and no data are available regarding long-term clinical outcome.[34] For the present time, long-term oxygen therapy is not recommended as a standard treatment for patients with heart failure and CSB.

13.4.1.3 Positive airway pressure (PAP) support

CPAP has been shown to be an effective and safe treatment for patients with acute pulmonary edema.[39] A multicenter trial (CANPAP) was conducted to evaluate the efficacy of CPAP in reducing mortality and morbidity associated with CSB in patients with HF.[40] The study revealed that CPAP had positive effects on oxygen saturation, left ventricular ejection fraction, 6 m walk distance, and resulted in a 53% reduction in AHI. However, no significant difference was found in transplant-free survival, rate of hospitalization, or quality of life. The study group advised against routine use of CPAP in patients with HF and CSB. However, in the post-hoc analysis of the study, patients with residual AHI < 15/hr on CPAP had improvement in both left ventricular ejection fraction and heart transplant–free survival.[41] Therefore, the investigators recommended that a 1 month trial of CPAP be continued only if AHI is suppressed significantly. Based on this study, CPAP should not be routinely used to manage CSB in patients with HF.

Adaptive servo-ventilation (ASV) is a new mode of automated pressure support that performs breath-to-breath analysis and delivers ventilatory needs accordingly, in order to prevent the hyperventilation that drives CSB. Figure. 2 shows a zoomed epoch (2 min) of a patient with CSB on ASV. Numerous observational studies have demonstrated the beneficial effect of ASV in CSB in patients with HF.[42] Nevertheless, randomized long-term controlled trials are needed to determine the long-term clinical efficacy of these devices. ResMed Company issued on May 2015 a serious safety concern during the preliminary primary data analysis from the SERVE-HF clinical trial.[43] The investigators reported increased risk of cardiovascular death with ASV therapy for patients with symptomatic chronic HF with reduced ejection fraction (45%) and CSA.[43]

13.5 SLEEP RELATED HYPOVENTILATION DISORDERS

Sleep related hypoventilation disorders, as defined by the most recent version of the AASM Manual (ICSD-3) are characterized by insufficient sleep related ventilation leading to abnormally elevated $PaCO_2$ during sleep.[1]

Sleep-related hypoventilation disorders include:[1]

- Obesity hypoventilation syndrome (OHS).
- Congenital central alveolar hypoventilation.
- Late-onset central hypoventilation with hypothalamic dysfunction.
- Idiopathic central alveolar hypoventilation.
- Sleep-related hypoventilation due to a medication or substance.
- Sleep-related hypoventilation due to a medical disorder.

In the sleep related hypoventilation disorders, other than OHS, daytime hypoventilation may or may not be present. If hypoventilation is present during wakefulness, it worsens during sleep in all of these disorders. Awake hypoventilation

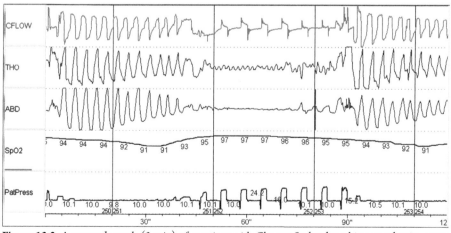

Figure 13.2. A zoomed epoch (2 min) of a patient with Cheyne-Stokes breathing on adaptive servo ventilation (ASV). ASV performs breath-to-breath analysis and delivers ventilatory needs accordingly. During periods of apnea or reduced flow, pressure support (PS) tries to compensate for the decrease of flow, while respecting the breathing frequency of the patient.

is defined as $PaCO_2$ > 45 mmHg.[1] In this chapter we will discuss OHS.

13.5.1 Obesity Hypoventilation Syndrome (OHS)

OHS is also called hypercapnic sleep apnea or sleep related hypoventilation associated with obesity.[1] Previously, it used to be called Pickwickian syndrome; however, the use of this term is discouraged now.

To diagnose OHS, the following criteria must be met:[1]

A. Presence of hypoventilation during wakefulness ($PaCO_2$ > 45 mmHg) as measured by arterial $PaCO_2$, end-tidal $PaCO_2$, or transcutaneous $PaCO_2$.

B. Presence of obesity (BMI > 30 kg/m^2).

C. Hypoventilation is not primarily due to lung parenchymal or airway disease, pulmonary vascular pathology, chest wall disorder (other than mass loading from obesity), medication use, neurologic disorder, muscle weakness, or a known congenital or idiopathic central alveolar hypoventilation syndrome.

Although arterial oxygen desaturation is a common finding in patients with OHS, it is not required to make the diagnosis of OHS. It is important to note that OSA often coexists with OHS, in those cases; the diagnosis of both OSA and OHS should be made.

13.5.1.1 History and Physical Examination

About 90% of patients with OHS have coexisting OSA; therefore, symptoms and many of the physical findings of OHS patients are similar to those in patients with OSA,[44] such as excessive daytime sleepiness, snoring, choking during sleep, morning headaches, fatigue, mood disturbance and impairments of memory or concentration.[44] However, when compared to eucapnic OSA patients, those with OHS tend to complain more often of shortness of breath.[44]

On examination, obesity is the main feature. The prevalence of OHS increases with the increase in BMI.[44] OHS patients tend to have crowded oropharynx and an increase in neck circumference.[44] They may also exhibit signs of corpulmonale or circulatory congestion including plethora, scleral injection, peripheral edema and a prominent pulmonic component of the second heart.[44]

13.5.1.2 Diagnostic Testing

OHS is a diagnosis of exclusion and therefore many diagnostic tests should be carried out to distinguish OHS from other disorders in which hypercapnia is a common finding such as pulmonary diseases, skeletal restriction, neuromuscular disorders, hypothyroidism or pleural pathology.[1,44] Tests should include ABG, pulmonary function tests, chest imaging, laboratory tests, electrocardiography (ECG), transthoracic echocardiogram and polysomnography. ABG sampling is a key test since hypercapnia is a fundamental feature of the disorder and is required to make the diagnosis. Usually, ABG reveals high $PaCO_2$ and a high bicarbonate level which reflect the chronic nature of the disease.

Serum bicarbonate level is a useful test in screening patients with morbid obesity or if there is difficulty or delay in obtaining ABG.[45,46] A serum bicarbonate of > 27 mEq/L has a 92% sensitivity for identifying awake hypercapnia in obese patients; however, the specificity is significantly lower.[45]

Pulmonary function tests (PFT) are essential to exclude other causes of hypercapnia such as chronic pulmonary diseases. Although PFT can be normal, it usually reveals mild-to-moderate restrictive pattern due to obesity.[46] The expiratory reserve volume is also significantly reduced in patients with significant obesity.[44,47] Chest radiographs, and if indicated chest computed tomography are usually performed to rule in or out other potential causes of hypercapnic respiratory failure.[44] ECG and transthoracic echocardiogram could show features of right heart strain, right ventricular hypertrophy, right atrial enlargement and elevated pulmonary artery pressure.[44] Other laboratory testing should include a complete blood count to rule out secondary erythrocytosis, and thyroid function test to rule out severe hypothyroidism.[44,48]

Polysomnography in patients with OHS may show oxygen desaturation and hypercapnia during sleep not related to obstructive apneas and hypopneas periods. Hypoventilation is usually more prominent during REM sleep compared to NREM sleep.[7]

If $PaCO_2$ can be monitored, it may also demonstrate an increase of more than 10 mmHg in $PaCO_2$ level during sleep, compared with levels during wakefulness.[44]

13.5.1.3 Management of obesity hypoventilation syndrome

Untreated OHS is associated with a high mortality rate, a reduced quality of life, and numerous morbidities, including hypertension, pulmonary hypertension, right heart failure, angina, and acute hypercapnic respiratory failure.[34,47]

The optimal management of patients with OHS requires a multidisciplinary approach including different medical and surgical subspecialties. Affected subjects require the contribution of internists and endocrinologists regarding their diabetes mellitus, hypertension, hyperlipidemia, heart failure, and hypothyroidism therapy; a dietician for weight reduction planning; a respirologist for respiratory failure management; and a surgeon for potential bariatric surgery when needed. Moreover, patients with OHS have higher rates of intensive care admission when compared to obese patients without hypoventilation, which obviously requires an expert input from medical intensivists regarding the management of acute or chronic respiratory failure episodes.[44]

Although there are no treatment guidelines for OHS, treatment approaches are based on reversing the underlying pathophysiology of OHS including the reversal of sleep-disordered-breathing, weight reduction, and treatment of comorbid conditions.

13.5.1.4 Weight Loss

Significant weight loss is desirable in patients with OHS and will lead to improvement in pulmonary physiology and function including improvement in alveolar ventilation and in nocturnal oxyhemoglobin saturation.[49] Furthermore, in patients with coexisting OSA, weight loss decreases the frequency of obstructive respiratory events.[50,51,52]

All patients with OHS should pursue lifestyle modifications to lose weight. If lifestyle modifications are not sufficient, bariatric surgery should be entertained. The benefits of weight loss appear to occur regardless of whether weight loss was due to lifestyle modification (i.e., diet, exercise) or surgery. However, it is important to realize that weight loss cannot be used as the sole initial treatment.[44]

13.5.1.5 Positive Airway Pressure Therapy (PAP)

Application of positive airway pressure is the mainstay of therapy for OHS. It seems reasonable to start with CPAP knowing that the majority of OHS patients have coexisting OSA. CPAP has been shown to be effective in a group of patients with stable OHS, especially in those with severe OSA.[53] There are no clear guidelines on when to start or switch to bi-level PAP (BPAP); however, BPAP should be strongly considered in patients with OHS without OSA, in patients with OHS and coexisting OSA if CPAP is insufficient and hypercapnia persists despite being on long-term CPAP or if they fail to tolerate CPAP. In addition, BPAP should be used in patients with OHS who experience acute-on-chronic respiratory failure.[47,54]

13.5.1.6 Average Volume-Assured Pressure-Support (AVAPS)

AVAPS is a hybrid mode of ventilation which combines pressure-support and volume-controlled modes of ventilation into one ventilation mode to ensure a more consistent tidal volume and hence minute volume ventilation.[47] Data seem to suggest that such assurance of tidal volume has advantages with regard to lower transcutaneous PCO_2 readings.[55] However, AVAPS did not provide further clinical benefits regarding sleep quality and health-related quality of life (HRQL).[55,56] Nevertheless, this does not preclude this mode of ventilatory support being indicated on an individual basis for patients failing fixed BPAP.

Treatment of OHS with PAP improves blood gases, this improvement could be achieved in 2-4 weeks. Therefore, early follow-up is important and should include repeat measurement of ABG with assessment of adherence to PAP.[53]

13.5.1.7 Oxygen Therapy

Patients with OHS commonly suffer from prolonged episodes of hypoxemia during sleep, in addition to daytime hypoxemia. Therefore oxygen therapy is needed if hypoxemia persists despite the relief of upper airway obstruction and hypoventilation with PAP therapy in order to prevent the long-term consequences of hypoxemia on pulmonary vasculature and other vital organs. However, it is important to keep in mind, that treatment with oxygen alone is inadequate and is not recommended as it does not reverse hypoventilation or airway obstruction on its own.[44,47]

13.5.1.8 Tracheostomy

Tracheostomy is rarely necessary and is generally reserved for patients who are not tolerant or not adherent to PAP therapy or patients who have difficulty in weaning them from invasive ventilation.[47]

13.5.1.8 Treatment of Comorbid Conditions and Complications

It is important to manage comorbid conditions that impair ventilation or reduce the ventilatory response to hypoxemia or hypercapnia, which are likely to worsen OHS such as chronic obstructive pulmonary disease (COPD) and hypothyroidism. In addition, complications associated with OHS should be investigated and managed appropriately.

ACKNOWLEDGMENTS

This work was supported by a grant from the Strategic Technologies Program of the National Plan for Sciences and Technology and Innovation in the Kingdom of Saudi Arabia.

KEY WORDS

- **obstructive sleep apnea**
- **central sleep apnea**
- **hypopnea**
- **Cheyne-Stokes respiration**
- **hypoventilation**
- **CPAP**
- **positive airway pressure**

REFERENCES

1. American Academy of Sleep Medicine. *International Classification of Sleep Disorders, 3rd ed.* American Academy of Sleep Medicine:Darien, IL, 2014.

2. Berry, R. B.; Gamaldo, C. E.; Harding, S. M.; Lloyd, R. M.; Marcus, C. L.; Vaughn, B. V. *The AASM Manual for the Scoring of Sleep and Associated Events: Rules, Terminology and Technical Specifications, Version 2.1.* for the American Academy of Sleep Medicine American Academy of Sleep Medicine: Darien, IL, 2014.

3. Peppard, P. E.; Young, T.; Barnet, J. H.; Palta, M.; Hagen, E. W.; Hla, K. M. Increased Prevalence of Sleep-Disordered Breathing in Adults. *Am. J. Epidemiol.* **2013**, *177*(9), 1006–1014.

4. Johns, M. W. A New Method for Measuring Daytime Sleepiness: The Epworth Sleepiness Scale. *Sleep.* **1991**, *14*(6), 540–545.

5. Myers, K. A.; Mrkobrada, M.; Simel, D. L. Does This Patient have Obstructive Sleep Apnea?: The Rational Clinical Examination Systematic Review. *JAMA.* **2013**, *310*(7), 731–741.

6. Friedman, M.; Hamilton, C.; Samuelson, C. G.; Lundgren, M. E.; Pott, T. Diagnostic Value of the Friedman Tongue Position and Mallampati Classification for Obstructive Sleep Apnea: A Meta-Analysis. *Otolaryngol. Head. Neck. Surg.* **2013**, *148*(4), 540–547.

7. Kushida, C. A.; Littner, M. R.; Morgenthaler, T.; Alessi, C. A.; Bailey, D.; Coleman, J. Jr. et al. Practice Parameters for the Indications for Polysomnography and Related Procedures: An Update for 2005. *Sleep.* **2005**. *28*(4), 499–521.

8. Collop, N. A.; Tracy, S. L.; Kapur, V.; Mehra, R.; Kuhlmann, D.; Fleishman, S.A.; et al. Obstructive Sleep Apnea Devices for Out-Of-Center (OOC) Testing: Technology Evaluation. *J. Clin. Sleep. Med.* **2011**, *7*(5), 531–548.

9. Ruehland, W. R.; Rochford, P. D.; O'Donoghue, F. J.; Pierce, F. J.; Singh, P.;Thornton A. T. The New AASM Criteria for Scoring Hypopneas: Impact on the Apnea Hypopnea Index. *Sleep.* **2009**. *32*(2), 150–157.

10. Bradley, T. D.; Floras, J. S. Obstructive Sleep Apnoea and its Cardiovascular Consequences. *Lancet.* **2009**, *373*(9657) 82–93.

11. Aldabal, L.; Bahammam, A. S. Metabolic, Endocrine, and Immune Consequences of Sleep Deprivation. *Open. Respir. Med. J.* **2011**, *5*, 31–43.

12. Epstein, L. J.; Kristo, D.; Strollo, P. J. Jr.;Friedman, N.; Malhotra, A.; Patil, S. P.; et al. Clinical Guideline for the Evaluation, Management and Long-Term Care of Obstructive Sleep Apnea in Adults. *J. Clin. Sleep. Med.* **2009**, *5*(3), 263–276.

13. Strohl, K. P.; Brown, D. B.; Collop, N.; George, C.; Grunstein, R.; Han, F.; et al. An Official American Thoracic Society Clinical Practice Guideline: Sleep Apnea, Sleepiness, and Driving Risk in Noncommercial Drivers. An Update

of a 1994 Statement. *Am. J. Respir. Crit. Care. Med.* **2013**, *187*(*11*), 1259–1266.

14. Randerath, W. J.; Verbraecken, J.; Andreas, S.; Bettega, G.; Boudewyns, A.; Hamans, E.; et al. Non-CPAP Therapies in Obstructive Sleep Apnoea. *Eur. Respir. J.* **2011**, *37*(*5*), 1000–1028.

15. Morgenthaler, T.I.; Kapen, S.; Lee-Chiong, T.; Alessi, C.; Boehlecke, B.; Brown, T.; et al. Practice Parameters for the Medical Therapy of Obstructive Sleep Apnea. *Sleep.* **2006**, *29*(*8*), 1031–1035.

16. Dixon, J. B.; Schachter, L. M.; O'Brien, P. E.; Jones, K.; Grima, M.; Lambert, G.; et al. Surgical vs Conventional Therapy for Weight Loss Treatment of Obstructive Sleep Apnea: A Randomized Controlled Trial. *JAMA.* **2012**, *308*(*11*), 1142–1149.

17. Pevernagie, D.A.; Stanson, A. W.; Sheedy, P. F.; Daniels, D. K.; Shepard, J. W. Jr. Effects of Body Position on the Upper Airway of Patients with Obstructive Sleep Apnea. *Am. J. Respir. Crit. Care. Med.* **1995**, *152*(*1*), 179–185.

18. Issa, F. G.; Sullivan, C. E. Alcohol, Snoring and Sleep Apnea. *J. Neurol. Neurosurg. Psychiatry.* **1982**, *45*(*4*), 353–359.

19. Al-Jawder, S. E.; Bahammam, A. S. Comorbid Insomnia in Sleep-Related Breathing Disorders: An Under-Recognized Association. *Sleep. Breath.* **2012**, *16*(*2*), 295–304.

20. Aurora, R .N.; Chowdhuri, S.; Ramar, K.; Bista, S. R.; Casey, K. R,; Lamm, C. I.; et al. The Treatment of Central Sleep Apnea Syndromes in Adults: Practice Parameters with an Evidence-Based Literature Review and Meta-Analyses. *Sleep.* **2012**, *35*(*1*), 17–40.

21. Ou, Q.; Chen, Y. C.; Zhuo, S. Q.; Tian, S. Q.; He, C. H.; Lu, X. L.; et al. Continuous Positive Airway Pressure Treatment Reduces Mortality in Elderly Patients with Moderate to Severe Obstructive Severe Sleep Apnea: A Cohort Study. *PloS. one.* **2015**, *10*(*6*), e0127775.

22. Campos-Rodriguez, F.; Martinez-Garcia, M. A.; de la Cruz-Moron, I.; Almeida-Gonzalez, C.; Catalan-Serra, P.; Montserrat, J. M. Cardiovascular Mortality in Women with Obstructive Sleep Apnea With or Without Continuous Positive Airway Pressure Treatment: A Cohort Study. *Ann. Intern. Med.* **2012**. *156*(*2*), 115–122.

23. Kryger, M. H.; Berry, R. B.; Massie, C. A. Long-Term Use of a Nasal Expiratory Positive Airway Pressure (EPAP) Device as a Treatment for Obstructive Sleep Apnea (OSA). *J. Clin. Sleep. Med.* **2011**, *7*(*5*), 449–453.

24. Barbe, F.; Duran-Cantolla, J.; Capote, F.; de la Pena, M.; Chiner, E.; Masa, J. F.; et al. Long-Term Effect of Continuous Positive Airway Pressure in Hypertensive Patients with Sleep Apnea. *Am. J. Respir. Crit. Care. Med.* **2010**. *181*(*7*), 718–726. Lanfranco, F.; Motta, G.; Minetto, M. A.; Baldi, M.; Balbo, M.; Ghigo, E.; et al. Neuroendocrine Alterations in Obese Patients with Sleep Apnea Syndrome. *Int. J. Endocrinol.* **2010**, *181*(*7*), 718–726.

25. Wozniak, D. R.; Lasserson, T. J.; Smith, I. Educational, Supportive and Behavioural Interventions to Improve Usage of Continuous Positive Airway Pressure Machines in Adults with Obstructive Sleep Apnoea. *Cochrane. Database. Syst. Rev.* **2014**, *1*, CD007736.

26. Wu, H.; Yuan, X.; Zhan, X.; Li, L.; Wei, Y. A Review of EPAP Nasal Device Therapy for Obstructive Sleep Apnea Syndrome. *Sleep. Breath.* **2014**, *1*, CD007736.

27. Berry, R. B.; Kryger, M. H.; Massie, C. A. A Novel Nasal Expiratory Positive Airway Pressure (EPAP) Device for the Treatment of Obstructive Sleep Apnea: A Randomized Controlled Trial. *Sleep.* **2011**, *34*(*4*), 479–485.

28. Kushida, C. A.; Morgenthaler, T. I.; Littner, M. R.; Alessi, C. A.; Bailey, D.; Coleman, J. Jr.; et al. Practice Parameters for the Treatment of Snoring and Obstructive Sleep Apnea with Oral Appliances: An Update for 2005. *Sleep.* **2006**, *29*(*2*), 240–243.

29. Smith, D. F.; Cohen, A. P.; Ishman, S. L. Surgical Management of OSA in Adults. *Chest.* **2015**, *147*(*6*), 1681–1690.

30. Van de Heyning, P. H.; Badr, M. S.; Baskin, J. Z.; Cramer Bornemann, M. A.; De Backer, W. A.; Dotan, Y.; et al. Implanted Upper Airway Stimulation Device for Obstructive Sleep Apnea. *Laryngoscope.* **2012**, *122*(*7*), 1626–1633.

31. Strollo, P. J. Jr.; Soose, R. J.; Maurer, J. T.; de Vries, N.; Cornelius, J.; Froymovich, O.; et al. Upper-Airway Stimulation for Obstructive Sleep Apnea. *N. Engl. J. Med.* **2014**, *370*(*2*), 139149.

32. Woodson, B. T.; Gillespie, M. B.; Soose, R. J.; Maurer, J. T.; de Vries, N.; Steward, D. L.; et al. Randomized Controlled Withdrawal Study of Upper Airway Stimulation on OSA: Short- and Long-Term Effect. *Otolaryngol. Head. Neck. Surg.* **2014**, *151*(*5*), 880–887.

33. Mason, M.; Welsh, E. J.; Smith, I. Drug Therapy for Obstructive Sleep Apnoea in Adults. *Cochrane. Database. Syst. Rev.* **2013**, *5*, CD003002.

34. AlDabal, L.; BaHammam, A.S. Cheyne-Stokes Respiration in Patients with Heart Failure. *Lung.* **2010**, *188*(*1*), 5–14.

35. Carmona-Bernal, C.; Ruiz-Garcia, A.; Villa-Gil, M.; Sanchez-Armengol, A.; Quintana-Gallego, E.; Ortega-Ruiz, F.; et al. Quality of Life in Patients with Congestive Heart Failure and Central Sleep Apnea. *Sleep. Med.* **2008**, *9*(*6*), 646–651.

36. Sin, D. D.; Fitzgerald, F.; Parker, J. D.; Newton, G.; Floras, J. S.; Bradley, T. D. Risk Factors for Central and Obstructive Sleep Apnea in 450 Men and Women with Congestive Heart Failure. *Am. J. Respir. Crit. Care. Med.* **1999**, *160*(*4*), 1101–1106.

37. Xie, A.; Skatrud, J.B.; Puleo, D. S.; Rahko, P. S.; Dempsey, J. A. Apnea-Hypopnea Threshold for CO2 in Patients with Congestive Heart Failure. *Am. J. Respir. Crit. Care. Med.* **2002**, *165*(*9*), 1245–1250.

38. Carmona-Bernal, C.; Quintana-Gallego, E.; Villa-Gil, M.; Sanchez-Armengol, A.; Martinez-Martinez, A.; Capote, F. Brain Natriuretic Peptide in Patients with Congestive Heart Failure and Central Sleep Apnea. *Chest.* **2005**, *127*(*5*), 1667–1673.

39. Vital, F. M.; Ladeira, M. T.; Atallah, A. N. Non-Invasive Positive Pressure Ventilation (CPAP or Bilevel NPPV) for Cardiogenic Pulmonary Oedema. *Cochrane. Database. Syst. Rev.* **2013**, *5*,CD005351.

40. Bradley, T. D.; Logan, A. G.; Kimoff, R. J.; Series, F.; Morrison, D.; Ferguson, K.; et al. Continuous Positive Airway Pressure for Central Sleep Apnea and Heart Failure. *N. Engl. J. Med.* **2005**, *353*(*19*), 2025–2033.

41. Arzt, M.; Floras, J. S.; Logan, A. G.; Kimoff, R. J.; Series, F.; Morrison, D.; et al. Suppression of Central Sleep Apnea by Continuous Positive Airway Pressure and Transplant-Free Survival in Heart Failure: A Post Hoc Analysis of the Canadian Continuous Positive Airway Pressure for Patients with Central Sleep Apnea and Heart Failure Trial (CANPAP). *Circulation.* **2007**, *115*(*25*), 3173–3180.

42. Javaheri, S.; Brown, L. K.; Randerath, W. J. Clinical Applications of Adaptive Servoventilation Devices: Part 2. *Chest.* **2014**, *146*(*3*), 858–868.

43. ResMed. *Important medical device warning.* 2015 [cited 2015 6/29/2015]; Available from: http://www.thoracic.org.au/imagesDB/wysiwyg/ServeHFDoctorLetter.pdf.

44. Al Dabal, L.; Bahammam, A. S. Obesity Hypoventilation Syndrome. *Ann. Thorac. Med.* **2009**, *4*(*2*), 41–49.

45. Mokhlesi, B.; Tulaimat, A.; Faibussowitsch, I.; Wang, Y.; Evans, A. T. Obesity Hypoventilation Syndrome: Prevalence and Predictors in Patients with Obstructive Sleep Apnea. *Sleep. Breath.* **2007**, *11*(*2*), 117–124.

46. BaHammam, A. S. Prevalence, Clinical Characteristics, and Predictors of Obesity Hypoventilation Syndrome in a Large Sample of Saudi Patients with Obstructive Sleep Apnea. *Saudi. Med. J.* **2015**, *36*(*2*), 181–189.

47. Bahammam, A. S.; Al-Jawder, S. E. Managing Acute Respiratory Decompensation in the Morbidly Obese. *Respirology.* **2012**, *17*(*5*), 759–771.

48. Bahammam, S. A.; Sharif, M. M.; Jammah, A. A.; Bahammam, A. S. Prevalence of Thyroid Disease in Patients with Obstructive Sleep Apnea. *Respir. Med.* **2011**, *105*(*11*), 1755–1760.

49. Aaron, S. D.; Fergusson, D.; Dent, R.; Chen, Y.; Vandemheen, K. L.; Dales, R. E. Effect of Weight Reduction on Respiratory Function and Airway Reactivity in Obese Women. *Chest.* **2004**, *125*(*6*), 2046–2052.

50. Guardiano, S. A.; Scott, J. A.; Ware, J. C.; Schechner, S. A. The Long-Term Results of Gastric Bypass on Indexes of Sleep Apnea. *Chest.* **2003**, *124*(*4*), 1615–1619.

51. Harman, E. M.; Wynne, J. W.; Block, A. J. The Effect of Weight Loss on Sleep-Disordered Breathing and Oxygen Desaturation in Morbidly Obese Men. *Chest.* **1982**, *82*(*3*), 291–294.

52. Nguyen, N. T.; Hinojosa, M. W.; Smith, B. R.; Gray, J.; Varela, E. Improvement of Restrictive and Obstructive Pulmonary Mechanics Following Laparoscopic Bariatric Surgery. *Surg. Endosc.* **2009**, *23*(*4*), 808–812.

53. Mokhlesi, B. Obesity Hypoventilation Syndrome: A State-of-the-Art Review. *Respir. Care.* **2010**, *55*(*10*), 1347–1362, discussion 1363–1365.

54. BaHammam, A. Acute Ventilatory Failure Complicating Obesity Hypoventilation: Update on a 'Critical Care Syndrome'. *Curr. Opin. Pulm. Med.* **2010**, *16*(*6*), 543–551.

55. Storre, J. H.; Seuthe, B.; Fiechter, R.; Milioglou, S.; Dreher, M.; Sorichter, S.; et al. Average Volume-Assured Pressure Support in Obesity Hypoventilation: A Randomized Crossover Trial. *Chest.* **2006**, *130*(3), 815–821.

56. Murphy, P. B.; Davidson, C.; Hind, M. D.; Simonds, A.; Williams, A. J.; Hopkinson, N. S. et al. Volume Targeted Versus Pressure Support Non-Invasive Ventilation in Patients with Super Obesity and Chronic Respiratory Failure: A Randomised Controlled Trial. *Thorax.* **2012**, *67*(8), 727–734.

SURGICAL OPTIONS FOR SLEEP APNEA

Adriana Martin, Tomasz Rogula, and Philip R. Schauer

14.1 INTRODUCTION

Obstructive sleep apnea (OSA) is a chronic disorder characterized by obstructive apnea, hypopnea and arousals leading to fragmented sleep that occurs secondary to intermittent complete or partial upper-airway collapse during sleep. This is a relatively common problem, and its prevalence in North American population varies between 9 and 43% in males, and 3 and 27% in females, depending in what classification criteria is used, the age and the body mass index (BMI) of the patient.[1]

Adriana Martin

1Universidade Federal do Rio Grande do Sul- Hospital de Clinicas de Porto Alegre, Rua Ramiro Barcelos, 2350, Porto Alegre, Brazil

Tomasz Rogula2
Bariatric and Metabolic Institute, Cleveland Clinic, 9500 Euclid Ave, M66-06, Cleveland, OH 44118, USA

Philip R. Schauer
Bariatric and Metabolic Institute, Cleveland Clinic, 9500 Euclid Ave, M66-06, Cleveland, OH 44118, USA

Corresponding author: Tomasz Rogula, MD, PhD, Bariatric and Metabolic Institute, Cleveland Clinic, 9500 Euclid Ave, M66-06, Cleveland, OH 44118, USA. Email: tomrogula@gmail.com

This prevalence has grown during the last decades, mostly due to the ongoing obesity epidemic.[2]

This pathological airway obstruction, instead of the physiological airway narrowing that occurs during sleep, is usually caused by an association of multiple risk factors. Risk factors and demographic correlations include: male gender, age, alcohol ingestion, morbid obesity, and large neck girth.[3, 4] Craniofacial and upper-airway soft-tissue abnormalities also have an important role in OSA occurrence. Patients with OSA are found to have smaller minimum airway area, thicker lateral pharyngeal wall, larger soft palate area, larger tongue, and larger amount of lateral and posterior subcutaneous fat.[5] Other anatomical abnormalities include: dimorphisms related to mandibular or maxillary size and position, narrowed nasal cavities, and tonsillar and adenoids hypertrophy. The latter, are especially important in childhood and may cause abnormal growth patterns of the lower face and jaw that in the future would predispose to OSA.[6]

As a consequence of this airway collapse, patients will present with important physiologic changes. Sleep fragmentation, oxyhemoglobin desaturation, hypercapnia, marked swings in intrathoracic pressure, fluctuations in blood pressure and heart rate, and increased sympathetic activity[3,7] are some changes that can lead to OSA's symptoms, complications, as well increased mortality.

The usual clinical presentation is complains of daytime sleepiness, nonrefreshing sleep, nocturia and morning headaches, witnessed loud snoring and breathing interruptions, decreased concentration and memory with increased irritability. OSA can predispose to a series of harmful long-term sequelae as hypertension, stroke, myocardial infarct, *cor pulmonale* and also can increase the risk of motor vehicle accidents.[7, 8]

In this scenario, it is important to stablish an accurate and prompt diagnosis of OSA and classify the disease's severity in order to decide which is the most appropriate treatment for each patient. There are three major ways of managing OSA: behavioral changes, medical interventions as continuous positive airway pressure (CPAP) and oral appliances, and surgical treatment.[4] Even though CPAP is the most widely used management to decrease symptoms and cardiovascular risk,[9] it is a treatment that only protects when used chronically. Problems with adhesion and compliance are its main issue. Apparently, up to 40% of the patients will discontinue CPAP after the first 3 months of use.[10] On the other hand, surgery offers a "short course" and apparently immediate curative intervention,[4] but the role of surgery in the treatment of OSA is still controversial and most of recommendations are based in case series.[11]

In this chapter we will focus on the surgical options for treating OSA that can be indicated depending on the severity of the disease, anatomical findings and patient preferences.

14.2 THE GOAL OF SURGICAL TREATMENT

There are three main surgical approaches for OSA treatment: by increasing the airway surface area, by bypassing the obstruction or by removing a specific pathological lesion.[3]

The desired outcomes are resolution of symptoms, mainly daytime sleepiness (assessed by the Epworth sleepiness scale (ESS)), improvement in quality of sleep and life, and normalization of apnea-hypopnea index (AHI) and oxyhemoblogin saturation levels in the postoperative period.[12] More long-term follow-up studies are needed to show reduction of cardiovascular risk and normalization of mortality rate with surgical approach.

A successful surgical outcome is defined by the Sher's criteria as a reduction of the AHI by at least 50% and to a value below 20 /h in a patient whose preoperative AHI was > 20 /h.[13]

14.3 INDICATIONS FOR SURGICAL TREATMENT

All indications for surgery in OSA are summarized in Table 14.1. Surgery can be indicated as a primary therapy in patients with OSA and severe obstructive anatomy that is surgically correctible. The main application of primary surgery is in pediatric OSA, where surgical resolution of the disease occurs in 80% of the patients.[14] Another primary indication for surgical intervention is life-threatening situations requiring emergent tracheostomy, as in the case of

severe arrhythmias and respiratory insufficiency.[12]

All other indications are as secondary
treatment, when first-line CPAP therapy
or oral appliances have failed improving
symptoms and objective parameters, are
not tolerated by the patient or the patient
has low compliance after 3 months trial.
Surgery can also be an adjuvant therapy
to improve outcomes with CPAP therapy
when obstructive anatomy or functional
deformities compromises firs-line therapy
efficacy.[7] Another adjuvant surgical option
available and effective in curing OSA is bariatric surgery for those with BMI > 40 or 35
kg/m[2] when associated comorbidities. The
patient´s option and preferences should always be considered and discussed.

Finally, surgical intervention is indicated only if the patient's general condition allows surgery and if a correctable
anatomical abnormality has been carefully
and diagnostically evaluated before the op-
eration.

14.4 PREOPERATIVE EVALUATION AND APPROACH

Preoperative approach includes evaluation
of patient's eligibility for surgery, by assessing any medical, psychological, or social
comorbidity that might affect surgical outcomes.[7]

The severity of OSA should be assessed
by polysomnography prior to surgery, it is
needed to aid in deciding which is the best
treatment option.

The Friedman staging system[15] is
a helpful preoperative tool assessed by
examination of the oral cavity with the
mouth open widely without the protrusion of the tongue. It evaluates palate position in relation to the tongue, tonsils size,
and BMI. It classifies the patient into four
stages that are correlated to rates of successful surgery and are directly related to
the severity of OSA.[16] Uvulopalatopharyngoplasty (UPPP) is indicated in patients
in stage I, since they have an 80% chance

Table 14.1 Summary of main indications for surgical approach in OSA

Indications for surgical intervention in OSA	
Primary	- Pediatric OSA.
	- Anatomical obstruction (e.g. benign or malignant tumor of the upper-airway).
	- Emergencies (e.g. arrhythmias and acute respiratory insufficiency).
Secondary	- Low effectiveness of CPAP therapy.
	- Intolerance to CPAP therapy (e.g. important CPAP-related side effects).
	- Low adherence to CPAP therapy.
Adjuvant	- Anatomical findings compromising CPAP use and efficacy. (e.g. complete concentric collapse of the soft palate).
	- Bariatric surgery indicated if BMI > 40 or 35 kg/m[2] if comorbidities.

of successful outcome. On the other hand, for patients with stage II or III disease the best approach is a multi-level surgery, since they have obstruction in more than one site. Lastly, patients in stage IV, BMI > 40 kg/m², should have bariatric surgery indicated.[17]

It is also recommended to perform a drug-induced sleep endoscopy (DISE) preoperatively to determine the anatomical site and pattern of the obstruction and/or collapse, particularly regarding hypopharyngeal and laryngeal obstruction. It is shown to be a good predictor of the likelihood of response to upper-airway surgery,[18] because it can help to decide which surgical technique would be more appropriate for the patient, individualizing each particular case. It is usually performed in the operating room, under a sedation technique that best approximates natural sleep pharmacologically. The nasopharyngoscope is introduced after the patient is sleeping and actively snoring, so the collapsible region is better visualized and diagnosed. Another option is an awake endoscopy performing Muller's maneuver. Although it is less helpful in predicting surgery's outcomes as DISE,[19] with a predicted efficacy of only 33%,[20] it can be performed in a clinic. The technique involves the patient performing a forced inspiratory effort while the examiner is watching with the fiber optic camera and searching for obstructions.

Studies suggest that computed tomography (CT) scanning and magnetic resonance imaging (MRI) are not necessary in the preoperative workup to help guide the surgical intervention in OSA. Their harmful and expensiveness do not overcome their benefit.[21] They can be helpful in diagnosing skeletal deformities, when that is the case.

It is a standard recommendation that patients be advised about surgical options, success rates, side effects, and possible complications before surgery. Also, alternative treatments available must be presented and explained for the patient, including its levels of effectiveness and success rates. The patient should always participate in the decision of which modality of treatment is the most adequate for him.[12]

14.5 TYPES OF SURGICAL PROCEDURES

There are multiple surgical options to treat OSA. The main ones are summarized in Table 14.2 and better described below.

14.5.1 Nasal Cavity Procedures

There are a variety of nasal procedures that can be done to improve the airflow through the nasal cavity. Septoplasty for those whose obstruction is due to a deviated septum, polypectomy for patients with nasal polyps, nasal turbinate reduction when turbinates are hypertrophic, or rhinoplsty when the nasal architecture is the causative factor.

Studies show an objective and subjective association between nasal obstruction and sleep-related breathing disorders.[22] Moreover, nasal obstruction can cause a statistically significant increase in the number of partial and total obstructive respiratory events.[23] Nevertheless, nasal surgery frequently fails to correct OSA.[22] Nasal surgery can improve quality of life, quality of sleep, and snoring. Still, it is not an effective treatment for OSA[24] because it is

Table 14.2 Summary of upper-airway procedures for OSA

Anatomic site	Procedure
Nasal cavity	Septoplasty Rhinoplasty Nasal turbinate reduction Nasal polypectomy
Nasopharynx and oropharynx	UPPP and modified UPP Laser assisted uvulopalatoplasty RFA of the soft palate Soft palate implants Adenoidectomy Tonsillectomy
Hypopharynx/laryngopharynx	RFA of the base of the tongue Genioglossus advancement Laser midline glossectomy Lingualplasty Lingual tonsillectomy Glossoplexia Hyoid suspension
Mandible and maxilla	Maxillomandibular advancement (MMA)
Trachea	Tracheostomy

not able to significantly decrease AHI or daytime sleepiness accessed by ESS postoperatively.[25]

On the other hand, a group of patients with OSA that can have positive results with nasal surgery are those patients who do not tolerate CPAP because of an obstructive nasal cavity. The main reason why patients find CPAP uncomfortable is because of nasal symptoms as nasal congestion, rhinorrhea, nasal dryness, and sneezing, due to pre-existing nasal obstruction or CPAP-induced rhinitis.[26] In this select group, nasal airway surgery can reduce CPAP pressure requirements, improve oxygen saturation, and, consequently, improve CPAP compliance and adherence.[27]

Summarizing, nasal surgery may not be recommended as a primary treatment and single intervention for OSA because it is not effective in eliminating OSA, but can be recommended as adjuvant treatment in patients with nasal obstruction and issues with CPAP therapy in order to improve CPAP compliance and efficacy.[24]

14.5.2 Nasopharynx and Oropharynx Procedures

14.5.2.1 Uvulopalatopharyngoplasty

UPPP is the most common surgical procedure performed in OSA patients to increase the oropharyngeal airspace. It involves resection of the uvula, the posterior soft palate, thetonsils, and closure of the tonsillar pillars. Modifications of this technique with uvula preservation have been described to minimize complications.[28] Even though UPP is the most frequent performed procedure, it does not normalize AHI when performed in patients with moderate to severe OSA.[14] A meta-analysis shows that AHI remain elevated in the postoperative period, with a mean of 29.8 /h.[29] Furthermore, no significant improvement in lowest oxygen desaturation indices (LSAT) is acquired.[4] On the other hand, subjective data reports improvement of daytime sleepiness after 3 and 12 months.[4] UPP seems to have good outcomes when performed in preselected patients with retropalatal obstruction according to their stage in the Friedman stage system. Patients in stage I (BMI < 40 kg/m^2, enlarged tonsils that can extend to the midline but a palate position that permits a good visualization of the uvula, soft palate, and tonsils) have a successful result in 80% of the UPP surgeries performed.[16]

Complication rate after surgery is 22% and the most commons are bleeding and infection.[4] Other side effects are difficulty swallowing, dry mouth, nasal regurgitation, taste disturbances, voice changes, and need for tracheostomy.[30] Another important issue with this procedure is that it may compromise future CPAP therapy that is often required since it is not effective in providing cure.[31] When proposing UPPP for those selected patients with retropalatal obstruction, potential benefits should be weighed against the risk of frequent long-term sideeffects.[22]

In summary, UPP is not an efficient procedure in curing OSA patients, but it can achieve good outcomes if indicated to previously selected patients in stage I of Friedman's.

14.5.2.2 Laser Assisted Uvulopalatoplasty

In this procedure incisions or trenches along both sides of the uvula are performed, and after that, a laser ablation of the uvula is performed. It is an outpatient surgical technique, performed under local anesthesia, designed to shorten the uvula and modify and tighten the soft palatal tissue.[29] It can be performed as a multistage or one-stage procedure. A randomized trial showed 21% reduction in mean AHI,[32] but not a curative reduction, reducing only from 18.6 to 14.7 /h. No improvement on daytime sleepiness or quality of life was found.[22, 29] An observational report found a 73% reduction in AHI when patients were carefully selected upon sleep endoscopy that determined the anatomical level of obstruction and their suitability for the procedure.[33]

Complications of this procedure are low, since it is a minimally invasive procedure, but can include bleeding, infections, and temporary loss of taste sense.[34]

Laser assisted uvulopalatoplasty has not proven to significantly improve either OSA severity, daytime sleepiness or quality of life domains.[4, 22] In patients carefully selected by nasoendoscopy, it may have apositive outcome, but randomized trials are needed to prove its efficiency.[33]

14.5.2.3 Temperature-Controlled Radiofrequency Tissue Ablation

This is a minimally invasive procedure that applies temperature controlled energy to target tissue, which ranges from 40 to 70 ºC, and cause coagulative necrosis. It can be directed to the soft palate, to the base of tongue or be a multi-level procedure combining both. It is a procedure performed under local anesthesia that shows an improvement in airway patency with low morbidity and low rate of complications.

Radiofrequency ablation (RFA) significantly reduces sleepiness symptoms, improves quality of live, decreases AHI and increases LSAT in patients with OSA syndrome, in short-term and long-term assessment.[29, 35] This data is based on non-controlled case series and there is no difference in symptoms when compared to CPAP therapy. Also, CPAP is known to have a greater decrease in polysomnographic parameters.[4, 22]

Summarizing, RFA is a relatively new, low-morbidity procedure that can be considered a valid treatment option for patients with mild to moderate OSA who refuse or are unable to tolerate CPAP.[12]

14.5.2.4 Soft Palatal Implants

Soft palatal implants are also a relatively new treatment option that is less invasive and have a low morbidity and complication rate.[29] It is a procedure performed under local anesthesia, were three Dacron rods are inserted into the soft palate, in order to increase the structural support of the soft palate. This implants, as well as the surrounding fibrosis, reduce the vibration of the soft palate and therefore the inspiratory airway resistance.

This procedure was found to decrease daytime sleepiness, decrease snoring intensity, and increase bed partner satisfaction, being a good option for snoring treatment.[36] Two controlled clinical trials were able to show a significant improvement in AHI in non-obese patients,[4, 36] with a reduction of AHI in 81% of the patients,

and to lower than 10 in 61%. In non-obese patients with mild obstructive sleep apnea, palatal implants may be an effective alternative therapy for those who do not want to use the first-line therapy or cannot tolerate it.[12]

14.5.2.5 Adenoidectomy and Tonsillectomy

Adenotonsillar hypertrophy is a well-defined risk factor for the development of OSA and is the most common etiology of OSA in children.[6] Adenotonsillectomy is the first-line treatment for OSA in pediatrics patients.[14] It is usually curative and is associated with a significant and consistent decrease in AHI, with a resolution rate of 100–73%, depending on the preoperative AHI.[37] Also, a study found a decrease in hypertension rate from 34 to 14%[38] and improvement in quality of life. Surgery is indicated when there is adenoid and/or tonsils' hypertrophy that leads topathologic upper-airway obstruction, diagnosed either by clinical signs and symptoms or by positive polysomnography.[22]

Despite significant improvements in respiratory parameters, there was evidence of residual OSA in 20–40% of the patients.[6] Obesity and AHI at the moment diagnosis may be the major determinants for worst surgical outcome.[39]

Complication rates are low, ranging from 7 to 1.4%, and include respiratory complications as pulmonary edema, hypoxemia and bronchosp asm; bleeding and fever.[40]

Adult patients with OSA who are either intolerant to CPAP or require high CPAP pressures to treat upper-airway obstruction due to tonsillar hypertrophy have favorable outcomes with tonsillectomy, with significant decrease in the nasal resistance and AHI values.[41]

Tonsillectomy may play a role similar to nasal surgery in increasing the compliance and efficacy of CPAP therapy in patients with tonsillar hypertrophy. It functions as an adjuvant therapy or even as an alternative secondary therapy for those that are not willing to use CPAP and have obstruction due to tonsillar hypertrophy.[22]

14.5.3 Retrolingual-Hypopharyngeal Procedures

In contrast to the soft palate's procedures, no established standard technique exists for surgical treatment of retrolingual and hypopharyngeal obstruction[12] and only low evidence studies show improvement in symptoms and polysomnographic parameters.

14.5.3.1 Base of the Tongue's Procedures

This procedures are performed to remove portions of the tongue, such as laser midline glossectomy, lingualplasty, lingual tonsillectomy, and glossopexia, have some case series report of improvement in AHI,[42] but they still may not be recommended as a single treatment option for patients with moderate to severe OSA.[22]

14.5.3.2 Hyoid Suspension

Hyoide suspension with palatal surgery performed previously or in the same setting also show a reduction of AHI in case series, but no controlled study comparing this intervention was performed yet.[42] Hyoid suspension aims to avoid the back positioning of the tongue during sleep. It is performed trough a horizontal incision in the midline of the neck to advance the hyoid bone anteriorly and inferiorly to the thyroid cartilage. It is best indicated for patients with OSA and retrolingual orhypopharyngeal obstruction.[22]

14.5.3.3 Genioglossus Advancement

This procedure aims to enlarge the hypopharyngeal air space bringing forward the base of tongue by advancement of the genial tubercle and genioglossus muscle.[22] This procedure can significantly improve the AHI as showed in 3 case reports[42] but no controlled trial have been performed yet. Nowadays, genioglossus advancement still may not be recommended as a single procedure for the surgical treatment of OSA, since its outcomes are not well defined.[22]

14.5.4 Maxilla and Mandible Procedures

14.5.4.1 Maxillomandibular Advancement

This procedure enlarges the velo-orohypopharyngeal airway without direct manipulation of the pharyngeal tissues as in the UPPP and the other procedures described.[43] It promotes the advancement of the anterior pharyngeal tissues attached to the maxilla, mandible, and hyoid bone by LeFort I osteotomy with maxilla and mandible repositioning. MMA has been shown to be the most effective surgical option for OSA and one of the most powerful therapeutic options to treat patients with severe forms of OSA when the obstruction includes the retrolingual region.[43]

This procedure seems to be as efficient as CPAP,[22] particularly in young

OSA population without excessive BMI or other comorbidities, and can be indicated for patients with moderate to severe OSA who cannot tolerate or who are unwilling to adhere to CPAP.[12] Hypopharyngeal and velo-orohypopharyngeal narrowing are common anatomic criteria used to indicate MMA.[29]

A reduction of 87% in previous AHI can be achieved.[29] A meta-analysis shows a decrease from 86 to 9 /h in the mean AHI. A significant improvement in the LSAT after surgery was also obtained. Furthermore, subjective symptoms are reported to decrease after the intervention. It has shown a significant improvement in systemic blood pressure, particularly in those with previously established hypertension.[44] Lastly, long-term results shows that it achieves long-term cure in most patients.[45]

MMA is a lengthy and technically challenging procedure and presents inherent risks of dental malocclusion and facial neurosensory deficits.[29] The most frequent complication of MMA is dehiscence of the surgical wound leading to injuries in the maxillary and mandibular trigeminal components of the somatic and proprioceptive sensitivity.[46]

14.5.5 Multi-Level Surgery

Sleep-endoscopy shows that 73% of patients have upper-airway collapse in more than one level of the pharynx.[21] Because of that, a multi-level approach to the surgical treatment has been increasingly used. This surgery involves heterogeneity of combined procedures that can approach the nasal cavity, soft palate, base of the tongue, hyoid, and jaw, usually in the same surgical session. The majority combines one palatal surgery, usually UPP, with a surgical procedure on the base of the tongue, usually performing RFA or hyoid suspension.[29]

When multi-level surgery is compared to isolate UPP it has a significantly higher success rate if patients are preselected based on Friedman classification or on a DISE that that shows collapse in more than one level.[15] Patients with moderate to severe OSA in stage II or III probably have associated hypopharyngeal obstruction and should have multi-level surgery indicated.[16] The success rate, including reduction of > 50% in AHI and ESS > 11 in a long-term follow up, in patients that underwent UPP and tongue base surgery (either genioglossus advancement, RFA or hyoid suspension) was 76–78%.[47]

A new protocolled procedure involving uvuloflap, tonsillectomy, modified hyoid suspension or RFA of the base of the tongue, and nasal surgery if nasal pathology present, was found to have a significant reduction in the AHI and in daytime sleepiness assessed by ESS, with a 50% rate of successful treatment.[48] When the same surgery was compared with the one with all procedures but no hyoid suspension, the hyoid advancement was found to be an important and a key part of the procedure to improve outcomes, mainly by decreasing AHI.[49]

Multi-level surgery can also be performed as a multi-phase or stepwise surgery. The majority of the studies use the Riley-Powell-Stanford surgical protocol, where phase I consists of UPP for those with retropalatal obstruction, genioglossus advancement with hyoid suspension for those with retrolingual obstruction or both procedures if both obstruction sites.

Phase II consists of MMA reserved for those failing phase I. A 95% long-term success rate compared with 61% after phase I was found.[50]

Multi-level surgery is a painful procedure that causes more discomfort than the others presented before. In one study patients needed analgesic treatment for 2 weeks after surgery. Other symptoms that patients complained were dysphagia and problems with the speech.[51]

14.5.6 Trachea Procedures

14.5.6.1 Tracheostomy

Tracheostomy is an option for treating OSA when all other therapies have failed, do not exist, are refused, or when an urgent management is needed.[12] Emergencies in OSA can be due to respiratory insufficiency or life-threatening arrhythmias. Cardiac arrhythmias are present in almost 50% of OSA patients and they are correlated with the number of apneic episodes and the severity of the hypoxic episodes.[52]

The objective of the procedure is to bypass the upper-airway obstruction or collapse region, by creating an opening in the trachea to ventilate the patient. Tracheostomy is an effective single intervention that significantly decreases AHI to cure levels, decreases sleepiness's symptoms, increases LSAT and decreases mortality in OSA patients.[53] It is also effective in reversing cardiac arrhythmias.[52]

Tracheostomy should only be recommended as last option for treating severe OSA or as life-saving procedure in emergent cases, since is an option that requires daily care and reduces the patient's autonomy.[12]

14.5.7 Adjuvant Surgical Option

14.5.7.1 Bariatric surgery

Obesity is an important and well described risk factor for OSA, and could be the primary cause of sleep apnea or the primary factor contributing to the severity of the disease.The prevalence of sleep apnea in obese patients is found to be as high as 80%,[54] and weight loss should be an important therapy goal for all OSA patients with associated obesity. The first option is always diet and lifestyle changes but it only achieves modest weight reduction, leading to persistence or recurrence of OSA. Between all the interventions to promote weight loss, bariatric surgery is known to be the one with best long-term and sustainable results. Also, bariatric surgery induce significant metabolic changes that are known as weight-independent mechanisms that helps to improve OSA after surgery, including adipokines effects, cytokine actions, altered gut hormonal release, and the improvement of insulin resistance.[55] OSA resolution or improvement can be achieve in 77–99% of the patients submitted to bariatric surgery, depending on the type of surgery performed, with mal-absorptive procedures being the most effective ones and purely restrictive procedures showing less effectiveness in some studies.[56]

Most of the surgeries for OSA described above have worst outcomes when performed in obese patients. In this population, bariatric surgery is a good alternative therapy for those failing CPAP, since airway surgery has not proved good outcomes.

Summarizing, bariatric surgery can supplement CPAP and improve its outcomes, or can completely resolve OSA, but for those, follow up with sleep study testing should be performed routinely, since these patients are at high risk of weight gaining and recurrence of the disease.

14.6 FOLLOW UP

Follow up should include postoperative surgery visit to evaluate wound healing, surgery complications, patient satisfaction, and also a general OSA evaluation.[7]

General sleep evaluation should be performed after an adequate healing period. The assessment should include: resolution of sleepiness, quality of life measures, patient and spousal satisfaction, approach the control of worsening risk factors, and if the patient is practicing adequate sleep hygiene.[7] Patients should undergo an objective measure of the presence and severity of OSA, that evaluate AHI and oxygen saturation, to compare with the previous values, and evaluate if the surgery objectives have been achieved.

Additionally, long-term follow up with a sleep specialist is recommended to early detect recurrence of disease.[7] It is not known yet all long-term results and some studies have important recurrence rates.

14.7 CONCLUSION

Airway surgery is a potential treatment option for OSA. The disadvantage with this management is that most of the literature is composed of cases reviews with little evidence-based results supporting it. Until now, surgery's outcomes have not overcome clinical management with CPAP therapy, and indications for surgery are still limited. Surgery may be indicated for patients with OSA who have failed or have poor adherence to CPAP therapy.

- Nasal surgery may be a good adjuvant treatment to improve CPAP compliance and efficacy in those with nasal obstruction.
- UPPP is the most common procedure performed, but its outcomes are not good in decreasing polysomnographic parameters. Some studies show that patients in stage I of Friedman are the ones with the best outcomes with UPPP, supporting that surgery can be indicated for those patients.
- Adenotonsillectomy is the first-line treatment for OSA in pediatrics patients and may work as adjuvant treatment in adults.
- MMA has been shown to be the most effective surgical option for OSA and one of the most powerful therapeutic options to treat patients with severe forms of OSA when the obstruction includes the retrolingual region.
- Multi-level surgery has good outcomes when performed in patients in stage II or III of Friedman's classification.
- Tracheostomy is the most effective option for treating OSA when all others therapies have failed, do not exist, are refused, or when an urgent management is needed.
- Bariatric surgery is an effective adjuvant therapy and, when indicated, has proven to successfully cure OSA.

With this variety of procedures, surgeons must individualize each patient, based on his gravity of the disease, symptoms, physical exam, and preferences. Minimally invasive procedures can be a good option for those that do not want a big surgery, but still have issues with CPAP. Always, patients should have a close follow up after surgery to a prompt diagnose of recurrence.

KEY WORDS

- surgery
- obstructive sleep apnea
- uvulopalatopharyngoplasty

REFERENCES

1. Peppard, P. E.; Young, T.; Barnet, J. H.; Palta, M.; Hagen, E. W.; Hla, K. M. Increased Prevalence of Sleep-disordered Breathing in Adults. *Am. J. Epidemiol.* **2013**, 177(9), 1006–1014.
2. Flegal, K. M.; Carroll, M. D.; Kit, B. K.; Ogden, C. L. Prevalence of Obesity and Trends in the Distribution of body Mass Index Among US Adults, 1999–2010. *JAMA* **2012**, 307(5), 491–497.
3. Young, T.; Skatrud, J.; Peppard, P. E. Risk Factors for Obstructive Sleep Apnea in Adults. *JAMA* **2004**, 291(16), 2013–2016.
4. Bridgman, S. A.; Dunn, K. M. Surgery for Obstructive Sleep Apnoea. *Cochrane Database Syst. Rev.* **2000**, (2), CD001004.
5. Schwab, R. J.; Gupta, K. B.; Gefter, W. B.; Metzger, L. J.; Hoffman, E. A.; Pack, A. I. Upper Airway and Soft Tissue Anatomy in Normal Subjects and Patients with Sleep-Disordered Breathing. Significance of the Lateral Pharyngeal Walls. *Am. J. Respir. Crit. Care Med.* **1995**, 152(5 Pt 1), 1673–1689.
6. Redline, S.; Amin, R.; Beebe, D.; Chervin, R. D.; Garetz, S. L.; Giordani, B.; Ellenberg, S.The Childhood Adenotonsillectomy Trial (CHAT): Rationale, Design, and Challenges of a Randomized Controlled Trial Evaluating a Standard Surgical Procedure in a Pediatric Population. *Sleep* **2011**, 34(11), 1509–1517.
7. Epstein, L. J.; Kristo, D.; Strollo, P. J.; Friedman, N.; Malhotra, A.; Patil, S. P. Adult Obstructive Sleep Apnea Task Force of the American Academy of Sleep Medicine. Clinical Guideline for the Evaluation, Management and Long-Term care of Obstructive Sleep Apnea in Adults. *J. Clin. Sleep. Med.* **2009**, 5(3), 263–276.
8. Shahar, E.; Whitney, C.W.; Redline, S.; Lee, E.T.; Newman, A.B.; Nieto, F.J. Sleep-Disordered Breathing and Cardiovascular Disease: Cross-Sectional Results of the Sleep Heart Health Study. *Am. J. Respir. Crit. Care Med.* **2001**, 163(1), 19–25.
9. He, J.; Kryger, M.H.; Zorick, F.J.; Conway, W.; Roth, T. Mortality and Apnea Index in Obstructive Sleep Apnea. Experience in 385 Male Patients. *Chest.* **1988**, 94(1), 9–14.
10. McArdle, N.; Devereux, G.; Heidarnejad, H.; Engleman, H.M.; Mackay, T.W.; Douglas, N.J. Long-Term Use of CPAP Therapy for Sleep Apnea/Hypopnea Syndrome. *Am. J. Respir. Crit. Care Med.* **1999**, 159(4 Pt 1), 1108–1114.
11. Elshaug, A.G.; Moss, J.R.; Southcott, A.M.; Hiller, J.E. Redefining Success in Airway Surgery for Obstructive Sleep Apnea: a Meta Analysis and Synthesis of the Evidence. *Sleep* **2007**, 30(4), 461–467.
12. Aurora, R.N.; Casey, K.R.; Kristo, D.; Auerbach, S.; Bista, S.R.; Chowdhuri, S. Practice Parameters for the Surgical Modifications of the Upper Airway for Obstructive sleep Apnea in Adults. *Sleep* **2010**, 33(10), 1408–1413.
13. Sher, A. E.; Schechtman, K. B.; Piccirillo, J. F. The Efficacy of Surgical Modifications of the Upper Airway in Adults with Obstructive Sleep Apnea Syndrome. *Sleep* **1996**, 19(2), 156–177.
14. Brietzke, S. E.; Gallagher, D. The Effectiveness of Tonsillectomy and Adenoidectomy in the Treatment of Pediatric Obstructive Sleep Apnea/Hypopnea Syndrome: A Meta-Analysis. *Otolaryngol. Head. Neck Surg.* **2006**, 134(6), 979–984.
15. Friedman, M.; Ibrahim, H.; Bass, L. Clinical Staging for Sleep-Disordered Breathing. *Otolaryngol. Head. Neck Surg.* **2002**, 127(1), 13–21.
16. Rodrigues, M. M.; Dibbern, R. S.; Goulart, C. W. K.; Palma, R. A. Correlation Between the Friedman Classification and the Apnea-Hypopnea Index in a Population with OSAHS. *Braz. J. Otorhinolaryngol.* **2002**, 76(5), 557–560.
17. Friedman, M.; Ibrahim, H.; Joseph, N. J. Staging of Obstructive Sleep Apnea/Hypopnea Syndrome: a Guide to Appropriate Treatment. *Laryngoscope* **2004**, 114(3), 454–459.
18. Koutsourelakis, I.; Safiruddin, F.; Ravesloot, M.; Zakynthinos, S.; Vries, N. Surgery for Obstructive Sleep Apnea: Sleep Endoscopy Determinants of Outcome. *Laryngoscope* **2012**, 122(11), 2587–2591.

19. Fernández-Julián, E.; García-Pérez, M. Á.; García-Callejo, J.; Ferrer, F.; Martí, F. Marco, J. Surgical Planning After Sleep Versus Awake Techniques in Patients with Obstructive Sleep Apnea. *Laryngoscope* **2014**, 124(8), 1970–1974.

20. Doghramji, K.; Jabourian, Z. H.; Pilla, M.; Farole, A.; Lindholm, R. N. Predictors of Outcome for Uvulopalatopharyngoplasty. *Laryngoscope* **1995**, 105(3 Pt 1), 311–314.

21. Gillespie, M. B.; Reddy, R. P.; White, D. R.; Discolo, C. M.; Overdyk, F. J.; Nguyen, S. A. A Trial of Drug-Induced Sleep Endoscopy in the Surgical Management of Sleep-Disordered Breathing. *Laryngoscope* **2013**, 123(1), 277–282.

22. Randerath, W. J.; Verbraecken, J.; Andreas, S.; Bettega, G.; Boudewyns, A.; Hamans, E. Non-CPAP Therapies in Obstructive Sleep Apnoea. *Eur. Respir. J.* **2011**, 37(5), 1000–1028.

23. Olsen, K. D.; Kern, E. B.; Westbrook, P. R. Sleep and Breathing Disturbance Secondary to Nasal Obstruction. *Otolaryngol. Head. Neck Surg.* **1981**, 89(5), 804–810.

24. Georgalas, C. The Role of the Nose in Snoring and Obstructive Sleep Apnoea: an Update. *Eur. Arch. Otorhinolaryngol.* **2011**, 268(9), 1365–1373.

25. Koutsourelakis, I.; Georgoulopoulos, G.; Perraki, E.; Vagiakis, E.; Roussos, C.; Zakynthinos, S. G. Randomised Trial of Nasal Surgery for Fixed Nasal Obstruction in Obstructive Sleep Apnoea. *Eur. Respir. J.* **2008**, 31(1), 110–117.

26. Hoffstein, V.; Viner, S.; Mateika, S.; Conway, J. Treatment of Obstructive Sleep Apnea with Nasal Continuous Positive Airway pressure. Patient Compliance, Perception of Benefits, and Side Effects. *Am. Rev. Respir. Dis.* **1992**, 145(4 Pt 1), 841–845.

27. Friedman, M.; Tanyeri, H.; Lim, J. W.; Landsberg, R.; Vaidyanathan, K.; Caldarelli, D. Effect of Improved Nasal Breathing on Obstructive Sleep Apnea. *Otolaryngol. Head. Neck Surg.* **2000**, 122(1), 71–74.

28. Kwon, M.; Jang, Y. J.; Lee, B.J.; Chung, Y.S. The Effect of Uvula-Preserving Palatopharyngoplasty in Obstructive Sleep Apnea on Globus Sense and Positional Dependency. *Clin. Exp. Otorhinolaryngol.* **2010**, 3(3), 141–146.

29. Caples, S. M.; Rowley, J. A.; Prinsell, J. R.; Pallanch, J. F.; Elamin, M. B.; Katz, S. G.; Harwick, J. D. Surgical Modifications of the Upper Airway for Obstructive Sleep Apnea in Adults: A Sys-tematic Review and Meta-Analysis. *Sleep* **2010**, 33(10), 1396–1407.

30. Franklin, K. A.; Anttila, H.; Axelsson, S.; Gislason, T.; Maasilta, P.; Myhre, K. I.; Rehnqvist, N. Effects and Side Effects of Surgery for Snoring and Obstructive Sleep Apnea-a Systematic Review. *Sleep* **2009**, 32(1), 27–36.

31. Mortimore, I. L.; Bradley, P. A.; Murray, J. A.; Douglas, N. J. Uvulopalatopharyngoplasty may Compromise Nasal CPAP Therapy in Sleep Apnea Syndrome. *Am. Rev. Respir. Dis.* **1996**, 154(6 Pt 1), 1759–1762.

32. Ferguson, K. A.; Heighway, K.; Ruby, R. R. F. A Randomized Trial of Laser-Assisted Uvulo-palatoplasty in the Treatment of Mild Obstructive Sleep Apnea. *Am. Rev. Respir. Dis.* **2003**, 167(1), 15–19.

33. Chisholm, E.; Kotecha, B. Oropharyngeal Surgery for Obstructive Sleep Apnoea in CPAP Failures. *Eur. Arch. Otorhinolaryngol.* **2007**, 264(1), 51–55.

34. Scierski, W.; Namysłowski, G.; Urbaniec, N.; Misiołek, M.; Namysłowska, K.; Pilch, J. Complication After Laser Assisted Uvulopalatoplasty in the Treatment of Snoring and Obstructive Sleep Apnea Syndrome. *Otolaryngol. Pol.* **2003**, 57(5), 675–678.

35. Baba, R. Y.; Mohan, A.; Metta, V. V. S. R.; Mador, M. J. Temperature Controlled Radiofrequency Ablation at Different Sites for Treatment of Obstructive Sleep Apnea Syndrome: a Systematic Review and Meta-Analysis. *Sleep. Breath.* **2005**, 19(3), 891–910.

36. Nordgård, S.; Stene, B. K.; Skjøstad, K. W.; Bugten, V.; Wormdal, K.; Hansen, N. V.; Midtlyng, T. H. Palatal Implants for the Treatment of Snoring: Long-Term Results. *Otolaryngol. Head. Neck Surg.* **2006**, 134(4), 558–564.

37. Mitchell, R. B. Adenotonsillectomy for Obstructive Sleep Apnea in Children: Outcome Evaluated by Pre- and Postoperative Polysomnography. *Laryngoscope* **2007**, 117(10), 1844–1854.

38. Lee, L. A.; Li, H. Y.; Lin, Y. S.; Fang, T. J.; Huang, Y. S.; Hsu, J. F.; Huang, C. G. Severity of Childhood Obstructive Sleep Apnea and Hypertension Improved After Adenotonsillectomy. *Otolaryngol. Head. Neck Surg.* **2015**, 152(3), 553–560.

39. Tauman, R.; Gulliver, T. E.; Krishna, J.; Montgomery-Downs, H. E.; O'Brien, L. M.; Ivanenko, A.; Gozal, D. Persistence of Obstruc-

tive Sleep Apnea Syndrome in Children After Adenotonsillectomy. *J.Pediatr.* **2006,** 149(6), 803–808.

40. Konstantinopoulou, S.; Gallagher, P.; Elden, L.; Garetz, S. L.; Mitchell, R. B.; Redline, S.; Marcus, C. L. Complications of Adenotonsillectomy for Obstructive Sleep Apnea in School-Aged Children. *Int. J. Pediatr. Otorhinolaryngol.* **2015,** 79(2), 240–245.

41. Nakata, S.; Miyazaki, S.; Ohki, M.; Morinaga, M.; Noda, A.; Sugiura, T.; Nakashima, T. Reduced Nasal Resistance After Simple Tonsillectomy in Patients with Obstructive Sleep Apnea. *Am. J. Rhinol.* **2007,** 21(2), 192–195.

42. Kezirian, E. J.; Goldberg, A. N. Hypopharyngeal Surgery in Obstructive Sleep Apnea: An Evidence-Based Medicine Review. *Otolaryngol. Head. Neck Surg.* **2006,** 132(2), 206–213.

43. Jaspers, G. W.; Booij, A.; de Graaf, J.; de Lange, J. Long-Term Results of Maxillomandibular Advancement Surgery in Patients with Obstructive sleep Apnoea Syndrome. *Br. J. Oral. Maxillofac. Surg.* **2013,** 51(3), e37–39.

44. Islam, S.; Taylor, C. J.; Ormiston, I. W. Effects of Maxillomandibular Advancement on Systemic Blood Pressure in Patients with Obstructive Sleep Apnoea. *Br. J. Oral. Maxillofac. Surg.* **2015,** 53(1), 34–38.

45. Li, K. K.; Powell, N. B.; Riley, R. W.; Troell, R. J.; Guilleminault, C. Long-Term Results of Maxillomandibular Advancement Surgery. *Sleep. Breath.* **2000,** 4(3), 137–140.

46. Holty, J.E. C.; Guilleminault, C. Maxillomandibular Advancement for the Treatment of Obstructive Sleep Apnea: a Systematic Review and Meta-Analysis. *Sleep Med. Rev.* **2010,** 14(5), 287–297.

47. Jacobowitz, O. Palatal and Tongue Base Surgery for Surgical Treatment of Obstructive Sleep Apnea: a Prospective Study. *Otolaryngol. Head. Neck Surg.* **2006,** 135(2), 258–264.

48. Verse, T.; Baisch, A.; Maurer, J. T.; Stuck, B. A.; Hörmann, K. Multilevel Surgery for Obstructive Sleep apnea: Short-Term Results. *Otolaryngol. Head. Neck Surg.* **2006,** 134(4), 571–577.

49. Baisch, A.; Maurer, J. T.; Hörmann, K. The Effect of Hyoid Suspension in a Multilevel Surgery Concept for Obstructive Sleep Apnea. *Otolaryngol. Head. Neck Surg.* **2006,** 134(5), 856–861.

50. Riley, R. W.; Powell, N. B.; Guilleminault, C. Obstructive Sleep Apnea Syndrome: a Review of 306 Consecutively Treated Surgical Patients. *Otolaryngol. Head. Neck Surg.* **1993,** 108(2), 117–125.

51. Baisch, A.; Hein, G.; Gössler, U.; Stuck, B. A.; Maurer, J. T.; Hörmann, K. Subjective Outcome After Multi-Level Surgery in Sleep-Disordered Breathing. *HNO* **2005,** 53(10), 863–868.

52. Tilkian, A. G.; Guilleminault, C.; Schroeder, J. S.; Lehrman, K. L.; Simmons, F. B.; Dement, W. C. Sleep-Induced Apnea Syndrome. Prevalence of Cardiac Arrhythmias and Their Reversal After Tracheostomy. *Am. J. Med.* **1977,** 63(3), 348–358.

53. Camacho, M.; Certal, V.; Brietzke, S. E.; Holty, J. E. C.; Guilleminault, C.; Capasso, R. Tracheostomy as Treatment for Adult Obstructive Sleep Apnea: a Systematic Review and Meta-Analysis. *Laryngoscope* **2014,** 124(3), 803–811.

54. Fredheim, J. M.; Rollheim, J.; Omland, T.; Hofsø, D.; Røislien, J.; Vegsgaard, K.; Hjelmesæth, J. Type 2 Diabetes and Pre-Diabetes are Associated with Obstructive Sleep Apnea in Extremely Obese Subjects: A Cross-Sectional Study. *Cardiovasc. Diabetol.* **2011,** 10, 84.

55. Ashrafian, H.; le Roux, C. W.; Rowland, S. P.; Ali, M.; Cummin, A. R.; Darzi, A.; Athanasiou, T. Metabolic Surgery and Obstructive Sleep Apnoea: The Protective Effects of Bariatric Procedures. *Thorax* **2012,** 67(5), 442–449.

56. Sarkhosh, K.; Switzer, N. J.; El-Hadi, M.; Birch, D. W.; Shi, X.; Karmali, S. The Impact of Bariatric Surgery on Obstructive Sleep Apnea: a Systematic Review. *Obes. Surg.* **2013,** 23(3), 414–423.

15

SLEEP AND CHRONIC KIDNEY DISEASE

Camila Hirotsu, Sergio Tufik, and Monica Levy Andersen

ABSTRACT

Sleep complaints are common in the population and affect well-being, productivity, and physical and mental capacity of individuals. Recently, there has been also a reduction in sleep duration due to higher socioeconomic pressure and 24/7 style of urban centers. However, both sleep disorders and sleep restriction constitute risk factors for hypertension, obesity and diabetes. These comorbidities are known as important contributors of chronic kidney disease (CKD), which is characterized by progressive and irreversible loss of kidney function. Patients with CKD often have sleep disorders such as obstructive sleep apnea, insomnia, and restless legs syndrome. Besides being related to poorer quality of life and develop early with the disease, sleep disturbances are being recognized as risk factors for increased morbidity and mortality in CKD patients. Moreover, recent findings point to a bidirectional relationship between sleep and kidney, as short sleep duration has been associated with significant reduction of renal function in the general population.

Camila Hirotsu
Department of Psychobiology, Universidade Federal de São Paulo, Rua Napoleão de Barros, 925, Vila Clementino, São Paulo, 04024-002, Brazil

Sergio Tufik
Department of Psychobiology, Universidade Federal de São Paulo, Rua Napoleão de Barros, 925, Vila Clementino, São Paulo, 04024-002, Brazil

Monica Levy Andersen
Department of Psychobiology, Universidade Federal de São Paulo, Rua Napoleão de Barros, 925, Vila Clementino, São Paulo, 04024-002, Brazil

Corresponding author: Camila Hirotsu, E-mail: milahirotsu@gmail.com

15.1 INTRODUCTION

Chronic kidney disease (CKD) is characterized by progressive and irreversible loss of kidney function and is increasingly acknowledged as a global public health problem, affecting 10% of adults and 35% of those over 70 years old (Coresh et al., 2007). Although this increasing prevalence of CKD has been attributed to the increased prevalence in type 2 diabetes, hypertension, and obesity, there is evidence that poor sleep quality and sleep disorders may contribute to the pathophysiology of CKD and its progression to kidney failure. Basic and clinical studies have shown that sleep restriction and sleep disturbances can promote and exacerbate the severity of well-known risk factors of CKD. In an animal model of CKD, the influence of uremic toxins led to changes in sleep architecture such as reversal of the sleep–wake cycle and sleep fragmentation (Hirotsu et al., 2010). To date, few studies have objectively characterized sleep in patients with CKD prior to end-stage kidney failure, but emerging evidence suggests a high prevalence of obstructive sleep apnea (OSA) with a singular clinical presentation (Nicholl et al., 2012), in which rostral fluid shift has an important influence on upper airway size (Elias et al., 2012). Indeed, OSA and nocturnal hypoxemia have been associated with kidney injury, suggesting that they contribute to the pathogenesis of continued loss in kidney function due to intra-renal hypoxia and activation of the renin-angiotensin system. Thus, in this chapter, we will provide a review about the important relationship between sleep and renal health, which is a relatively new topic that is growing and needs more attention from both researchers and clinicians in Sleep Medicine (Fig. 15.1).

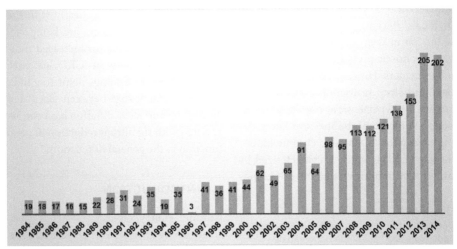

Figure 15.1. Scientific production in the field of "Sleep and Kidney" in the last 3 decades. Source of data: Pubmed.

15.2 CHRONIC KIDNEY DISEASE

The kidneys are dynamic organs that play vital functions in the body such as homeostatic, excretory, endocrine, and regulatory. The homeostatic function is responsible for the maintenance of electrolytes and the acid–base balance. The excretory function allows the removal of catabolites such as creatinine, urea, uric acid, phosphates, and sulfates, among others. Endocrine and hormone regulating functions are related to the production of vitamin D, erythropoietin, renin, and the catabolism of several other hormones. Above all, the kidney is primarily an excretory organ and depends on the glomerular filtration process. In this process, blood is filtered through the glomerular capillaries, and the water and its proteinaceous and non-proteinaceous constituents of low molecular weight are transferred to the Bowman's capsule, while the blood cells and high molecular weight proteins are retained (Kretzler et al., 2001). Throughout life, it naturally occurs a decrease in renal function due to loss of glomerular filtration capacity and decreased renal blood flow, which may vary between individuals (Turin and Hemmelgarn, 2011).

However, when the performance or number of nephrons, the functional unit of the kidney, is reduced considerably, morphological and functional compensatory responses occur. These changes are represented by hypertrophy and hyperplasia of the remaining nephrons, which are able to multiply their work compass to avoid a reduction in the glomerular filtration rate (Brenner et al., 1982). However, when the loss of renal function compromises most of the functioning nephrons and glomerular filtration rate is reduced by about 30 to 50%, progression to CKD occurs inevitably (Fine et al., 1998; Terzi et al., 1998). According to estimates of the World Health Organization, CKD affects one in 10 adults worldwide. In 2004, the number of dialysis patients in the U.S. was approximately 497,934 according to the U.S. Renal Data System. Despite the important implications of CKD as a global public health problem, the understanding about the factors responsible for its development and progression is still incipient (Eckardt et al., 2013). Also, there is an interindividual variability in the progression of the disease poorly understood.

Hypertension appears to be a contributing factor to glomerulosclerosis in different kidney diseases, significantly contributing to cardiovascular morbidity and mortality in CKD patients. In 1998, Foley et al. demonstrated that even in the early stages of CKD, these patients had a higher risk of developing cardiovascular disease compared to the general population. Many studies have shown that lowering blood pressure in CKD can reduce the loss of renal function and that the blockade of the renin-angiotensin-aldosterone system has renoprotective effects (Parving et al., 1987; Klag et al., 1996). Cardiovascular disease is the leading cause of mortality among patients with CKD. Advanced age, male gender, high low-density lipoprotein (LDL)-cholesterol levels, low high-density lipoprotein (HDL)-cholesterol levels, sedentary lifestyle, menopause, smoking, and family history of cardiovascular disease are independent factors well known to increase the risk of cardiovascular events (Kannel et al., 1961). However, there are other risk factors intrinsically associated with CKD, such as albuminuria, hiperhomocisteinemia, anemia, calcium, and phosphorus abnormalities, fluid overload, electrolyte imbalances, inflammation, thrombogenic factors, oxidative stress, and, in particular, sleep disorders.

15.3 SLEEP AND CHRONIC KIDNEY DISEASE

15.3.1 Sleep Quality and Renal Function

During sleep, there is a physiological reduction in the renal function, resulting in lower production of electrolytes urine excretion, glomerular filtration rate, and renal blood flow. In addition to the influence of sleep, the activities of kidneys also present circadian variations, although the mechanisms that control these daily fluctuations are still unknown. Our group has demonstrated a close relationship between sleep pattern and the CKD progression in an animal model. Rats with CKD presented a fragmented sleep, with increased number of awakenings, which in turn was associated with hypertension and creatinine retention (Hirotsu et al., 2010). Furthermore, CKD animals showed a reversal of the sleep–wake cycle in the later stages of the disease, suggesting that sleep quality may reflect, at least in part, the functioning of the kidneys (Hirotsu et al., 2010).

In hemodialysis patients, a significant correlation was found between sleep quality and C-reactive protein, body mass index, and albumin (Emami Zeydi et al., 2014). The quality of sleep in hemodialysis patients was also analyzed between the ratio of pro-inflammatory cytokines showing that high levels of these cytokines were related to poor sleep quality in these individuals (Taraz et al., 2013).

The first changes related to sleep in patients with CKD were described in 1959, when it was shown that uremic subjects had a reversal pattern of the sleep–wake cycle in addition to an atypical response to sedatives (Schreiner, 1959). In 1967, a study using electroencephalography showed that CKD patients had reduced total sleep time and rapid eye movement (REM) sleep in association with excessive number of arousals (Mises et al., 1967). Since then, several studies using questionnaires confirmed the high prevalence of sleep complaints, both in individuals undergoing hemodialysis and in peritoneal dialysis (Holley et al., 1992; Parker, 2003; Unruh et al., 2008). Compared to the general population, the prevalence of sleep disorders is considerably higher in renal patients. About 80% of dialysis patients with CKD report sleep disorders, which impair their quality of life and functional status (Gusbeth-Tatomir et al., 2007). These disorders are complex and involve difficulty sleeping and waking up, sleep fragmentation, nightmares, restless leg syndrome (RLS), sleep apnea, excessive daytime sleepiness, insomnia, among others.

Emerging evidence shows a high prevalence of sleep disorders in patients with CKD in non-dialysis phase (Turek et al., 2012), supporting the hypothesis that sleep has a key role in glomerular filtration process. However, studies using sleep questionnaires have shown different estimates of the prevalence of sleep disorders in patients with CKD, ranging from 14 to 85% (Kurella et al., 2005; De Santo et al., 2008; Kumar et al., 2010), with some inconsistencies. For example, Cohen et al. (2007) found no significant difference in the sleep pattern of CKD patients compared to general outpatients. In another analysis, De Santo et al. (2008) assessed the prevalence of sleep disorders in patients with recent CKD diagnosis through questionnaires. In this study, approximately 90% of patients showed some subclinical sleep disorder, indicating that even in the early stages of the disease the changes in sleep are already present, functioning as possible biomarkers of disease.

The sleep disorders are independent predictors of poor quality of life and can affect physical and psychological parameters. Patients with sleep disorders have a higher perception of disease severity and a pessimistic view of their health status (Iliescu et al., 2003). Recent studies have evaluated the relationship between sleep and life quality in CKD patients, finding high levels of poor sleep quality, especially in patients at advanced stages of the disease (Han and Kim, 2014). Parvan et al. (2013) found that 83.3% of hemodialysis treatment subjects had poor sleep in association with mental, physical, and sexual function impairments. Another study evaluated the health-related quality of life in a sample of patients who underwent different therapies for CKD: hemodialysis or kidney transplant, being the last one associated with corticosteroids and immunosuppressants. The findings showed that only patients treated with hemodialysis had worse physical well-being, occupational functioning, spiritual fulfillment, and physical functioning compared to the controls. In relation to transplanted patients, hemodialysis obtained higher scores of depressive symptoms (Martínez-Sanchis et al., 2015). Furthermore, each 1 h of sleep duration reduction was significantly associated with worse levels of glomerular filtration rate (Petrov et al., 2014).

The relationship between sleep disorders and CKD appears to be mediated by the effect of uremic toxins in the central nervous system (CNS), but this central action is not yet fully elucidated in CKD patients (Fein et al., 1987). Studies have shown that 50% of CKD patients treated with hemodialysis or peritoneal dialysis report insomnia compared to 12% in a control population (Sabbatini et al., 2002). Another study showed that non-apnea sleep disorders (NASD) were risk factors for the development of CKD in a large population-based retrospective cohort. In this study, patients with NASD had a 14% increased risk of developing CKD, whereas patients with insomnia had a 13% increased risk of

subsequent CKD compared with the non-NASD cohort (Huang et al., 2015). About 30–69% of CKD patients with end-stage renal disease RLS and periodic leg movement (PLM) disorders during sleep (Benz et al., 2000). In a study carried out in 2010 in Greece, it was observed a high RLS index in hemodialysis patients compared with the general population (26.6 vs 3.9%) (Stefanidis et al., 2013). However, findings suggest that the prevalence of RLS in CKD looks pretty close to that found in the general population after a diagnosis made by an expert and not just by questionnaire. This observation raised the question of whether the RLS in CKD has a causal relationship or is a confounding factor associated with other discomforts in the lower limbs (Calviño et al., 2015). Consequently, due to the great potential of sleep disruption promoted by sleep disorders, it is not surprising to find that about 50–66% of patients with end-stage renal disease also report the consequent excessive daytime sleepiness (Walker et al., 1995; Hanly et al., 2003). Although polysomnography is the gold standard method for the evaluation of sleep disorders, the literature is still scarce regarding the systematic evaluation of objective sleep related to CKD.

15.3.2 Sleep Apnea and CKD: a New Risk Factor

Substantial evidence suggests that inadequate sleep duration and poor sleep quality promote the development and exacerbate the severity of three major risk factors for CKD: type 2 diabetes, hypertension, and obesity (Knutson et al., 2009; Spiegel et al., 2009; Aronsohn et al., 2010; St-Onge et al., 2011). Therefore, sleep should have a direct role in the development and progression of CKD. Under physiological conditions, hormones that regulate body fluid and blood pressure are highly modulated by the sleep–wake cycle (Brandenberger et al., 1994; Charloux et al., 1999). Moreover, the sleep fragmentation related to OSA is associated with an increase in the activity of the sympathetic nervous system, which has an adverse impact on renal function (Turek et al., 2012). Indeed, increasing evidence points that OSA may be directly related to the development and progress of CKD.

Patients with OSA are at increased risk for CKD and end-stage renal disease compared to the general population (Lee et al., 2015). Fleischmann et al. (2010) evaluated the sleep of 158 CKD patients using polysomnography and found high rates of OSA (80–94%). Another study revealed a prevalence of 54.3% of sleep-disordered breathing and 30% of PLM in non-dialysis CKD patients (Markou et al., 2006). Individuals with glomerular filtration rate between 45 and 89 mL/min/1.73 m² had a higher risk of developing OSA as compared to those with glomerular filtration rate greater than 90 mL/min/1.73 m². This result suggests a bidirectional relationship between sleep and kidney: Loss of renal function can directly contribute to the development of OSA and this, in turn, could be added to other factors, accelerating the progression of CKD.

Specific factors associated with the pathophysiology of OSA, for example, intermittent hypoxia and re-oxygenation stimulate the formation of reactive oxygen species, promoting inflammation, oxidative stress, and systemic endothelial dysfunction (Yamauchi and Kimura, 2008;

Lavie, 2009); conditions that by themselves may affect renal function (Fassett et al., 2011). A prospective cohort study with 6 years follow-up showed that endothelial dysfunction and inflammation are predictors of the arterial stiffness development in subjects without CKD (Van Bussel et al., 2011). There is a strong association between OSA and arterial stiffness, and the treatment for OSA with continuous positive airway pressure (CPAP) device reduced significantly the arterial stiffness (Tsioufis et al., 2007; Doonan et al., 2011).

Kinebuchi et al. (2004) evaluated 27 patients with OSA before and after treatment with CPAP. After one week of treatment, no change was found in renal physiology; however, renal plasma flow was increased, suggesting a possible improvement of kidney function mediated by vasodilation promoted by CPAP treatment. Recently, Koga et al. (2013) evaluated the renal function of 38 men diagnosed with OSA by polysomnography. In this study, after 3 months of CPAP treatment individuals showed a significant reduction in circulating concentrations of creatinine as well as an increase in glomerular filtration rate. Taken together, these data suggest that improving sleep may work as an adjuvant treatment for CKD and help delay its progression, leading to a better quality of life and disease prognosis.

15.3.3 Sleep Restriction and Kidney Function

In addition to the high prevalence of sleep disorders, sleep deprivation is common in today's society due to socioeconomic, social and health issues and may lead to a chronic sleep deficit state even in CKD patients. Lack of sleep by itself has an impact on health and leads to premature death, cardiovascular disease, and diabetes (Ayas et al. 2003; Zhang et al., 2014). Recently, a retrospective cross-sectional survey used data from the national health interview survey and analyzed the prevalence of self-reported kidney failure (Salifu et al., 2014). The authors found that CKD frequency was higher in both short sleepers (sleep duration ≤ 6 h/night) and long sleepers (≥ 8 h/night) when compared to those who slept in the range of 7 h of sleep per night (Salifu et al., 2014). Moreover, a significant association between short sleep duration and proteinuria was reported among Japanese people aged 20–65 years who did not have CKD at baseline, providing the notion that short sleep duration might serve as a possible marker of kidney impairment among those who are at relatively low risk of cardiovascular disease (Yamamoto et al., 2012).

There is a growing body of evidence that sleep disorders are common in patients with CKD. The question of whether sleep duration is independently associated with CKD, however, has been little studied. It is important to answer this question since sleep duration has been associated with conditions that are associated with CKD and could be a key prognostic factor of CKD.

15.4 CONCLUSIONS

There is evidence that the prevalence of sleep disorders is greater in individuals with CKD compared to the general population. It is plausible to think that renal disease can lead to the development of sleep disorders, due to uremic toxins that accumulates in the body, including the CNS. Sleep disorders arising from CKD have a direct impact on quality of life, since they are able to worse patient survival and aggravate the clinical picture. In addition, changes in sleep pattern and the lifestyle of CKD patients may be associated with sleep deprivation, which in turn can affect cardiometabolic function, which is already changed in CKD.

Currently, CKD is considered a growing challenge for both patients and health-care budgets. Recent research suggests that sleep disorders may contribute to the pathogenesis of CKD, which provides the opportunity to have an impact on this important clinical problem. Thus, the recent advances on this topic from the perspective of sleep medicine may provide direction for further research that will advance the field of nephrology. This could include specific strategies such as optimizing sleep duration and quality as well as treating sleep disorders like OSA, to reduce the prevalence of CKD and delay its progression.

KEY WORDS

- **sleep**
- **renal function**
- **chronic kidney disease**
- **hypertension**

REFERENCES

Aronsohn, R. S.; Whitmore, H.; Van Cauter, E.; Tasali, E. Impact of Untreated Obstructive Sleep Apnea on Glucose Control in Type 2 Diabetes. *Am. J. Respir. Crit. Care. Med.* **2010,**181 (5), 507–513.

Ayas, N. T.; White, D. P.; Al-Delaimy, W. K.; Manson, J. E.; Stampfer, M. J.; Speizer, F. E.; Patel, S.; Hu, F.

B. A Prospective Study of Self-Reported Sleep Duration and Incident Diabetes in Women. *Diabetes. Care.* **2003,** 26 (2), 380–384.

Benz, R.L.; Pressman, M. R.; Hovick, E. T.; Peterson, D. D. Potential Novel Predictors of Mortality in End-Stage Renal Disease Patients with Sleep Disorders. *Am. J. Kidney Dis.* **2000,** 35 (6), 1052–1060.

Brandenberger, G.; Follenius, M.; Goichot, B.; Saini, J.; Spiegel, K.; Ehrhart, J.; Simon, C. Twenty-Four-Hour Profiles of Plasma Renin Activity in Relation to the Sleep-Wake Cycle. *J. Hypertens.* **1994,** 12 (3), 277–283.

Brenner, B. M.; Meyer, T. W.; Hostetter, T. H. Dietary Protein Intake and the Progressive Nature of Kidney Disease: the Role of Hemodynamically Mediated Glomerular Injury in the Pathogenesis of Progressive Glomerular Sclerosis in Aging, Renal Ablation, and Intrinsic Renal Disease. *N. Engl. J. Med.* **1982,** 307 (11), 652–659.

Calviño, J.; Cigarrán, S.; Lopez, L. M.; Martinez, A.; Sobrido, M. J. Restless Legs Syndrome in Non-Dialysis Renal Patients: is It Really That Common? *J. Clin. Sleep Med.* **2015,** 11 (1), 57–60.

Charloux, A.; Gronfier, C.; Lonsdorfer-Wolf, E.; Piquard, F.; Brandenberger, G. Aldosterone Release During the Sleep-Wake Cycle in Humans. *Am. J. Physiol.* **1999,** 276 (1 Pt 1), E43–49.

Cohen, S. D.; Patel, S. S.; Khetpal, P.; Peterson, R. A.; Kimmel, P. L. Pain, Sleep Disturbance, and Quality of Life in Patients with Chronic Kidney Disease. *Clin. J. Am. Soc. Nephrol.* **2007,** 2 (5), 919–925.

Coresh, J.; Selvin, E.; Stevens, L. A.; Manzi, J.; Kusek, J. W.; Eggers, P.; Van Lente, F.; Levey, A. S. Prevalence of Chronic Kidney Disease in the United States. *JAMA* **2007,** 298 (17), 2038–2047.

De Santo, R. M.; Bartiromo, M.; Cesare, C. M.; Cirillo, M. Sleep Disorders Occur Very Early in Chronic Kidney Disease. *J. Nephrol.* **2008,** 21 Suppl 13, S59–65.

Doonan, R. J.; Scheffler, P.; Lalli, M.; Kimoff, R. J.; Petridou, E. T.; Daskalopoulos, M. E.; Daskalopoulou, S. S. Increased Arterial Stiffness in Obstructive Sleep Apnea: a Systematic Review. *Hypertens. Res.* **2011,** 34 (1), 23–32.

Eckardt, K. U.; Coresh, J.; Devuyst, O.; Johnson, R. J.; Köttgen, A.; Levey, A. S.; Levin, A. Evolving Importance of Kidney Disease: from Subspecialty to Global Health Burden. *Lancet* **2013,** 382 (9887), 158–169.

Elias, R.M.; Bradley, T.D.; Kasai, T.; Motwani, S.S.; Chan, C.T. Rostral Overnight Fluid Shift in End-Stage Renal Disease: Relationship with Obstructive Sleep Apnea. *Nephrol. Dial. Transplant.* **2012,** 27(4):1569–1573.

Emami Zeydi, A.; Jannati, Y.; Darvishi Khezri, H.; Gholipour Baradari, A.; Espahbodi, F.; Lesani, M.; Yaghoubi, T. Sleep Quality and Its Correlation with Serum C-reactive Protein Level in Hemodialysis Patients. *Saudi. J. Kidney Dis. Transpl.* **2014,** (4), 750–755.

Fassett, R. G.; Venuthurupalli, S. K.; Gobe, G. C.; Coombes, J. S.; Cooper, M. A.; Hoy, W. E. Biomarkers in Chronic Kidney Disease: a Review. *Kidney Int.* **2011,** 80 (8), 806–821.

Fein, A. M.; Niederman, M. S.; Imbriano, L.; Rosen, H. Reversal of Sleep Apnea in Uremia by Dialysis. *Arch. Intern. Med.* **1987,** 147 (7),1355–1356.

Fine, L. G.; Orphanides, C.; Norman, J. T. Progressive Renal Disease: the Chronic Hypoxia Hypothesis. *Kidney Int. Suppl.* **1998,** 65, S74–78.

Fleischmann, G.; Fillafer, G.; Matterer, H.; Skrabal, F.; Kotanko, P. Prevalence of Chronic Kidney Disease in Patients with Suspected Sleep apnoea. *Nephrol. Dial. Transplant.* **2010,** 25 (1), 181–186.

Foley, R. N.; Parfrey, P. S.; Sarnak, M. J. Cardiovascular Disease in Chronic Renal Disease: Clinical Epidemiology of Cardiovascular Disease in Chronic Renal Disease. *Am. J. Kidney Dis.* **1998,** 32 (5 Suppl 3), S112–119.

Gusbeth-Tatomir, P.; Boisteanu, D.; Seica, A.; Buga, C.; Covic, A. Sleep Disorders: a Systematic Review of an Emerging Major Clinical Issue in Renal Patients. *Int. Urol. Nephrol.* **2007,** 39 (4), 1217–1226.

Han, S. J.; Kim, H. W. Quality of Sleep in Predialysis Patients with Chronic Kidney Disease. *International Journal of Bio-Science and Bio-Technology* **2014,** 6 (5), 101–110.

Hanly, P. J.; Gabor, J. Y.; Chan, C.; Pierratos, A. Daytime Sleepiness in Patients with CRF: Impact of Nocturnal Hemodialysis. *Am. J. Kidney Dis.* **2003,** 41 (2), 403–410.

Hirotsu, C.; Tufik, S.; Bergamaschi, C. T.; Tenorio, N. M.; Araujo, P.; Andersen, M. L. Sleep Pattern in an Experimental Model of Chronic Kidney Disease. *Am. J. Physiol. Renal. Physiol.* **2010,** 299 (6), F1379–1388.

Holley, J. L.; Nespor, S.; Rault, R. A Comparison of Reported Sleep Disorders in Patients on Chronic Hemodialysis and Continuous Peritoneal Dialysis. *Am. J. Kidney Dis.* **1992,** 19 (2), 156–161.

Huang, S. T.; Lin, C. L.; Yu, T. M.; Yang, T. C.; Kao, C. H. Nonapnea Sleep Disorders and Incident Chronic Kidney Disease: a Population-Based Retrospective Cohort Study. *Medicine (Baltimore)* **2015**, 94 (4), c429.

Iliescu, E. A.; Coo, H.; McMurray, M. H.; Meers, C. L.; Quinn, M. M.; Singer, M. A.; Hopman, W. M. Quality of Sleep and Health-Related Quality of Life in Haemodialysis Patients. *Nephrol. Dial. Transplant.* **2003**, 18(1), 126–132.

Kannel, W. B.; Dawber, T. R.; Kagan, A.; Revotskie, N.; Stokes, J 3rd. Factors of Risk in the Development of Coronary Heart Disease--Six Year Follow-up Experience. The Framingham Study. *Ann. Intern. Med.* **1961**, 55, 33–50.

Kinebuchi, S.; Kazama, J. J.; Satoh, M.; Sakai, K.; Nakayama, H.; Yoshizawa, H.; Narita, I.; Suzuki, E.; Gejyo, F. Short-Term Use of Continuous Positive Airway Pressure Ameliorates Glomerular Hyperfiltration in Patients with Obstructive Sleep Apnoea Syndrome. *Clin. Sci. (Lond)* **2004**, 107 (3), 317–322.

Klag, M. J.; Whelton, P. K.; Randall, B. L.; Neaton, J. D.; Brancati, F. L.; Ford, C. E.; Shulman, N. B.; Stamler, J. Blood Pressure and End-Stage Renal Disease in Men. *N. Engl. J. Med.* **1996**, 334 (1), 13–18.

Knutson, K. L.; Van Cauter, E.; Rathouz, P. J.; Yan, L. L.; Hulley, S. B.; Liu, K.; Lauderdale D. S. Association Between Sleep and Blood Pressure in Midlife: the CARDIA Sleep Study. *Arch. Intern. Med.* **2009**, 169 (11), 1055–1061.

Koga, S.; Ikeda, S.; Yasunaga, T.; Nakata, T.; Maemura, K. Effects of Nasal Continuous Positive Airway Pressure on the Glomerular Filtration Rate in Patients with Obstructive Sleep Apnea Syndrome. *Intern. Med.* **2013**, 52 (3), 345–349.

Kretzler, M.; Teixeira, V. P.; Berger, T.; Blattner, S. M.; Unschuld, P. G.; Cohen, C. D.; Schlöndorff, D. Altering Glomerular Epithelial Function in Vitro Using Transient and Stable Transfection. *J. Nephrol.* **2001**, 14,:211–219.

Kumar, B.; Tilea, A.; Gillespie, B. W.; Zhang, X.; Kiser, M.; Eisele, G.; Finkelstein, F.; Kotanko, P.; Levin, N.; Rajagopalan, S.; Saran, R. Significance of Self-Reported Sleep Quality (SQ) in chronic Kidney Disease (CKD): the Renal Research Institute (RRI)-CKD Study. *Clin. Nephrol.* **2010**, 73 (2), 104–114.

Kurella, M.; Luan, J.; Lash, J. P.; Chertow, G. M. Self-Assessed Sleep Quality in Chronic Kidney Disease. *Int. Urol. Nephrol.* **2005**, 37 (1), 159–165.

Lavie, L. Oxidative Stress--a Unifying Paradigm in Obstructive sleep Apnea and Comorbidities. *Prog. Cardiovasc. Dis.* **2009**, 51 (4), 303–312.

Lee, Y. C.; Hung, S. Y.; Wang, H. K.; Lin, C. W.; Wang, H. H.; Chen, S. W.; Chang, M. Y.; Ho, L. C.; Chen, Y. T.; Liou, H. H.; Tsai, T. C.; Tseng, S. H.; Wang, W. M.; Lin, S. H.; Chiou, Y. Y. Sleep Apnea and the Risk of Chronic Kidney Disease: a Nationwide Population-Based Cohort Study. *Sleep* **2015**, 38 (2), 213–221.

Markou, N.; Kanakaki, M.; Myrianthefs, P.; Hadjiyanakos, D.; Vlassopoulos, D.; Damianos, A.; Siamopoulos, K.; Vasiliou, M.; Konstantopoulos S. Sleep-Disordered Breathing in Nondialyzed Patients with Chronic Renal Failure. *Lung* **2006**, 184 (1), 43–49.

Martínez-Sanchis, S.; Bernal, M. C.; Montagud, J. V.; Abad, A.; Crespo, J.; Pallardó L. M. Quality of Life and Stressors in Patients with Chronic Kidney Disease Depending on Treatment. *Span. J. Psychol.* **2015**, 18, E25.

Mises, J.; Lerique-Koechlin, A.; Rimbot, B. The Electroencephalogram in Renal Insufficiency. *Rev. Neurol. (Paris)* **1967**, 117 (3), 524.

Nicholl, D. D.; Ahmed, S. B.; Loewen, A. H.; Hemmelgarn, B. R.; Sola, D. Y; Beecroft, J. M.; Turin, T. C.; Hanly, P. J. Clinical Presentation of Obstructive Sleep Apnea in Patients with Chronic Kidney Disease. *J. Clin. Sleep Med.* **2012**, 8(4), 381–387.

Parker, K. P. Sleep Disturbances in Dialysis Patients. *Sleep Med. Rev.* **2003**, 7 (2), 131–143.

Parvan, K.; Lakdizaji, S.; Roshangar, F.; Mostofi, M. Quality of Sleep and Its Relationship to Quality of Life in Hemodialysis Patients. *J. Caring. Sci.* **2013**, 2 (4), 295–304.

Parving, H. H. Antihypertensive Drugs and Diabetes Mellitus. *Dan. Med. Bull.* **1987**, 34 Suppl 1, 16–17.

Petrov M. E.; Kim Y.; Lauderdale D. S.; Lewis C. E.; Reis J. P.; Carnethon M. R.; Knutson K. L.; Glasser S. P. Objective Sleep, a Novel Risk Factor for Alterations in Kidney Function: the CARDIA Study. *Sleep Med.* **2014**, 15 (9), 1140–1146.

Sabbatini, M.; Minale, B.; Crispo, A.; Pisani, A.; Ragosta, A.; Esposito, R.; Cesaro, A.; Cianciaruso, B.; Andreucci, V. E. Insomnia in Maintenance Haemodialysis Patients. *Nephrol. Dial. Transplant.* **2002**, 17 (5), 852–856.

Salifu, I.; Tedla, F.; Pandey, A.; Ayoub, I.; Brown, C.; McFarlane, SI.; Jean-Louis, G. Sleep Duration and Chronic Kidney Disease: Analysis of the National Health Interview Survey. *Cardiorenal. Med.* **2014**, 4(3–4), 210–216.

Schreiner, G. E. Mental and Personality Changes in the Uremic Syndrome. *Med. Ann. Dist. Columbia* **1959,** 28 (6), 316–323.

Spiegel, K.; Tasali, E.; Leproult, R.; Van Cauter, E. Effects of Poor and Short Sleep on Glucose Metabolism and Obesity Risk. *Nat. Rev. Endocrinol.* **2009,** 5(5), 253–261.

Stefanidis, I.; Vainas, A.; Dardiotis, E.; Giannaki, C. D.; Gourli, P.; Papadopoulou, D.; Vakianis, P.; Patsidis, E.; Eleftheriadis, T.; Liakopoulos, V.; Pournaras, S.; Sakkas, G. K.; Zintzaras, E.; Hadjigeorgiou, G. M. Restless Legs Syndrome in Hemodialysis Patients: an Epidemiologic Survey in Greece. *Sleep Med.* **2013,** 14 (12), 1381–1386.

St-Onge, M. P.; Roberts, A. L.; Chen, J.; Kelleman, M.; O'Keeffe, M.; RoyChoudhury, A.; Jones, P. J. Short Sleep Duration Increases Energy Intakes but Does Not Change Energy Expenditure in Normal-Weight Individuals. *Am. J. Clin. Nutr.* **2011,** 94 (2), 410–416.

Taraz, M.; Khatami, M. R.; Hajiseyedjavadi, M.; Farrokhian, A.; Amini, M.; Khalili, H.; Abdollahi, A.; Dashti-Khavidaki, S. Association Between Antiinflammatory Cytokine, IL-10, and Sleep Quality in Patients on Maintenance Hemodialysis. *Hemodial. Int.* **2013,** 17 (3), 382–390.

Terzi, F.; Burtin, M.; Friedlander, G. Early Molecular Mechanisms in the Progression of Renal Failure: Role of Growth Factors and Protooncogenes. *Kidney. Int. Suppl.* **1998,** 65, S68–73.

Tsioufis, C.; Thomopoulos, K.; Dimitriadis, K.; Amfilochiou, A.; Tousoulis, D.; Alchanatis, M.; Stefanadis, C.; Kallikazaros, I. The Incremental Effect of Obstructive Sleep Apnoea Syndrome on Arterial Stiffness in Newly Diagnosed Essential Hypertensive Subjects. *J. Hypertens.* **2007,** 25 (1), 141–146.

Turek, N. F.; Ricardo, A. C.; Lash, J. P. Sleep Disturbances as Nontraditional Risk Factors for Development and Progression of CKD: Review of the Evidence. *Am. J. Kidney Dis.* **2012,** 60 (5), 823–833.

Turin, T. C; Hemmelgarn, B. R. Change in Kidney Function Over Time and Risk for Adverse Outcomes: is an Increasing Estimated GFR Harmful? *Clin. J. Am. Soc. Nephrol.* **2011,** 6 (8), 1805–1806

Unruh, M. L.; Sanders, M. H.; Redline, S.; Piraino, B. M.; Umans, J. G.; Chami, H.; Budhiraja, R.; Punjabi, N. M.; Buysse, D.; Newman, A. B. Subjective and Objective Sleep Quality in Patients on Conventional Thrice-Weekly Hemodialysis: Comparison with Matched Controls from the Sleep Heart Health Study. *Am. J. Kidney Dis.* **2008,** 52 (2), 305–313.

Van Bussel, B. C.; Schouten, F.; Henry, R. M.; Schalkwijk, C. G.; de Boer, M. R.; Ferreira, I.; Smulders, Y. M.; Twisk, J. W.; Stehouwer, C. D. Endothelial Dysfunction and Low-Grade Inflammation are Associated with Greater Arterial Stiffness Over a 6-Year Period. *Hypertension* **2011,** 58 (4), 588–595.

Walker, S.; Fine, A.; Kryger, M. H. Sleep Complaints are Common in a Dialysis Unit. *Am. J. Kidney Dis.* **1995,** 26(5), 751–756.

Yamamoto, R.; Nagasawa, Y.; Iwatani, H.; Shinzawa, M.; Obi, Y.; Teranishi, J.; Ishigami, T.; Yamauchi-Takihara, K.; Nishida, M.; Rakugi, H.; Isaka, Y.; Moriyama, T. Self-Reported Sleep Duration and Prediction of Proteinuria: a Retrospective Cohort Study. *Am. J. Kidney Dis.* **2012,** 59 (3), 343–355.

Yamauchi, M.; Kimura, H. Oxidative Stress in Obstructive Sleep Apnea: Putative Pathways to the Cardiovascular Complications. *Antioxid. Redox. Signal.* **2008,** 10 (4),755–768.

Zhang, J.; Wang, C.; Gong, W.; Peng, H.; Tang, Y.; Li, C. C.; Zhao, W.; Ye, Z.; Lou, T. Association Between Sleep Quality and Cardiovascular Damage in Pre-Dialysis Patients with Chronic Kidney Disease. *BMC Nephrol.* **2014,** 15, 131.

16

SLEEP DISORDERS IN PARKINSON'S DISEASE AND DEMENTIA

Luciano Ribeiro Pinto Jr. and Andrea Frota Bacelar Rêgo

ABSTRACT

The causes of dementia are: Alzheimer's Disease, frontotemporal lobar degeneration, Dementia with Lewy body (DLB) disease, Multiple System Atrophy (MSA), Huntington disease, Prion disease and Parkinson disease. Sleep disturbances (SD) in Parkinson disease (PD) were recognized as early as 1817, by James Parkinson. PD, Multiple system atrophy and DLB are synucleinopathies. Progressive supranuclear palsy (PSP), however, results from the deposit of the tau protein forming tangles within the glial cells. Sleep-related problems in dementias and Parkinson disease can be divided into disturbances of sleep and disturbances of wakefulness.

The sleep disorders in degenerative dementias are: sleep architecture alterations; changes in the circadian sleep-wake rhythm, insomnia, sleep fragmentation, increased periods of wakefulness during the night, sundowning syndrome, sleep-related breathing disorders, periodic limb movement disorder (PLM), restless leg syndrome (RLS) and parasomnias like REM sleep behavior disorder (RBD).

Polysomnography and Multiple Sleep Latency (MSLT) are valuable tests in patients with PD and dementias.

Luciano Ribeiro Pinto Jr.
Neurology and Sleep Medicine, Brazilian Academy of Neurology, Universidade Federal de São Paulo Hospital Alemão Oswaldo Cruz, São Paulo (SP), Brazil

Andrea Frota Bacelar Rêgo
Neurology and Sleep Medicine, Brazilian Academy of Neurology, Universidade Federal de São Paulo Hospital Alemão Oswaldo Cruz, São Paulo (SP), Brazil

Corresponding author: Luciano Ribeiro Pinto Jr., E-mail: lucianoribeiro48@gmail.com

16.1 PARKINSON'S DISEASE

16.1.1 Introduction

Parkinson's disease (PD) is a progressive neurodegenerative illness that affects about 1 million patients in North America, 1% of middle-aged patients, with 85% sporadic and 15% genetic cases.[1]

The physical aspects of PD such as tremor, rigidity, and postural imbalance, have traditionally been regarded as the most important features of the disease, and have understandably received the most attention in both research and clinical practice. Many of the non-motor aspects of PD are common, and significantly affect the daily lives of individuals. Better treatment for these aspects of the illness could produce an important reduction in suffering.[2]

One of the major non-motor aspects affected is sleep. Sleep disturbances (SD) in PD were recognized as early as 1817, by James Parkinson[3] in his classic monograph in which he noted that: "The sleep becomes much disturbed. The tremulous motions of the limbs occur during sleep, and augment until they awaken the patient, and frequently with much agitation and alarm and at the last, constant sleepiness, with slight delirium".

16.1.2 Neuroanatomy and Pathology of Parkinsonism

Parkinsonism is caused by an impaired functioning of the extrapyramidal system within the basal ganglia. Parkinsonism is also observed in other related neurodegenerative diseases: (1) patients with multiple-system atrophy (MSA), in whom other systems, such as the cerebellar system, the upper motor neurons, and the autonomic system, may be affected; (2) in patients with dementia with Lewy bodies (DLB), in whom cognitive functions are markedly impaired at an early stage; (3) in patients with progressive supranuclear palsy (PSP), in whom speech, eye movements, and postural balance are deficient.[1]

PD, MSA and DLB are synucleinopathies (i.e. disorders linked to the abnormal accumulation of the alpha-synuclein protein-forming Lewy bodies included in the neurons); PSP, however, results from the deposit of the tau protein forming tangles within the glial cells.[4]

Although the main lesion in PD is the massive loss of nigro-striatal neurons, the loss of neurons may affect as many as 50% of other monoamine-containing neurons, such as: (1) dopamine neurons in the ventral tegmental area; (2) norepinephrine neurons in the locus coeruleus and vagus nerve dorsal nucleus; and (3) serotonin neurons in the dorsal raphe. Also damaged are cholinergic neurons in the Meynert basal nucleus, neocortex, hippocampus, Edinger-Westphal, and pedunculopontine nuclei; peptidergic neurons, including neuropeptide Y neurons in the spinal cord, somatostatin neurons in the hippocampus and cortex, substance P neurons in the medulla; and cholecystokinin and met-enkephalin neurons in the basal ganglia. Nishino S. et al.[5] have shown that the levels of the hypothalamic neuropeptide, hypocretin, which are abnormally low in the cerebrospinal fluid of patients with primary narcolepsy-cataplexy, may also be markedly decreased within the cerebral lateral ventricle of patients with severe PD. The occurrence of lesions in non-dopaminergic neurons may explain the dopamine-insensitive symptoms.[1]

16.1.3 Neuroanatomy of Sleep in Parkinson's Disease

The anatomical basis of SD in PD is not fully understood, but it likely it probably involves degeneration of both dopaminergic and non-dopaminergic systems. SD are primarily due to the progressive disease process impairing thalamocortical arousal and affecting sleep-regulating centers in the brainstem.[6] Their degeneration leads to the disruption of basic rapid eye movement (REM) and non-REM sleep architecture, manifesting as insomnia, parasomnias, and hallucinations.[7] The pedunculopontine nucleus and the retro-rubral nucleus have strong influences on REM atonia and phasic generator circuitry and have been implicated in the pathogenesis of REM behavior disorder (RBD). A flip-flop-switch pattern of regulation of sleep–wake cycle has been proposed by Saper[9], suggesting that the brain can be either 'off' (asleep by activating the ventrolateral preoptic area, the sleep promoter) or 'on' (in quiet wakefulness with the activation of the tuberomamillary nucleus (TMN), the wake-promoting area along with locus coerulius (LC) and the raphe nuclei (DRN)). The internal rhythm between the two switches is regulated by the anterior hypothalamic suprachiasmatic nucleus (SCN; the 'biological clock'). Hypocretin 1 (orexin), a hypothalamic neuropeptide[5], is now thought to have a complex relationship with the dopaminergic systems in the basal ganglia and may function as an external regulator of the flip-flop switch promoting wakefulness.[7] In PD, dopaminergic dysfunction and neuronal degeneration can destabilize this switch and its regulators, promoting rapid transitions to sleep.

Secondary causes are nocturnal disease manifestations, and side effects of pharmacological treatment.

The relative frequency of certain SD in the neurodegenerative disorders characterized by alfa-synucleine deposition compared to the tauopathies suggests differences in the anatomic areas affected related to sleep (Table 16.1).[10] RBD is more frequently associated with synucleinopathies, in MSA and DLB, RBD symptoms are more likely to precede other disease manifestations, and occur at a younger age than in PD, suggesting that the more extensive pathological changes in these disorders may involve brainstem areas controlling REM sleep at an earlier stage in the disease. In Alzheimer's disease (AD) and PSP, RBD is rare.[11]

NOMENCLATURE

AD	Alzheimer's Disease	REM	Rapid Eye Movement
DLB	Dementia with Lewy Bodies	REMw/oA	REM sleep without Atonia
EDS	Excessive Daytime Sleepiness	RBD	REM Behavior Disorder
ESS	Epworth Sleepiness Scale	RLS	Restless Leg Syndrome
MSA	Multiple-System Atrophy	SA	Sleep Attack
MSLT	Multiple Sleep Latency Test	SD	Sleep Disturbances
PD	Parkinson's Disease	SOREMP	Sleep Onset in REM Period
PSG	Polysomnography	PDSS	Parkinson's Disease Sleep Scale
PSP	Progressive Supranuclear Palsy	PLMS	Periodic Limb Movements of Sleep

	REM behavior disorder/REM sleep without atonia	Sleep apnea	PLMD	EDS
Parkinson's disease	+ +	+ +	+ +	+ + +
MSA	+ + +	+ + + +	+ + + +	+ + +
DLB	+ + +		+ + +	+ + + +
PSP	+ +	+/−	?	+
AD	+/−	+ +	?	+ +

Table 1. Relative Frequency of Sleep Disorders in Neurodegenerative Diseases. Multiple System Atrophy (MSA); Dementia with Lewy Bodies (DLB); Progressive Supranuclear Palsy (PSP); Alzheimer's Disease (AD).[10]

16.1.4 Clinical Presentation of SD in PD

The recognition and clinical diagnosis of SD in PD patients may be an important issue to further understand the pathophysiology of the disease itself. SD may develop several years before the onset of clinically evident PD and may therefore be considered the first clinical manifestation of PD; moreover, SD also occur during the course of PD in combination with some clinical phenotypes or with a more severe progression of the disease.[12]

The neurology community has only recently realized that excessive daytime sleepiness (EDS) can also be a disabling and potentially dangerous symptom in patients with parkinsonism.[13]

Non-motor features (which include autonomic nervous system dysfunction, disorders of cognition and mood, psychosis, pain, loss of smell, gastrointestinal (GI) dysfuntions, and fatigue), appear early in the course of PD, and contribute to EDS. All of these symptoms have significant adverse effects on the quality of life of both patients and caregivers and require proper identification and treatment.[14]

Sleep-related problems in PD can be divided into disturbances of sleep and disturbances of wakefulness.[6] Disturbances of sleep include nocturnal akinesia, insomnia, sleep fragmentation with increased periods of wakefulness during the night, nocturia, restless leg syndrome (RLS), periodic limb movement syndrome (PLMS), RBD, sleep apnea, and parasomnias. Disturbances of wakefulness include EDS and "sleep attacks" (SA). With aging, there is disruption of normal sleep architecture and alterations in the circadian rhythm leading to impaired nocturnal sleep and EDS. These

problems are accentuated in PD patients, with 60%–90% having these symptoms which increase with disease duration.[15] SD in PD are associated with disease progression, older age, male sex, rigid-akinetic phenotype, and severe motor dysfunction.[12]

The most common clinical feature is RBD, a parasomnia first reported by Schenck *et al.* in 1986[16], which is characterized by the loss of normal skeletal muscle atonia during REM sleep and associated with excessive motor activity while dreaming. Patients enact their dreams which can be vivid or unpleasant, and partners report vocalisations (talking, shouting, and vocal threats) and abnormal movements (arm/leg jerks, falling out of bed, and violent assaults). EDS and RBD may be harbingers of PD and other synucleinopathies, such as MSA[17], and thus already present in the premotor phase of the disease. Schenck *et al.* reported that in 11 of 29 men (38%) 50 years or older in whom idiopathic RBD was diagnosed, a parkinsonian disorder was identified after a mean interval of 3.7 ± 1.4 years following the diagnosis of RBD and 12.7 ± 7.3 years after the onset of RBD.[18]

Insomnia is defined as an almost nightly complaint of an insufficient amount of sleep, divided into difficulty in falling asleep (sleep initiation), staying asleep (sleep maintenance), and awakening too early in the morning. Although all three problems occur in patients with PD, sleep maintenance difficulties are the most common, affecting up to 74–88% of patients.[19] Obviously, many patients with PD who complain of insomnia actually have other primary SD such as RBD, PLMS, or RLS. Furthermore, insomnia may be related to factors such as depression, poor sleep hygiene, nocturia, pain, dystonia, akinesia, difficulty turning in bed, and reactions to medications.[2]

RLS also known as Willis-Ekbom disease (WED), is a sleep-related movement disorder characterized by an uncontrollable urge to move the legs, usually accompanied by unpleasant sensations that are typically partially or totally relieved by movement. Symptoms occur in a circadian pattern, with onset usually in the evening hours when lying in bed to sleep.

PLMS are rhythmic moving or jerking of the limbs during sleep. Both of these disorders may interfere with the quantity and quality of sleep. RLS symptoms have been shown to appear before the onset of PD in a variable percentage of patients (ranging from 0.6%[20] to 50% of cases[21]). PD patients with RLS have been shown to have a longer duration of PD symptoms, more severe PD disability, a greater degree of cognitive dysfunction, and a longer duration of antiparkinsonian therapy than those without RLS.[20]

Sleep apnea is characterized by a deficit in breathing drive in the brain (CSA; central sleep apnea) or a problem with the passage of air through the breathing passages (OSA; obstructive sleep apnea). As breathing becomes more difficult or ceases a decrease in blood oxygen level results, which in turn results in sufficient awakening to restore breathing, which may occur hundreds of times a night. Consequently, the patient experiences little deep restorative sleep at night and extreme daytime sleepiness.[19] Sleep apnea was more frequent and severe in the most disabled patients with PD.[22] Sleep apnea may co-exist with RLS or PLMS or RBD.[23] This is important to

diagnose as these patients need specific and targeted treatment.

In patients with MSA, respiratory disturbances during sleep are very common and they may develop stridor during sleep arising from vocal cord abductor paralysis.[24]

The term SA, was used initially to characterize irresistible episodes of unwanted sleep in patients with primary narcolepsy. It has been recently re-defined as 'an event of sudden, irresistible, and overwhelming sleepiness that occurs without sleepiness and is not preceded by being sleepy. Severe SA in patients treated with the new nonergotic dopamine agonists pramipexole and ropinirole, night-time sleep troubles (including sleep deprivation caused by nocturnal pain and akinesia) and sleep fragmentation caused by sleep apnea and PLM are also a potential cause of EDS.[1] The disease itself may eventually affect the sleep-wake system, as suggested by the report of narcolepsy-like sleepiness in patients with PD (Narcolepsy type 2).[25,33] In contrast to narcolepsy, however, cataplexy has not been described in PD and a recent study suggested that there is little overlap between these disorders.[25]

It is also clear that dopaminergic medications and particularly dopamine agonists can have a complex effect on sleep. Sometimes these medications cause insomnia, and their sedative properties may contribute to daytime sleepiness. In other situations, they improve the quality of sleep by improving nocturnal immobility.[26] Therefore, dopaminergic medications can either improve or worsen sleep in PD patients. Although both SD and hallucinations may be influenced by the dopaminergic treatment, results on the association between RBD and hallucinations and antiparkinsonian therapy are conflicting[27] and may be explained by a common dysfunction of the REM sleep regulatory system. Patients with SD, older age, more severe motor and cognitive dysfunctions are more likely to develop hallucinations over the years.

16.1.5 Epidemiology of SD in PD

The prevalence of SD in PD is difficult to ascertain due to the heterogeneity of patients, differences of age, parkinsonian motor dysfunction, dyskinesia, pain, nocturia, nightmares, dopaminergic and non-dopaminergic medications, and cognitive impairment. SD have been reported in up to 90% of PD patients and their frequency increases with advancing disease.[12]

There was no sex difference for difficulty initiating sleep.[6] The occurrence of insomnia remains high (more than 50% of patients) during follow-up, because insomnia frequently fluctuates over time in individual patients.[28] Van Hilten et al.[29] observed that female patients experienced more difficulty maintaining sleep (87.5%) and excessive dreaming (68.4%) than males (64% and 31.6%, respectively).

EDS is more frequent and more severe in patients with PD than in controls, and epidemiological studies of PD have shown that men are more prone than women to suffer from sleepiness.[30] Case-controlled epidemiological studies conducted in various countries consistently found higher Epworth Sleepiness Scale (ESS) scores and a higher percentage (16–74%) in patients with PD than in controls.[1]

The estimated prevalence of RLS in PD patients ranged widely (from 0%[20] to 50%[21]) but was higher than in controls (0.5–6.3).

Apnea has been found in as many as 50% of patients with PD. Snoring and apneic episodes also may be up to three times more common in PD (12%) than in the general population.[22]

The percentage of patients with PD who have experienced SA varies from 1–27%, a percentage either higher[31] or the same as controls[32], whereas 1–21% of patients with PD have reported experiencing SA while driving.[1]

This narcolepsy-like phenotype (Narcolepsy type 2)[33] was found in up to 39% of sleepy PD patients with EDS[10] and 70% of patients with PD suffering from severe hallucinations.[34] RBD or REM sleep without atonia (REMw/oA) is present in 25–50% of PD patients[35], and an even greater percentage of MSA and DLB. RBD or REMw/oA was extremely prevalent in patients with narcolepsy-like sleepiness.

In a cross-sectional clinic survey of 289 consecutive PD patients, the presence of RBD was associated with an almost 3-fold increased risk of hallucinations.[36] A longitudinal study over 8 years showed RBD to be highly associated with hallucinations.[37]

16.1.6 Diagnosis of SD in PD

An evaluation of a sleep problem in a patient with PD involves taking a careful history, neurological examination, use of questionnaires and, sometimes, polysomnography (PSG) is necessary to make the correct differential diagnosis.

The most widely used scale is the ESS[38], which scores the tendency to fall asleep (from 0 to 3) during eight everyday situations. It ranges from 0 to 24. Abnormal somnolence is considered as a value in ESS greater than 10 or [2] a mean sleep latency of less than eight 8min in the Multiple Sleep Latency Test (MSLT).[39]

The Parkinson's Disease Sleep Scale (PDSS), with an extended spectrum of nocturnal disabilities and easier use for patients, is the first formal instrument for quantifying sleep problems in PD. It is a reliable, valid, precise, and potentially treatment-responsive tool for measuring sleep disorders in PD[8] (Table 16.2).

The PDSS-2[40], now in its second edition, consists of 15 questions about various sleep and nocturnal disturbances which are rated by the patients using one of five categories, from 0 (never) to 4 (very frequent). The PDSS-2 total score ranges from 0 (no disturbance) to 60 (maximum nocturnal disturbance). A gradual increase of sleep disturbance with PD disease severity could be observed by applying PDSS-2. It can be divided into specific factors that reflect the diversity and complexity of the sleep problem in PD patients: first, PD specific nocturnal motor symptoms such as akine-

Table 2. The Parkinson's Disease Sleep Scale (PDSS-2).[40]

Parkinson's Disease Sleep Scale (PDSS-2)

Please rate the severity of the following based on your experiences during the past week (7 days). Please make a cross in the answer box

	Very often (This means 6 to 7 days a week)	Often (This means 4 to 5 days a week)	Sometimes (This means 2 to 3 days a week)	Occasionally (This means 1 day a week)	Never
1) Overall, did you sleep well during the last week?	0	1	2	3	4
2) Did you have difficulty falling asleep each night?	4	3	2	1	0
3) Did you have difficulty staying asleep?	4	3	2	1	0
4) Did you have restlessness of legs or arms at nights causing disruption of sleep?	4	3	2	1	0
5) Was your sleep disturbed due to an urge to move your legs or arms?	4	3	2	1	0
6) Did you suffer from distressing dreams at night?	4	3	2	1	0
7) Did you suffer from distressing hallucinations at night (seeing or hearing things that you are told do not exist)?	4	3	2	1	0
8) Did you get up at night to pass urine?	4	3	2	1	0
9) Did you feel uncomfortable at night because you were unable to turn around in bed or move due to immobility?	4	3	2	1	0
10) Did you feel pain in your arms or legs which woke you up from sleep at night?	4	3	2	1	0
11) Did you have muscle cramps in your arms or legs which woke you up whilst sleeping at night?	4	3	2	1	0
12) Did you wake early in the morning with painful posturing of arms and legs?	4	3	2	1	0
13) On waking, did you experience tremor?	4	3	2	1	0
14) Did you feel tired and sleepy after waking in the morning?	4	3	2	1	0
15) Did you wake up at night due to snoring or difficulties with breathing?	4	3	2	1	0

sia, early morning dystonia, tremor during waking period at night, PLM, restless behavior, and motor symptoms probably due to RBD; second, PD-specific nocturnal non-motor symptoms like hallucinations, confusional states, pain, muscle cramps, difficulties in breathing with snoring, and immobility; third, sleep-specific disturbances like insomnia, sleep maintenance, unrestored sleep in the morning, getting up at night to pass urine and the overall quality of sleep in the patient's rating.

PSG is the "gold standard" method used to evaluate sleep disorders and provides detailed information about actual sleep status. It can detect the co-occurrence of sleep apnea, RLS, PLMS, and RBD. Figures 16.1 and 16.2 show examples of PLM and REMw/oA, respectively. PSG findings of RBD include excessive chin muscle tone and limb jerking during REM. MSLT is helpful to quantify the severity of EDS, identifying multiple sleep onset in REM periods (SOREMPs) in patients with PD. Because it resembled the sleep pattern observed in patients with primary narcolepsy without cataplexy, we defined the occurrence of two or more SOREMPs during MSLT as a 'narcolepsy-like phenotype.'[33]

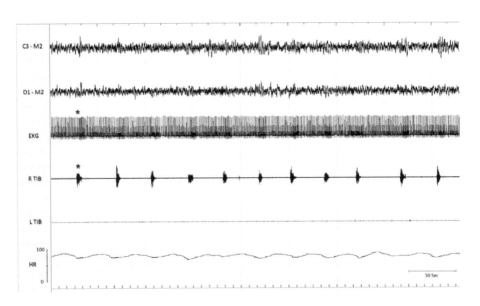

Figure 1. PLM with arousals in PD.

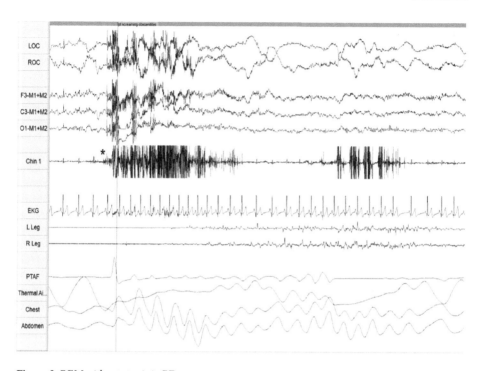

Figure 2. REM without atonia in PD.

16.1.7 Treatment of SD in Parkinson's Disease

The main points that need to be addressed in the treatment of SD in patients with PD are; the introduction of proper sleep hygiene, eliminating unnecessary sedative medications, using the lowest dose of do-paminergic medication that provides satisfactory clinical control, improving sleep fragmentation and early morning function, identifying (PSG) and treat sleep disorders such as sleep apnea and PLM, and counselling patients on the risks of daytime sleepiness and sudden sleep episodes. Patients with EDS should not drive a motor vehicle until this problem has been corrected.[6] A management strategy for SD in PD is shown in Table 16.3.[8]

A dose as small as 0.25 mg of clonazepam may reduce RBD episodes and prevent sleep-related injuries. Melatonin was effective at doses ranging from 3 mg to 12 mg[41], and pramipexole also improves RBD[42], The serotonergic antidepressants have been shown to increase the risk of RBD and should be avoided in susceptible individuals.[43]

The treatment of PD hallucinations involves discontinuation of drugs including anticholinergic agents, anxiolytics, centrally active pain medications, antidepressants, and other centrally active drugs. The atypical antipsychotics such as clozapine or quetiapine reduce hallucinations without worsening PD.[44]

Table 3: Management Strategies for Symptoms Contributing to Nocturnal.

Insomnia symptoms
Fragmented sleep with difficulty in sleep onset and sleep maintenance
Avoidance of alcohol, caffeine and tobacco
Increase in day-time physical activity and exposure to daylight
Psychological therapies: relaxation training, cognitive therapies and biofeedback
Short-acting benzodiazepines: clonazepam, temazepam and diazepam
Non-benzodiazepine hypnotics: zopiclone
Tricyclic antidepressants: amitriptyline (may help nocturia)
Motor symptoms
Fidgeting, painful cramps and posturing, tremor, sleep akinesia, RLS-type symptoms
Use of satin bed sheets and bed straps
Sustained dopaminergic stimulation:
CR levodopa ± COMT inhibitor
Long-acting dopamine agonists, e.g. cabergoline
Nocturnal apomorphine infusion
Combination of day-time apomorphine and evening cabergoline (dual-agonist therapy)
Practical measures to aid bioavailability of dopaminergic medications
Avoidance of high-protein meals at night
Domperidone if delayed gastric emptying
REM behaviour disorder
Clonazepam
Pramipexole
Melatonin
Neuropsychiatric symptoms
Distressing dreams, hallucinations and depression
Consider alternative diagnosis: MSA, LBD and PSP
Hallucinations
Drug-induced: optimise therapy
Atypical neuroleptics: Quetiapine
Depression
Amitriptyline;
noradrenaline re-uptake inhibitors;
dopamine agonists, e.g. pramipexole
Panic attacks
During 'on' periods: alprazolam and lorazepam
During 'off' periods: CR levodopa ± COMT inhibitor; cabergoline; apomorphine infusion
Any time: sertraline, fluoxetine and paroxetine
*Urinary symptoms*Nocturia;Incontinence of urine - inability to move during 'off ' phase
Reduction of evening fluid intake
Emptying bladder before bed
Use of condom catheters/bedside commode
If associated with postural hypotension, head-up tilt of bed
Low-dose amitriptyline
Possible role for D1/D2 agonists, e.g. cabergoline, pergolide and apomorphine
If associated with detrusor instability: oxybutinin and tolterodine
If associated with morning hypotension: desmopressin nasal spray; avoidance of
evening diuretics, antihypertensives and vasodilators

If alterations in dopaminergic medications fail to help EDS, or if narcolepsy type 2 is diagnosed, the addition of a wakefulness-promoting agent like modafinil should be considered. Modafinil appears to exert its effects specifically on the hypothalamus sleep-wake system. It should be started at a dose of 100 mg and increased to 200–400 mg per day as necessary.[45] Side effects include insomnia, head pains, and depression. Depression should be evaluated and treated accordingly.

The nocturnal administration of apomorphine has been found to increase nocturnal sleep time and reduce night-time movements in PD.[46] Deep brain stimulation surgery has also been demonstrated to improve sleep quality and mitigate sleep fragmentation.[47]

Further areas of research are now also focusing on adenosine A2A receptor antagonist, sodium oxybate and caffeine to promote wakefulness. Studies conducted worldwide report an inverse association between caffeine/coffee consumption and the risk of developing PD, in men.[48]

For nocturia, the patients should be encouraged to avoid diuretics such as tea or coffee at bedtime and the use of a desmopressin nasal spray could be beneficial.

Sustained-release levodopa/benserazide significantly improved night-time akinesia in patients with PD and distressing nocturnal symptoms.[8]

Practice points[12]

- RBD and EDS may be the heraldic non-motor manifestation of PD, as neuropathological changes in brainstem nuclei involved in sleep-wake cycle regulation may occur early in the course of the disease.
- The prevalence of SD during the course of the disease progressively increases with disease duration. SD are associated with PD with more rapid progression, older age, male sex, rigid-akinetic phenotype, and severe motor dysfunction.
- There is a strict relationship between SD and other non-motor symptoms of PD, such as hallucinations and cognitive decline.
- EDS generally correlates with disease severity and high dose of levodopa equivalents.
- SA have been associated with all (ergot and non-ergot) dopamine agonists, and sometimes with levodopa alone and may disappear after reducing or changing the dopamine agonist.
- The risk of SA increases with increasing scores on ESS and adjunction of a dopamine agonist. Patients should be asked not to drive during treatment changes.
- 50% of patients with PD with EDS have an obstructive sleep apnea syndrome.
- Two out of five patients with PD with EDS have a narcolepsy-like phenotype.
- PSG and MSLT are valuable tests in patients with PD with EDS.

16.2 DEMENTIAS

16.2.1 Introduction

There is a bidirectional relationship between sleep and dementia.[49] Degenerative dementias are characterized by a progressive decline in cognitive function which has a major impact on an individual's ability to take part in social activities and on their daytime functioning (Table 16.4)

(DSM-V).[50] There are three main types of degenerative dementias: AD, Lewy body dementia and fronto-temporal dementia (Table 16.5). SD found in dementia are sometimes similar to those of normal elderly individuals in the same age group, but are more accentuated.[51]

Table 16.4. Neurocognitive Domains Affected in Neurocognitive Disorders According to the DSM- V.

1. Complex attention
2. Executive function
3. Learning and memory
4. Language
5. Perceptual-motor
   visual perception
   visuo-constructional
   perceptual-motor
   praxis
   gnosis
6. Social cognition

Table 16.5. Causes of Dementia.

A. Degenerative disease
1. Alzheimer's Disease
2. Frontotemporal lobar degeneration
3. Synucleipathies
   a) Parkinson Disease
   b) Lewy body disease
   c) Multiple System Atrophy
4. Huntington disease
5. Prion disease
B. Vascular dementia
C. Another medical conditions: deficiency of vitamin B12 and folate, hypothiroidism, depression, infectious disease, syphilis, tuberculosis, HIV, encephalitis, normal pressure hydrocephalus, tumors, subdural hematoma, substance/medication use, drug intoxication, alcohol, and metabolic disorders

The anatomical structures involved in the maintenance of wakefulness and sleep induction are extremely complex. The sleep trigger depends on the brain stem and cortical and subcortical regions. In addition, several neurotransmitters are involved in the process as homeostatic and circadian factors. Neurotransmitters that promote wakefulness include acetylcholine, monoamines, glutamate and hypocretin, while those associated with the induction and maintenance of sleep include gamma-amino butyric acid (GABA), adenosine and serotonin.[52]

Dementias are characterized by progressive impairment of multiple neuronal systems. AD is characterized by the formation of extracellular plaques of amyloid-ß (Aß) protein and intracellular neurofibrillary tangles containing hyperphosphorylated microtubule associated protein tau.[53,54] The main affected regions in dementia are: the nucleus basalis of Meynert, the entorhinal cortex, the hippocampus, the lateral hypothalamus, the amygdala, the raphe nuclei, the magnocellular reticular formation, the coeruleus nucleus, the limbic areas, and the associative cortex.[7] These regions have in common their development in hominids, and low efficiency in the clearance of Ca^{2+}, being in general more susceptible to neurotoxic and oxidative stress.[55] The fact that all these areas of the central nervous system can be affected in dementia explains the existence of the widespread sleep problems affecting patients. Disruptions in these regions can influence all the sleep disorders which dementia patients may exhibit: circadian rhythm disorders, difficulty in falling and staying asleep, multiple awakenings, respiratory disorders, periodic limb movement

disorder, and parasomnias such as REM behavioural disorder. Changes in sleep architecture, such as a reduction in REM sleep due to the effects of dementia on the cholinergic system, can also be observed. The cholinergic neurons of the nucleus basalis of Meynert and the basal forebrain are also active and are involved in the generation of REM sleep.[56] The nucleus basalis of Meynert is one of the first structures affected early in AD. This nucleus sends acetylcholine profusely to the cortex and the hippocampus. While EEG desynchronization during wakefulness is mainly produced by noradrenergic and serotonergic stimuli, during REM sleep it is generated by glutamatergic stimulation of the thalamus and cholinergics from the gigantocellular tegmental field, the septal nucleus, and the nucleus basalis of Meynert. The relationship between REM sleep and AD may not be random, but functional, since REM sleep is related to learning processes, which are compromised in dementia.[57] The nucleus basalis of Meynert activates the cortex and suppresses slow rhythms and timing of the reticulo-thalamic system. It has been suggested that in AD there is a reduction in the percentage of REM sleep, with increased theta and delta bands in the spectral analysis.[57,58] These changes in brain wave frequency can constitute an objective parameter to distinguish AD from normal aging.[59]

Many studies have shown that drugs which act on the cholinergic system can affect REM sleep. Studies with rivastigmine and galantamine have shown an increase in REM sleep. Donepezil may cause an increase in the percentage of REM sleep alongside cognitive improvement.[60] Other parameters of REM sleep such as latency, number of periods, atony, and REM density are not consistently affected, possibly due to the lower dependence of these variables on cholinergic neurons in the basal forebrain.[56]

Compromise of the SCN is associated with a significant impairment of the sleep-wake cycle, which is what many patients with dementia have. In AD there is an absence of physiological rhythmic expression of the genes related to biological rhythm. In addition, there is a progressive reduction in the production of melatonin by the pineal gland. Melatonin protects neuronal cells from the toxicity mediated by the β-amyloid via (through) its antioxidant and anti-amyloid properties, furthermore melatonin attenuates tau hyperphosphorylation. Melatonin regulates circadian rhythm, eliminates free radicals, improves immunity, and inhibits the oxidation of biomolecules.[53] It seems that the altered function of the brain stem pathways associated with sleep promotes a specific regional disintegration of the cortical-subcortical network, increasing the susceptibility for AD. The formation of a toxic state within this network, linked to the disruption of the sleep-wake cycle, could lead to a failure of synaptic repair with an increase in the transmission of pathogenic proteins and the neurodegenerative process.[60] Synaptic changes seem to play an important role in the pathogenesis of AD, and can promote the transfer of pathogenic β-amyloid molecules, with consequent dissemination of proteins such as hyperphosphorylated tau. An important synaptic phase change occurs due to the disruption of the circadian sleep-wake cycle.[61] Many sleep disorders that occur in AD can be attributed to this disruption of circadian cycles due to atrophy of the suprachiasmatic nucleus[62] and this disruption increases with disease progression. Changes in the rhythm of melatonin secretion can alter the sleep of patients with AD and exogenous melatonin administration, or the stimulation of its production through morning phototherapy, can help correct this.[63]

16.2.2 Sleep Disorders in Degenerative Dementias

A summary of sleep disorders in degenerative dementias is given in Table 16.6.

Table 16.6. Sleep Disorders in Degenerative Dementias.

1. Sleep architecture alterations
2. Changes in the circadian sleep-wake rhythm
3. Insomnia
4. Sundowning syndrome
4. Sleep-related breathing disorders
5. Periodic limb movement disorder
6. REM sleep behavior disorder

16.2.2.1 Sleep architecture disorders

Changes in sleep architecture in dementia are similar to those which occur in normal aging, but with greater intensity. There is a reduction in total sleep time and sleep efficiency, sleep-phase advance, a reduction in the amplitude of circadian cycles, a reduction in the percentage of slow wave sleep, and especially a reduction in the percentage of REM sleep.[64-67]

16.2.2.2 Changes in the circadian sleep-wake rhythm

The effect on the circadian sleep-wake rhythm is one of the most common and disabling symptoms in AD. Night walking, daytime naps and prolonged awakenings during the night, which are accentuated as the disease progresses, are all common in AD. Nocturnal restlessness is the symptom of dementia that is considered to cause most trouble to patient caregivers, and is one of the main causes of patients' admission to specialized care institutions.

16.2.2.3 Sundowning syndrome

One type of nocturnal restlessness is known as sundowning syndrome and is characterized by increased mental confusion and agitation in the evening, and increased awakenings during the night with consequent daytime sleepiness.

Many patients also suffer from episodes of unrest related to other phases of the circadian cycle, especially on awakening.[68] This agitation that occurs in the late afternoon and early evening is associated with chronobiological changes. Due to the disruption of the circadian cycle caused by atrophy of the suprachiasmatic nucleus, some patients reverse their cycle, with day sleepiness and restlessness at night.[62]

16.2.2.4 Insomnia

Insomnia is a common symptom in dementias. Other complicating conditions, such as nocturia due to prostatic hypertrophy in men and low urethral resistance in post-menopausal women, can ultimately lead to multiple night-time awakenings.[69] Environmental and psychological factors, such as social isolation, material impoverishment, lack of exposure to sunlight, and limited physical activity, can aggravate the insomnia of patients with dementia.[69]

Insomnia can also be complicated by the use of psychotropic medications, cardiovascular treatments, analgesics, infectious diseases, and chronic pain. Abuse of benzodiazepines (BZD) in the treatment of these patients is common. BZD induce tolerance and dependence, increase the risk of falls and worsen sleep apnea. There seems to be an association between BZD use in the elderly and being at higher risk of developing AD.[70]

16.2.2.5 Sleep-related breathing disorders

The association between AD and sleep apnea has often been studied. Sleep apnea has been linked to impairment of memory and executive functions. Compromised bulbar respiratory centers could be the causal factor of sleep apnea.[71] The association between dementia and apnea may be linked to a genetic base common to these two conditions.[72]

16.2.2.6 Periodic limb movement disorder

Periodic limb movements increase with aging and dementia. Periodic limb movement disorder can be one of the causes of sleep fragmentation and daytime sleepiness related to aging.[73] It is estimated that about 45% of individuals over 65 years of age have these kinds of movements, with most being asymptomatic. Many mechanisms present in aging appear to increase the frequency of periodic limb movements and, in parallel, restless legs syndrome. An important central mechanism is the dopaminergic deficit, which can be successfully corrected by the administration of L-dopa or dopamine agonists.

Systemic factors, such as uremia and iron deficiency, also contribute to the increased frequency of periodic limb movements, operating on central mechanisms. In the case of iron deficiency, the effect is mediated through the dopaminergic system, since iron is an important cofactor in the D2 dopamine receptor. Proprioceptive changes in the peripheral nervous system are also involved in the initiation of periodic limb movements, explaining its exacerbation in spinal canal stenosis and peripheral neuropathies. Local triggering factors, such as vascular insufficiency of the lower limbs and the osteoarthrosis, may act by this mechanism.[74]

16.2.2.7 REM sleep behavior disorder

RBD is characterized by the occurrence of REM sleep with a lack of the muscular atony that characterizes this stage of sleep. The result is that the patient experiences dreams, which are often stereotypical, described as being persecutory nightmares, often with violence and fighting. These can have tragic consequences as the patient may suffer serious accidents or unintentionally harm their bedmate.[3]

RBD is common in neurodegenerative disorders, especially synucleipathies like PD and Dementia with Lewy bodies[75–79] and may precede other symptoms of the disease.[67,80–82] RBD in dementia is related to impairment of mesopontine cholinergic neurons and locus coeruleus neurons. RBD may occur due to loss of cholinergic neurons in the basal forebrain and bridge, or the loss of their control, could lead to both clinical pictures of REM without atony, due to the intrusion of REM sleep during wakefulness.

16.3 TREATMENT

The treatment of insomnia in patients with dementia and in the elderly in general follows similar principles. Initially non-pharmacological treatments such as the physical exercise and phototherapy, good sleep hygiene, increased exposure to morning sunlight, daytime sleep restriction, and adjustment of circadian rhythm.[82] The importance of physical activity should not be underestimated, mild exercise during the day is enough to improve sleep consolidation in many patients with dementia. The patients should be kept awake during the day by being offered social activities and daytime naps should be avoided. Exposure to sunlight can be replaced by artificial phototherapy with intensity ≥ 2500 lux, depending on the time of exposure.

When there is a need for a pharmacological treatment for the purpose of inducing sleep, a selective GABA-A agonist such as zolpidem, should be used.

Treatment of nocturnal restlessness in patients with dementia depends on the cause. Most cases of nocturnal restless are probably due to circadian rhythm disorders. The drugs most commonly used for the treatment of nocturnal restlessness are the atypical antipsychotics, which in addition to reducing nocturnal restlessness can improve sleep. BZD should be avoided because of their risk of dependence, worsening of respiratory disorders and increased risk of accidents.

Some authors, based on the role of melatonin on dementia, suggest the use of melatonin in addition to phototherapy, in order to regulate a disturbed circadian rhythm.[53,63,84,85] Some studies suggesting that doses of 2 mg at bedtime were effective in the treatment of nocturnal restlessness and confusion, probably by acting on the chronobiological rhythms of patients with dementia.[62] RBD is a cause of nighttime agitation, especially in Lewy body dementia. Clonazepam is used as first-line therapy for RBD and melatonin for second-line therapy.[53,63,81,86]

Various other drugs have been used including antiepileptics, antidepressants, lithium carbonate, selegiline and beta-blockers. The clinical causes of nocturnal restlessness are varied and include medication, medical procedures, malnutrition, and infections.

When sleepiness caused by sleep apnea impairs the quality of life of patients, treatment with CPAP may be an option. The option of treating sleep apnea with continuous positive airway pressure devices (CPAP) in patients with dementia depends on the etiology of the dementia.[87,88] The use of CPAP can reduce the cognitive and behavioral deterioration of patients with AD.[88] The use of drugs which enhance cholinergic transmission such as donepezil, may improve sleep apnea in patients with AD.[71] The treatment of PLMS and RLS includes improved sleep hygiene and increased physical activity. The use of substances which may worsen the clinical picture, such as antidepressants should be investigated. The replacement of iron may be necessary. When these measures are not sufficient, the use of levodopa and dopamine agonists such as pramipexole can be effective, but with a risk of triggering or aggravating nocturnal restlessness.

In addition to the treatment of insomnia, nocturnal restlessness, changes to the intrinsic circadian rhythm and sleep disorders such as sleep apnea, PLMS, and RBD, obviously the treatment of sleep disorders in dementia implies the proper treatment of the source of these conditions, mainly through the use of medicines with an anticholinesterase action, with a consequent improvement not only in cognition, but also in sleep.[60] Sedating antidepressants are used when depression and its accompanying SD coexist with dementia. Antipsychotics are usually administrated to control the behavioral and neuropsychiatric manifestations of AD and sometimes for insomnia treatment after failure of all other measures (Table 16.7).[89]

Table 16.7. Management of Sleep Disorders in Degenerative Dementia.

1. Pharmacological treatment

    Melatonin

    Acetylcholinesterase inhibitors

    Antypsychotics

    Hypnotics

    Antidepressants

2. Non-pharmacological treatment

    Cognitive behavioral therapy

    Social strategies

KEY WORDS

- **sleep disorder**
- **Parkinson disease**
- **Alzheimer dementia**
- **rapid eye movement sleep behavior disorder**

REFERENCES

1. Arnulf, I. Excessive Daytime Sleepiness in Parkinsonism. *Sleep. Med. Rev.* **2005,** 9, 185–200.
2. Menza,M; Dobkin, R; Marin, H; Bienfait, K. Sleep Disturbances in Parkinson's Disease. *Mov. Disord.* **2010,** Vol. 25, Suppl. 1, 117–122.
3. Parkinson J. An Essay on the Shaking Palsy. London: Whittingham and Rowland for Sherwood. Neely and Jones, 1817.
4. Hardy,J; Gwinn-Hardy, K. Genetic Classification of Primary eurodegenerative Disease. *Science* **1998,** 282, 1075–1079.
5. Nishino, S; Ripley, B; Overeem, S; Lammers, GJ; Mignot, E. Hypocretin (orexin) Deficiency in Human Narcolepsy. *Lancet* **2000,** 355, 39–40.
6. Salawu,F.; Olokoba, A. Excessive Daytime Sleepiness and Unintended Sleep Episodes Associated with Parkinson's Disease. *Oman. Med. J.* **2015,** Vol. 30, No. 1, 3–10
7. Rye, D.B; Jankovic, J. Emerging Views of Dopamine in Modulating Sleep/ake State from an Unlikely Source: PD. *Neurology.* **2002,** 58, 341–6.
8. Dhawan,V; Healy, D.G; Pal, S; Chaudhur,R. Sleep-Related Problems of Parkinson's Disease. *Age Ageing.* **2006,** 35, 220–228.
9. Saper, C.B. *et al.* The Sleep Switch: The Hypothalamic Control of Sleep and Wakefulness. *Trends. Neurosci.* 2001, 24, 726–731.
10. Comella, C.L. Sleep Disorders in Parkinson's Disease: An Overview. *Mov. Disord.* Vol. 22, Suppl. 17, **2007,** 367–373
11. Jellinger, K.A. Lewy Body-Related Alpha-Synucleinopathy in the Aged Human Brain. *J. Neural. Transm.* **2004,** 111,1219–1235.
12. Zoccolella, S.; et al. Sleep Disorders and the Natural History of Parkinson's Disease:The Contribution of Epidemiological Studies. *Sleep. Med. Rev.* 15; **2011,** 41–50.
13. Olanow, C.; Schapira, A.; Roth, T. Waking Up to Sleep Episodes in Parkinson's Disease. *Mov. Disord.* **2000,** 15, 212–215.
14. Adler, C.H.; Thorpy, M.J. Sleep Issues in Parkinson's Disease. *Neurology.* **2005** Jun, 64(12) Suppl 3,12–20.
15. Trenkwalder, C. Sleep Dysfunction in Parkinson's Disease. *Clin. Neurosci.* **1998,** 5(2), 107–114.
16. Schenck, C.H.; Mahowald, M.W. REM Sleep Behavior Disorder: Clinical, Developmental, and Neuroscience Perspectives 16 Years after Its Formal Identification in Sleep. *Sleep* **2002,** 25, 120–138.
17. Postuma, R.B.; Gagnon, JF; Vendette, M; Fantini, ML; Massicotte-Marquez, J; Montplaisir, J. Quantifying the Risk of Neurodegenerative Disease in Idiopathic REM Sleep Behavior

Disorder. *Neurology.* **2009** Apr, 72(15),1296–1300.

18. Schenck, C.H.; Bundlie, S.R.; Mahowald, MW. Delayed Emergence of a Parkinsonian Disorder in 38% of 29 Older men Initially Diagnosed with Idiopathic Rapid Eye Movement Sleep Behavior Disorder. *Neurology* **1996**, 46, 388–393.

19. Factor, A.S.; McAlarney, T.; Sanchez-Ramos, J.R.; Weiner, W.J. Sleep Disorders and Sleep Effect in Parkinson's Disease. *Mov. Disord.* **1990**, 5, 280–285.

20. Nomura, T.; Inoue, Y.; Nakashima, K. Clinical Characteristics of Restless Legs Syndrome in Patients with Parkinson's Disease. *J. Neurol. Sci.* **2006**, 250, 39–44.

21. Loo, H.V., Tan E.K. Casecontrol Study of Restless Legs Syndrome and Quality of Sleep in Parkinson's Disease. *J. Neurol. Sci.* **2008**, 266,145–149.

22. Oerlemans W.G.; de Weerd, A.W. The revalence of Sleep Disorders in Patients with Parkinson's Disease: A Self-Reported, Community- Based Survey. *Sleep. Med.* **2002**, 3,147–149.

23. Arnulf,I.; Konofal, E.; Merino-Andreu, M. *et al.* Parkinson's Disease and Sleepiness – An Integral Part of PD. *Neurology* **2002**, 58, 1019–1024.

24. Silber, M.H.; Levine, S. Stridor and Death in Multiple System Atrophy. *Mov. Disord.* **2000**, 15, 699–704.

25. Baumann, C.; Ferini-Strambi, L.; Waldvogel, D.; Werth, E.; Bassetti, C.L. Parkinsonism with Excessive Daytime Sleepiness—a Narcolepsy-Like Disorder? *J. Neurol.* **2005**, 252,139–145.

26. Sinforiani, E.; Pacchetti, C.; Zangaglia, R.; Pasotti, C.; Manni, R.; Nappi. G. REM Behavior Disorder, Hallucinations and Cognitive Impairment in Parkinson's Disease: a Two-Year Follow Up. *Mov. Disord.* **2008**, 23,1441–1445.

27. Happe, S.; Berger, K.; FAQT Study Investigators. The Association of Dopamine Agonists with Daytime Sleepiness, Sleep Problems and Quality of Life in Patients with Parkinson's Disease e a Prospective Study. *J. Neurol.* **2001**, 248,1062–1067.

28. Gjerstad ,M.D.; Alves, G.; Wentzel-Larsen, T.; Aarsland, D.; Larsen, J.P. Excessive Daytime Sleepiness in Parkinson Disease: Is It the Drugs Or the Disease? *Neurology* **2006**, 67,853–858.

29. Van Hilten, J.J.; Weggeman, M.; van der Velde, E.A.; Kerkhof, G.A.; van Dijk, J.G.; Roos, R.A.C. Sleep, Excessive Daytime Sleepiness and Fatigue in Parkinson's Disease. *J. Neural. Transm. Park. Dis. Dement. Sect.* **1993**, 5(3), 235–244.

30. Paus, S.; Brecht, H.M.; Koster, J.; Seeger, G.; Klockgether, T.; Wullner, U. Sleep Attacks, Daytime Sleepiness, and Dopamine Agonists in Parkinson's Disease. *Mov. Disord.* **2003**, 18, 659–667.

31. Tan, E.; Lum, S.; Fook-Chong, S.; Teoh, M.; Yih Y.; Tan L.; Tan, A.; Wong, M. Evaluation of Somnolence in Parkinson's Disease:Comparison with Age- and Sex-Matched Controls. *Neurology.* **2002**, 58,465–468.

32. Montastruc, J.; Brefel-Courbon, C.; Senard, J.; Ferreira, J.; Rascol, O. Sleep Attacks and Antiparkinsonian Drugs: A Pilot Prospective Pharmacoepidemiologic study. *Clin. Neuropharmacol.* **2001**, 24,181–183.

33. American Academy of Sleep Medice (AASM). *International Classifications of Sleep Disorders* (ICDS), 3ª Ed; 2014.

34. Arnulf, I.; Bonnet, A.M.; Damier, P.; Bejjani, P.B.; Seilhean, D.; Derenne, J.P.; Agid, Y. Hallucinations, REM Sleep and Parkinson's Disease. *Neurology* **2000**, 55, 281–288.

35. Gagnon, J.F.; Bedard, M.A.; Fantini, M.L. et al. REM Sleep Behavior Disorder and REM Sleep Without Atonia in Parkinson's Disease. *Neurology* **2002**, 59, 585–589.

36. Pacchetti, C.; Manni, R.; Zangaglia. R. et al. Relationship Between Hallucinations, Delusions, and Rapid Eye Movement Sleep Behavior Disorder in Parkinson's Disease. *Mov. Disord.* **2005**, 20, 1439–1448.

37. Onofrj, M.; Thomas, A.; D'Andreamatteo, G. et al. Incidence of RBD and Hallucination in Patients Affected by Parkinson's Disease: 8-Year Follow-up. *Neurol. Sci.* **2002**, 23 (Suppl 2), 91–94.

38. Johns, M.H. A New Method for Measuring Daytime Sleepiness: the Epworth Sleepiness Scale. *Sleep* **1991**, 14, 540–545.

39. Johns, M.W. Sensitivity and Specificity of the Multiple Sleep Latency Test (MSLT), the Maintenance of Wakefulness Test and the Epworth Sleepiness Scale: Failure of the MSLT As a Gold Standard. *J. Sleep. Res.* **2000**, 9, 5–11.

40. Trenkwalder, C.; Kohnen, R.; Ho, B.; Metta, V.; Sixel-Doring, F.; Frauscher, B; lsmann, J.H.; Martinez-Martin, P.; Chaudhuri, K.R. Parkinson's Disease Sleep Scale—Validation of the Revised Version PDSS-2. *Mov. Disord.* **2011**, Vol. 26, No. 4, 644–652.

41. Boeve, B.F.; Silber, M.H.; Ferman, T.J. Melatonin for Treatment of REM Sleep Behavior Disorder in Neurologic Disorders: Results in14 Patients. *Sleep. Med.* **2003**, 4, 281–284.

42. Fantini, M.L.; Gagnon, J.F.; Filipini ,D.; Montplaisir, J. The Effects of Pramipexole in REM Sleep Behavior Disorder. Neurology **2003**, 61,1418–1420.

43. Winkelman, J.W.; James, L. Serotonergic Antidepressants are Associated with REM Sleep Without Atonia. *Sleep* **2004**, 27, 317–321.

44. Friedman, J.H.; Fernandez, H.H. Atypical Antipsychotics in Parkinson- Sensitive Populations. *J. Geriatr. Psychiatry. Neurol.* **2002**, 15,156–170.

45. Adler, C.H.; Caviness. J.N.; Hentz, J.G *et al.* Randomised Trial of Modafinil for Treating Subjective Daytime Sleepiness in Patients with Parkinson's Disease. *Mov Disord.* **2003**, 18, 287–293.

46. Priano, L.; Albani, G.; Brioschi. A. et al. Nocturnal Anomalous Movement Reduction and Sleep Microstructure Analysis in Parkinsonian Patients During 1-Night Transdermal Apomorphine Treatment. *Neurol. Sci.* **2003**, 24, 207–208.

47. Hjort, N.; Ostergaard. K.; Dupont, E. Improvement of Sleep Quality in Patients with Advanced Parkinson's Disease Treated with Deep Brain Btimulation of the Subthalamic Nucleus. *Mov. Disord.* **2004**,19, 196–199.

48. Postuma, R.B.; Lang, A.E.; Munhoz, R.P.; Charland, K.; Pelletier, A.; Moscovich, M. et al. Caffeine for Treatment of Parkinson Disease: A Randomized Controlled Trial. *Neurology* **2012** Aug, 79(7), 651–658.

49. Ju, Y-ES.; Lucey, B.P.; Holtzman, D. M. Sleep and Alzheimer Disease Pathology a Bidirectional Relationship. *Nat. Rev. Neurol.* **2014**, 10, 115–119.

50. American Psychiatric Association. *Diagnostic and Statistical Manual of Mental Disorders.* 5th ed. Washington, DC: American Psychiatric Association, 2013.

51. Moraes W.; Pinto, Jr. L.R. Transtorno do Sono nas Demências e na Doença de Alzheimer. In *Sono e Seus Transtornos – do Diagnóstico ao Tratamento*; Pinto Jr. L.R., Ed.; Atheneu: São Paulo, 2012; p 189.

52. Andersen M.; Pinto, Jr. L.R.; Tufik, S. O Sono Normal. In *Sono e Seus Transtornos – do Diagnóstico ao Tratamento*; Pinto Jr. L. R., Ed.; Atheneu: São Paulo, 2012; p 1.

53. Lin, L.; Huang, Q-X.; Yang, S-S.; Chu, J.; Wang, J-Z.; Tian, Q. Melatonin in Alzheimer's Disease. Int. J. Mol. Sci. 2013, 14, 14575–14593.

54. Lucey, B. P.; Bateman, R.J. Amyloid-β Diurnal Pattern: Possible Role in Alzheimer's Disease Pathogenesis. *Neurobiol. Aging* **2014**, 35, S29–S34.

55. Heininger, K. A Unifying Hypothesis of Alzheimer's Disease. II Pathophysiological Processes. *Hum. Psychopharm. Clin.* **1999**, 14, 525–581.

56. Montplaisir, J.; Petit, D.; Lorrain, D.; Gauthier, S.; Nielsen, T. Sleep in Alzheimer's Disease: Further Considerations on the Role of Brainstem and Forebrain Cholinergic Populations in Sleep-Wake Mechanisms. **1995**, *Sleep.* 13,145–148.

57. Christos, G.A. Is Alzheimer's Disease Related to a Deficit or Malfunction of Rapid Eye Movement (REM) Sleep. *Med. Hypotheses.* **1993**, 41, 435–439.

58. Petit, D.; Montplaisir, J.; Lorrain, D.; Gauthier, S. Spectral Analysis of the Rapid Eye Movement Sleep Electroencephalogram in Right and Left Temporal Regions: A Biological Marker of Alzheimer's Disease. *Ann. Neurol.* **1992**, 32,172–176.

59. Hassainia, F.; Petit, D.; Nielsen, T.; Gauthier, S.; Montplaisir, J. Quantitative EEG and Statistical Mapping of Wakefulness and REM Sleep in the Evaluation of Mild to Moderate Alzheimer's Disease. *Eur. Neurol.* **1997**, 37, 219–224.

60. Moraes, W. A. S.; Poyares, D. R.; Guilleminault, C.; Ramos, L. R.; Bertolucci, P.H.; Tufik, S. The Effect of Donepezil on Sleep and REM Sleep EEG in Patients with Alzheimer Disease: A Double-Blind Placebo-Controlled Study. *Sleep* **2006**, 29,199–205.

61. Clark, C. N.; Warren, J. D. A Hypnic hypothesis of Alzheimer's Disease. Neurodegener Dis. **2013**, 12,165–176.

62. Swaab, D. F.; Fliers, E.; Partiman, T. S. The Suprachiasmatic Nucleus of the Human Brain in Relation to Sex, Age, and Senile Dementia. *Brain Res.* **1985**, 342, 37–44.

63. Girardin, J.; Zizi, F.; Gizycki, H.; Taub, H. Effects of Melatonin in two Individuals with Alzheimer's Disease. *Percept. Mot. Skills.* **1998**, 87, 331–339.

64. Richardson, G. S.; Carskadon, M. A.; Orav, E. J; Dement, W. C. Circadian Variation of Sleep

Tendency in Elderly and Young Adult Subjects. *Sleep.* **1982,** 5 (Suppl 2), S82–94.

65. Aharon-Peretz, J.; Masiah, A.; Pillar, T.; Epstein, R.; Tzischinsky, O.; Lavie, P. Sleep-Wake Cycles in Multi-Infarct Dementia and Dementia of the Alzheimer Type. *Neurology* **1991,** M 41, 1616–1619.

66. Bliwise, D. L.; Carroll, J. S.; Lee, K. A. et al. Sleep and Sundowning in Nursing Home Patients with Dementia. *Psychiatry. Res.* **1993,** 48, 227–292.

67. Lucchesi, L. M.; Pradella-Hallinan, M.; Lucchesi, M.; Moraes, W.A. Sleep in Psychiatric Disorders. Rev. Bras. Psiquiatr. **2005,** 27 Suppl 1, 27–32.

68. Johnson, J. Delirium in the Elderly: Incidence, Diagnosis, Management, and Functional Status in General Medical Patients [abstract]. *Gerontologist* **1987,** 27, 243A.

69. Foley, D. J.; Monjan, A. A.; Brown, S. L.; Simonsick, E. M.; Wallace, R. B.; Blazer, D. G. Sleep Complaints Among Elderly Persons: An Epidemiologic Study of Three Communities. *Sleep* **1995,** 18, 425-432.

70. Gage, S. B.; Moride, Y.; Ducruet, T et al. Benzodiazepine Use and Risk of Alzheimer's Disease: Case-Control Study. *BMJ.* **2014,** 349, 1–10.

71. Presti, M. F.; Schmeichel, A. M.; Low, P. A.; Parisi, J. E.; Benarrochi, E. M. Degeneration of Braistem Respiratory Neurons in Dementia with Lewy Bodies. *Sleep* **2014,** 37, 373–378.

72. Moraes, W.; Poyares, D.; Sukys-Claudino, L.; Guilleminault, C.; Tufik S. Donepezil Improves Obstructive Sleep Apnea in Alzheimer Disease: A Double-Blind Placebo-Controlled Study. *Chest* **2008,** 133, 677–83.

73. Feinberg, I. Changes in Sleep Cycle Patterns with Age. *J. Psychiatr. Res.* **1974,** 10, 283–306.

74. Rongve, A.; Boeve, B. F.; Aarsland, D. Frequency and Correlates of Caregiver-Reported Sleep Disturbances in a Sample of Persons with Early Dementia. *J. Am. Geriatr. Soc.* **2010,** 58, 480–486.

75. Shenck, C.H.; Bundlie, S.R.; Ettinger, M.G. Chronic Behavioral Disorder of Human REM Sleep: A New Category of Parasomnia. *Sleep.* **1986,** 9, 293–308.

76. Turner, R. S.; Chervin, R. D.; Frey, K. A.; Minoshima, S.; Kuhl, D. E. Probable Diffuse Lewy Body Disease Presenting as REM Sleep Behavior Disorder. *Neurology* **1997,** 49, 523–527.

77. Boeve, B. F.; Silber, MI-I.; Ferman, U. et al. REM Sleep Behavior Disorder and Degenerative Dementia. *Neurology* **1998,** 51, 363–370.

78. Boeve, B.F.; Silber, M.H.; Ferman, T.J. et al. Clinicopathologic Correlations in 172 Cases of Rapid Eye Movement Sleep Behavior Disorder with or Without a Coexisting Neurologic Disorder. *Sleep Med.* **2013,** 14, 754–762

79. Jennum, P.; Mayer, G.; Postuma, R. Morbidities in Rapid Eye Movement Sleep Behavior Disorders. *Sleep. Med.* **2013,** 14, 782–787.

80. Iranzo, A.; Fernandez-Arcos, A.; Tolosa, E. et al. Neurodegenerative Disorder Risk in Idiopathic REM Sleep Behavior Disorder: Study in 174 Patients. *PLoS One.* **2014,** 9, e89741.

81. Trotti, L. M. REM Sleep Behavior Disorder in Older Individuals: Epidemiology, Pathophysiology, and Management. *Drugs Aging* **2010,** 27, 457–470.

82. Claassen, D. O.; Josephs, K. A.; Ahlskog, J. E.; Silber, M. H.; Tippmann-Peikert, M.; Boeve, B. F. REM Sleep Behavior Disorder Preceding Other Aspects of Synucleipathies bu Up to Half Century. *Neurology* **2010,** 75, 494–499.

83. Shub, D.; Darvishi, R.; Kunik, M. E. Non-Pharmacologic Treatment of Insomnia in Persons with Dementia. *Geriatrics* **2009,** 64, 22–26.

84. Salami, O.; Lyketsos, C.; Rao, V. Treatment of Sleep Disturbance in Alzheimer's Dementia. *Int. J. Geriatr. Psychiatry* **2011,** 26, 771–782.

85. Hanford, N.; Figueiro, M. Light therapy and Alzheimer's disease and related dementia: past, present, and future. J. Alzheimers Dis. **2013,** 33, 913-922.

86. Anderson, K. N.; Jamieson, S.; Graham. A. J.; Shneerson, J. M. REM Sleep Behaviour Disorder Treated with Melatonin in a Patient with Alzheimer's Disease. *Clin. Neurol. Neurosurg.* **2008,** 110, 492–495.

87. Diomedi, M.; Placidi, F.; Cupini, L. M. et al. Cerebral Hemodynamic Changes in Sleep Apnea Syndrome and Effect of Continuous Positive Airway Pressure Treatment. *Neurology* **1998,** 51, l051–1056.

88. Ancoli-Israel, S. Sustained Use of CPAP Slows Deterioration of Cognition, Sleep, and Mood in Patients with Alzheimer's Disease and Obstructive Sleep Apnea: A Preliminary Study. *J Clin. Sleep Med.* **2009,** 5, 305–309.

89. Peter-Derex, L; Yammine, P; Bastuji, H; Croisile, B. Sleep and Alzheimer's Disease . *Sleep Med. Rev.* **2015,** 19, 29–38.

17

NARCOLEPSY

Lawrence Scrima and Todd J. Swick

ABSTRACT

Narcolepsy means "to be seized by sleep" and is a relatively rare neurological sleep disorder that is the bane of those afflicted, yet provides a window on the neuromechanisms and functions of sleep. This chapter reviews and examines narcolepsy characteristics, primarily excessive daytime sleepiness (EDS), cataplexy (sudden loss of muscle tone during strong emotions) and disrupted nocturnal sleep, as well as hypnagogic/hypnogogic hallucinations; incidence, differentiating tests, and appropriate treatment options. Fundamental and promising research is reviewed on the causes, treatments, and neuromechanisms of narcolepsy and cataplexy. The impact of narcolepsy on day-to-day living has substantial morbidity and reductions in health-related quality of life, whereas optimal treatment can have positive life altering benefit. Physicians' lack of understanding of narcolepsy is reviewed, as well as their obligations to patients and the public. Sources of information for patients are provided.

Lawrence Scrima
Sleep Expert Consultants, LLC, 15011 E Arkansas Dr., B, Aurora CO 80012, USA
Fellow, American Academy of Sleep Medicine, Darien, IL, USA

Todd J. Swick
Fellow, American Academy of Neurology, Minneapolis, MN, USA
Fellow, American Academy of Sleep Medicine, Darien, IL, USA
School of Medicine, University of Texas Health Sciences Center-Houston, Houston, TX, USA
Neurology and Sleep Medicine Consultants, 7500 San Felipe, Houston, TX 77063, USA6
North Cypress Medical Center Sleep Disorders Center, 21214 NW Freeway Cypress, TX 77429, USA
Apnix Sleep Diagnostics Laboratories, Houston, TX, USA

Corresponding author: Lawrence Scrima,
E-mail: scrimasleepdoc@msn.com

17.1. INTRODUCTION

Narcolepsy is life-long, neurologic, sleep disorder characterized by: excessive daytime sleepiness (EDS), cataplexy, hypnagogic/hypnogogic hallucinations (both spelllings are used), sleep paralysis, and disrupted nocturnal sleep (DNS) (the narcolepsy "pentad").[1] From a mechanistic standpoint, the syndrome represents a dysregulation of the wake and sleep (rapid eye movements (REM or R) and non-REM(NREM or NR)) states.[2] It affects approximately 0.05% of the general population, with incidence rates depending on ethnicity and geographical location.[3,4] Narcolepsy is associated with significant and substantial morbidity and reductions in health-related quality of life (QOL).[5]

It has been 138 years since Westphal, in 1877, first published a description of cases of severe daytime sleepiness and sudden attacks of muscle weakness triggered by emotions.[6] In 1880, Gélineau named the condition "narcolepsy" (from the Greek word "narcosis" meaning drowsiness and "lambanein" meaning to seize, or take).[6] Following these initial case descriptions, the attempt to categorize the condition has gone through several iterations. In 1902, Lowenfeld was the first to identify that cataplexy was a separate component of narcolepsy and was independent of the sleep attacks.[7] In 1916, Henneberg was the first to name attacks of muscle weakness, "cataplectic inhibition"[6] and, in 1926, Adie changed that term to "cataplexy", from the Latin word "cataplessa", which means, "to strike down with fear".[8] In 1934, Daniels described hypnagogic sleep paralysis and noted that narcoleptics have lower pulse, basal metabolism, and body temperature than normal.[9] In 1957, Yoss and Daly described the clinical "tetrad" which included hypersomnia, cataplexy, hypnagogic hallucinations, and sleep paralysis.[10] A fifth component, disrupted night time sleep, though often noted, was only recently added to form the current pentad of symptoms that make up narcolepsy.[11]

Kleitman hypothesized that the carotid sinus seems implicated in the etiology of narcolepsy and cataplexy.[12] More recently, this hypothesis was extended to include the pressor–depressor neuro-regulatory mechanism as involved in triggering cataplexy, pressor trigger of cataplexy (PTC),[33] given several prerequisites, that is, baroreceptors reset by low blood pressure, REM sleep deprivation and approximately 90% hypocretin (Hcrt) cell loss, with growing research support, as recently reviewed.[13,14]

Narcolepsy is considered an "orphan disease" in the USA, that is, affecting less than 200 000 individuals, however this may be an underestimate due to a high number of individuals that have been misdiagnosed or continue to go undiagnosed.[3] In western countries (North America and Europe), narcolepsy has an estimated prevalence of 40/100 000 inhabitants.[3] However, narcolepsy is not uniform in its distribution. The highest reported prevalence is in Japan and the lowest is in Israel (native born Israelis as opposed to those who have emigrated to Israel).[15] The incidence of narcolepsy is also difficult to pin down,[33] and recently has been estimated to be approximately 1.4/100 000 person-years.[3] There is no sex preference and the age of onset has a peak in adolescence at 14.7 years with a second smaller peak at age 35.[16] The youngest onset of narcoleptic symptoms has been described in a child at age 2 and in the case of a Hcrt gene mutation at age 6 months.[17]

Hypersomnolence or EDS, defined as an irrepressible urge or need to sleep or daytime lapses into sleep (sleep attacks), usually occurs as the first independent symptom of narcolepsy. The onset of cataplexy vis-a-vis the onset of daytime sleepiness is highly correlated to age of onset. In the Stanford University Sleep Disorders Clinic study of 100 patients between the ages of 14–23, cataplexy was present simultaneously with EDS in 49% of the cases. Onset of cataplexy occurred in 41% of the subjects between 6 months to 2 years following the onset of EDS. In 4% of the subjects, cataplexy developed between 2 and 6 years after the onset of EDS. Cataplexy preceded the onset of sleepiness by 6 months to 3 years in 6% of the subjects.[18]

The research into the pathophysiology of narcolepsy has been greatly accelerated following the discovery of a canine model[19] with evidence of monogenetic autosomal recessive transmission.[20] In 1998, the discovery of the orexin/Hcrt system revolutionized research in the control of sleep/wake states.[21,22] Within 1 year, researchers identified the cause of narcolepsy in dogs with the identification of an autosomal recessive mutation of the Hcrt receptor 2 gene. In 2000, a seminal paper was published that established the pathophysiology of human narcolepsy/cataplexy as the critical reduction in cerebrospinal fluid (CSF) Hcrt/orexin[23] and the near total absence of Hcrt/orexin producing cells in the postero-lateral hypothalamus[24] sparing melanin-concentrating hormone neurons which are co-localized with the Hcrt/orexin cells in the hypothalamus.[25] Thus, the loss of Hcrt/orexin neurons is selective and given the high association between narcolepsy with cataplexy and the presence of human leucocyte antigen (HLA) DQB1*06:02, an antigen responsible for immune regulation, the most likely explanation is that the loss of the Hcrt/orexin cells comes about as a result of an "autoimmune attack" directly on these neurons.

17.2. DIAGNOSTIC CLASSIFICATIONS

The clinical manifestations of narcolepsy, include EDS (with or without inadvertent sleep attacks), cataplexy, hypnagogic hallucinations, sleep paralysis, and disrupted nocturnal sleep. The diagnostic nosology has undergone several changes since the first description of the "narcolepsy tetrad" by Yoss and Daly in 1957. In 1979 the Association of Sleep Disorder's Centers (the predecessor to the American Academy of Sleep Medicine(AASM)) published the first edition of Diagnostic Classification of Sleep and Arousal Disorders and organized the sleep disorders into symptomatic categories. Narcolepsy was considered a unitary disease and no distinction was made for those with cataplexy and those without. In 1990, the International Classification of Sleep Disorders (ICSD) was published with input from the international sleep societies (American Sleep Disorders Association, European Sleep Research Society, the Japanese Society of Sleep Research, and the Latin American Sleep Society). In 2005, the ICSD underwent updates and modifications with publication of version 2 (ICSD-2)[26,27] and in 2014, the third edition of ICSD was released.[1] The classification of narcolepsy has been revised with each update. In the 2005 second edition of the ICSD, the nosology split narcolepsy into four categories: narcolepsy with cataplexy, narcolepsy without cataplexy, narcolepsy due to medical condition and narcolepsy, unspecified. All require the presence of EDS on a daily basis for at least 3 months.

The essential diagnostic tests for all narcolepsy categories were listed including, an overnight polysomnogram (PSG) to exclude other sleep disorders that could cause daytime sleepiness (e.g., obstructive sleep apnea syndrome, periodic limb movements of sleep, insufficient sleep, circadian rhythm disorders, etc.), followed by a multiple sleep latency test (MSLT), demonstrating a mean sleep latency of < 8 min and two or more sleep onset REM periods (SOREMP). The PSG/MSLT test should be done after documentation of 2 weeks of "regular" sleep. The MSLT should not be carried out if the PSG demonstrated < 6 h of sleep or if there was significant sleep disordered breathing or moderate to severe periodic limb movements. The determination of HLA status was mentioned in the context that a positive HLA DQB1*06:02 genotype is not diagnostic of narcolepsy, but might be useful to exclude a diagnosis in "very" selected cases or when a secondary etiology is suspected.[27]

The diagnosis of narcolepsy without cataplexy requires the same objective sleep testing as narcolepsy with cataplexy, with the exception that there is no clinical evidence of cataplexy. In cases where there is EDS, no cataplexy and negative PSG/MSLT tests, the diagnosis defaults to idiopathic hypersomnia (IH) with or without long sleep time (Table 17.1).[28]

In 2014, the ICSD 3rd edition changed the classification of narcolepsy into Narcolepsy Type 1 (Narcolepsy with cataplexy, and either low or absent CSF Hcrt-1 or a positive PSG/MSLT) and Narcolepsy Type 2 (narcolepsy without cataplexy and a positive PSG/MSLT). [See section on PSG/MSLT below for definition of a positive result.]

The designation was made to help differentiate those patients exhibiting narcolepsy with cataplexy and presumed Hcrt deficiency from patients with narcolepsy conditions that either do not have the associated Hcrt loss and/or there is no

Table 17.1. ICSD-3 AASM 2014[1] criteria for narcolepsy with and without cataplexy.

ICSD Dx Criteria:

Type I. Narcolepsy w Cataplexy

A. Complaint of EDS almost daily for ≥ 3 months

B. Cataplexy sudden muscle weakness triggered by emotions

C. Confirmed by Polysomnographic findings ≥ 6 hrs :
     sleep onset latency <8 min & REM Latency ≤15 min
     & MSLT findings: mean sleep latency <8 min, ≥2 SOREMPs   or: from the
previous night PSG: a REM sleep latency of < 15 min
                                        + 1 REM Nap on MSLT

   Or: Hypocretin-1 levels, < 110 pg/mL a third of normal 90% CI

D. Not due to other sleep, medical, neurological or mental disorders, and not due to medications, or substance use disorder.

Type 2. W/out Cataplexy: A + C + D

Type 3. Due to Medical Condition: A + B or C + D w Med/Neuro

ICSD = International Classification of Sleep Disorders; EDS = excessive daytime sleepiness;
MSLT = Multiple Sleep Latency Test; SOREMP = sleep-onset REM period.
Amer Acad Sleep Med 2014), Intern Classif. Sleep Disorders, 3rd ed.

clinically evident cataplexy. If an individual has EDS and low or absent CSF Hcrt-1(≤ 110 pg/ml or 1/3 of normal), that individual is classified as having Narcolepsy Type 1, even if they do not exhibit cataplexy.[1] The thinking was that patients with low or absent CSF Hcrt, EDS and / or cataplexy comprise a specific disease population with a unitary etiology and reproducible clinical and polysomnographic features (Fig. 17.1).[1]

As noted previously, the onset of symptoms and the time of the accurate diagnosis is usually significantly delayed. It has been postulated that the delay results from lack of diagnostic consideration (symptoms of narcolepsy may be suggestive of other conditions for which there is greater awareness) and the presence of cataplexy should not be the sole criteria for establishing the diagnosis.[29,30] In 2014, a published survey of 1000 US adults, 300 primary care physicians (PCPs) and 100 sleep medicine specialists (36% self-designated as "board

certified").[31] Overall, only 62% of the sleep specialists and 24% of the PCPs considered themselves to be very or extremely knowledgeable about narcolepsy and furthermore, only 42% of the sleep specialists and 9% of the PCPs felt "very"" or "extremely" comfortable diagnosing the disorder. Of greater concern was only 22% of the sleep specialists and 7% of the PCPs were able to identify all five of the key symptoms. The conclusion was that substantial gaps exist in recognizing narcolepsy, not only among general practitioners, but the training and knowledge base of sleep physicians is very poor.[31]

17.2.1. Diagnosis

Narcolepsy is a life-long disorder once it develops. The chief presenting complaint is excessive sleepiness[1,32,33] for at least 3 months, which can be screened in the office using the Epworth Sleepiness Scale[34] and checking for a history of excessive

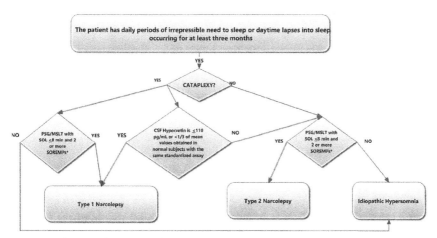

* The Sleep Onset REM period (SOREMP) can occur during the overnight PSG and then only one additional SOREMP on the MSLT is required for a positive evaluation

Figure 17.1. Diagnostic decision flow diagram for narcolepsy based on ICSD-3.[1]

sleepiness, such as falling asleep at inappropriate times, for example, while driving, waiting for a light to change, as a passenger during a short car ride, or during business meetings. The medical history should also assess for use of central nervous system (CNS)-depressant medications (including antidepressants), sleep apnea, shift work, insufficient sleep, and other potential causes of hypersomnolence.[1] Often narcolepsy patients have additonal sleep disorders, such as periodic limb movement disorder (PLMD), REM sleep behavior disorder (RBD), and/or obstructive sleep apnea (OSA), requiring appropriate treatment(s)[1]. Narcolepsy most often begins with the symptom of EDS between the ages of 15 and 25, rarely before age 5, and occasionally in senior adults[1]. Typically, cataplexy develops within 5–10 years after the onset of hypersomnolence in narcolepsy patients.[30,35,36]

Table 17.2. Differential diagnosis of central hypersomnias.[1]

DIFFERENTIAL DIAGNOSIS OF CENTRAL HYPERSOMNIAS

- Chronic sleep deprivation
- Sleep Apnea
- Narcolepsy with Cataplexy
- Narcolepsy without Cataplexy
- Secondary narcolepsy (due to a primary neurologic or medical condition)
- Idiopathic Hypersomnia (with and without long sleep time)
- Recurrent hypersomnia (Kleine-Levin syndrome)
- Restless Legs Syndrome/PLMDs
- REM sleep behavior disorder and associated conditions
  - Parkinson disease
  - PSP (progressive supranuclear palsy)
  - Alzheimer's
  - Multiple Sclerosis
  - Stroke
- Myotonic Dystrophy
- Prader-Willi syndrome

Table 17.3. Signs and symptoms of narcolepsy/cataplexy[1].

- Classic Pentad:
  - Excessive Daytime Sleepiness
  - Cataplexy
  - Hypnogogic hallucinations
  - Sleep paralysis
  - Sleep fragmentation

(Only about 10% of patients exhibit all 5 features concurrently)

However, in a small percent of patients, cataplexy or one or more of the other REM sleep related phenomena occur as the first symptom[1]. Patients with the chief complaint of EDS, in whom narcolepsy is a strong diagnostic possibility, have to be evaluated in a rigorous manner. A full history, including information to identify REM-related phenomena such as cataplexy, sleep paralysis, and hypnagogic hallucinations, needs to be thoroughly evaluated. The only confirmatory test is a CSF analysis of low or absent Hcrt-1 (≤ 110 pg/ml or 1/3 of normal), however this is not a readily available clinical test and as such one has to rely on sleep testing, the overnight PSG followed by a MSLT to diagnose or rule out other sleep disorders and narcolepsy[1].

17.2.2 Objective Evaluation

17.2.2.1. Use of Polysomnography

The overnight PSG is used to diagnose or rule-out sleep disorders that can directly cause EDS by fragmenting and reducing total nocturnal sleep (e.g., due to moderate to severe OSA, moderate to severe PLMDs, etc.).1 It needs to be emphasized that narcolepsy has significant comorbidity with OSA and PLMD, as well as with RBD.2,14 One of the most important aspects of the overnight PSG is to establish that the patient had at least 6 h of sleep prior to the MSLT, in that inadequate or fragmented sleep just prior to the MSLT, can give spurious results, in terms of both the sleep onset latency of the naps and the presence of SOREMP (Table 17.4).[37]

Table 17.4. Methodology for the overnight PSG & MSLT.

PSG:

A. To differentiate wake and sleep stages:
    a. Electroencephalogram (EEG): 2 frontal leads (right and left), 2 central leads (right and left) and 2 occipital leads (right and left).
    b. Electroocculograms (EOG, eye movements)
    c. Submental electromyogram (EMG);
B. Evaluate for primary sleep disorders (e.g., sleep apnea, periodic limb movements of sleep, parasomnias, etc.):
    a. Pulse oximetry (SpO2) to monitor oxygen saturation
    b. Electrocardiogram lead II (EKG/ECG),
    c. Airflow at the nose and mouth (pressure gauge and/or thermistor)
    d. Thoracic and abdomen effort
    e. Leg EMG (to monitor right and left leg movement)
    f. Snoring sensor
    g. Body position sensor (to record sleeping positions).

ICSD 3rd Edition. AASM 2014[1]. Iber C et al. The AASM Manual for the scoring of sleep and associated events. AASM 2007.[32]
MSLT: To differentiate wake and sleep stages a. Electro-encephalogram (EEG): 2 frontal leads (right, left), 2 central leads (right and left) and 2 occipital leads (right, left) b. Electro-occulograms (EOG, eye movements) c. Submental electro-myogram (EMG) d. Electro-cardiogram (ECG)
ICSD 3rd Edition. AASM 2014[1]. Arand D et al. Clinical use of the MSLT and MWT. Sleep 2005, 28(1) 123-144.[37]

17.2.2.2. Use of Multiple Sleep Latency Test

The purpose of the MSLT is to ascertain the degree of sleepiness the patient exhibits over the course of their waking day and to establish the presence or absence of REM onset sleep. The MSLT is a series of daytime nap tests, starting 1.5–3 h after an overnight PSG test, with each nap opportunity in a comfortable darkened bedroom, lasting 20–35 min, repeated five times, every 2 h, and each nap ending after 20 min of no sleep or 15 min after sleep onset without any evidence of REM sleep. For an MSLT diagnosis of hypersomnolence with "narcolepsy," the average latency to sleep onset must be 8 min or less, and at least two out of the five MSLT naps demonstrating REM sleep (SOREMP) starting within 15 min of the first epoch of sleep or one SOREMP on the overnight PSG test, with one SOREMP on the daytime MSLT as revised recently in 2014.[1,38]

If OSA is diagnosed during the PSG (as a prelude to the MSLT), then the MSLT should be cancelled (unless information about the degree of EDS is necessary). The OSA should be treated for 6–8 weeks (or more) to obtain a valid MSLT test result (compliance and efficacy with the OSA treatment modality needs to be verified).

If the patient's EDS, remits with treatment of OSA, then the likelihood of narcolepsy is decreased, which would be supported by a MSLT result negative for narcolepsy and perhaps also for EDS.

Also, the patient should keep a sleep–wake-symptom log for 1–2 weeks prior to the PSG/MSLT test, to ensure that there is adequate sleep and that the sleep/wake times are consistent and allowing the sleep clinic to schedule the PSG to start and end close to the patient's usual times (wrist worn actigraphy can easily substitute for sleep logs, but availability and insurance coverage is limited). A urine drug screen obtained at the conclusion of the MSLT naps may be useful to rule out unreported CNS drug use that would also invalidate the diagnostic MSLT.

It is sometimes necessary to repeat the MSLT test, because of the aforementioned confounding issues, to confirm the diagnosis of narcolepsy, especially, if the patient was using CNS medication, was exceptionally nervous, or did not have sufficient sleep (6–7 h or more) during the prior PSG test, or if there is one or more untreated comorbidities, for example, OSA.

Patients with sleep apnea or depression can sometimes exhibit false positive or false negative MSLT results, due to prior REM sleep deprivation (e.g., frequent apneas occurring during REM sleep can lower the REM sleep threshold), and also REM sleep suppressing CNS medications, such as antidepressants, and barbiturates. Sedatives can cause false positive shortened sleep latencies on the MSLT (thus the suggestion that a urine drug screen be obtained following the conclusion of the MSLT, to be certain there were no nefarious actions to influence the outcome of the test).

Excessive sleepiness due to OSA typically remits after the patient receives optimal continuous positive airway pressure (CPAP) treatment, but if excessive sleepiness persists with OSA treated, then the patient also could have narcolepsy or other disorders that promote this symptom.

Ideally, if the patient is taking a CNS active medication(s), they should stop using them for at least 2 weeks (i.e., at least five times the half-life of the medication(s) since some have very long half-lives and may metabolize even slower among senior citizens), prior to the PSG/MSLT test, to allow REM sleep to rebound and re-stabilize and to prevent a false positive or negative result.[1] Since CNS-depressant and/or anti-depressant medication(s) can promote hypersomnolence, PLMD and some REM sleep-related symptoms (e.g., RBD),[2,14,33] some patients with these symptoms may have improvement with withdrawal of these medications, after washout and rebound of REM sleep have finished; sometimes observed in preparation of the MSLT.

Besides EDS and cataplexy, symptoms of narcolepsy can include one or more of the following: sudden daytime "sleep attacks", with or without REM sleep or other REM sleep phenomena, that is, sleep paralysis, hypnagogic hallucinations, and/or RBD, and poor sleep maintenance. REM sleep-related phenomena can occur in patients without narcolepsy, but in narcoleptics, especially with cataplexy, they tend to be more frequent and intense, with greater propensity for REM sleep intrusion into wakefulness.[14]

Narcoleptics have a strong pressure for REM sleep[3,14,33,57] evidenced by frequent short REM onset latency from sleep onset (typically ≤ 15 min). In Figure 17.2, a 24 h hypnograph or a typical normal versus an untreated narcoleptic wake and sleep

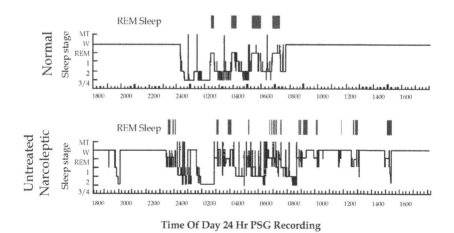

Time Of Day 24 Hr PSG Recording

Adapted from Rogers et al. Sleep. 1994;17:590.

Figure 17.2. Normal versus Untreated-Narcoleptic 24 h hypnograph.[39]

stages is displayed and contrasted. Normal adults have their first REM episode of the night approximately 90 min after sleep onset, with REM sleep cycling approximately every 90 min and each REM episode trending to be longer as the night progresses. In contrast, narcolepsy patients exhibit an abnormal predominance of REM sleep during the first half of nocturnal sleep and often experience intense, vivid, story-like dreaming during their REM sleep periods, that can also occur during daytime short naps or sleep attacks.[1,13,14,33]

Patients with narcolepsy can also have OSA, especially if they are overweight (it has been estimated that there is a 20–25% comorbidity as defined by an AHI > 10).[40] Central sleep apnea comorbidity is also possible, especially in elderly patients, or those with a history of head injury, stroke, use of opiates, or living at higher elevations.[14,41] There is also research suggestive of a high comorbidity with PLMDs.[42]

The diagnosis of narcolepsy without cataplexy requires the same objective sleep testing as narcolepsy with cataplexy, with the exception there is no clinical evidence of cataplexy. As such, the MSLT becomes the critical test to differentiate Type 2 Narcolepsy from IH. The reliability of the MSLT to accurately diagnose narcolepsy rests on the mean sleep onset latencies and the identification of SOREMP, previously described. However, a recent study reported poor inter-test stability in clinical practice, evaluated in tertiary sleep clinics.[43] Diagnoses changed in half the patients over an average of 4 years, raising the question of how much reliance should we place on the MSLT when we don't really know the etiology of narcolepsy without cataplexy, vis-a-vis IH. Moreover, MSLTs are easily

compromised, if all of the requirements of a valid MSLT are not met.

Besides chronic or intermittent hypersomnolence, intrusion of REM sleep related phenomena into wakefulness is the most defining symptom of narcolepsy, i.e., cataplexy, hypnagogic hallucinations, and sleep paralysis.[1] Narcoleptics have a strong pressure for REM sleep, evidenced by frequent short latency from sleep onset (typically ≤ 15 min) to the first REM sleep period (normally it takes about 90 min), a predominance of REM sleep during the first half of nocturnal sleep (normally REM sleep dominates during the last half of nocturnal sleep, see Fig. 17.2), and often having REM sleep and story-like dreaming occurring during short naps or sleep attacks. [1,13,14,32,33,44]

Although most OSA patients have EDS and some also have a strong pressure for REM sleep (short latency to REM sleep and intense dreaming), the REM sleep abnormalities in OSA patients usually normalize with effective treatment for OSA (e.g., with CPAP treatment), unless narcolepsy or other comorbid medical condition(s) are present, for example, depression, use of CNS medications and so forth. Cataplexy and sleep paralysis share the same profound generalized muscle atonia that is a defining feature of normal REM sleep. Hypnagogic hallucination is a term used to describe fleeting, usually visual, but also can be other sensory perceptions or mentation, not story-like or emotional, during the normal transition to sleep, and can be accompanied by a sensation of falling (perhaps due to the muscle relaxation that occurs at sleep onset) that may prompt a startle response and arousal. In contrast to "normal" hypnagogic hallucinations, those of narcoleptics have abrupt on-set, are more intense, visually vivid, more story-like, bizarre, long lasting and can be more frightening, with a feeling like it is really happening.[1] Most likely, the narcoleptic hypnagogic hallucination is a REM sleep dream breaking into wakefulness, causing difficulty distinguishing it from awake reality.[1,44,45,46,47,48] The narcoleptic hypnagogic hallucination and SOREMP can include the perception of sleep paralysis, since muscle atonia accompanies REM sleep[1], adding to the frightening effect, that can be abruptly broken with touch or with continued effort to move. Hypnopompic hallucinations occur emerging from sleep, in people with or without narcolepsy, and are more likely to have dream story-like mentation and may include sleep paralysis, as the person emerges from REM sleep.[1] Therefore, intense, dream-like hypnagogic hallucinations and sleep paralysis at the start of sleep is more likely a differentiating symptom of narcolepsy than hypnopompic hallucinations with or without sleep paralysis at the end of the night.

17.2.3. Genetic Predisposition and Other Physiological, Behavioral and Etiological Considerations

Narcolepsy in humans, unlike other species (e.g., dogs where it is an autosomal recessive trait) is not necessarily a genetic disorder.[33] The risk of a first degree relative of a patient having narcolepsy/cataplexy is 1–2%. However, a risk of 1–2% is 10–40 times higher than the prevalence observed in the general population, strongly suggesting the presence of predisposing genetic factors.[49] These predisposing factors are most likely related to the strong association

of narcolepsy with cataplexy (Narcolepsy Type 1) with the HLA subtypes DR2/DRB1*15:01 and DQB1*06:02.[50] Familial forms of narcolepsy with cataplexy are quite rare and probably represent only 1–2% of all cases of narcolepsy. Even less common are families with more than two affected individuals.[51–53]

Further evidence that narcolepsy with or without cataplexy is not genetically determined comes from the twin studies looking for concordance in monozygotic (MZ) twins. In a large Finnish study, almost 14 000 MZ and same-sex dizygotic (DZ) twin pairs were evaluated for narcolepsy with or without cataplexy. Three individuals were found and each of them was discordant DZ with a negative family history.[54] In several studies of MZ twins,[49] there was a concordance of only 25–30%, emphasizing the importance and presence of environmental or triggering factors.[50,51]

Narcolepsy with cataplexy has long been thought to have an autoimmune pathogenesis. As described above, the disorder has an extremely tight association with HLA-DQB1*06:02. The HLA genes, also called major histocompatibility (MHC) genes are located on human chromosome 6. The HLA system has been shown to be an integral part of the immune system response to infection as well as strongly associated with autoimmune diseases. Epidemiologic studies have increasingly recognized the association between exposure to infectious illnesses, such as influenza (i.e. H1N1) and streptococcus, and narcolepsy- cataplexy. Following the 2009 pandemic of H1N1, there was an apparent increase in onset of narcolepsy, particularly in young children. In China, where there was on ongoing research to study the epidemiology of narcolepsy/cataplexy, there was a 3–5 times increase in the number of children with narcolepsy compared to previous years. The peak was noted to occur 4–6 months after the peak of the H1N1 infections.[55] In Sweden, there was an increase in children diagnosed with narcolepsy/cataplexy following vaccination for the H1N1 strain of influenza.[56] There has been evidence to suggest that streptococcal infections are a significant environmental trigger for narcolepsy.[57]

The most commonly reported medical causes of narcolepsy-cataplexy (secondary narcolepsy) are tumors or sarcoidosis of the hypothalamus, autoimmune or paraneoplastic disorders associated with anti-Ma-2 or anti-aquaporin-4 antibodies, multiple sclerosis, myotonic dystrophy, Prader–Willi syndrome, Parkinson disease, and head trauma.[1,50] Behavioral "causes" most cited include abrupt change in sleep–wake patterns or sustained sleep deprivation. Other medical and behavioral conditions have been reported as occurring as a prelude to the onset of narcolepsy in a case series of 100 narcoleptics.[58]

The impact of narcolepsy on patients, who are untreated or sub-optimally treated, highlights the importance of developing treatments that target all aspects of the narcolepsy/cataplexy pentad; that is, normalization of sleep, including redistribution and normal timing of both NREM and REM sleep to allow for healthy physiological, emotional, and cognitive stability.[13,14,53–59] Before receiving optimal treatment, narcolepsy patients describe severe hardships in both physical and psychological spheres.[13,14,33,58-60] They describe difficulty coping with continuous or intermittent exhaustion, mental fogginess (colloquially termed "narco-fog"[60]), and experiencing feelings of being "out of it" unable to sustain reading or concentration. Many individuals with narcolepsy/cataplexy seek social isolation because of the need to constrain their emotions in order to avoid cataplexy, embarrassment (ridicule especially for pediatric patients), and accidents. Some liken untreated narcolepsy to "feeling like not having slept for 48 hours" or "living in the twilight zone"[61,62] Therefore untreated or inadequately treated narcolepsy can be severely disabling, compromising daily functioning and individual potential.[62]

More serious consequences of living with narcolepsy involve falling asleep or having cataplexy during a business meeting, while cooking, crossing a street, or while driving. A recent study found that patients with central hypersomnia reported more car crashes (in the previous 5 years), a risk that was confirmed in both treated and untreated subjects (Untreated, OR = 2.21 95% CI = $[1.30–3.76]$, Treated OR = 2.04 95% CI = $[1.26–3.30]$).[61] However, the risk of car accidents of patients treated for at least 5 years was no different than for healthy subjects (OR = 1.23 95% CI = $[0.56–2.69]$).[63]

17.2.4. Hypocretin (Orexin) Cell Loss

As noted previously, the loss of Hcrt (orexin) cells is the hallmark of the pathology of narcolepsy/cataplexy (cataplexy appears to typically emerge when approximately 90% Hcrt cell loss occurs, associated with CSF Hcrt-1 of ≤ 110 pg/ml or about 1/3 of normal Hcrt CSF level).[63,1] The Hcrts (Hcrt-1 and Hcrt-2) are a pairs of neuropeptides that among other physiologic attributes promote wakefulness.[24]

In the absence of these wake-promoting neuropeptides, chronic dysregulation of the sleep–wake and REM/NREM cycles occurs. It has been postulated that the symptoms of narcolepsy are associated with a strong pressure for REM sleep.[33,13,64] This hypothesis is now supported by the finding of > 60% increase in histamine neurons, theorized to be involved in the decrease of Hcrt cells in narcolepsy patients.[65] It has been theorized that the triggering events of human narcolepsy may involve a proliferation of histamine-containing cells as a residual effect of an increased histamine expression causing Hcrt cell loss in addition to antigens(s) as triggering an auto immune response, for example, HIN1 influence, HIN1 vaccination or streptococcal infection/exposure in the pathophysiology of narcolepsy/cataplexy.[66]

The Hcrt neurons are tonically active during the waking state, decreasing their activity during NREM sleep and demonstrating the least activity during REM sleep.[67] Hcrt cells activate neurons that promote wakefulness and inhibit sleep-activating neurons of the ventrolateral preoptic nucleus. Recent animal research with mice is supportive of the theory that Hcrt neurons are necessary for circadian control of REM sleep.[68]

17.2.5. Cataplexy

Cataplexy is defined as an "episodic loss of muscle tone with preserved consciousness and triggered by emotions".[1,69] Cataplexy can be difficult to diagnose because of its extreme clinical variability and expression. The clinical symptoms can vary not only from patient to patient but

within individuals over time. The presentation can be age specific, for example, facial weakness with slurred speech is more common in children, whereas appendicular weakness with a broader range of emotional triggers is seen in adolescence and adults. In general, cataplexy attacks range from partial muscle weakness to complete paralysis. In most cases the weakness is bilateral, but case reports have been published that describe strictly unilateral weakness.[70] Cataplexy tends to diminish over time in terms of both frequency of attacks and severity of the weakness,[71,72,73] perhaps in part because patients learn to avoid situations likely to trigger the muscle weakness or by keeping a "tight rein" over their emotions or at the onset of cataplexy to abort it, for example, by concentrating on neutral thoughts.[33,13]

A recent review[72] of behavioral and cognitive techniques have been used and reported to assist in the treatment of narcolepsy with positive benefits, utilizing similar techniques as used in the treatment of insomnia, that is, cognitive behavioral treatment (CBT) applied to narcolepsy, i.e., CBT-narcolepsy or CBT-N, as summarized below.

A series of multicomponent behavioral-cognitive strategies have been used, including techniques to produce "sleep satiation" for example, scheduled naps; hypnosis to reduce frequency of sleep paralysis attacks and symptoms of sleepiness; lucid dreaming to reduce sleep paralysis and hypnagogic hallucinations; cognitive therapy, including advice on problem solving, sleep and wake cycle control analysis and scheduling of productivity for circadian peak levels of alertness; muscle relaxation to reduce cataplexy and sleep attacks (most importunately while driving); as well as dietary guidelines, with admonitions not to eat in the middle of the night and to more tightly schedule meals.[74] These techniques and similar strategies have been reported for several decades[33] and are very useful to use, for example, given that cataplexy is often triggered by sleep deprivation and strong emotions, learning emotional control techniques can help prevent or abort some attacks. Sleepiness is well known to be in part decreased with proper diet and timing of meals. Therefore, such behavior training can be utilized as part of the patient's comprehensive treatment program.

Cataplexy attacks of muscle weakness affects all skeletal muscle groups (with the exception of the diaphragm and the extra-ocular muscles), but its most frequent target involves the muscles of the face, throat, head, and neck. The clinical picture is one of slurred speech (dysarthria) phasic twitching of facial muscles, jaw tremor, jaw dropping, and head bobbing. The next group of affected muscles includes the shoulder girdle and muscles of the upper extremities. Approximately 30% of the time, muscles of the lower extremities are involved. It has been estimated that 50% of cataplexy attacks have independent partial muscle attacks as well as complete paralysis.[73]

17.2.6. Triggers of Cataplexy

Cataplexy, as noted previously, is usually precipitated by strong positive emotions (e.g., laughter, excitement, elation, surprise, competitive sports, and sexual orgasm). Less often, negative emotions can trigger the episodes. These include, fright, anger, frustration, and stress. Fatigue can also trigger cataplexy. However, it should be noted that for an estimated 50% of patients with narcolepsy/cataplexy, episodes of muscle weakness can occur without any obvious emotional trigger,[1] although it is conceivable, for example, that fleeting funny or other intense thoughts might cause some attacks that the patient does not later recall. A significant number of pediatric patients (< 18 years of age) with narcolepsy/cataplexy exhibit complex abnormal motor behavior that occurs close to disease onset (daytime sleepiness) that do not meet the classical definition of cataplexy. The behaviors include hypotonic events (called "negative motor phenomena") such as, head drop, ptosis, tongue protrusion, drooping facial muscles, and generalized hypotonia. There are also "positive motor phenomena" including, eyebrow raising, perioral and tongue movements, facial grimacing, head/neck swaying, dyskinetic and dystonic limb movements, and stereotyped motor behavior. It is thought that these motor abnormalities at some point will evolve into the more typical cataplexy motor phenomena.[75]

Almost everyone has heard the expression that we get "weak with laughter". This is an example of a normal physiologic response to laughter characterized by suppression of the Hoffman reflex.[76] In fact, orexin neurons increase their firing rate in response to strong emotions. In orexin deficient narcoleptics, presumably there is destabilization of the motor control system which leads to more frequent episodes of muscle weakness and intensifies the degree of weakness (i.e., cataplexy).[14,77] Physiological studies cited in these publications include one by Pickering and Sleight of 1977, which demonstrated that chronically induced low blood pressure causes the baroreceptors to reset and become hyper-reactive to sudden increases in blood pressure (as well as the reverse seen in chronic hypertension), promoting a larger than normal pressor response. The PTC theory takes into account, animal studies reviewed[14] that report the area receiving this pressor response, the solitarii nuclei, has collaterals to other areas of the solitarii nuclei that promote a depressor response to lower blood pressure and have indirect input via connections to the gigantocellular field, which in turn projects to and receives projections from the locus coeruleus. The locus coeruleus and sub lateral dorsal nucleus (SLD), just ventral to the locus coeruleus are reported to be involved in the atonia neuromechanism of REM sleep. The locus coeruleus has also been reported to elicit a pressor response when stimulated and to modulate the depressor response mechanism in animals, and there is some evidence that there are projections of the locus coeruleus that terminate near the solitarii nuclei. Therefore, once there is sufficient Hcrt cell loss, EDS promotes low blood pressure, that causes the baroreceptors to reset and become hypersensitive to sudden increases in blood pressure, triggered by strong emotions, then the blood pressure neuroregulatory mechanism will be activated to lower blood pressure, but also may trigger the atonia neuromechanism of REM sleep, which continues until blood pressure has been stabilized and /or the pressure for REM sleep has been decreased by SOREMP.[13,14]

Among more recent reports documenting low blood pressure in narcoleptics, is a retrospective study of 554 narcoleptics off all medication, from the early modafinil clinical trials, where especially in the non-overweight group (males < 180 lbs., females < 160 lbs.), there was notable lower blood pressures, up to 1 SD (standard deviation) and typically approximately 0.5 SD below the norm for both males and females.[78] A subset of 23 patients from the 554 narcolepsy patients from one site, with only a few overweight subjects, were reported to have systolic and diastolic blood pressure means of 1.0 SD below the norms.[79] In another pilot study, an increase in blood pressure and pulse transit time during cataplexy was reported.[80] A clinical trial inspired by the PTC theory was done with a blood pressure stabilizing medication, propranolol at high doses, that was reported effective in treating cataplexy, but most narcoleptics could not tolerate the side effects for long-term treatment.[81] Collectively, these studies provide some empirical support for the PTC theory.[13,33]

From studies on the limbic system, there is evidence that patients with cataplexy have altered neuronal responses to positive emotion, as demonstrated using neuroimaging studies, especially the amygdala (a key brain region for regulation of emotional activity). The amygdala exhibits a reduced threshold for neuronal activation using functional MRI (fMRI). Positron emission tomography (PET) scans showed increased metabolic activity during cataplexy in the bilateral precentral and postcentral gyri and the primary somatosensory cortex and a marked decrease in activity in the hypothalamus.[82–84]

Evidence that patients with cataplexy have altered neuronal responses to positive emotions arising within the limbic system also has been demonstrated, using neuroimaging technique studies. It has been demonstrated that the amygdala (a key brain region for regulation of emotional activity) exhibits a reduced threshold for neuronal activation, using fMRI, PET scans, and showed increased metabolic activity during cataplexy in the bilateral precentral and postcentral gyri and the primary somatosensory cortex and a marked metabolic decrease in the hypothalamus.[82,83,84] Neurons in the medial prefrontal cortex innervate the lateral hypothalamus and parts of the amygdala, and also have been implicated in promoting cataplexy,[64] for example, thinking of something funny or sad might trigger cataplexy in a narcoleptic.

17.2.7. Neurology of Narcolepsy/Cataplexy

As has been repeatedly stated, Type 1 narcolepsy (narcolepsy with cataplexy) is most likely due to abolition or near complete loss of Hcrt/orexin. One of the primary goals of ongoing research is to identify the factor(s) that initiate the destruction of the Hcrt neurons. We have already covered the idea that genetic susceptibility probably plays a significant role in that triggers such as microbiologic exposure in the form of the an infectious agent (streptococcus or H1N1 influenza) or an attenuated form of the infectious agent such as that found in vaccines start a cascade that results in an auto immune attack specifically directed against the Hcrt neurons.

Hcrts/orexins are stimulatory neurotransmitters and are potent wake-promoting agents; when absent there is loss of the stability of the sleep–wake system and loss of normal NREM and REM cycling. The best model is that of the inherently stable flip-flop sleep switch in which orexin provides the governing stability to maintain one state over the other until normal physiologic processes cause a properly timed and executed change in states. In the absence of this governor, an unstable system is present causing frequent state changes from sleep to wake and back again with unstable REM sleep intruding into wakefulness and NREM sleep where it should be absent.[85,86,87]

17.2.8. Treatment Considerations and Options

It is strongly advised to consult the current Physician's Desk Reference or the pharmaceutical company's product safety information for each drug description and the full list of contraindications, adverse experiences (AEs), interactions, risk of interactions, dosing and withdrawal information, before prescribing treatment(s) for narcolepsy, in light of each patient's medical history, to avoid disease contraindications and drug interactions.[2] The summary Tables 17.5 and 17.6 list available drugs and a promising new drug in Phase 3 clinical trials for narcolepsy with and without cataplexy.

Table 17.5. Medications available & in phase 3 clinical trial for treatment of narcolepsy with & without cataplexy.

Drug	FDA approval (narcolepsy indication)	EMA approval (narcolepsy indication)	Treatment guidelines
Sodium oxybate (Xyrem®)	Yes (EDS & cataplexy)	Yes (narcolepsy with cataplexy)	Cataplexy, EDS, and disrupted nighttime sleep; option for hypnagogic hallucinations and sleep paralysis
Amphetamine salts (Adderall® but not Adderall XR®)	Yes*	No	EDS
Methylphenidate HCL (Ritalin® but not Concerta®, Methylin®, Equasym XL®)	Yes (narcolepsy general indication)	Yes **	EDS
Modafinil (Provigil®)	Yes (EDS)	Yes (promote wakefulness in narcolepsy)	EDS
Armodafinil (Nuvigil®)	Yes (EDS)	No	Developed subsequent to the guidelines
Selegiline (Eldepryl®, Zelapar®)	No	No	Cataplexy and EDS
Mazindol®	No	No	EDS and cataplexy
Pitolisant®	No	Submitted to EMA (narcolepsy)	EDS
Lisdexamfetamine (Vyvanse®)	No	No	EDS
Dexmethylphenidate (Focalin®)	No	N	EDS
Phenylalanine derivative (JZP-110)	No recommendation	No	EDS in Phase 3 clinical trials

Adapted and updated from: Thorpy M, Dauvilliers Y. Clinical and practical considerations in the pharmacologic management of narcolepsy. Sleep Medicine. 2015;16;10.

EDS = Excessive daytime somnolence* narcolepsy general indication;

**immediate release only (narcolepsy with or without cataplexy in adults when modafinil is ineffective in children over the age of 6.

Table 17.6. Contraindications & potential drug-drug interactions of current medication therapies & in phase 3 clinical trials for narcolepsy.

Pharmaceutical agent	Disease contraindication	Drug-drug interactions
Selective serotonin reuptake inhibitors (SSRIs)	Use with caution in patients with renal or hepatic impairment; seizure disorders, morbid obesity, restless legs syndrome RLS); periodic limb movements of sleep (PLMDS) and uncontrolled hypertension	MAOIs; SNRI's, TCAs Valerian Root, alcohol; drugs that inhibit, induce or metabolized by cytochrome P450 pathways and drugs affecting hemostasis.
Selective noradrenergic reuptake inhibitors (SNRIs)	Glaucoma; RLS; PLMDS	MAOIs; SNRI's, TCAs Valerian Root, alcohol; drugs that inhibit, induce or metabolized by cytochrome P450 pathways and drugs affecting hemostasis.
Tricyclic antidepressants (TCAs)	Glaucoma; RLS; PLMDS and seizures	Alcohol; MAOIs; potential interactions with drugs that inhibit, induce, or are metabolized by specific cytochrome P450 pathways; SSRIs; SNRIs.
Amphetamines	Myocardial ischemia, cardiac arrhythmias, hypertension, anxiety, bipolar disorder, glaucoma; substance abuse disorder, RLS, PLMDS	Insulin, antihistamines, antihypertensives, MAOIs; TCAs;
Methylphenidate (Ritalin®)	Myocardial ischemia, cardiac arrhythmias, hypertension, anxiety, bipolar disorder, glaucoma; substance abuse disorder, RLS, PLMDS and Tourette's syndrome	Coumarin-type anticoagulants; antihypertensives, MAOIs; and TCAs.
Modafinil/Armodafinil (Provigil®/Nuvigil®)	NONE SPECIFIED	Oral contraceptives exhibit decreased serum concentrations because of metabolic induction and require higher doses for pregnancy prevention; drugs that inhibit, induce or metabolized by cytochrome P450 pathways.
Selegiline (Eldepryl®, Zelapar®)	NONE SPECIFIED	Dextromethorphan, meperidine, MAOIs; SSRIs; SNRIs, TCAs

Sodium Oxybate (Xyrem®)	Succinic semialdehyde dehydrogenase deficiency (SSDD) an inborn error of metabolism; patients with congestive heart failure require close monitoring; uncontrolled hypertension; impaired renal function; major depressive disorder; and untreated obstructive sleep apnea.	Sedative hypnotics; alcohol; and divalproex sodium.
Mazindol	May cause valvular heart disease	Antihistamines; antihypertensives; MAOIs; TCAs.
Pitolisant	NONE SPECIFIED	Antihistamines.
Phenylalanine derivative (JZP-110)	History of phenylketonuria or hypersensitivity to phenylalanine	-

Adapted and updated from: Thorpy M, Dauvilliers Y. Clinical and practical considerations in the pharmacologic management of narcolepsy. Sleep Medicine. 2015;16;10.

17.2.9. Stimulant and Wake Promoting Drugs to Treat Hypersomnolence (EDS)

All narcoleptics have hypersomnolence, and is typically the first symptom that needs to be treated for safety, productivity, and QOL reasons. Currently the drugs of choice to treat excessive sleepiness are the Class IV wake-promoting medications: modafinil (Provigil®) (approved by the Food and Drug Administration (FDA) and European Medicines Agency (EMA)) a racemic mixture and its R-enantiomer armodafinil (Nuvigil®) (approved by the FDA). Prior to 1998, when modafinil was released in USA, the mainstay treatment for EDS were the Class II CNS stimulates, for example, amphetamine.[2] The first amphetamine to treat "narcolepsy" was introduced in 1935 under the trade name Benzedrine®which was the racemic mixture of α-methylphenethylamine. Several years later, the generic designation was named "amphetamine" by the Council on Pharmacy and Chemistry of the American Medical Association.

In 1937, the dextroversion (d- isomer of α-methylphenethylamine) was synthesized under the trade name Dexedrine® (dextroamphetamine). In the 1970s, it was found that the d-isomer was more effective as an alerting agent (and as a drug to treat *attention deficit hyperactivity disorder* (ADHD)), than the l-isomer or the racemic mixture. Today, the only available l-isomer is in the mixed salts/mixed enantiomers amphetamine (MES-amphetamine) which consists of a 3:1 enantiomeric mixture d-amphetamine: l-amphetamine salts that is available in both immediate release (Adderall®, generic) and extended-release (Adderall XR®, generic).[88]A more recent addition to the amphetamine class stimulants is lisdexamphetamine, which is an amphetamine prodrug (Vyvanse®) and is only approved for treatment of ADHD and binge-eating disorder.[89]

Methamphetamine (Desoxyn®), which is closely related to amphetamine, but has greater central versus peripheral effects than amphetamine, presumably due to its greater lipophilicity, has been shown to be effective in the treatment of EDS in narcoleptics.[90]

Methylphenidate (Ritalin®) is a CNS stimulant derived from piperidine, structurally similar to amphetamine, and acts as a norepinephrine-dopamine reuptake inhibitor. Methylphenidate, compared to amphetamines, has a shorter C_{max} and T_{max}, overall efficacy, in terms of its ability to increase alertness, but because it has a better tolerance and side-effect profile, it has gained wide use in the therapy of EDS.[91]

The extensive side effects and development of tolerance of the stimulants Class II scheduled drugs, especially the amphetamines, represent their major limitations. Some of most notable AEs include irritability, agitation, headache, and peripheral sympathetic stimulation, hyperthyroidism, glaucoma, high blood pressure, among a number of other AEs. These AEs are usually dose related. Tolerance develops in many patients, which leads to the need for higher doses to maintain efficacy, but also an increase in AEs and addiction.[92] Disease contraindications for amphetamines are structural cardiac abnormalities or other serious heart problems and glaucoma. The drug interactions of concern are insulin, antihistamines, anti-hypertensives, monoamine oxidase inhibitors (MAOIs); and tricyclic antidepressants (TCAs).[2]

17.2.10. The Class IV Scheduled Wake Promoting Drugs

Modafinil and armodafinil have been studied more rigorously for the treatment of EDS in patients with narcolepsy than the older amphetamines. These drugs are usually grouped with the amphetamines, but they have a more limited mechanism of action. It appears that instead of flooding the CNS with dopamine (DA), there is a more selective increase of DA in the synaptic clefts and terminals, secondary to blocking dopamine transporter (DAT). This leads to the reduction in frequency and severity of irritability and agitation and to some degree less sympathetic stimulation. Of particular note is the drug interaction for modafinil and armodafinil, both increase the metabolism of oral contraceptives and it is recommended that women of childbearing potential should use alternative or concomitant form of contraception; also there is potential interaction with drugs that inhibit, induce or are metabolized by cytochrome P450 pathways.[2] Noteworthy AEs include headache, diarrhea, dry mouth, nasal and throat congestion, nervousness and anxiety, dizziness, back pain, and difficulty sleeping, among others.

Armodafinil is the dextro-enantiomer component of modafinil. Clinically it has a longer effect in terms of allowing better alertness after 6–7 h of drug ingestion, but interestingly enough the actual half-life and C_{max} is similar to modafinil. Typical AEs include headache, mild anxiety and nervousness, and insomnia, among others.[93] Amphetamines (including methylphenidate) have mild REM sleep suppressing properties and when withdrawn there is rebound of both sleepiness and REM sleep. It is preferable to prescribe the long acting (or extended release formulations) stimulants to be taken first thing in the morning, than relegating the short acting stimulants to a prn basis because the serum peaks and troughs generated by multiple dosing of the short acting stimulants, creates more sympathetic stimulation between doses and as such there is an overall increase in side effects.[2]

Gamma-hydroxybutyrate (γ-hydroxybutyrate or GHB), formulated as sodium oxybate (SXB), now available by prescriptions as Xyrem®, is the only drug approved to treat both cataplexy (approved in 2002) and EDS (approved in 2005) by the FDA.[76,94] SXB has an FDA-mandated disease contraindication, succinic semialdehyde dehydrogenase deficiency, an inborn error of metabolism, causing accumulation of endogenous (and if taken therapeutically) GHB, due to the defect in the Krebs cycle. There is also a cautionary warning for patients with heart failure, hypertension, or impaired renal function; SXB is not to be used with sedative hypnotics, divalproex sodium, or alcohol. Indeed patients, who drink alcohol, should not take SXB afterward, as together they augment sedation and increase the risk for regurgitation and aspiration. Notable AEs include nausea, dizziness, vomiting, bedwetting, and diarrhea.

SXB is the sodium salt of GHB and is only available in the United States as Xyrem®. GHB has a history of being used recreationally and illegally (as one of the "date rape drugs"), and as such has a bifurcated FDA classification. If an individual has a legitimate (legal) prescription, the drug is classified as Class III. However if GHB or SXB is in the possession of an individual who does not have a legitimate or legal prescription, the classification is Class 1. GHB has been used and abused recreationally and illegally (as a "street" drug, typically in very much higher doses than SXB is prescribed for narcolepsy patients, and combined with alcohol and/or other drugs with serious side effects, in-

cluding, seizures, confusion, psychomotor impairment, withdrawal symptoms, stupor, coma, and some deaths have been reported). However in patients treated with SXB or GHB for their narcolepsy, there has been no evidence that patients suffered any significant withdrawal signs or symptoms when the drug was abruptly discontinued.[95,96,99]

Since the introduction of SXB, there is an ongoing risk evaluation and mitigation strategy (REMS). The evidence of any abuse, misuse, or diversion of SXB has been studied and has been reported to be extremely low. [104,105]

17.2.11. Cataplexy

Clinical management of cataplexy is limited by etiologic uncertainty, as well as by challenges in the clinical identification and diagnosis of the underlying narcolepsy; diagnosis is often delayed by many years after initial symptom onset.[29] In particular, there are no standard measures for identifying or assessing cataplexy, increasing the challenges not only of its identification but also of its evaluation during treatment. Thus, the recognition of cataplexy has primarily been based on clinical interview and patient self-reports. With evaluation of cataplexy during clinical trials or in the clinical setting, relying on patient recall and/or diaries, the manner of assessment often varies and may not always include a full description of frequency and severity. Although self-administered cataplexy questionnaires have been developed for eliciting information on occurrence, [106,107] they may be limited by a high burden of administration (one of the validated questionnaires consists of 51 questions[106]) and/or used only as a screening tool, since their sensitivity to treatment effects has not been demonstrated. However, despite these challenges, the pharmacologic management of cataplexy has a long history and a potentially promising future.

Although no longer recommended as the first line treatment for cataplexy, TCAs such as clomipramine, imipramine, desipramine, and protriptyline have been used as a therapeutic approach for narcolepsy for many years, and their efficacy for cataplexy was reported as early as 1960.[108,109] Since then, evidence from case reports and small open-label studies has further supported their beneficial effects for improving cataplexy,[2] however they do have a number of disease contraindications, drug interactions and side effects that are often problematic. It is important to note that these drugs have never been evaluated in larger and more formal randomized, placebo-controlled, double-blind clinical trials, nor FDA approved for treating narcolepsy or cataplexy. Although the mechanism of action of these drugs for cataplexy remains unknown, it has been thought that it may be related to their augmentation of noradrenergic signaling and/or the relative suppression of REM sleep.[110]

17.2.12. Current Therapeutic Approaches

SXB is the only medication currently indicated and FDA approved for the treatment of cataplexy associated with narcolepsy.[111] Treatment recommendations for cataplexy are included as part of the overall guidelines for the treatment of narcolepsy issued by the AASM[112] and European Federation of Neurological Societies (EFNS).[113] Both of these guidelines recommend several drugs as being of potential benefit for the treatment of cataplexy, but only SXB is recommended as a first-line treatment for cataplexy, based on high levels of evidence obtained from randomized, controlled clinical trials. Additionally, the guidelines also recommend SXB as a first-line therapy for treatment of EDS in narcolepsy, for which it is indicated and FDA approved. The AASM guidelines also recommend SXB as a standard treatment for disrupted sleep (but not FDA-approved nor European Medicines Agency-approved for the treatment of this symptom). Both guidelines further suggest that SXB be considered for hypnagogic hallucinations and/or sleep paralysis and/or for cases of disrupted nighttime sleep, but it is not specifically indicated for these symptoms (not FDA-approved nor European Medicines Agency-approved for the treatment of these symptoms), since the level of evidence for such indications is lower than for cataplexy and EDS.[111,112,113]

As mentioned in the guidelines, the evidentiary basis for therapeutic alternatives to SXB for the treatment of cataplexy is limited. Commonly used alternatives include TCAs, particularly clomipramine, and newer antidepressants such as SSRIs, the SNRI venlafaxine, and the norepinephrine reuptake inhibitor reboxetine (not available in USA). The monoamine oxidase type B (MAO-B) inhibitor selegiline is also noted for its efficacy in cataplexy, although both guidelines warn of limitations due to its safety profile including the potential for drug and food interactions. The European guidelines identify TCAs as the most effective of the alternatives to SXB for cataplexy, whereas the American guidelines do not state any general preferences among the second-line drug classes.[112,113]

Overall, the currently available treatments for cataplexy act symptomatically, and there is no evidence suggesting that they target the Hcrt/orexin system.[96]

SXB has been shown to enhance sleep in a number of ways, that will be detailed below.[97,101] Although the mechanism of action of SXB is unknown, it is hypothesized that its therapeutic effects on cataplexy and EDS are thought to be mediated through $GABA_B$ receptor activity, potentially impacting noradrenergic and dopaminergic neuronal function as well as that of thalamocortical neurons.[97] However, a clinical comparison with the prototypical $GABA_B$ agonist baclofen (a racemic mixture of R- and S-isomers of baclofen) showed that, while SXB reduced EDS and cataplexy, baclofen had no effect on these narcolepsy symptoms[134]. These results suggest that the efficacy of SXB may derive from mechanisms more complex than direct $GABA_B$ agonism. A further consideration is the fact that GHB satisfies criteria as a neurotransmitter, complete with CNS GHB receptors, and hence has been hypothesized to be a sleep neurotransmitter or neuromodulator.[101] As such, the benefit of GHB on cataplexy and EDS may be also, or in part indirect, by improving sleep and decreasing the pressure for sleep and REM sleep.[1100, 101, 114]

In 1989, the first placebo-controlled, complete cross-over, counter-balance, double-blind clinical trial of GHB (at 25 mg/kg hs and repeated 3 h later) versus placebo was reported, each treatment given for 4 weeks ,with six washout nights between treatments, in 20 diagnosed narcoleptics with cataplexy (10 males and 10 females)[99], followed by a continuation open label study, lasting for up to 14 months. Each subject had baseline, first night and last night PSGs and also a fixed MSLT of 20 min, providing seminal results[99-101,114] that would later be essentially replicated by another research group[102] and also by multi-center clinical trials with the SXB formulation.[120-122] The first double-blind study of GHB treatment versus placebo treatment documented the following results from daily logs: GHB significantly decreased cataplexy ($p < 0.03$), decreased subjective arousals from sleep ($p < 0.04$) and also noteworthy, abrupt withdrawal of GHB did not cause cataplexy rebound, but cataplexy did gradually return.[99]

The sleep data from this double-blind study documented the following PSG results from GHB treatment versus placebo treatment: GHB significantly increased delta sleep ($p < 0.05$), decreased stage 1 sleep ($p < 0.02$), decreased awakenings ($p < 0.01$), decreased stage shifts ($p < 0.01$), and also noteworthy, did not decrease REM sleep.[115.] The modified MSLT (fixed 20 min nap) data, from this double-blind study, done after 28 nights of GHB and again after 28 nights of placebo, yielded the following results, GHB vs Placebo: GHB: significantly increased stage 0 ($p < 0.05$), in female participants significantly decreased naps with REM sleep ($p = 0.02$), and for both male and female participants yielded a strong trend for increased sleep onset latency mean ($p < 0.08$).[114]

The 4–14 month follow-up on 12 of the 20 double-blind narcoleptics, who continued to keep daily cataplexy logs, documented continued efficacy, on GHB 25 mg/kg hs and repeated 3 h later, consistently having cataplexy ≤ 1/day versus the original mean baseline of about 3/day.[98] The 12 participants did a follow-up PSG[115] and MSLT.[100] The 4–14 month follow-up PSG revealed continued significant increase in sleep stage N3 ($p < 0.05$). The 4–14 month follow-up MSLT results (off methylphenidate from 5 PM the night before until after the last nap) revealed significant further improvement compared to the double-blind MSLT results: longer sleep onset latency ($p < 0.05$), increased awake time during naps ($p < 0.01$), decreased number of REM sleep naps ($p < 0.01$), and decreased REM sleep minutes ($p < 0.02$), compared to their baseline. These patients not only logged significant decrease in daytime cataplexy with nightly GHB use,[98] but also objective decrease in NREM sleep and REM sleep pressure[100], though still diagnostic of narcolepsy. An up to 5 year follow-up on 34 patients (16 followed for 3–5 years), for cataplexy efficacy with daily logs kept at least 2 of 4 weeks), where all had on average ≤ 1 cataplexy event/day, reported cataplexy to be noticeably milder, shorter lasting and more controllable.[103] Further continuation of this trial for up to 14 years and adding more narcoleptic patients, revealed that GHB remained efficacious for almost all patients (noteworthy: 4 insufficient therapeutic effect, 3 AE: seizure, heart attack, combined treatment, 9 cataplexy remitted or became too mild to require GHB), and of the 65 narcoleptic patients (12 from the original double-blind study) at 25 mg/kg (also several patients at either 32.5 mg/kg or 40 mg/kg) dosing at bedtime and repeated 3–4 hours later.[116] Moreover, 36 of the 65, responded to a survey that revealed: 83% had "life-altering" positive impact, 14% had moderate positive impact and 1% had slight positive impact; 97% reported mild to hardly noticeable cataplexy, 86% had mild to hardly noticeable excessive sleepiness, 91% reported slight to hardly noticeable sleep paralysis or hypnagogic hallucinations.[116]

The mechanism of action of the antidepressants in narcolepsy with cataplexy (none of which are FDA approved) is generally related to their ability to inhibit reuptake of monoamines, with a high correlation between receptor affinity and potency for their effects on cataplexy. [71,117]

The TCAs are nonspecific monoamine reuptake inhibitors, with effects that promote the availability and activity of serotonin, noradrenaline, and, for some agents, DA. The TCAs most commonly used in narcolepsy—clomipramine, imipramine, desipramine, and protriptyline—exert, in addition to monoamine reuptake inhibition, anticholinergic actions that may contribute to their effect on cataplexy, but may also cause important side effects such as constipation, dry mouth, blurred vision, and alterations in cardiac conduction, tardive dyskinesia, loss of libido, insomnia, hypersomnolence, impotence, suicide, violent impulses, and RBD.

The efficacy of SSRIs in reducing cataplexy has been shown to derive from their adrenergic reuptake inhibition. However, very selective SSRIs such as escitalopram or fluoxetine are usually not as effective for cataplexy as venlafaxine, an SNRI that is used mainly because it has demonstrated a short onset of action. Abrupt withdrawal of TCAs, SSRIs or SNRIs often precipitate rebound cataplexy, once wash-out is complete.[118,71,119] The MAO-B inhibitor selegiline increases availability of monoamine neurotransmitters, including DA, noradrenaline, and serotonin, but its use is limited by potential side effects, as also noted in the narcolepsy treatment guidelines.[112,113]

As suggested by the high level of evidence used to support its recommendation as a first-line therapy, SXB demonstrated efficacy for treatment of cataplexy and daytime sleepiness in multiple randomized, double-blind, placebo-controlled clinical trials.[94,99,102,,111,120,121] In particular, the administration of nightly SXB produced a decrease in the reported frequency of cataplexy attacks and this change was significant across doses ($P = 0.0021$); the median percent change in the number of weekly cataplexy attacks across doses was significant ($P < 0.03$); a decrease of approximately 70% after 4 weeks of treatment at the 9 g/night dose was most significant ($P=0.0008$).[120] No rebound cataplexy was observed in patients who had their SXB abruptly discontinued at any of the studied doses, nor in any of the reported studies.[99,111-,120,121,94] Long-term SXB studies showed that the therapeutic effect on cataplexy was maintained for 12 months and up to 4 years.[122,123] The earlier GHB studies also reported therapeutic effects up to 5 years[103] and 14 years.[116] Two systematic reviews and meta-analyses of SXB provided further support for the robustness of its efficacy for reducing cataplexy episodes, with the greatest effects observed at the highest dose of 9 g/night, and the lower doses of 6 g and 7.5 g were also highly significant from 1–12 months.[64,124] Both the 3 g and 4.5 g are reported to be significant after the 2nd month, through the rest of the 12 month interval,[122] and essentially replicated the results from first double blind[99] and continuation GHB studies.[98,103,116]

These meta-analyses also reported that SXB was generally well tolerated, with most adverse events of mild-to-moderate severity. Furthermore, a retrospective analysis of SXB in a pediatric population, suggested that in children and adolescents, aged 4 through 16 years, treatment significantly reduced the median frequency of cataplexy episodes from 38/week to < 1/week ($n = 14$; $p < 0.001$), and reduced cataplexy severity relative to baseline ($p < 0.001$) (SXB is not indicated for use in patients < 18 years old).[125]; pediatric clinical trials with SXB are in progress.

17.2.13. Brief Case Report

A memorable narcolepsy patient, seen in 1995, had been on antidepressants for cataplexy, for several years.[14] He reported that the antidepressant he was taking had stopped working and that he had developed RBD to the point that his wife would not sleep in the same bed with him. After taking GHB for a few months and also treating his OSA, his cataplexy and RBD both gradually decreased and stopped, except for some very mild and short cataplexy twinges, on rare occasions. His remission from cataplexy and RBD continued for several years until he was lost to follow-up.

17.3. DRUG DEVELOPMENT

17.3.1. Histamine H₃ Receptor Antagonism Using Direct Receptor Antagonists or Inverse Agonists

In addition to the catecholamines, tuberomammillary histaminergic neurons play a crucial role in maintenance of wakefulness, but remain active during cataplexy, helping to preserve consciousness.[66,126] Indeed, these neurons appear to be increased in narcolepsy, perhaps as a compensatory response to Hcrt/orexin loss and the resulting deficit in excitatory adrenergic drive.[127] Pitolisant is an inverse agonist of the histamine H_3 autoreceptor, which theoretically reduces H_3 inhibitory activity below basal rates and thereby functions more effectively than H_3 antagonists to activate histaminergic neuronal activity in the brain and promote wakefulness. While it reduced EDS in patients with narcolepsy in a small clinical trial,[128] a post hoc analysis of another clinical trial showed that pitolisant for 8 weeks resulted in statistically significant reductions in the rate of cataplexy from baseline compared with placebo ($p < 0.05$), and was "non-inferior" relative to modafinil in terms of improvement in daytime sleepiness.[129] Pitolisant was also shown to slightly improve cataplexy severity and frequency in a case series of four teenagers with narcolepsy. [130] Based on these preliminary results, a phase 3 randomized, controlled trial was initiated to specifically evaluate the effects of pitolisant on cataplexy as a primary outcome compared with placebo (ClinicalTrials.gov identifier NCT01800045).

17.3.2. Hypocretin/Orexin Replacement Therapy

Compensating for Hcrt/orexin deficiency through the use of Hcrt/orexin peptide supplementation or cell replacement therapies may provide a rational approach to narcolepsy with cataplexy therapy. Potential techniques include: 1) delivery of Hcrt/orexin peptides via intranasal, in-travenous, intracisternal, or intracerebroventricular modes; 2) use of prodrugs or agonists; or 3) by genetic engineering or cell replacement techniques.[96] While these techniques are still in their early stages of development, the few available human studies have shown potential benefits for sleep and wakefulness, but the effects on cataplexy were not evaluated.[131] However, in a canine model of narcolepsy, repeated systemic administration of Hcrt-1/orexin-A consolidated waking and sleeping periods and abolished cataplexy completely for periods of ≥ 3 days.[132,133] Further evaluation of these techniques may confirm the benefits of this therapeutic approach, and may also provide insight into the underlying mechanisms of narcolepsy and cataplexy.

17.3.3. GABA$_B$ Agonist

Baclofen, a GABA$_B$ agonist that has been suggested to improve nighttime sleep in patients with several neurologic conditions including narcolepsy, has not demonstrated efficacy for cataplexy or EDS associated with narcolepsy.[134] However, a more recent study in a mouse model of narcolepsy showed that R-baclofen, an enantiomer with a 3-fold higher affinity for GABA$_B$ receptors than the clinically available racemate, had greater efficacy than placebo in reducing cataplexy-like activity and REM sleep disturbances.[135]

17.3.4. Immunomodulation

The autoimmune hypothesis of narcolepsy with cataplexy provides a rationale for use of immunomodulation therapy,[96] which is exemplified by experimental use of intravenous immunoglobulin therapy, although it has only been evaluated in case studies in adults and children. While benefits were not consistently demonstrated in all studies, several studies did report that, if administered shortly after disease onset, intravenous immunoglobulin therapy may be effective in reducing narcolepsy symptoms including cataplexy and may have long-term benefits. These initial results suggest that further, more formal evaluation of intravenous immunoglobulin may be warranted.[136–142]

17.3.5. New Wake Promoting Agent

A potential new wake promoting agent (JZP-110), that indirectly enhances dopaminergic and adrenergic neurotransmission has completed phase 2 clinical trials with encouraging results on both subjective measures (clinical global impression, patient global impression, and Epworth Sleepiness Scale) and objective measures (multiple wakefulness testing or MWT). JZP 110 is now in multi-center phase 3 efficacy clinical trials.[143–145]

17.4. CONCLUSION

The story of narcolepsy/cataplexy is really about mind, body, and environment effects on neurophysiology of sleep (made more understandable by sleep, neuroscience, and psychology research) as they influence the sleep–wake states and what happens when critical elements of the circadian rhythm are perturbed. The discovery of the Hcrt/orexin system has enabled researchers and clinicians to better conceptualize and understand the intricacies of the regulatory system that controls wakefulness and sleep, as well as the oscillations NREM and REM sleep cycles.

Sleep researchers are in the process of discerning the relationship between the immune system and the loss of Hcrt neurons that currently defines narcolepsy with cataplexy. This relationship will hopefully shape our diagnostic and therapeutic options in the future. In the meantime, optimal treatment for narcolepsy often includes a wake-promoting medication, such as modafinil or armodafinil or methylphenidate and once cataplexy develops, SXB. One question that needs to be answered through additional research is whether SXB use might delay or prevent cataplexy from developing, once the narcolepsy diagnosis is established by one or more positive MSLT tests, since it is very efficacious in treating cataplexy as well as hypersomnolence in patients with 90% or more Hcrt cells dead. Neither SXB, modafinil, nor armodafinil is FDA-approved for pediatric narcolepsy as yet, even though the problem typically starts in early teenage years. Fortunately, clinical trials with SXB for pediatric narcolepsy-cataplexy are currently underway.

Narcolepsy remains a chronic neurologic disorder that has a disproportionate degree of burden on its afflicted, because it is underdiagnosed and even when evaluated by "specialists" remains, for the most part, misdiagnosed, inappropriately or undertreated. The sleep medical community has a responsibility to learn more about and understand this fascinating condition, diagnose it and treat it appropriately.

Footnote: Physicians must be aware that laws vary between states and might require reporting EDS that may impair driving.[4] The treating physician has a patient and public responsibility to make recurring clinical assessments of the patient's overall risk for unsafe driving, document driving recommendations, and precautions. Treating physicians should report patients who fail to comply with their prescribed treatment(s), particularly those high-risk patients, such as airline pilots, train, truck, taxi, limousine, bus, or other public transportation drivers, and those jobs involving public safety or excellent alertness maintenance.

Resources for the patients

Narcolepsy Network, Inc. - http://www.narcolepsynetwork.org/ -(888)-292-6522

Wake up Narcolepsy - http://www.wakeupnarcolepsy.org/ -(978) 751-3693

National Sleep Foundation - http://www.sleepfoundation.org/-(703) 243-1697

Narcolepsy Institute - http://www.narcolepsyinstitute.org/ -(718) 920-6799

KEY WORDS

- narcolepsy type 1
- narcolepsy type 2
- cataplexy
- sleep paralysis
- hypnagogic/hypnogogic hallucinations
- hypnopompic hallucinations
- disrupted nocturnal sleep (DNS)
- idiopathic hypersomnolence (IH)
- polysomnography (PSG)

- multiple sleep latency test (MSLT)
- multiple wakefulness test (MWT)
- hypocretin (Hcrt)/orexin (OXR) cells
- OXR-1 CSF levels
- rapid eye movement (REM or R) sleep
- sleep onset REM period (SOREMP)
- prevalence
- etiology
- neuromechanisms
- treatment options
- diagnosis
- management

REFERENCES

1. *International Classification of Sleep Disorders, 3rd ed.* Darien, IL: American Academy of Sleep Medicine; 2014.

2. Thorpy M.; Dauvilliers Y. Clinical and Practical Considerations in the Pharmacologic Management of Narcolepsy. *Sleep Med.* **2015**,*16*, 9–18.

3. Longstreth W.; Koepsell T.; Ton T.; Hendrickson A.; van Belle G. The Epidemiology of Narcolepsy. *Sleep.* **2007**,*30(1)*,13–26.

4. Frauscher B.; Ehrmann L.; Mitterling T., et al. Delayed Diagnosis, Range of Severity, and Multiple Sleep Comorbidities: A Clinical and Polysomnographic Analysis of 100 Patients of the Innsbruck Narcolepsy Cohort. *J. Clin. Sleep. Med.* **2013**,*9(8)*,805–812.

5. Ozaki A.; Inoue Y.; Nakajima T., et al. Health-Related Quality of Life Among Drug-Naïve Patients with Narcolepsy with Cataplexy, Narcolepsy Without Cataplexy, and Idiopathic Hypersomnia Without Long Sleep Time. *J. Clin. Sleep Med.* **2008**,*4(6)*,572–578.

6. Schenck C.; Bassetti C.; Arnulf I.; Mignot E. English Translations of the First Clinical Reports on Narcolepsy and Cataplexy by Westphal and Gélineau in the Late 19th Century, with Commentary. *J. Clin. Sleep Med.* **2007**,*3(3)*,301–311.

7. Lowenfeld L. Uber Narkolepsie. *Munch. Med. Wochenschr.* **1902**, *49*,1041–1045.

8. Adie W. Idiopathic Narcolepsy: A Disease Sui Generis; with Remarks on the Mechanism of Sleep. *Brain.* **1926**,*49(3)*,257–306.

9. Daniels L. Narcolepsy. *Medicine.* **1934**,*13(1)*,1–122.

10. Yoss R.; Daly D. Criteria for the Diagnosis of the Narcoleptic Syndrome. *Proc. Staff. Meet. Mayo. Clin.* **1957**,*32(12)*,320–328.

11. Roth T.; Dauvilliers Y.; Mignot E., et al. Disrupted Nighttime Sleep in Narcolepsy. *Sleep Med.* **2013**,*9(9)*,955–965.

12. Kleitman N. *Sleep and Wakefulness;* University of Chicago Press:Chicago, IL, 1963; pp 233–242.

13. Scrima L. Dreaming Epiphenomena of Narcolepsy. In *Sleep Medicine Clinics;* Pagel J., Ed.; W.B. Saunders:Philadelphia, PA, 2010; Vol. 5, pp 261–275.

14. Scrima L. Narcolepsy. In *Primary Care Sleep Medicine a Practical Guide,* 2nd ed.; Pagel J., Pandi-Perumal R., Eds.; Springer : New York, NY, 2014; pp 269–281.

15. Lavie P.; Peled R. Narcolepsy is a Rare Disease in Israel. *Sleep.* **1987**,*10*, 608–609.

16. Dauvilliers Y.; Montastruc J.; Molinari N., et al. Age at Onset of Narcolepsy in Two Large Populations of Patients in France and Quebec. *Neurology.* **2001**,*57*, 2029–2033.

17. Guilleminault C.; Pelayo R. Narcolepsy in Prepubertal Children. *Ann Neurol.* **1998**,*43*, 135–142.

18. Guilleminault C.; Lee J.; Arias V. Cataplexy. In *Narcolepsy and Hypersomnia*; Bassetti C., Billiard M., Mignot E., Eds.; Informa Healthcare USA, Inc. :New York, NY, 2007; Vol. 220, pp 50–51.

19. Mitler M.; Boysen B.; Campbell L.; Dement W. Narcolepsy-Cataplexy in a Female Dog. *Exp. Neuro.* **1974**,*45(2)*,332–340.

20. Foutz A.; Mitler M.; Cavalli-Sforza L.; Dement W. Genetic Factors in Canine Narcolepsy. *Sleep.* **1979**,*1(4)*,413–422.

21. De Lecea L.; Kilduff T.; Peyron C., et al. The Hypocretins: Hypothalalamus-Specific Peptides with Neuroexcitatory Activity. *Proc. Natl. Acad. Sci.USA.* **1998**,*95*, 322–327.

22. Sakurai T.; Amemiya A.; Ishii M., et al. Orexins and Orexin Receptors: A Family of Hypothalamic Neuropeptides and G Protein-Coupled Receptors that Regulate Feeding Behavior. *Cell.* **1998**,*92*, 573–585.

23. Nishino S.; Ripley B.; Overeem S.; Lammers G.; Mignot E. Hypocretin (Orexin) Deficiency in Human Narcolepsy. *Lancet.* **2000**, *355(9197)*, 39–40.

24. Thannickal T.; Moore R.; Nienhuis R., et al. Reduced Number of Hypocretin Neurons in Human Narcolepsy. *Neuron.* **2000**,*27*, 469–474.

25. Crocker A.; Espana R.; Papadopoulou M., et al. Concomitant Loss of Dynorphin, NARP, and Orexin in Narcolepsy. *Neurology.* **2005**,*65*, 1184–1188.

26. Thorpy M. Classification of Sleep Disorders. *Neurotheraputics.* **2012**,*9*, 687–701.

27. American Academy of Sleep Medicine.*The International Classification of Sleep Disorders. Diagnostic & Coding Manual,* 2nd ed. American Academy of Sleep Medicine:Westchester, IL, 2005.

28. Hypersomnia of central origin. In *International Classification of Sleep Disorders,* 3rd ed.;Sateia M., Ed. American Academy of Neurology:Westchester, IL, 2005; pp 78–93.

29. Thorpy M.; Krieger A. Delayed Diagnosis of Narcolepsy: Characterization and Impact. *Sleep Med.* **2014**,*15*, 502–507.

30. Scrima L. Lag Time between Onset of Excessive Sleepiness and Cataplexy in Narcolepsy Patients. *Sleep Res.* **1991**,*20*, 328.

31. Rosenberg R.; Kim A. The AWAKEN Survey; Knowledge of Narcolepsy Among Physicians and the General Public. *Postgrad. Med.* **2014**,*126(1)*,1–8.

32. Iber C.; Ancoli-Israel S.; Chesson A., et al. *The AASM Manual for the Scoring of Sleep and Associated Events.* Westchester, IL: American Academy of Sleep Medicine; 2007.

33. Scrima L. An Etiology of Narcolepsy Cataplexy and a Proposed Cataplexy Neuromechanism. *Int. J. Neurosci.* **1991**,*15*, 69–86.

34. Johns M. A New Method for Measuring Daytime Sleepiness: The Epworth Sleepiness Scale. *Sleep.* **1991**,*14(6)*,540–545.

35. Sasai T.; Inoue Y.; Komada Y.; Sugiura T.; Matsushima E. Comparison of Clinical Characteristics among Narcolepsy with and without Cataplexy and Idiopathic Hypersomnia without Long Sleep Time, Focusing on HLA-DRB1*1501/DQB1*0602 Finding. *Sleep Med.* **2009**,*10*, 961–966.

36. Guilleminault C. Narcolepsy. In *Narcolepsy;* Guilleminault C., Dement W., Passouant P.,

Eds.; Spectrum Publications: Holliswood, NY,1976.

37. Arand D.; Bonnet M.; Hurwitz T.; Mitler M.; Rosa R.; Sangal R. The Clinical Use of the MSLT and MWT. *Sleep.* **2005**,*28(1)*,123–144.

38. Sateia M. International Classification of Sleep Disorders-Third Edition: Highlights And Modifications. *Chest.* **2014**,*146(5)*,1387–1394.

39. Rogers A.; Aldrich M.; Caruso C. Patterns Of Sleep And Wakefulness In Treated Narcoleptic Subjects. *Sleep.* **1994**,*17(7)*,590–597.

40. Sansa G.; Iranzo A.; Santamaria J. Obstructive Sleep Apnea in Narcolepsy. *Sleep Med.* **2010**,*11*, 93–95.

41. Scrima L.; Hoddes E.; Johnson F.; Cardin R.; Thomas E.; Hiller F. Effect of High Altitude on a Patient with Obstructive Sleep Apnea. *Sleep Res.* **1987**,*16*, 427.

42. Hartman P.; Scrima L. Muscle Activity in the Legs (MAL) Associated with Frequent Arousals in Narcoleptics and OSA Patients. *Clin. Electroencephalogr.* **1986**,*17*, 181–186.

43. Trotti L.; Staab B.; Rye D. Test-Retest Reliability of the Multiple Sleep Latency Test in Narcolepsy without Cataplexy and Idiopathic Hypersomnia. *J. Clin. Sleep.Med.* **2013**,*9(8)*,789–795.

44. Nielsen T. Mentation During Sleep: the NREM/REM Distinction. In *Handbook of behavioral state control cellular and molecular mechanisms;* Lydic R., Baghdoyan H., Eds.; CRC Press : Boca Raton, FL, 1999; pp 101–128.

45. Ness R. The Old Hag Phenomenon as Sleep Paralysis: A Biocultural Interpretation. *Cult. Med. Psychiatry.* **1978**,*2*, 15–39.

46. Dodet P.; Chavez M.; Leu-Semenescu S.; Golmard J.; Arnulf I. Lucid Dreaming in Narcolepsy. *Sleep.* **2015**,*38(3)*,487–497.

47. Hufford D. *The Terror that Comes in the Night.* University of Pennsylvania Press: Philadelphia, PA; 1982.

48. Wamsley E.; Donjacour C.; Scammell T.; Lammers G.; Stickgold R. Delusional Confusion of Dreaming and Reality in Narcolepsy. *Sleep.* **2014**,*37(2)*,419–422.

49. Mignot E. Genetic and Familial Aspects of Narcolepsy. *Neurology.* **1998**, *50* (Suppl. 1),S16-S22.

50. Mignot E.; Hayduk R.; Black J.; Grumet F.; Guilleminault C. HLA DQB1*0602 is Associated with Cataplexy in 509 Narcoleptic Patients. *Sleep.* **1997**,*20(11)*,1012–1020.

51. Bassetti C. Spectrum of Narcolepsy. In *Narcolepsy and Hypersomnia;* Bassetti C., Billiard M., Mignot E., Eds.; Informa Healthcare USA, Inc.:New York, NY, 2007; Vol. 220, pp 97–108.

52. Nevsimalova S.; Mignot E.; Sonka K.; Arrigoni J. Familial Aspects of Narcolepsy-Cataplexy in the Czech Republic. *Sleep.* **1997**,*20(11)*,1021–1026.

53. Guilleminault C.; Mignot E.; Grumet F. Familial Patterns of Narcolepsy. *Lancet.* **1989**,*335*,1376–1379.

54. Hublin C.; Kaprio J.; Partinen M., et al. The Prevalence of Narcolepsy: An Epidemiological Study of the Finnish Twin Cohort. *Ann Neurol.* **1994**,*35(6)*,709–716.

55. Han F.; Lin L.; Warby S., et al. Narcolepsy Onset is Seasonal and Increased Following the 2009 H1N1 Pandemic in China. *Ann Neurol.* **2011**,*70*, 410–417.

56. Szakacs A.; Darin N.; Hallbook T. Increased Childhood Incidence of Narcolepsy in Western Sweden after H1N1 Influenza Vaccination. *Neurology.* **2013**,*80*, 1315–1321.

57. Aran A.; Lin L.; Nevsimalova S., et al. Elevated Anti-Streptococcal Antibodies in Patients with Recent Narcolepsy Onset. *Sleep.* **2009**,*32(8)*,979–983. 58. Scrima L.; Miller B. Pre-Onset Medical and Sleep History of 100 Narcoleptics. *Sleep.* **1999**, *22*(Suppl.1), S155.

59. Mamelak M. Narcolepsy and Depression and the Neurobiology of Gammahydroxybutyrate. *Prog. Neurobiol.* **2009**,*89*, 193–219.

60. Narcolepsy Public Meeting on Patient-Focused Drug Development, FDA White Oak Campus, Silver Spring, MD, September 24, 2013; CDER, Food and Drug Administration: Silver Spring, MD, 2013; pp 72–101.

61. Ingravallo F.; Gnucci V.; Pizza F.; Vignatelli L., et al. The Burden of Narcolepsy with Cataplexy: How Disease History and Clinical Features Influence Socio-Economic Outcomes. *Sleep Med.* **2012**,*13*, 1293–1300.

62. Goswami M.; Thorpy M. Commentary on the Article by Ingravallo et al. "The burden of narcolepsy with cataplexy: how disease history and clinical features influence socio-economic outcomes" Sleep Medicine 2012; 13:1293–1300. *Sleep Med.* **2013**,*14*, 301–302.

63. Pizza F.; Jaussent I.; Lopez R., et al. Car Crashes and Central Disorders of Hypersomnolence: A French Study. *PloS ONE.* **2015**,*10(6)*,1–14.

64. Alshaikh M.; Tricco A.; Tashkandi M.; Mamdani M.; Straus S.; BaHammam A. Sodium Oxybate for Narcolepsy with Cataplexy: Systematic Review and Meta-Analysis. *J. Clin. Sleep. Med.* **2012**,*8(4)*,451–458.

65. Joshi J.; Thannickal T.; McGregor R., et al. Greatly Increased Numbers of Histamine Cells in Human Narcolepsy with Cataplexy. *Ann Neurol.* **2013**,*74*, 786–793.

66. Burgess C.; Scammell T. Narcolepsy: Neural Mechanisms of Sleepiness and Cataplexy. *J Neurosci.* **2012**,*32(36)*,12305–12311.

67. Swick T. The Neurology of Sleep 2012. In *Biology of Sleep*; Teofilo Lee-Chiong J., MD, Ed.; W.B. Saunders:Philadelphia, PA, 2012; Vol. 7, pp 399–415.

68. Kantor S.; Mochizuki T.; Janisiewicz A.; Clark E.; Nishino S.; Scammell T. Orexin Neurons are Necessary for the Circadian Control of REM Sleep. *Sleep.* **2009**,*32(9)*,1127–1134.

69. Overeem S.; Mignot E.; van Dijk J.; Lammers G. Narcolepsy: Clinical Features, New Pathophysiologic Insights, and Future Perspectives. *J. Clin. Neurophysiol.* **2001**,*18(2)*,78–105.

70. McCarty D. A Case of Narcolepsy with Strictly Unilateral Cataplexy. *J. Clin. Sleep. Med.* **2010**,*6(1)*,75–76.

71. Dauvilliers Y.; Siegel J.; Lopez R.; Torontali Z.; Peever J. Cataplexy–Clinical Aspects, Pathophysiology and Management Strategy. *Nat. Rev. Neurol.* **2014**,*10*, 386–395.

72. Dauvilliers Y.; Billiard M.; Montplaisir J. Clinical Aspects and Pathophysiology of Narcolepsy. *Clin. Neurophysiol.* **2003**,*114(11)*,2000–2017.

73. Overeem S.; van Nues S.; van der Zande W.; Donjacour C.; van Mierlo P.; Lammers G. The Clinical Features of Cataplexy: A Questionnaire Study in Narcolepsy Patients with and without Hypocretin-1 Deficiency. *Sleep Med.* **2011**,*12*, 12–18.

74. Agudelo H.; Correa U.; Sierra J.; Seithikurippu R.; Pandi-Perumal R.; Schenck C. Cognitive Behavioral Treatment for Narcolepsy: Can it Complement Pharmacotherapy? *Sleep Sci.* **2014**,*7*, 30–42.

75. Plazzi G.; Pizza F.; Palaia V., et al. Complex Movement Disorders at Disease Onset in Childhood Narcolepsy with Cataplexy. *Brain.* **2011**,*134*, 3477–3489.

76. Overeem S.; Lammers G.; Dijk D. Weak with Laughter. *Lancet.* **1999**,*354*, 838.

77. Burgess C.; Peever J. A Noradrenergic Mechanism Functions to Couple Motor Behavior with Arousal State. *Curr. Biol.* **2013**,*23(18)*,1719–1725.

78. Scrima L.; Garlick I.; Victor Y.; Miller B. Narcolepsy Patient's Blood Pressure in Higher and Lower Weight Groups. *Sleep.* **1998**, *21* (Suppl.), 53.

79. Scrima L.; Garlick I.; Victor Y., et al. Narcolepsy Patinets' Blood Pressure and Pulse. *Sleep Res.* **1996**,*25*, 366.

80. Scrima L.; Hartman P.; Anderson J.; Winters R. Cataplexy Related to Blood Pressure and Transit Time Changes. *Soc. Neurosci.* **1984**,*10(2)*,832.

81. Meier-Ewert K.; Matsubayashi K.; Benter L. Propranolol: Long-Term Treatment in Narcolepsy-Cataplexy. *Sleep.* **1985**,*8(2)*,95–104.

82. Ponz A.; Khatami R.; Poryazova R., et al. Abnormal Activity in Reward Brain Circuits in Human Narcolepsy with Cataplexy. *Ann Neurol.* **2010**,*67(2)*,190–200.

83. Reiss A.; Hoeft F.; Tenforde A.; Chen W.; Mobbs D.; Mignot E. Anomalous Hypothalamic Responses to Humor in Cataplexy. *PloS ONE.* **2008**,*3(5)*,e2225.

84. Dauvilliers Y.; Comte F.; Bayard S.; Carlander B.; Zanca M.; Touchon J. A Brain PET Study in Patients with Narcolepsy-Cataplexy. *J. Neurol. Neurosurg. Psych.* **2010**,*81(3)*,344–348.

85. Saper C.; Chou T.; Scammell T. The Sleep Switch: Hypothalamic Control of Sleep and Wakefulness. *Trends Neurosci.* **2001**,*24*, 726–731.

86. Espana R.; Scammell T. Sleep Neurobiology from a Clinical Perspective. *Sleep.* **2011**,*34*, 845–858.

87. Saper C.; Scammell T.; Lu J. Hypothalamic Regulation of Sleep and Circadian Rhythms. *Nature.* **2005**,*437*,1257–1263.

88. Heal D.; Smith S.; Gosden J.; Nutt D. Amphetamine, Past and Present–A Pharmacological and Clinical Perspective. *J. Psychopharm.* **2013**,*27(6)*,479–496.

89. Pennick M. Absorption of Lisdexamfetamine Dimesylate and its Enzymatic Conversion to D-Amphetamine. *Neuropsychiatr. Dis. Treat.* **2010**,*24*, 317–327.

90. Mitler M.; Hajdukovic R.; Erman M. Treatment of Narcolepsy with Methamphetamine. *Sleep.* **1993**,*16(4)*,306–317.

91. Challman T.; Lipsky J. Methylphenidate-Its Pharmacology and Uses. *Mayo. Clin. Proceed.* **2000**,*75*, 711–721.

92. Wise M.; Arand D.; Auger R.; Brooks S.; Watson N. Treatment of Narcolepsy and other Hypersomnias of Central Origin. An American Academy of Sleep Medicine Review. *Sleep.* **2007**,*30(12)*,1712–1727.

93. Schwartz J.; Roth T.; Drake C. Armodafinil in the Treatment of Sleep/Wake Disorders. *Neuropsychiatr. Dis. Treat.* **2010**,*6*, 417–427.

94. The Xyrem International Study Group. A Double-Blind, Placebo-Controlled Study Demonstrates Sodium Oxybate is Effective for the Treatment of Excessive Daytime Sleepiness in Narcolepsy. *J. Clin. Sleep. Med.* **2005**,*1(4)*,391–397.

95. US Xyrem Multicenter Study Group. The Abrupt Cessation of Therapeutically Administered Sodium Oxybate (GHB) Does Not Cause Withdrawal Symptoms. *J. Toxicol. Clin. Toxicol.* **2003**,*41(2)*,131–135.

96. Mignot E.; Nishino S. Emerging Therapies in Narcolepsy-Cataplexy. *Sleep.* **2005**,*28(6)*,754–763.

97. Pardi D.; Black J. γ-Hydroxybutyrate/Sodium Oxybate Neurobiology, and Impact on Sleep and Wakefulness. *CNS Drugs.* **2006**,*20(12)*,993–1018.

98. Scrima L.; Hartman P.; Nowack W.; Johnson F.; Hiller F. Long-Term Effectiveness of Gamma Hydroxybutyrate in Treating Cataplexy and Sleepiness in Narcolepsy Patients. *Sleep Res.* **1989**,*18*, 77.

99. Scrima L.; Hartman P.; Johnson FJ, Hiller F. Efficacy of Gamma-Hydroxybutyrate Versus Placebo in Treating Narcolepsy-Cataplexy: Double-Blind Subjective Measures. *Biol. Psych.* **1989**,*26(4)*,331–343.

100. Scrima L.; Johnson F.; Thomas E.; Hiller F. Effects of Gamma-Hydroxybutyrate (GHB) on Multiple Sleep Latency Test (MSLT) in Narcolepsy Patients: A Long Term Study. *Sleep Res.* **1990**,*19*, 290.

101. Scrima L.; Hartman P.; Johnson F.; Thomas E.; Hiller F. The Effects of Gamma Hydroxybutyrate on the Sleep of Narcolepsy Patients: A Double Blind Study. *Sleep.* **1990**,*13(6)*,479–490.

102. Lammers G.; Arendt J.; Declerck A.; Ferrari M.; Schouwink G.; Troost J. Gammahydroxybutyrate and Narcolepsy: A Double-Blind Placebo-Controlled Study. *Sleep.* **1993**,*16(3)*,216–220.

103. Scrima L. Gamma-hydroxybutyrate (GHB) Treated Narcolepsy Patients Continue to Report Cataplexy Controlled for up to Five (5) Years. *Sleep Res.* **1992**,*21*, 262.

104. Wang Y.; Swick T.; Carter L.; Thorpy M.; Benowitz N. Safety Overview of Postmarketing and Clinical Experience of Sodium Oxybate (Xyrem): Abuse, Misuse, Dependence, and Diversion. *J. Clin. Sleep Med.* **2009**,*5(4)*,365–371.

105. Wang Y.; Swick T.; Carter L.; Thorpy M.; Benowitz N. Updates and Correction to Previously Published Data. *J. Clin. Sleep Med.* **2011**,*7(4)*,1–2.

106. Anic-Labat S.; Guilleminault C.; Kraemer H.; Meehan J.; Arrigoni J.; Mignot E. Validation of a Cataplexy Questionnaire in 983 Sleep-disorders Patients. *Sleep.* **1999**,*22(1)*,77–87.

107. Moore W.; Silber M.; Decker P., et al. Cataplexy Emotional Trigger Questionnaire (CETQ)-A Brief Patient Screen to Identify Cataplexy in Patients with Narcolepsy. *J. Clin. Sleep Med.* **2007**,*3(1)*,37–40.

108. Akimoto H.; Honda Y.; Takahashi K. Pharmacotherapy in Narcolepsy. *Dis. Nerv. Syst.* **1960**,*21*, 704–706.

109. Hishikawa Y.; Ida H.; Nakai K.; Kaneko Z. Treatment of Narcolepsy with Imipramine (Tofranil) and Desmethylimipramine (Pertofran). *J. Neuro. Sci.* **1963**,*3*, 453–461.

110. Vignatelli L.; D'Alessandro R.; Candelise L. Antidepressant Drugs for Narcolepsy (Review). *Cochrane Database Syst. Rev.* **2010** *(6)*, 1–27.

111. Xyrem (sodium oxybate) oral solution [prescribing information]. Jazz Pharmaceuticals, Inc.: Palo Alto, CA, 2014.

112. Morgenthaler T.; Kapur V.; Brown T., et al. Practice Parameters for the Treatment of Narcolepsy and Other Hypersomnias of Central Origin. *Sleep.* **2007**,*30(12)*,1705–1711.

113. Billiard M.; Bassetti C.; Dauvilliers Y., et al. EFNS Guidelines on Management of Narcolepsy. *Eur. J. Neurol.* **2006**,*13*, 1035–1048.

114. Scrima L.; Johnson F.; Thomas E.; Hiller F. Effect of Gamma-Hydroxybutyrate (GHB) on Multiple Sleep Latency Test (MSLT) Narcolepsy Patients: A Double Blind Study. *Sleep Res.* **1990**,*19*, 287.

115. Scrima L.; Johnson F.; Hiller F. Long-Term Effect of Gamma-Hydroxybutyrate on Sleep in Narcolepsy Patients. *Sleep Res.* **1991**,*20*, 330.

116. Scrima L.; Skakich-Scrima S.; Miller B. The Efficacy of Gamma-Hydroxybutyrate (GHB) in

Narcoleptics: 9 Years. *Sleep.* **2000**, *23*(Abstract Supplement 2), A 293.

117. Nishino S.; Fruhstorfer B.; Arrigoni J.; Guilleminault C.; Dement W.; Mignot E. Further Characterization of the Alpha-1 Receptor Subtype Involved in the Control of Cataplexy in Canine Narcolepsy. *J. Pharmacol. Exp. Ther.* **1993**,*264*(*3*),1079–1084.

118. Wang J.; Greenberg H. Status Cataplecticus Precipitated by Abrupt Withdrawal of Venlafaxine. (Case Report). *J. Clin. Sleep Med.* **2013**,*9*(*7*),715–716.

119. Mignot E. A Practical Guide to the Therapy of Narcolepsy and Hypersomnia Syndromes. *Neurotheraputics.* 2012.

120. US Xyrem Multicenter Study Group. A Randomized, Double Blind, Placebo-Controlled Multicenter Trial Comparing the Effects of Three Doses of Orally Administered Sodium Oxybate with Placebo for the Treatment of Narcolepsy. *Sleep.* **2002**,*25*(*1*),42–49.

121. The Xyrem International Study Group. Further Evidence Supporting the Use of Sodium Oxybate for the Treatment of Cataplexy: A Double-Blind, Placebo-Controlled Study in 228 Patients. *Sleep Med.* **2005**,*6*, 415–421.

122. US Xyrem Multicenter Study Group. A 12-month, Open-Label, Multicenter Extension Trial of Orally Administered Sodium Oxybate for the Treatment of Narcolepsy. *Sleep.* **2003**,*1*, 31–35.

123. US Xyrem Multicenter Study Group. Sodium Oxybate Demonstrates Long-Term Efficacy for the Treatment of Cataplexy in Patients with Narcolepsy. *Sleep Med.* **2004**,*5*, 119–123.

124. Boscolo-Berto R.; Viel G.; Montagnese S.; Raduazzo D.; Ferrara S.; Dauvilliers Y. Narcolepsy and Effectiveness of Gamma-Hydroxybutyrate (GHB): A Systematic Review and Meta-Analysis of Randomized Controlled Trials. *Sleep Med Rev.* **2012**,*16*, 431–443.

125. Mansukhani M.; Kotagal S. Sodium Oxybate in the Treatment of Childhood Narcolepsy–Cataplexy: A Retrospective Study. *Sleep Med.* **2012**,*13*, 606–610.

126. John J.; Wu M.; Boehmer L.; Siegel J. Cataplexy-Active Neurons in the Hypothalamus: Implications for the Role of Histamine in Sleep and Waking Behavior. *Neuron.* **2004**,*42*(*4*),619–634.

127. Valko P.; Gavrilov Y.; Yamamoto M., et al. Increase of Histaminergic Tuberomammil-

lary Neurons in Narcolepsy. *Ann Neurol.* **2013**,*74*(*6*),794–804.

128. Sundvik M.; Panula P. Interactions of the Orexin/Hypocretin Neurones and the Histaminergic System. *Acta Physiologica.* **2015**, *213*, 321–333.

129. Dauvilliers Y.; Bassetti C.; Lammers G., et al. Pitolisant Versus Placebo or Modafinil in Patients with Narcolepsy: A Double-Blind, Randomised Trial. *Lancet Neurol.* **2013**,*12*, 1068–1075.

130. Inocente C.; Arnulf I.; Bastuji H., et al. Pitolisant, an Inverse Agonist of the Histamine H3 Receptor: An Alternative Stimulant for Narcolepsy-Cataplexy in Teenagers with Refractory Sleepiness. *Clin. Neuropharm.* **2012**,*35*(*2*),55–60.

131. Baier P.; Hallschmid M.; Seeck-Hirschner M.; Weinhold S.; et al. Effects of Intranasal Hypocretin-1 (Orexin A) on Sleep in Narcolepsy with Cataplexy. *Sleep Med.* **2011**,*12*, 941–946.

132. Fujiki N.; Yoshida Y.; Ripley B.; Mignot E.; Nishino S. Effects of IV and ICV Hypocretin-1 (Orexin A) in Hypocretin Receptor-2 Gene Mutated Narcoleptic Dogs and IV Hypocretin-1 Replacement Therapy in a Hypocretin-Ligand-Deficient Narcoleptic Dog. *Sleep.* **2003**,*26*(*8*),953–959.

133. Siegel J. Hypocretin Administration as a Treatment for Human Narcolepsy. *Sleep.* **2003**,*26*(*8*),953–959.

134. Huang Y.; Guilleminault C. Narcolepsy: Action of Two gamma-Aminobutyric Acid Type B Agonists, Baclofen and Sodium Oxybate. *Ped. Neuro.* **2009**,*41*, 9–16.

135. Black S.; Morairty S.; Chen T., et al. GABAB Agonism Promotes Sleep and Reduces Cataplexy in Murine Narcolepsy. *J. Neuro. Sci.* **2014**,*34*(*19*),6485–6494.

136. Fronczek R.; Verschuuren J.; Lammers G. Response to Intravenous Immunoglobulins and Placebo in a Patient with Narcolepsy with Cataplexy. *J. Neurol.* **2007**,*254*(*11*),1607–1608.

137. Plazzi G.; Poli F.; Franceschini C., et al. Intravenous High-Dose Immunoglobulin Treatment in Recent Onset Childhood Narcolepsy with Cataplexy. *J. Neurol.* **2008**,*255*(*10*),1549–1554.

138. Lecendreux M. Clinical Efficacy of High-Dose Intravenous Immunoglobulins Near the Onset of Narcolepsy in a 10-year-old Boy. *J. Sleep Res.* **2003**,*12*, 347–348.

139. Dauvilliers Y.; Carlander B.; Rivier F.; Touchon J.; Tafti M. Successful Management of Cataplexy

with Intravenous Immunoglobulins at Narcolepsy Onset. *Ann Neurol.* **2004**,*56(6)*,905–908.

140. Dauvilliers Y. Follow-Up of Four Narcolepsy Patients Treated with Intravenous Immunoglobulins. *Ann Neurol.* **2006**,*60(1)*,153.

141. Dauvilliers Y.; Abril B.; Mas E.; Michel F.; Tafti M. Normalization of Hypocretin-1 in Narcolepsy after Intravenous Immunoglobulin Treatment. *Neurology.* **2009**,*73*, 1333–1334.

142. Knudsen S.; Biering-Sorensen B.; Kornum B., et al. Early IVIg Treatment Has No Effect on Post-H1N1 Narcolepsy Phenotype or Hypocretin Deficiency. *Neurology.* **2012**,*79*, 102–103.

143. Black J.; Swick T.; Feldman N., et al. Oral JZP-110 (ADX-N05) for the Treatment of Excessive Daytime Sleepiness in Adults with Narcolepsy: Results of a Randomised, Double-Blind, Placebo-Controlled Trial. *J. Sleep Res.* **2014**,*23*, 32–33.

144. Bogan R.; Feldman N.; Emsellem H., et al. Effect of Oral JZP-110 (ADX-N05) Treatment on Wakefulness and Sleepiness in Adults With Narcolepsy. *Sleep Med.* 2015.

145. Scrima L.; Emsellem H.; Becker P., et al. Patient Global Impression of Change Correlates With Clinical Global Impression of Change in a Clinical Trial of JZP-110 for the Treatment of Narcolepsy. *Sleep.* **2015**, *23*(Abstract Supplement), A266.

18

FATIGUE MANAGEMENT

Tatyana Mollayeva and Colin M. Shapiro

ABSTRACT

Fatigue is a complex nonspecific symptom that is commonly experienced after a period of mental or physical activity, but can also be the result of other causes where energy expenditure exceeds restorative processes. When excessive, fatigue affects every aspect of one's life, decreasing physical and mental capacity, changing roles and relationships, and causing social isolation. Research in this area has provided foundational knowledge on the mechanisms of fatigue, its correlates, and management. Despite the complexity of the symptom, and the need for comprehensive investigation for its identification and determination of etiology, several targeted interventions have been described with encouraging results. The continuous and multifactorial nature of fatigue calls for a patient-centered, flexible, and broad approach to its management.

Tatyana Mollayeva
Toronto Rehabilitation Institute-University Health Network, 550 University Avenue, Rm 11120, Toronto, Ontario M5G 2A2. Tel: 416-597-3422 ext 7848; Fax: 416-946-8570

Colin M. Shapiro
Toronto Western Hospital, 399 Bathurst Street, Rm 7MP421, Toronto, Ontario M5T 2S8. Tel: 416-603-5800 ext 5160; Fax: 416-603-5292

Corresponding author: Tatyana Mollayeva, E-mail: tatyana.mollayeva@utoronto.ca

18.1. INTRODUCTION

There is no universally accepted conceptual framework for fatigue.[1] The term originates from the Latin word *fatigo*, meaning "to tire,"[2] and has multiple definitions. The Stedman Medical Dictionary's[2] main definitions for fatigue are: 1) a state following mental or bodily activity, characterized by a diminished capacity for and efficiency of work and feelings of weariness, sleepiness, or irritability; it can likewise be the result of other causes where energy expenditure exceeds restorative processes and can be experienced within a single organ, and 2) a feeling of boredom and weariness as a result of lack of stimulation, monotony, or disinterest in one's setting. In many cases, however, fatigue arises not in connection with a medical cause, physical or mental exertion, or boredom, and in these cases diagnostic uncertainty is unavoidable.[3] Also uncertain is whether fatigue represents a primary state or a composite of symptoms that may vary depending on the etiology of the underlying condition. In other words, fatigue may represent a surface phenomenon of multiply determined physiological and pathological processes which contribute to a final common pathway of a negative perceptual feeling state (Fig. 18.1). This chapter focuses largely on generic theories of fatigue and its management, developed by scientists from various disciplines whilst attempting to understand the fatigue experienced by both healthy and diseased populations.

18.2. THE EXPERIENCE OF FATIGUE ACROSS LIFESPAN

The lack of a universally accepted definition of fatigue, in both the physiological and pathological contexts, makes it difficult to analyze and compare epidemiological studies on the topic. Nevertheless, several general trends have been noted. First, most people experience fatigue across the lifespan, suggesting fatigue is a continuous construct related to diminished performance in response to either physical or cognitive tasks.[4,5] In a study of 611 general practice attenders, just 10 percent of men and 18 percent of women reported no fatigue, and over 10 percent of both males and females felt fatigue most of the time.[6] In the historical "Lundby Study," fatigue was reported to be the most common symptom in one-third of women and a fifth of men.[7] A population-based study conducted in Germany suggested age and sex differences in relation to fatigue frequency and severity in the general population.[8] Females were significantly more affected than males, and there was a strong and nearly linear age dependency, with the oldest age group having the highest fatigue

scores.[8] Similarly, in the general population in Colombia, women were more affected by fatigue than men, and there was an almost linear age trend, with higher mean scores reported for subjects that were older and those with chronic diseases.[9] In accordance with prior findings with regards to sex differences, a US study of the general population found women reported higher levels of fatigue than men, with results also yielding a main effect for age: younger participants gave significantly higher fatigue ratings.[10] Chronic disabling fatigue has recently emerged as affecting children, adolescents and younger adults.[11-13] From 11 years of age onward, young people have similar rates and types of chronic fatiguing illnesses as adults.[14]

The contextual understanding of the experience of fatigue in various age groups and populations comes from discussion with children, adolescents, parents, workers, and persons at the end of their lives.[15-17] Until recently, we assigned numerous connotations to the perceived state of fatigue such as lack of energy, weakness, fatigability, tiredness, boredom, and so forth. A recent study provides evidence that the taxonomy of tiredness, fatigue, and exhaustion are discrete domains that show the progression of an adaptive response across a continuum.[18] Thus, *tiredness* was described as an inconvenience with slight impact on mental, emotional, and functional aspects of daily activity; the effects were variable and manageable. The *fatigue* domain was associated with an increased sensitivity to hardship and necessary limitations. There was evidence that regular daily activities had become more difficult to perform, interfering with functional and social roles, and raising worry and frustration. Finally, *exhaustion* was characterized by an inability and disinterest in managing functional and social roles by planning and developing means to perform the required activities. People in this state described themselves as being attending only to immediate challenges and requirements. Researchers suggested that the experience of fatigue consists of 1) a mental challenge, including emotional effects and cognitive awareness of the debility and mental tenacity, and 2) a physical challenge, including limitations in leisure activities and functional roles. The described experience of fatigue highlights it as a socially constructed concept, and where mental or physical demands exceed present performance capacity.[19] Limited energy reserves and inability to undertake valued daily activities are key manifestations of excessive fatigue. Energy, vitality, and strength were commonly identified as fundamental to maintenance of health and well-being across age groups. Half of the over 100 individuals who took part in interviews about their beliefs with respect to health, reported fatigue as a central construct of health, where being healthy meant having the energy to work.[20] One of the emerging neuroscience topics in energy metabolism, energy regulation and energy balance is the function of sleep. In the following paragraph we provide an overview of the current state of knowledge on the relationship between sleep and energy balance.

18.3. THEORIES OF SLEEP AS IT RELATES TO ENERGY CONSERVATION AND ALLOCATION

While recent *energy conservation* strategies for managing fatigue focus on rearranging home and work environments to minimize obstacles in performing tasks and to consolidate similar tasks, pacing, priority-setting, and reassigning tasks, the theory of *sleep function*, as it concerns conservation of energy, is an important discussion.

Several models illustrating this relationship have been proposed.[21-24] Berger and Phillips' model suggests energy is preserved because all biological functions are slowed during sleep.[21] Other models propose that sleep is a time for recovery and recuperative processes to take place.[22,23] The recent *energy allocation model* of sleep function provides a new theory based on the temporal organization of energy acquisition and utilization.[24] Through the lens of this model, the underlying function of sleep is inextricably linked to waking, as investment into biological processes is differentially shared between the states in a manner that optimizes energy utilization to maximize lifetime reproductive success. It is the predictable cycling of sleep and wakefulness that allows for completion of biological activities to be coordinated between these behavioral states.[24] Utilizing this theory, it is possible to explain fluctuations in performance capacity across the daily cycle of the circadian clock (that is, fatigue is greater during the biological night than day) and sleep deprivation and time awake affects performance capacity, and thereby contributes to the experience of fatigue.

Several longitudinal studies in various clinical and non-clinical populations have discussed the relationship between sleep quality and energy/vitality and fatigue.[25,26] More recently, studies on symptom clusters have revealed that sleep dysfunction and fatigue commonly co-occur as part of a cluster of three or more symptoms. [27-31] The most commonly presenting cluster in patients with unipolar major depression, for example, included anxiety, fatigue, insomnia, and pain.[27] This observation provides a logical framework for a structured treatment strategy for fatigue to deliver optimal care. Moreover, it suggests that diverse behavioral responses to internal and external stimuli are coordinated in unison by the elaborate networks across brain structures, and therefore a sophisticated and structured treatment strategy is necessary to deal with any one of the components of a cluster in etiology.

18.4. FATIGUE CAUSES AND RELATED FACTORS

The complexity of fatigue calls for investigation of its possible causes for its management (Fig. 18.1).[32] Current differential diagnoses encompassing fatigue include vascular, infectious, neuroplastic/neurogenic, psychogenic (e.g., depression, sleep disorder, anxiety, and so forth), autoim-

mune, toxic, endocrine, sedentary lifestyle, nutrition-related (e.g., iron and vitamin B deficiency), and drug-related,[33] each of which can be effectively managed through application of a targeted intervention. If etiology remains undetermined, current best practices include reassurance and frequent follow-up, supportive counseling, behavioral therapy, encouragement to maintain maximal function, and comprehensive work-up to ensure exclusion of side-effects of drugs.[33] Follow-up of patients referred to hospital with unexplained fatigue (that is, uncertain cause) found that most patients had made a functional recovery by 2 years after initial clinic attendance.[34] Impaired functioning was associated with certain patient characteristics such as a belief in a viral cause of the illness, limited exercise, change in or departure from employment, particular diets (that is, avoidance of certain foods), alcohol consumption, being a member of a patients' self-help organization, and having an emotional disorder. In the following paragraphs, we provide an overview of the mechanisms and manage-

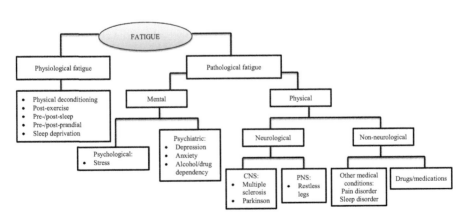

Figure 1: Pathways to fatigue. Modified from Mollayeva et al. (2014)[32]

ment strategies for fatigue in the general population.

18.4.1. Fatigue and Inflammation

Generally, inflammation is a critical bodily response against foreign invaders (that is, infection or injury); chronic or exaggerated inflammation, however, is linked to health decline, implicated in significant medical disorders including cardiovascular disease, diabetes, arthritis, certain cancers, and age-related decline.[35,36] These medical disorders are recognized as being associated with fatigue, even in the absence of an ongoing active disease process.[37] Each of these disorders has also been reported to be associated with poor sleep, highlighting the possibility of common neural and behavioral pathways (that is, stress and inflammation) among them.

The relationship between stress and inflammation has been described in humans and animal models. Pro-inflammatory cytokines increase in response to acute stressors, and the magnitude of inflammation observed in response to stressors was not found to be predicted by perceived stress.[38] This observation supports the notion of self-perceived assessment of symptoms, fatigue included, as a poor predictor of bodily responses to acute stressors.[39] Higher levels of circulating inflammatory markers have been observed in young and middle-aged adults with major depression than in controls with no psychiatric history.[40] Such findings were also reported in older adults.[41] Physiological mechanisms linking stress, depression, and inflammation have been proposed through stress-induced activation of the hypothalamic-pituitary-adrenal (HPA) axis resulting in release of cortisol, which signals back to the hypothalamus and pituitary producing anti-inflammatory effects on the cytokine-producing cells acutely; extended exposure to elevated cortisol results in insensitivity in the HPA axis and cytokine-producing cells, provoking production of pro-inflammatory cytokines.[42] Cytokines affect the brain via transfer through 1) the blood-brain barrier, 2) stimulation of the efferent vagal nerves, and 3) by reducing the availability of tryptophan (that is, a precursor of serotonin).[43] It is well documented that cytokines produce various sickness behaviors (e.g., fatigue, lethargy, impaired sleep, and withdrawal[44]) and can mimic symptoms of depression.[45] Research with emphasis on physiological mechanisms to benefit fatigue outcomes has grown dramatically in recent years. A number of pharmacological trials have been conducted in both clinical and non-clinical populations, proposing use of progesterone derivates,[46,47] corticosteroids,[48,49] anabolic drugs,[50,51] cannabinoids,[52,53] nonsteroidal anti-inflammatory drugs,[54,55] and cytokine modulation drugs.[56] However, there remain controversies with regards to the above pharmacological approaches, particularly with their appropriateness for long-term management.[57] Moreover, challenges in defining the pathophysiological mechanisms of fatigue have left much still to be understood before implication of such interventions.

18.4.2. Challenges in Studying the Physiological Mechanisms of Fatigue

Appropriate assessment is of utmost importance in studies seeking to elucidate physiological pathways linking manifestations of fatigue with endocrine, emotional, and somatic human functions. This is particularly relevant to the assessment of fatigue in persons who have experienced it continuously over a prolonged period of time, for example, in persons with chronic illnesses and those of older age. It is conceivable that not only participation in research is reduced in such individuals,[58] but the ability to concentrate and report information about one's experienced symptomatology may also be impacted.[59] Consequently, the most extreme manifestations of fatigue are likely to be under-reported and their impact likely reduced by proxy respondents. Objective assessment of muscle fatigability (that is, weakness) through tests of functional capacity (that is, dynamometers) are reported to be useful in the clinical setting due to their potential to provide relevant information (that is, the index of muscle activity) using a single or repetitive motor task. However, such measurements are subject to natural variability due to the menstrual cycle,[60] current health conditions, physical and mental work capacity, and the number of pain localizations.[61] In addition, the issue of motivation is encapsulated in any measurement of a voluntary task.

An actigraph, measuring periods of activity and inactivity, has an advantage over the dynamometer, specifically due to its capacity to provide longitudinal assessment during which combined measurements of self-perceived fatigue, circadian rhythms, blood sampling, and so forth, may provide valuable insight into the individual's underlying pathophysiology. Studies using actigraphy indicate that disruptions to circadian rhythms may be associated with both fatigue and depression,[62] and that the association between fatigue and circadian disruption is independent of depression.[63] Consistency in the day- to-day patterns of rest and activity was related to lower fatigue scores and depression as assessed by actigraph.[64]

Testing *the vagal nerve hypothesis* for the substrate of fatigue initiation and maintenance in humans is problematic.[65] A gold standard to provide direct evidence to support the hypothesis may be dissection of cadavers of patients who suffered with chronic fatigue and immunohistochemical staining for inflammation within the vagus nerve.[65] Given the difficulty of dissecting out all possible infection locations in the long and highly branched vagus nerve, other models and approaches should be considered.

The *tryptophan* (that is, amino acid precursor for serotonin (5-HT)) *dysregulation hypothesis* arises from studies of exercise-induced fatigue. Sustained exercise was shown to elevate central 5-HT,[66,67] which, decades ago was linked to 'lethargy' and diminished 'muscular tone' in animal models.[68,69] The revised *central fatigue hypothesis* suggests that an increase in the central ratio of serotonin to dopamine is associated with feelings of tiredness and lethargy, accelerating the onset of fatigue, whereas a low ratio favors improved performance through the maintenance of motivation and arousal.[70] Convincing evidence for the role of dopamine in the development of fatigue comes from work investigating physiological responses to amphetamine use. Other central catecholamines have yet to be studied with respect to links to exercise capacity during exercise in temperate conditions. Additional evidence implicates dysregulation of 5-HT in disrupted hypothalamic-pituitary function.[71] In patients with chronic fatigue, mild hypercortisolism has been observed, with the proposed explanation being improperly functioning hypothalamic neurons producing cortico-trophin-releasing hormone.[72] Testing this hypothesis requires detailed study of the HPA function in matched samples of controls and persons with chronic fatigue. The basal secretion of hormones may vary, thus provocative endocrinologic challenges to investigate the ability of the system to respond to a stimuli is key. Particular attention should be given to special precautions, patient preparation, timing of the test, and dose of medication.[73]

The existing experimental research on the mechanisms of fatigue is full of assumptions. Thus, it is conceivable that other unidentified mechanisms are involved and yet to be uncovered. There are many poorly understood phenomena which are clearly fertile for research. Among them are persons with chronic fatigue who will recover. Are there neurophysiological or behavioral signatures that can be detected at early fatigue onset and that are harbingers of recovery? Given the complexity and intertwined relationship of neural and behavioral processes in humans, can fatigue mechanisms be explained by a biological or a behavioral approach alone, or is a multilevel integrative analysis necessary?

18.4.3. Interventions for Fatigue

18.4.3.1. Exercise for fatigue

Fatigue can interfere with performance of activities of daily living. Its detrimental effect on routine activities can discourage a person's adherence to these activities, and propel them to seek the help of others or even completely avoid these activities on a regular basis.[74] If fatigue interferes with physical (that is, manual tasks) or cognitive (that is, concentration or ability to direct attention) demands of a job, the consequent effect on one's ability to continue with gainful employment can cause social isolation and feelings of a sense of loss, withdrawal from those who could be most supportive, initiation of the grief process, which can manifest itself in many forms, including loss of appetite, depression, disturbed sleep, and fatigue.[75-77] Fatigue may also interfere with maintenance of self-care such as preparation of food, which then lead to malnutrition and weight loss, decreased muscle mass and loss of functional ability.[78] For the past several decades, exercise has been recognized as a promising strategy for managing chronic fatigue.[79] Such non-competitive and rhythmical activities as walking, cycling, swimming, dancing, and running have been shown to produce psychological benefits.[80] Exercise 30–45 min in duration at least three times per week has been advocated as having physiological benefits, linked directly to the enhancement of functional capacities of the cardiovascular, pulmonary and muscular systems.[79,80] With training, the relative stress experienced with performing certain activities shown to be lower, and the physiological reserve higher.[81] A systematic review of randomized controlled trials in 1518 participants with chronic fatigue reported exercise generally benefitted participants and persons felt less fatigued following exercise therapy, with no evidence suggesting exercise therapy may worsen outcomes.[80] A positive effect with respect to sleep, physical function, and self-perceived general health was also observed.[80]

18.4.3.2. Sleep intervention for fatigue

A recent observational study of active duty military personnel with obstructive sleep apnea (OSA) and adherence to positive airway pressure reported improvement in fatigue and energy, depressive symptoms, sleepiness, emotional well-being, and social functioning.[82] A six-week mindfulness meditation intervention directed toward sleep disturbance in older adults showed significant improvement in insomnia symptoms, depression symptoms, fatigue interference, and fatigue severity compared to the sleep hygiene education intervention.[83] A randomized controlled trial of cognitive behavioral therapy for insomnia (CBT-I) in older adults with sleep maintenance insomnia[84] found that participants in the CBT-I group reported large reductions in fatigue, daytime sleepiness, and impaired daily feelings and functioning immediately following treatment, reductions significantly greater than those reported in the waitlisted control group; these effects remained at three-month follow-up.

A stress management intervention directed toward psychosocial adaptation for women with early-stage breast cancer reported greater improvements in sleep quality scores compared to controls as well as greater reductions in fatigue-related daytime interference, though there were no significant differences in changes in fatigue intensity.[85]

The potential value of schedule optimization for reducing errors due to fatigue (that is, currently used in aviation and by the US Military and Federal Transporta-tion Association) has been investigated in medical settings.[86] Researchers have found that the predicted impairment from fatigue on the night shift can be significantly reduced by development of countermeasures. Implementing the day shift and modifying the night shift has the potential to eliminate fatigue impairment.

The *MORE Energy* (*mHealth*) intervention to reduce fatigue showed that counseling and advice about exposure to daylight, sleep, exercise, and nutrition was effective with regards to fatigue management in employees with irregular flight schedules and circadian disruption.[87]

18.4.3.3. Nutritional support for fatigue

Experiencing fatigue for a prolong period of time can interfere with food preparation. Moreover, fatigue as a result of various medical chronic disorders can lead to poverty, and inadequate access to food, starvation, and other impairments associated with poor nutritional status. It is conceivable that adequate early nutrition intervention and support may help patients improve prognosis. Several studies of fibromyalgia patients describing nutritional deficiencies demonstrated the benefits of specific diet and nutritional supplementation in terms of managing pain, fatigue, poor sleep quality, and mood.[88] A study of fatigue in patients scheduled for elective colorectal surgery, reported that one-third were at nutrition risk and that an acceptable intake of dietary protein was not achieved during the first 3 days of hospitalization, which was shown to be associated with recovery.[89] Riccio and Rossano recently investigated whether dietary habits and lifestyle influence the course of symptoms arising from multiple sclerosis (MS).[90] They systematically showed that dietary factors and lifestyle may exacerbate or ameliorate symptoms, including fatigue, in both relapsing-remitting MS and in primary-progressive MS by controlling metabolic and inflammatory cellular pathways as well as commensal gut microbiota. Researchers suggested high-calorie diets, with high sodium, sugar and animal product intake, together with a sedentary activity level increases inflammation and over time drives cellular metabolism toward biosynthetic pathways including those of proinflammatory molecules. Other effects of such a diet are dysbiotic gut microbiota, alteration of intestinal immunity, and low-grade systemic inflammation. Conversely, exercise and low-calorie diets with high vegetable, fruit, legume, fish, prebiotic, and probiotic intake act to upregulate oxidative metabolism, decrease synthesis of proinflammatory molecules, and uphold healthy symbiotic gut microbiota. Similarly, research has shown that nutritional intervention with anti-inflammatory foods and dietary supplements can alleviate side effects of immune-modulatory drugs in MS and the symptoms of chronic fatigue syndrome.[90] Bringing this knowledge into clinical practice and providing nutritional guidance and physical activity opportunities to persons with chronic fatigue may presently be the only method for managing fatigue with non-maleficent prospects. A population-based study was conducted in a Japanese working population with three major dietary styles (that is, healthy, westernized breakfast, and traditional Japanese).[91] The healthy diet, characterized by high intakes of vegetables, mushrooms, potatoes, seaweed, soy products, and eggs, was found to be associated with a decreased prevalence of difficulty falling asleep; the traditional Japanese diet, with high intakes of fish, shellfish, *natto (that is, soybeans fermented with* hay bacillus*)*, and low intake of red meat, was associated with a decreased but non-significant prevalence of poor sleep quality among non-shift workers. While the pathophysiological mechanisms of fatigue were not discussed here, the results intuitively highlight effective means by which one can avoid energy deficits and maintain positive energy balances. Asymmetries within either can at least partially explain multiple challenges we face when dealing with disorders of energy storage and their manifestations.

18.4.4. Future Perspectives

Fatigue requires a multidisciplinary management approach starting at the very early stages since its occurrence. Given the non-specificity of the fatigue symptom, investigation of fatigue in the person presenting with it must begin with a comprehensive understanding of the experience of "feeling fatigued". Next, the cause of the fatigue must be determined and a diagnosis established. Although the pathophysiology of fatigue is likely to be multifactorial, the goal is to determine whether the fatigue is caused by a correctable factor (that is, depression, endocrine dysfunction, deconditioning, poor sleep, and so forth) or otherwise, so that suitable interventions are applied. In a systematic review on fatigue in patients with traumatic brain injury (TBI), an algorithm for study fatigue in TBI was proposed. We have adapted this framework for the general management of fatigue (Fig. 18.2).

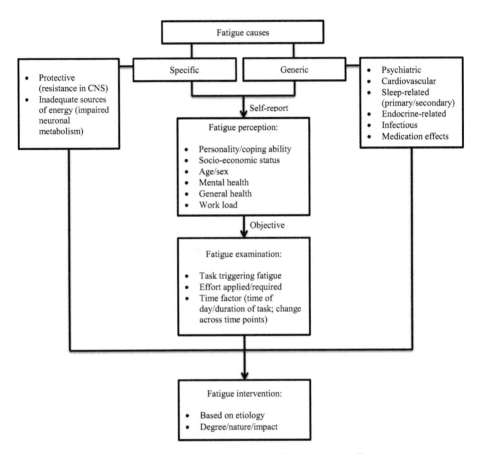

Figure 2: Model for investigating fatigue. Modified from Mollayeva et al, 2014[32]

18.5. CONCLUSIONS

Fatigue affects both healthy and diseased populations. The factors that cause fatigue and the mechanisms responsible for its manifestation, persistence, or amelioration are still under study. Multiple hypotheses have been proposed in the general population as well as in certain disorders associated with high fatigue burden. Commonly, in persons presenting fatigue of uncertain etiology, the symptom declines over time. However, in some cases fatigue can persist. Clinicians seeing patients in the acute stag- es of the symptom's onset should be aware that degree of recovery and functional impairment may be associated with certain patient characteristics. Medication effects, personal factors such as coping ability, physical deconditioning, stress level, and time should be investigated and interventions focusing on the individual's personal circumstances should guide clinical decision-making when dealing with fatigue. A caveat to this point is that each patient has his/her own legitimacy with regard to a particular symptom or problem, and consequently, requires personalized help.

KEY WORDS

- **fatigue**
- **sleep**
- **energy**
- **theories**
- **interventions**
- **management**

REFERENCES

1. Jason, L.A.; Evans, M.; Brown, M.; Porter, N. What is Fatigue? Pathological and Nonpathological Fatigue. *PM R.* **2010,** 2(5), 327–331.

2. Thomas Lathrop Stedman. *Stedman's Medical Dictionary,* 27th ed.; Baltimore: Lippincott Williams & Wilkins, 2000; pp 193.

3. Lloyd, A.R.; Meer, J.W. The Long Wait for a Breakthrough in Chronic Fatigue Syndrome. *B.M.J.* **2015,** 350, h2087.

4. Zijlstra, F.R.; Cropley, M.; Rydstedt, L.W. From Recovery to Regulation: An Attempt to Reconceptualize 'Recovery from Work'. *Stress Health* **2014,** 30(3), 244–252.

5. Moller, H.J. Neural Correlates of "absence" in Interactive Simulator Protocols. *Cyberpsychol. Behav.* **2008,** 11(2), 181–187.

6. David, A.; Pelosi, A.; McDonald, E.; Stephens, D.; Ledger, D.; Rathbone, R.; Mann, A. Tired, Weak, or in Need of Rest: Fatigue Among General Practice Attenders. *B.M.J.* **1990,** 301 (6762), 1199–1202.

7. Hagnell, O.; Gräsbeck, A.; Ojesjö, L.; Otterbeck, L. Mental Tiredness in the Lundby Study: Incidence and Course over 25 Years. *Acta. Psychiatr .Scand.* **1993,** 88(5), 316–321.

8. Schwarz, R.; Krauss, O., Hinz, A. Fatigue in the General Population. Onkologie. **2003,** 26(2): 140–144.

9. Hinz, A.; Barboz, C.F.; Barradas, S.; Körner, A.; Beierlein, V.; Singer, S. Fatigue in the General Population of Colombia - Normative Values for the Multidimensional Fatigue Inventory MFI-20. *Onkologie.* **2013,** 36(7–8), 403–407.

10. Junghaenel, D.U; Christodoulou, C.; Lai, J.S.; Stone, A.A. Demographic Correlates of Fatigue in the US General Population: Results from the Patient-Reported Outcomes Measurement In-formation System (PROMIS) Initiative. *J. Psychosom. Res.* **2011,** 71(3), 117–123.

11. Crawley, E.; Hughes, R.; Northstone, K.; Tilling, K.; Emond, A.; Sterne, J.A. Chronic Disabling Fatigue at Age 13 and Association with Family Adversity. *Pediatrics* **2012,** 130(1), e71–79.

12. Barkmann, C.; Braehler, E.; Schulte-Markwort, M.; Richterich, A. Chronic Somatic Complaints in Adolescents: Prevalence, Predictive Validity of the Parent Reports, and Associations with Social Class, Health Status, and Psychosocial Distress. *Soc. Psychiatry Psychiatr. Epidemiol.* **2011,** 46(10), 1003–1011.

13. Jones, J.F.; Nisenbaum, R.; Solomon, L.; Reyes, M.; Reeves, W.C. Chronic Ffatigue Syndrome and Other Fatiguing Illnesses in Adolescents: A Population-Based Study. *J. Adolesc. Health.* **2004,** 35(1), 34–40.

14. Farmer, A.; Fowler, T.; Scourfield, J.; Thapar, A. Prevalence of Chronic Disabling Fatigue in Children and Adolescents. *Br. J. Psychiatry.* **2004,** 184, 477–481.

15. Varni, J.W.; Beaujean, A.A.; Limbers, C.A. Factorial Invariance of Pediatric patient Self-Reported Fatigue Across Age and Gender: A Multigroup Confirmatory Factor Analysis Approach Utilizing the PedsQL™ Multidimensional Fatigue Scale. *Qual. Life. Res.* **2013,** 22(9), 2581–2594.

16. Frone, M.R.; Tidwell, M.C. The Meaning and Measurement of Work fatigue: Development and Evaluation of the Three-Dimensional Work Fatigue Inventory (3D-WFI). *J. Occup. Health. Psychol.* **2015,** 20(3), 273–288.

17. Kirshbaum, M.N.; Olson, K.; Pongthavornkamol, K.; Graffigna, G. Understanding the Meaning of Fatigue at the End of Life: An Ethnoscience Approach. *Eur. J. Oncol. Nurs.* **2013,** 17(2), 146–153.

18. Olson, K. A New Way of Thinking about Fatigue: A Reconceptualization. *Oncol. Nurs. Forum.* **2007,** 34(1): 93–99.

19. Lian, O.S.; Bondevik, H. Medical Constructions of Long-Term Exhaustion, Past and Present. *Sociol. Health. Illn.* **2015,** 37, 920–935 [Epub ahead of print].

20. Pierret, J. The Social Meaning of Health: Paris, the Essone, the Herault. In: *The Meanings of Illness. Anthropology, History and Sociology*; Auge, M., Herzlich, C., Eds.; Luxembourg: Harwood Academic Publishers, 2012; pp 175–206.

21. Berger, R.J.; Phillips, H.N. Energy Conservation and Sleep. *Behav. Brain. Res.* **1995,** 69(1–2), 65–73.

22. Oswald, I. Sleep as Restorative Process: Human Clues. *Prog. Brain. Res.* **1980,** 53, 279–288.

23. Mignot, E. Why We Sleep: The Temporal Organization of Recovery. *PLoS. Biol.* **2008,** 29, 6(4): e106.

24. Schmidt, M.H. The Energy Allocation Function of Sleep: A Unifying Theory of Sleep, Torpor, and Continuous Wakefulness. *Neurosci. Biobehav. Rev.* **2014,** 47, 122–153.

25. Ward-Ritacco, C.L.; Adrian, A.L.; O'Connor, P.J.; Binkowski, J.A.; Rogers, L.Q.; Johnson, M.A.; Evans, E.M. Feelings of Energy are Associated with Physical Activity and Sleep Quality, but not Adiposity, in Middle-Aged Postmenopausal Women. *Menopause* **2015,** 22(3), 304–311.

26. Choi, J.; Tate, J.A.; Hoffman, L.A.; Schulz, R.; Ren, D.; Donahoe, M.P.; Given, B.A.; Sherwood, P.R. Fatigue in Family Caregivers of Adult intensive Care Unit Survivors. *J. Pain. Symptom. Manage.* **2014,** 48(3), 353–363.

27. Woods, N.F.; Ismail, R.; Linder, L.A.; Macpherson, C.F. Midlife Women's Symptom Cluster Heuristics: Evaluation of an iPad application for Data Collection. Menopause **2015,** 22, 1058–1066 [Epub ahead of print].

28. Lin, S.Y.; Stevens, M.B. The Symptom Cluster-Based Approach to Individualize Patient-Centered Treatment for Major Depression. *J. Am. Board Fam. Med.* **2014,** 27(1), 151–159.

29. Vincent, A.; Hoskin. T.L.; Whipple, M.O.; Clauw, D.J.; Barton, D.L.; Benzo, R.P.; Williams, D.A. OMERACT-Based Fibromyalgia Symptom Subgroups: An Exploratory Cluster Analysis. *Arthritis. Res. Ther.* **2014,** 16(5), 463.

30. Ho, S.Y.; Rohan, K.J.; Parent, J.; Tager, F.A.; McKinley, P.S. A Longitudinal Study of Depression, Fatigue, and Sleep Disturbances as a Symptom Cluster in Women with Breast Cancer. *J. Pain. Symptom. Manage.* **2015,** 49(4), 707–715.

31. Khaleghipour, S.; Masjedi, M.; Kelishadi, R. Circadian Type, Chronic Fatigue, and Serum IgM in the Shift Workers of an Industrial Organization. *Adv. Biomed. Res.* **2015,** 4:61.

32. Mollayeva. T.; Kendzerska, T.; Mollayeva, S.; Shapiro, C.M.; Colantonio, A.; Cassidy, J.D. A Systematic Review of Fatigue in Patients with Traumatic Brain Injury: The Course, Predictors and Consequences. *Neurosci. Biobehav. Rev.* **2014,** 47: 684–716.

33. Toronto Notes. Fatigue, FM 27; Family Medicine, 2012; pp 176.

34. Sharpe, M.; Hawton, K.; Seagroatt, V.; Pasvol, G. Follow up of Patients Presenting with Fatigue to an Infectious Diseases Clinic. *B.M.J.* **1992,** 305(6846), 147–152.

35. Black, P.H.; Garbutt, L.D. Stress, Inflamation and Cardiovascular Disease. *J. Psychosom. Res.* **2002,** 52(1), 1–23.

36. Ishihara, K.; Hirano, T. Il-6 in Autoimmune Disease and Chronic Inflamatory Proliferative Disease. *Cytokine Growth Factor Rev.* **2002,** 13(4-5), 357–368.

37. Wessely, S.; Hotopf, M.; Sharpe, M. *Chronic Fatigue and Its Syndromes.* Oxford University Press: UK, **1998**; , pp 250–276.

38. Brydon, L.; Edwards, S.; Mohamed –Ali, V.; Steoptoe, A. Socio-Economic Status and Stress-Induced Increases In Interleukin -6. *Brain Behav. Immun.* **2004,** 18(3), 281–290.

39. Christian, L.M.; Stoney, C.M. Social Support Versus Social Evaluation: Unique Effects on Vascular and Myocardial Response Patterns. *Psychosom. Med.* **2006,** 68(6), 914–921.

40. Maes, M.; Meltzer, H.Y.; Bosmans, E.; Bergmans, R.; Vandoolaeghe, E.; Ranjan, R.; Desnyder, R. Increased Plasma Concentrations of Interleukin-6, Soluble Interleukin-6, Soluble Interleukin-2 and Transferrin Receptor in major Depression. *J. Affect. Disord.* **1995,** 34(4), 301–309.

41. Thomas, A.J.; Davis, S.; Morris, C.; Jackson, E.; Harrison, R.; O'Brien, JT. Increase in Interleukin-1beta in Late-Life Depression. Am. *J. Psychiatry.* **2005,** 162(1), 175–177.

42. Christian, L.M.; Deichert, N.T.; Gouin, J.P.; Graham, J.E.; Kiecolt-Glauser, J.K; Chapter 64. Psychological Influences of Neuroendocrine and Immune Outcomes. *Handbook of Neuroscience for the Behavioural Sciences,* John Wiley Sons: Hoboken, New Jersey, 2009; Vol 2 pp 414-426.

43. Hansen, M.K.; Taishi, P.; Chen, Z.; Krueger, J.M. Vagotomy Blocks the Induction of Interleukin-1beta (IL-1beta) mRNA in the Brain of Rats in Response to Systemic IL-1beta. *J. Neurosci.* **1998,** 18(6), 2247–2253.

44. Taishi, P.; Chen, Z.; Obál, F. Jr.; Hansen, M.K.; Zhang, J.; Fang, J.; Krueger, J,M. Sleep-Associated Changes in Interleukin-1beta mRNA in the

Brain. *J. Interferon Cytokine Res.* **1998**, 18(9), 793–798.

45. Dantzer, R.; Kelley, K.W. Twenty Years of Research on Cytokine-Induced Sickness Behavious. *Brain. Behav. Immun.* **2007**, 21(2), 153–160.

46. Noreika, D.; Griškova-Bulanova, I.; Alaburda, A.; Baranauskas, M.; Grikšienė, R. Progesterone and Mental Rotation Task: Is There Any Effect? *Biomed. Res. Int.* **2014**, 741–758.

47. Caruso, D.; Abbiati, F.; Giatti, S.; Romano, S.; Fusco, L.; Cavaletti, G.; Melcangi, R.C. Patients Treated for Male Pattern Hair with Finasteride Show, after Discontinuation of the Drug, Altered Levels of Neuroactive Steroids in Cerebrospinal Fluid and Plasma. *J. Steroid. Biochem. Mol. Biol.* **2015**, 146, 74–79.

48. Judson, M.A.; Chaudhry, H.; Louis, A.; Lee, K.; Yucel, R. The Effect of Corticosteroids on Quality of Life in a Sarcoidosis Clinic: The Results of a Propensity Analysis. *Respir. Med.* **2015**, 109(4), 526–531.

49. Schwartz, E.; Holtorf, K. Hormone Replacement therapy in the Geriatric Patient: Current State of the Evidence and Questions for the Future. Estrogen, Progesterone, Testosterone, and Thyroid Hormone Augmentation in Geriatric Clinical Practice: Part 1. *Clin. Geriatr. Med.* **2011**, 27(4), 541–559.

50. White, H.D.; Brown, L.A.; Gyurik, R.J.; Manganiello, P.D.; Robinson, T.D.; Hallock, L.S.; Lewis, L.D.; Yeo, K.T. Treatment of Pain in Fibromyalgia Patients with Testosterone Gel: Pharmacokinetics and Clinical Response. *Int. Immunopharmacol.* **2015**, 27, 249–256 [Epub ahead of print].

51. Valassi, E., Brick, D.J., Johnson, J.C., Biller, B.M., Klibanski, A., Miller, K.K. Effect of Growth Hormone Replacement Therapy on the Quality of Life in Women with Growth hormone Deficiency Who have a History of Acromegaly Versus Other Disorders. *Endocr. Pract.* **2012**, 18(2), 209–218.

52. Dlugos, A.M.; Hamidovic, A.; Hodgkinson, C.A.; Goldman, D.; Palmer, A.A.;de Wit H. More Aroused, Less Fatigued: fatty Acid Amide Hydrolase Gene Polymorphisms Influence Acute Response to Amphetamine. *Neuropsychopharmacology* **2010**, 35(3), 613–622.

53. Vermersch, P. Sativex(*) (Tetrahydrocannabinol + Cannabidiol), an Endocannabinoid System Modulator: Basic Features and Main Clinical Data. *Expert. Rev. Neurother.* **2011**, 11(4 Suppl), 15–19.

54. Shaygannejad, V.; Janghorbani, M.; Ashtari, F.; Zakeri, H. Comparison of the Effect of Aspirin and Amantadine for the Treatment of Fatigue in Multiple Sclerosis: A Randomized, Blinded, Crossover Study. *Neurol. Res.* **2012**, 34(9), 854–858.

55. Wingerchuk, D.M.; Benarroch, E.E.; O'Brien, P.C.; Keegan, B.M.; Lucchinetti ,C.F.; Noseworthy, J.H.; Weinshenker, B.G.;Rodriguez, M. A Randomized Controlled Crossover Trial of Aspirin for Fatigue in Multiple Sclerosis. *Neurology* **2005**, 64(7), 1267–1269.

56. Tyring, S.; Gottlieb, A.; Papp, K.; Gordon, K.; Leonardi, C.; Wang, A.; Lalla, D.; Woolley, M.; Jahreis, A.; Zitnik, R.; Cella, D.; Krishnan, R. Etanercept and Clinical Outcomes, Fatigue, and Depression in Psoriasis: Double-Blind Placebo-Controlled Randomised Phase III Trial. *Lancet* **2006**, 367(9504), 29–35.

57. Mücke, M.; Mochamat; Cuhls, H,; Peuckmann-Post, V.; Minton, O.; Stone, P.; Radbruch, L. Pharmacological Treatments for Fatigue Associated with Palliative Care. *Cochrane Database. Syst. Rev.* **2015**, 5:CD006788.

58. Wanger, T.; Foster, N.R.; Nguyen, P.L.; Jatoi, A. Patients' Rationale for Declining Participation in a Cancer-Associated Weight Loss Study. *J. Cachexia Sarcopenia Muscle.* **2014**, 5(2), 121–125.

59. Sloan, J.A.; Aaronson, N.; Cappelleri, J.C.; Fairclough, D.L.; Varricchio, C.; Clinical Significance Consensus Meeting Group.Assessing the Clinical Significance of Single Items Relative to Summated Scores. *Mayo Clin. Proc.* **2002**, 77, 479–487.

60. Sarwar, R.; Niclos, B.B.; Rutherford, O.M. Changes in Muscle Strength, Relaxation Rate and Fatiguability During the Human Menstrual Cycle. *J. Physiol.* **1996**, 493 (1), 267–272.

61. Rentzsch, M.; March, S.; Swart, E. Association of Hand Grip Strength with Subjective Health and Work Ability: Results of the Baseline Survey of the LidA Study. *Gesundheitswesen* **2015**, 77(4), e85–90.

62. Gander, P.H.; Mulrine, H.M.; van den Berg, M.J.; Smith, A.A.; Signal, T.L.; Wu, L.J.; Belenky, G. Effects of Sleep/Wake History and Circadian Phase on Proposed Pilot Fatigue Safety Performance Indicators. *J. Sleep Res.* **2015**, 24(1), 110–119.

63. Gander, P.H.; Mulrine, H.M.; van den Berg, M.J.; Smith, A.A.; Signal, T.L.; Wu, L.J.; Belenky, G. Pilot Fatigue: Relationships with Departure and Arrival Times, Flight Duration, and Direction. *Aviat. Space Environ. Med.* **2014**, 85(8): 833–840.

64. Akerstedt, T.; Axelsson, J.; Lekander, M.; Orsini, N.; Kecklund, G. Do Sleep, Stress, and Illness Explain Daily Variations in Fatigue? A Prospective Study. *J. Psychosom. Res.* **2014**, 76(4), 280–285.

65. VanElzakker, M.B. Chronic Fatigue Syndrome from Vagus Nerve Infection: A Psychoneuroimmunological Hypothesis. *Med. Hypotheses.* **2013**, 81(3), 414–423.

66. Gaddum, J.H.; Vogt, M. Some Central Actions of 5-hydroxytryptamine and Various Antagonists. *Br. J. Pharmacol. Chemother.* **1956**, 11(2): 1751–1759.

67. Nikiforuk, A. Targeting the Serotonin 5-HT7 Receptor in the Search for Treatments for CNS Disorders: Rationale and Progress to Date. *CNS Drugs* **2015**, 29(4), 265–275.

68. Yamamoto, T., Azechi, H., Board, M. Essential Role of Excessive Tryptophan and Its Neurometabolites in Fatigue. *Can. J. Neurol. Sci.* **2012**, 39(1): 40–47.

69. Meeusen, R.; Watson, P. Amino Acids and the Brain: Do They Play a Role in "central fatigue"? *Int. J. Sport Nutr. Exerc. Metab.* **2007**, 17 Suppl, S37–46.

70. Meeusen, R.; Watson, P.; Hasegawa, H.; Roelands, B.; Piacentini, M.F. Central Fatigue: the Serotonin Hypothesis and Beyond. *Sports Med.* **2006**, 36(10), 881–909.

71. Renan, M.J. Is Hypercortisolaemia a Factor in Chronic Fatigue Syndrome? *Horm. Metab. Res.* **2003**, 35(4): 201–203.

72. Nees, F., Richter, S., Lass-Hennemann, J., Blumenthal, T.D., Schächinger, H. Inhibition of Cortisol Production by Metyrapone Enhances Trace, but Not Delay, Eyeblink Conditioning. *Psychopharmacology (Berl).* **2008**, 199(2): 183–190.

73. Babic, K.; J.Yeo, K.T.; Weiss, R.E. *Endocrine Testing Protocols: Hypothalamic Pituitary Adrenal Axis*. In: De Groot, L.J., Beck-Peccoz, P., Chrousos, G., Dungan, K., Grossman, A., Hershman, J.M., Koch, C., McLachlan, R., New, M., Rebar, R., Singer, F., Vinik, A., Weickert, M.O., Eds.; MDText.com: South Dartmouth (MA), **2000–2014**. Endotext [Internet]. Retrieved Feb 10, 2016 at http://www-ncbi-nlm-nih-gov.myaccess.library.utoronto.ca/books/NBK278940/

74. Ross, S.D.; Estok, R.P.; Frame, D.; Stone, L.R.; Ludensky, V.; Levine, C.B. Disability and Chronic Fatigue Syndrome: A Focus on Function. *Arch. Intern. Med.* **2004**, 164(10), 1098–1107.

75. Feldthusen, C.; Björk, M.; Forsblad-d'Elia, H.; Mannerkorpi, K.; University of Gothenburg Centre for Person-Centred Care (GPCC). Perception, Consequences, Communication, and Strategies for Handling Fatigue in Persons with Rheumatoid Arthritis of Working Age--a Focus Group Study. *Clin. Rheumatol.* **2013**, 32(5): 557–566.

76. Northcott, S.; Hilari, K. Why Do People Lose Their Friends after a Stroke? *Int. J. Lang. Commun. Disord.* **2011**, 46(5), 524–534.

77. Toblin, R.L.; Riviere, L.A.; Thomas, J.L.; Adler, A.B.; Kok, B.C.; Hoge, C.W. Grief and Physical Health Outcomes in U.S. Soldiers Returning from Combat. *J. Affect. Disord.* **2012**, 136(3), 469–475.

78. Storfer-Isser, A.; Musher-Eizenman, D. Measuring Parent Time Scarcity and Fatigue as Barriers to Meal Planning and Preparation: Quantitative Scale Development. *J. Nutr. Educ .Behav.* **2013**, 45(2), 176–182.

79. Maughan, D.; Toth, M. Discerning Primary and Secondary Factors Responsible for Clinical Fatigue in Multisystem Diseases. *Biology (Basel).* **2014**, 3(3), 606–622.

80. Larun, L.; Brurberg, K.G.; Odgaard-Jensen, J.; Price, J.R. Exercise Therapy for Chronic Fatigue Syndrome. Cochrane Database. *Syst. Rev.* **2015**, 2:CD003200.

81. Pereira, V.H.; Gama, M.C.; Sousa, F.A.; Lewis, T.G.; Gobatto, C.A.; Manchado-Gobatto, F.B. Complex Network Models Reveal Correlations among Network Metrics, xercise Intensity and Role of Body Changes in the Fatigue Process. *Sci. Rep.* **2015**, 5:10489.

82. Mysliwiec, V.; Capaldi, V.F. 2nd; Gill, J.; Baxter, T.; O'Reilly, B.M.; Matsangas, P.; Roth, B.J. Adherence to Positive Airway Pressure Therapy in U.S. Military Personnel with Sleep Apnea Improves Sleepiness, Sleep Quality, and Depressive Symptoms. *Mil. Med.* **2015**, 180(4), 475–482.

83. Black, D.S.; O'Reilly, G.A.; Olmstead, R.; Breen, E.C.; Irwin, M.R. Mindfulness Meditation and Improvement in Sleep Quality and Daytime Im-

pairment among Older Adults with Sleep Disturbances: A Randomized Clinical Trial. *JAMA Intern. Med.* **2015**, 175(4), 494–501.

84. Lovato, N.; Lack, L.; Wright, H.; Kennaway, D.J. Evaluation of a Brief Treatment Program of Cognitive Behavior Therapy for Insomnia in Older Adults. *Sleep* **2014**, 37(1), 117–126.

85. Vargas, S.; Antoni, M.H.; Carver, C.S.; Lechner, S.C.; Wohlgemuth, W.; Llabre, M.; Blomberg, B.B.; Glück, S.; DerHagopian, R.P. Sleep Quality and Fatigue after a Stress Management Intervention for Women with Early-Stage Breast Cancer in Southern Florida. *Int. J .Behav. Med.* **2014**, 21(6), 971–981.

86. McCormick, F.; Kadzielski, J.; Evans, B.T.; Landrigan, C.P.; Herndon, J.; Rubash, H. Fatigue Optimization Scheduling in Graduate Medical Education: Reducing Fatigue and Improving Patient Safety. *J. Grad. Med. Educ.* **2013**, 5(1), 107–111.

87. van Drongelen, A.; Boot, C.R.; Hlobil, H.; Twisk, J.W.; Smid, T.; van der Beek, A.J. Evaluation of an mHealth Intervention Aiming to Improve Health-Related Behavior and Sleep and Reduce Fatigue among Airline Pilots. *Scand. J. Work Environ. Health.* **2014,** 40(6), 557–568.

88. Rossi, A.; Di Lollo, A.C.; Guzzo, M.P.; Giacomelli, C.; Atzeni, F.; Bazzichi, L.; Di Franco, M. Fibromyalgia and Nutrition: What News? *Clin. Exp. Rheumatol.* **2015,** 33(1 Suppl 88), S117–125.

89. Gillis, C.; Nguyen, T.H.; Liberman, A.S.; Carli, F. Nutrition Adequacy in Enhanced Recovery after Surgery: A single Academic Center Experience. *Nutr. Clin. Pract.* **2015**, 30(3), 414–419.

90. Riccio, P.; Rossano, R. Nutrition Facts in Multiple Sclerosis. *ASN Neuro.* **2015,** February 18, 7, pii: 1759091414568185.

91. Kurotani, K.; Kochi, T.; Nanri, A.; Eguchi, M.; Kuwahara, K.; Tsuruoka, H.; Akte, S., Ito, R.; Pham, N.M.; Kabe, I.; Mizoue, T. Dietary Patterns and Sleep Symptoms in Japanese Workers: The Furukawa Nutrition and Health Study. *Sleep Med.* **2015**, 16(2), 298–304.

19

FORENSIC AND LEGAL ASPECTS OF SLEEP MEDICINE

Kenneth Buttoo and Christian Guilleminault

ABSTRACT

Neuroscientific research has expanded our knowledge in sleep medicine. The growth of sleep medicine has the potential to make a significant contribution to legal doctrine and practice if the relationship is properly understood. Sleep medicine has extended into the following areas with legal implications: harm done during partial arousals from sleep; accidents or errors in judgment caused by sleepiness behind the wheel or in the workplace; and disability determinations caused by sleepiness-induced work impairment.

Governments have a legitimate interest in protecting public and occupational health and safety in their respective societies. It is not surprising that drowsiness-related risks have been addressed as a matter of law and public policy. Violent behaviors during sleep refer to a broad range of behaviors from simple dream enactment to complex behaviors that may have serious or even lethal consequences. Guidelines for the medical-legal evaluation of such behaviors have been developed and are evolving.

Kenneth Buttoo
601 Harwood Avenue South, Ajax, ON L1S 2J5, Canada

Christian Guilleminault
Stanford Sleep Medicine Center, 450 Broadway, St Pavilion C, 2nd Fl MC 5704, Redwood City, CA, USA

Corresponding author: Kenneth Buttoo, E-mail: kbuttoo@hotmail.com

FORENSIC AND LEGAL ASPECTS OF SLEEP MEDICINE

SLEEP DISORDERS ASSOCIATED WITH VIOLENCE

Prevalence and Risk Factors

Ohayon et al. reported the prevalence of sleep-related violence at 2% in the general population (1). This percentage was felt to be an overestimate as the study was conducted through telephone interviews based on a standard questionnaire and did not involve a sleep specialist. Harmful behavior has been reported in 59% of patients with sleep terrors and sleepwalking who were consecutively recruited at a sleep clinic (2). Violence with nocturnal wanderings of different etiologies was reported in 70% of patients (3). In both studies, these proportions may be overestimated as patients with sleep-related violence behavior are more likely to consult sleep clinics. Most complex behaviors observed in the course of parasomnias and seizures are benign. Why and in what circumstances violent behavior emerges has yet to be defined. The most consistent risk factor identified in this respect is male gender. Violent behavior and injury in arousal disorders is 1.6 – 2.8 times more common in males (4). About 97% of injuries and 80% of potentially lethal behaviors in rapid eye movement sleep behavior disorder (RDB) occur in males (5). Ictal and peri-ictal aggression is more common in males (6).

The emotional state and dream content which have been shown to be predominately unpleasant and aggressive in RBD and arousal disorders most likely play a role in the occurrence of sleep violence (7). Interaction with the sleeper, namely provocation by noise and touch, is an additional factor that has been shown to be related to violence in arousal disorders and sleep-related seizures.

Sleep dissociation is linked to the occurrence of sleep-related violence. The states of wakefulness (W), rapid eye movement (REM) sleep, and non-rapid eye movement (NREM) sleep are very complex phenomena. The determining properties of each state usually cycle in a predictable and uniform manner resulting in the behavioral appearance of a single prevailing state (8). Factors involved in state generation are complex and include a wide variety of neurotransmitters which act upon the multiple neural networks.

Sleep is most likely a fundamental property of numerous neuronal groups, rather than a phenomenon that requires the whole brain (9). State dissociations are the consequence of timing or switching errors in the normal process of the dynamic reorganization of the CNS as it moves from one state of being to another. Elements of one state persist or are recruited erroneously into another state, often with fascinating and dramatic consequences (8). These state dissociations can occur spontaneously or as the result of neurologic dysfunction or medication administration (8).

Genetic predisposition has been well established in studies of twins (10) and of families (11). However, the mode of transmission has not been clear. Ongoing studies suggest that the DBQ1*05 allele may distinguish sleepwalkers from comparison subjects (12). Maturational factors account for the higher prevalence of arousal disorders in childhood compared with adulthood. There is a strong association between RBD and neurodegenerative disorders (13).

The above information may help to explain the difficulties juries have expressed when faced with sleep-related violence. Juries may be perplexed in trying to understand how it is possible for an accused person to orient well in space, walk, climb, and drive without harm and yet not recognize the face of a familiar person whom the sleepwalker is attacking. Studies of the neuroanatomy of the visual system have established that pathways involved in guiding movement and in facial recognition are different

The pathway for visually guided (14) movement terminates in the posterior parietal cortex while that for recognizing faces and assigning meaning to visual stimuli terminates in the midtemporal cortex *(15)*. Broughton reported that sleepwalkers have a greater delay in reaction to visual stimuli when they are aroused from slow-wave sleep than do comparison subjects *(16)*. There are other apparent perceptional difficulties during sleepwalking episodes. In sleep walking violence, the sleep walkers do not appear to hear their victims cry out, nor do they register pain when they hurt themselves. Highly complex motor activity occurs in the absence of awareness, suggesting a deactivation of cortical structures for judgement and consciousness. A series of complex behavior sequences can be generated by brainstem structure alone, without the involvement of cortical regions *(17)*. These brainstem centers are called central pattern generators and control stereotyped innate motor behavior necessary for survival, such as locomotion, swimming, sexual activity, and other rhythmic motor sequences.

Perhaps the most vexing question for juries is that of motivation, since these attacks appear to be unpremeditated, to take place without awareness during the event, and to be followed by retrograde amnesia and remorse. Forensic reports often note extensive sleep deprivation before an aggressive event and suggest that the sleep loss was secondary to psychological stress *(18)*. Perhaps the sleepwalker's integrated goal-directed behavior noted by Bonkalo *(19)* represents a response to the waking stress. Aggression is modulated by the temporal (i.e. amygdala) and frontal lobes (i.e. orbitofrontal and medial prefrontal areas) *(20)*. Lesions in these areas might lead to unwarranted aggressive behavior. Impaired emotional recognition of faces,

errors in odour identification, and disadvantageous decisions in gambling tests can result from dysfunction in the medial prefrontal and orbitofrontal region during wakefulness *(21)*. Exaggerated amygdala activity and diminished activation of the orbitofrontal cortex in response to faces expressing anger and failure to demonstrate coupling between these two structures can result in intermittent explosive disorders during wakefulness *(22)*. A plausible hypothesis would have the amygdala and the orbitofrontal lobe act on subcortical systems mediating reactive aggression *(23)*. The role of the amygdala is to up or down regulate their response to threat, and the orbitofrontal cortex exerts its modulating activity in response to social cues. Therefore, during parasomnias, unwarranted aggressive behaviors are favored by the unrestrained influence of the amygdala that is no longer kept in balance by the hypoactivity of prefrontal associative cortices. A single photon emission computed tomography study performed during an episode of sleepwalking documented hyper perfusion of the posterior cingulate cortex and cerebellar vermis in addition to decreased cerebral blood flow in the frontal and parietal association cortices *(24)*. A recent intracerebral electroencephalogram (EEG) study showed that during confusional arousals, the motor and cingulate cortices are activated and display the same activity as during wakefulness, whereas the frontoparietal associative cortices exhibit an enhancement of delta activity characteristic of sleep *(25)*. Therefore, disorders of arousal share characteristics of both sleeping and waking states resulting from the selective activation of thalamocortical pathways implicated in the control of complex motor and emotional behavior and from hypoactivation of other thalamo-corital pathways including those projecting to the frontal lobes (26).

Classification

Parasomnias are undesirable events that accompany sleep. They usually involve complex, seemingly purposeful and goal-directed behaviors which presumably are performed with some personal meaning to the individual at the time, despite the illogical and unsound nature of the behaviors enacted outside the conscious awareness of the individual (27). Parasomnias involve sleep-related behaviors and experiences over which there is no conscious deliberate control. Parasomnias can be divided as follows: **Disorders of Arousal (From NREM Sleep)**

Confusional arousals
Sleepwalking
Sleep Terrors
Sleep-related eating disorder

Parasomnias Usually Associated With REM Sleep

REM sleep behavior disorder

Other Parasomnias

Sleep-related dissociative disorder
It is not uncommon for an individual to have two or three disorders of arousal. At times REM sleep behavior disorder is observed in association with NREM partial arousal disorders in which case the newly recognized parasomnia overlap can be diagnosed (27).

Disorders of Arousal (From NREM Sleep)

Four basic drive states - sleep, sex, feeding, and aggression – have been documented to emerge in pathologic form with the disorders of arousal, manifesting as sleep-related abnormal sexual behaviors, sleep-related eating disorders, and sleep violence (27).

Confusional arousals consist of mental confusion or confusional behavior which occurs during an arousal or awakening from nocturnal sleep or daytime nap which is not better explained by other disorders or medication use (27). If an individual leaves the bed and starts walking in the course of a confusional arousal, the disorder is referred to as sleepwalking. In sleep terrors the arousal is characterized initially by a cry or a loud scream that is accompanied by autonomic nervous system and behavioral manifestation of intense fear. In addition, the patient will experience difficulty in arousal, mental confusion on arousal, amnesia, or potentially dangerous behaviors (27). Different arousal disorders may coexist in the same individual and, not infrequently, an episode may start as one arousal disorder and evolve into another (that is, sleep terror evolving into sleepwalking) (28).

Disorders of arousal have been associated with sleep-related violence. Homicide, attempted homicide, filicide, suicide and inappropriate sexual behaviors have been reported in this setting (29). Pressman found aggressive behavior occurred in different ways in confusional arousals, sleepwalking, and sleep terrors (30). Pressman also found that 100% of confusional arousals, 81% of sleep terrors, and 40 - 90% of sleepwalking cases were associated with provocations including noise, touch, and/ or close proximity (28).

In confusional arousals, violence is usually elicited when individuals were awakened from sleep by someone else (19). In sleep walking, violent behavior tended to occur when the sleepwalking episode was already underway and the individual was approached by another person or accidentally encountered someone else (31). Violence related to sleep terrors appears to be a reaction to a concrete, frightening image that the individual can subsequently describe (32). These distressing, non-narrative dreamlike mentations are usually a single image that is less complex than the narrative sequence seen in REM behaviors (33).

Sleep-related sexual activity is frequently associated with violence. Sexual behaviors during sleep, termed sexsomnia, involve the repertoire of explicit vocalization with sexual content, violent masturbation, and complex sexual activities including oral sex and vaginal or anal intercourse. In over 90% of cases, sexsomnia occurs during confusional arousals; the remainder cases are related to sleepwalking, REM behavior disorders, and seizures (34). Among the 31 patients reviewed by Schenck and coworkers, 45% displayed assaultive behavior, 29% had sex with minors, and 36% sustained legal consequences from their sexual sleep-related behavior (34). Eighty percent were men who initiated sexual intercourse, whereas women preferentially engaged in masturbation or displayed sexual vocalization. Complete amnesia occurred in all reported cases of sexsomnia (34).

Sexsomnia often includes sexual arousal with autonomic activation, for example, nocturnal erection, vaginal lubrication, nocturnal emission, dream orgasm (wet dream), sweating, and cardiorespiratory response. Ejaculation has been reported in some cases (35). Sexsomnia without sexual arousal has also been reported, and this report may hinder correct diagnosis.

Parasomnias Usually Associated With REM Sleep

REM sleep behavior (RBD) results from REM sleep without atonia, confirmed on electromyography (EMG) with abnormal behavior either injurious or disruptive, captured on video in the absence of epilepsy and other medical disorders (27). RBD is associated with altered dream content and acting out dream mentation. Patients' eyes are usually closed in contrast to arousal disorders where the eyes are usually reported as being open. They appear to interact with their dream rather than with their immediate environment. The behaviors are often brief and, if awakened, the individual is oriented with a recollection of a dream that corresponds to the behaviour (36). Dreams of patients with idiopathic and secondary RBD have a more aggressive content compared to dreams of normal subjects, despite lower levels of daytime aggressiveness (37). In RBD associated with Parkinson's disease, motor activity may be vigorous and fast compared to the hypokinesia observed during wakefulness (38). The restored motor control during REM sleep in association with loss of REM sleep atonia and the action-loaded dream may account for the injury potential of this disorder. RBD is more prevalent in men and after the age of 50. In 45% of cases it represents the first manifestation of a neurodegenerative disorder including Parkinson's disease, multiple system atrophy, and Lewy body dementia (39). Sleep-related injuries to the patient or to the bed partner have been reported in 32 to 69% of cases (40). Attempted assault of sleep partners has been reported in 64% of cases with injuries in 3% (40).

Other Parasomnias

Sleep-related Dissociative Disorders

Sleep-related dissociative disorders can emerge throughout the sleep period and occur without the conscious awareness of the individual. They are a disruption of the usually integrated functions of consciousness, memory, identity, or perception of the environment (41). In contrast to parasomnias of NREM and REM sleep, dissociative disorders occur during well-established EEG wakefulness either at the transition from wakefulness to sleep or within several minutes after awakening from Stage N1 or N2 NREM sleep or from REM sleep (27). An alpha EEG rhythm with abnormal behaviors emerging shortly after an arousal from NREM sleep is not necessarily diagnostic of a dissociative disorder since disorders of arousal may also have an alpha rhythm (27). However, the lag time between EEG arousal and behavioral arousal can usually distinguish the two conditions. With sleep-related arousals, the behaviors emerge almost immediately after EEG arousal. With sleep-related dissociative disorders, there is often a lag-time of 15 - 60 seconds between EEG arousal and behavioral activation (27).

Fictitious Disorders

Fictitious disorders are diagnosed when the patient deliberately produces or falsifies a presentation of an illness with the purpose of either assuming the sick role or causing a sick role in another person (by proxy) (42). In contrast to fictitious disorder, malingering is not a mental illness. Intentionally produced symptoms or signs are motivated by external incentives for the behavior. Symptoms can be referred to the sleep period but when observed always occur during well-established EEG wakefulness.

Establishing a Diagnosis of Parasomnias Associated With Violence

History

The initial step is to obtain a complete history from both the patient and, if possible, from the bed partner. Information on sleep/wake habits, comorbidities, in particular neurological and psychiatric conditions, should be obtained. Drug intake history should also be noted. This history taking should be followed by a general physical, neurological, and psychiatric examination. As various sleep disorders, such as sleepwalking, often fail to occur in a laboratory setting, home videos documenting nocturnal episodes may be helpful (43).

Polysomnography

An extensive polygraphic study with a multichannel scalp EEG, electromyographic monitoring of all four extremities, and continuous audiovisual recording is necessary for diagnosis. Video EEG polysomnography is superior to standard polysomnography for the evaluation of parasomnias because of the increased capability to identify and localize EEG abnormalities and to correlate behavior with EEG and polysomnography (44). Only RBD requires video-polysomnographic documentation.

Sleep architecture is essentially normal in patients with arousal disorders (45). However, increased number of arousals or fragmentation of slow wave sleep, in particular in the first NREM-REM sleep episode, can occur (46). Hypersynchronous delta activity consisting of bilateral rhythmic delta waves occurring for 10 - 20 seconds during slow wave sleep is considered a typical but non-specific EEG marker for arousal disorder. It can occur in 66% of normal arousals (45). Abnormalities of the cyclic alternating pattern representing infra-slow oscillations observed on EEG between phases of sleep promoting slow oscillations and wake-promoting cortical arousals have been documented in chronic sleepwalkers. These include a decrease in phase A1 and an increase in phase A2 and A3 (47). Polysomnography recordings are useful in excluding coexistent sleep disorders. These may act as a trigger for parasomnias and require specific treatment. To establish the diagnosis of rapid eye movement sleep behavior disorder, polygraphic documentation of increased phasic muscle activity and loss of normal muscle atonia during REM sleep is necessary (27). Video-polysomnography typically shows rapid twitches of the extremities or more complex movements representing dream enacting behavior occurring during REM sleep. Although, on awakening, the individual can frequently recall a vivid dream

corresponding to the behavior displayed, the presence or absence of remembered dream imagery does not reliably differentiate between disorders of arousal and RBD (7).

SLEEP-RELATED VIOLENCE AND THE LAW

Linking Brain, Mind and Behavior

Neuroscience involves the study of relationships between the brain, mental activity, and behavior. The law has a keen interest in the mind, and, therefore, in the concept of intention which lies at the heart of the attribution of both moral responsibility and legal liability in the law of torts and of crimes. Many of the mental processes - thinking, sensing, memory, and consciousness - are helpful in explaining the relationship between brain and mind. The mappings from brain activity to mental process and from mental process to behavior remain complex and poorly understood (48).

What neuroscience can tell us about the mind and about the mind's relationship to law depends on one's concept of mind. What options does the law have for the concept of mind?

The first concept is the classic Cartesian dualism. Under this concept, the mind is thought to be some type of non-material (i.e., non-physical) activity that is a part of the person and is somehow in causal interaction with the person's body. It is the source and location of the person's mental life – thoughts, beliefs, sensations, and conscious experiences. The task is to understand how the non-physical entity known as "mind" causally interacts with the physical brain and body of a person. This concept was later repudiated by neuroscientists and neurolegalists (49).

The second concept is that the mind is identical to the brain. Under this concept, the mind is a material (i.e., physical) part of the human being - the brain - that is distinct from but in causal interaction with the rest of the organism. The brain is the subject of the person's mental properties (the brain thinks, feels, intends, and knows) and is the location of the person's conscious experiences. The second concept still keeps intact the same logical, formal structures of the mind as an entity that interacts with the body, replacing the Cartesian soul by the brain.

The third concept holds that the mind is not an entity or substance at all. To have a mind is to possess a certain array of rational powers and action. The roots of this conception are in Aristotle. Under the Aristotelian concept, the mind is not a part of the person that causally interacts with the person's body. It is the mental powers, abilities, and capacities possessed by humans. Likewise, the ability to see is not a part of the eye that interacts with other parts of the physical eye. Under this concept, the question of the mind's location in the body makes no sense, just as the location of eyesight within the eye makes no sense. This Aristotelian idea is also materialist/physicalist in an important sense: if you took away or changed the physical structures in the brain, the mind would go away or change as well. A properly working brain is necessary for having a mind, but the mind is not identical with the brain. The criteria for the ascription of mental attributes to human beings are constituted by their manifold behaviors; it is people who think,

feel, intend, and know (not parts of their brains). Under the third concept, because certain structures may be necessary to exercise various capacities or to engage in certain behaviors, neuroscience may contribute greatly by identifying these necessary conditions as well as by showing that because of injury or deformity, a person lacks them.

Challenges Presented By Applying Sleep Medicine to Law

Sleep medicine and law are very different disciplines. There are differences of language with many terms employed by both scientists and lawyers having different meaning, e.g., automatism. Neuroscience alone cannot answer questions of relevance to the law. It must be used in conjunction with other disciplines such as behavioral genetics, psychology, behavioral sciences, and sociology (50). Neuroscience can reveal some but, crucially, not all of the conditions necessary for behavior and awareness (51).

In modern societies, ideas about responsibility are linked to the extent to which people choose to act in a certain way and their ability to have acted otherwise. In the jurisprudence of punishment and responsibility, distinctions are drawn between acts that are (fully) justified, acts that are (fully) excused, and acts that merely merit some mitigation; and then there are questions about which acts fall into which of these categories.

In order to convict the accused of a crime, the prosecution must prove that the accused did what is alleged in the charge and that he/she did so with a particular mental element (mens rea) such as intention or recklessness (52). If a person pleads the insanity defense, the main thrust of the defense is that the accused should not be held responsible in criminal law at all. It follows that, in some cases, the accused might be trying to show that, because of his/her mental disorder, he/she was incapable of forming the mental element. In other cases, the accused might be trying to show that, although he/she did have the mental element as alleged in the charge, the reason the accused had same was on account of his/her mental disorder (52). The insanity defense is set out in the M'Naghten's rules as laid down by the House of Lords in 1843, " To establish a defence on the ground of insanity, it must be clearly proved that, at the time of the committing of the act, the party accused was labouring under such a defect of reason, from disease of the mind, as not to know the nature and quality of the act he was doing; or, if he did know or, that he did not know he was doing what was wrong." (53).

It is for the defendant to prove, on the balance of probabilities that he/she is insane within the test. If the test is met, the defendant is found not guilty by reason of insanity which is known as the special verdict (52). As a person found not guilty by reason of insanity has not been convicted of any crime, the person cannot be sentenced (52). Following a special verdict, the crown court has the power to make an absolute discharge (that there will be no further action), a supervisor order, or an order to detain the person in hospital possibly with the restriction that he/she is not to be released until permission is given by the secretary of state (52). Permission for release will depend on the person's mental health and the risk to the public that he/

she poses. Expert evidence when a mental condition defense is raised, such as in violent sleep cases, needs to address two questions: culpability and future risk. The statutory special verdict of not guilty by reason of insanity arose out of a need for social control of potentially dangerous individuals. Insanity is a legally defined concept rather than a medical diagnosis or classification (42). It does not require any existing psychiatric illness or permanent impairment. Additionally, the presence of a mental illness itself does not necessarily lead to a plea of insanity if the person is considered legally responsible for his/her behavior at the time of the guilty act (42). This plea is usually submitted following extensive psychological and psychiatric reports of the accused, determining whether the person's concept of right and wrong was impaired at the time of the act.

Applying this concept to a sleep-related violent offence can be problematic for a variety of reasons. Firstly, there are several degrees of consciousness, with numerous transitions between normal and pathological sleep (54). Secondly, there is currently no diagnostic tool that enables the diagnosis of an underlying sleep disorder with absolute certainty, and even if there were one, the presence of a sleep disorder does not necessarily establish causal link to committed act and cannot replicate the mental element associated with the alleged criminal act.

A sleep-related violent offence must be distinguished from a simulation. Criteria for establishing the putative role of an underlying sleep disorder in a specific violent act.

Presence of an underlying sleep disorder:

Presence of solid evidence supporting the diagnosis, previous occurrence of similar episodes

Characteristics of the act:

Occurs on awakening or immediately after falling asleep, abrupt onset and brief duration, impulsive, senseless, without apparent motivation, lack of awareness of individual during event

Victim:

Coincidentally present, possible arousal stimulus

On return to consciousness:

Perplexity, horror, no attempt to escape, amnesia for event

Presence of precipitating factors:

Attempts to awaken the subject, intake of sedative/hypnotic drugs, prior sleep deprivation, alcohol or drug intoxication precluding the use of disorder of arousal in forensic cases

In addition to a medical evaluation, an extensive forensic-psychiatric evaluation is required in this setting aimed at identifying a plausible offence mechanism based on the defendant's behavioral pattern and personality prolife. In the presence of a plausible wakeful offence mechanism, the assumption of a sleep-related violent offence becomes more unlikely (54).

The Defense of (Sane) Automatism

If a person totally lacked control of his/her body at the time of the offence, and if that lack of control was the person's fault, then he or she may not be guilty and may be acquitted (52). This is referred to as the defense of automatism. When automatism results from an external factor which is unforeseen and unlikely to recur (for example, a concussive blow or poisoning), the defense of non-insane automatism can be raised. The distinction between internal and external events is an attempt by the courts to address the policy issues of individuals who are potentially dangerous not being subject to social control if there is a plain acquittal (52).

This distinction has led to several difficulties. The first is the possibility that something that is judged to be an external factor is actually a circumstance that is likely to recur. Diabetics may suffer excessively high blood sugar levels (hyperglycemia) or excessively low blood sugar levels (hypoglycemia). Both states may be caused by external factors (alcohol or insulin) or by internal factors (lack of food or insufficient insulin). A diabetic who, without fault, fails to take insulin and then commits an allegedly criminal act would be treated as insane (52). In contrast, a diabetic who took insulin in accordance with a medical prescription would be acquitted if they were an automaton at the time of committing an allegedly criminal act.

The second issue is deciding whether the condition has an internal or external cause given that most medical conditions consist of one or more predispositions and precipitants. Sleepwalking precipitated by obstructive sleep apnea will be generally very responsive to continuous positive airway pressure (CPAP) treatment but would be treated by the courts as an internal cause and, therefore, insanity (42).

After careful examination of many legal reports from the American and Canadian courts, Schenck et al. criticized the legal notion that sleepwalking with a high rate of recurrence be considered an "insane automatism"(55). According to these authors, this conclusion would grant an open-ended confinement in mental hospitals for those who commit violent acts during a state of somnambulism if the existence of a high probability of recurrence was determined. Furthermore, the law would not, at any moment, grant those people specialized treatment in a sleep medicine center or consider that treatment could successfully change patients' legal status (converting it to a "non-insane automatism") once the possibility of recurrence was reduced (56). By the same token, no provision for accountability is made for individuals who, having been advised of their risk factors for initiating a sleep automatism, would deliberately seek exposure to such situations in order to commit a crime and offer a favorable, legal defense of "non-insane automatism" (55). Thus, the aforementioned authors not only suggest that parasomnia with continued danger be considered a "non-sane automatism" but also introduce the concept of "intermittent state-dependent recurring danger" (55). Such a concept helps determine the future accountability of someone who has just been acquitted by using the parasomnia state defense. Through clinical assessment, possible trigger factors for that parasomnia would be identified, and the defendant would be advised to avoid such triggers. Otherwise, sleepwalking would be unacceptable as a defense argument if another violent episode of parasomnia due to the same trigger should recur in the future.

Nevertheless, the recklessness of engaging in a risky behavior should be firmly determined. For instance, in a setting where it is already known that sleep deprivation for a certain subject is a trigger factor for automatisms and where this individual is a caregiver and is required on occasion to spend his/her precious sleep hours caring for someone gravely ill, if this person were to present a violent episode of parasomnia, he/she should not be considered accountable for such acts (55).

Sleep Medicine and the Court

The admissibility of evidence in both criminal and civil trials is subject to complex rules. The admissibility of neuroscientific evidence would be subject to both the general rules of evidence and the particular rules for scientific evidence (50). The Law Commission (UK) has recently recommended that such evidence should be admitted only if it is reliable. It has proposed a multi-factor test to determine reliability (50).

Admissibility of Expert Scientific Evidence

Tests for admitting expert scientific evidence in civil cases and in criminal cases seem, in practice, to turn on whether it is necessary to draw on experts, whether the evidence will assist the court, and, in particular, whether the evidence is reliable. It should be noted that the term "reliability" here used means that the evidence can be relied upon to make a decision, rather than the scientific sense of repeatable across a number of experiments.

The Law Commission's (UK) report on the admission of expert evidence in criminal cases outlines proposals to include a new reliability test (50). A long list of generic factors that bear on reliability includes the following:

- quality of the data on which the opinion is based and the validity of the methods by which these data were obtained;
- whether any material upon which the opinion is based has been reviewed by others with
- relevant expertise (for instance, in peer-reviewed publications) and the views of others on that material;
- whether the opinion is based on material falling outside the expert's own field of expertise;
- completeness of the information which was available to the expert, and whether the expert took account of all relevant information in arriving at the opinion including information as to the context of any facts to which the opinion relates;
- whether there is a range of expert opinion on the matter in question, and, if there is, where in the range the opinion lies and whether the expert's preference for the opinion proffered has been properly explained; and
- whether the expert's methods followed established practice in the field, and, if they did not, whether the reason for the divergence has been properly explained.

The application of neuroscientific data to law is not a straightforward matter.

Neuroscientific assessment of patients may be invalid as the inevitable time lag between the crime and the assessment may make it impossible to infer the state of the brain at the time the crime was committed. Neuroimaging techniques measure changes in the brain activity and may detect a correlation between some particular type of behavior and brain activity. Such a correlation whether it is between brain structure or brain activity and behavior does not amount to reliable evidence of causation (50).

There is a systematic relationship between particular brain areas and particular functions. For example, the brain areas associated with vision are located in the occipital lobe. Consistent relationships between brain structure and a mental process are not straight forward "one-to-one" links. Instead, a particular brain structure may be involved in many (but not all) mental processes, and a particular mental process will often involve several (but not all) brain areas. This "many-to-many" mapping of mental processes onto brain areas or structures makes it difficult, if not impossible, to infer particular mental processes from the observation of activity in a particular area (50). The presence of brain activity in a particular area may result from several different mental processes, for example, those of pain perception, arousal, and affect (57). The fallacy of reverse inference refers to the misguided and incorrect attempt to conclude from observation of activity in an area that a particular mental process was taking place (75).

Ideas about causality in neuroscience are not necessarily the same as those that operate in the legal sphere. In law, where it is claimed that event X caused outcome Y, then it must at least be shown that X satisfies the so-called "but-for test" (50). If Y would not have occurred but-for X, then X is at least a significant part of the causal story leading to Y. However, this does not establish that X is a sufficient cause of Y or that X should be singled out as the cause of Y where Y is the outcome of concurrent or cumulative conditions. Additional investigation would be necessary to establish the significance of X.

SLEEP-RELATED ACCIDENTS AND THE LAW

Sleepiness and Traffic Safety

There is a lack of awareness among the public, physicians, and authorities in general of the problem posed by sleepy drivers. It is estimated that in the United States there are approximately 4,800 fatal truck crashes each year and many more nonfatal crashes. In one study, the National Transportation Safety Board (NTSE) found that fatigue plus alcohol or drugs accounted for a large proportion of fatal-to-the-driver accidents (58).

Sleepiness results from the sleep component of the circadian cycle of sleep and wakefulness, restriction of sleep and/or interruption or fragmentation of sleep. The loss of one night's sleep can lead to extreme short-term sleepiness, while habitually restricting sleep by 1 or 2 hours a night can lead to chronic sleepiness (59).

Critical aspects of driving impairment associated with sleepiness are reaction time, vigilance, attention, and information processing. Sleepiness reduces optimum reaction times. Moderately sleepy people can have a performance-impairing increase in reaction time that will hinder stopping in time to avoid a collision (60). Performance on attention-based tests declines with sleepiness, including increased periods of non-responding or delayed responding (61). Processing and integrating information takes longer; the accuracy of

short-term memory decreases; and performance declines (60).

Risks For Drowsy-Driving Accidents

Chronic predisposing factors and acute situational factors are recognized as increasing the risk of drowsy driving and related crashes. These factors include sleep loss; driving patterns, including driving between midnight and 6am; driving substantial miles each year and/or a substantial number of hours each day; driving in the midafternoon hours (especially for older persons); driving for longer times without taking a break; use of sedating medications, especially prescribed anxiolytic hypnotics, tricyclic anti-depressants, and some anti-histamines; untreated or unrecognized sleep disorders, especially sleep apnea and narcolepsy; and the consumption of alcohol which interacts with and adds to drowsiness. These factors have cumulative effects. A combination of them substantially increases crash risk (59).

Sleep Restriction or Loss

Short duration of sleep appears to have the greatest negative effects on alertness (62). Although the need for sleep varies among individuals, sleeping 8 hours per 24 hour period is common, and 7 to 9 hours is needed to optimize performance (63). Experimental evidence shows that sleeping less than four consolidated hours per night impairs performance on vigilance tests (64). Acute sleep loss, even the loss of one night of sleep, results in extreme sleepiness (65). The effects of sleep loss are cumulative (66). Regularly losing one to two hours of sleep a night can create a "sleep debt" and lead to a chronic sleepiness over time. Only sleep can reduce sleep debt. People whose sleep was restricted to four to five hours per night for one week need two full nights of sleep to recover vigilance, performance, and normal mood (67). Both external and internal factors can lead to a restriction in the time available for sleep. External factors include work hours, job and family responsibilities, and school bus or school opening times. Internal factors could represent life-style choices - sleeping less to have more time to work or socialize and taking medication that interrupts sleep (59).

Sleep Fragmentation

Sleep disruption and fragmentation cause inadequate sleep and can negatively affect functioning (60). Sleep fragmentation can have internal and external causes. The primary internal cause is illness, including untreated sleep disorders. Externally, disturbances such as noise, restless spouse, or job-related duties (for example, workers who are on call) can interrupt and reduce the quality and quantity of sleep. Studies of commercial vehicle drivers present similar findings. The National Transportation Safety Board concluded that the critical factors in predicting crashes related to sleepiness were the duration of the most recent sleep period, the amount of sleep in the previous 24 hours, and fragmented sleep patterns (59). Circadian pacemaker regularly produces feelings of sleepiness during the afternoon and evening, even among people who are not sleep-deprived (60). Shift work also can disturb sleep by interfering with circadian sleep patterns. Investigators have demonstrated that cir-

cadian phase disruptions caused by rotating shift work are associated with lapses of attention, increased reaction time, and decreased performance (68). Working the night shift, overtime, or rotating shifts is a risk for drowsy driving that may be both chronic and acute. A higher reported frequency of driving drowsy was associated with working a rotating shift, working a greater number of hours per week, and driving more frequently for one's job (69). In the British study (70), respondents said that working the night shift led to sleepiness while driving, and in many studies a majority of shift workers admit having slept involuntarily on the night shift.

A study of hospital nurse staff working around the clock found higher levels of sleepiness and crashes following on-call periods (71). In a survey of hospital nurses, night nurses and rotators were more likely than nurses on other shifts to report nodding off at work and at the wheel and having had a driving mishap on the way home from work (72).

Driving Patterns

Driving patterns, including both time of day and amount of time driven, can increase crash risk. Drowsy-driving crashes occur predominately after midnight with a smaller peak in the mid-afternoon (69). Studies of commercial drivers have a similar pattern. The biology of sleep-wake cycle predicts sleepiness during this time period which is a circadian sleepiness peak and a usual time of darkness. Other driving time patterns that increase risk include driving a larger number of miles each year and a greater number of hours each day (69). Also, driving a longer time without taking a break or, more often, driving for three hours or longer increases crash risk *(70)*.

Sleep Medications

The Sleep in America Poll *(73)* indicated that 15% of the 1,010 subjects interviewed reported using prescription hypnotics (8%) or over the counter sleep aids (10%). Among those subjects reporting daytime sleepiness, 22% used sleep medicine regularly.

Increased traffic accident risks for benzodiazepine users are supported by evidence from on-the-road driving studies *(74)*. In the on-the-road driving test during normal traffic, subjects operate an instrumented vehicle over a highway circuit of 100 km. Subjects are instructed to drive with a steady-lateral position within the right traffic lane while maintaining a constant speed (95 km/h). The primary parameter of the test is the standard deviation of lateral position (cm) indexing the weaving of the car.

Most benzodiazepine hypnotics produce driving impairment exceeding the common legal limits for driving when compared to blood alcohol concentrations *(74)*. See chart below.

Effects of Hypnotics on Driving Ability

BAC >0.10%
Loprazolam 2 mg*
Flurazepam 30 mg*

BAC >0.08%
Flurazepam 15 mg*

BAC >0.05%

Flunitrazepam 2 mg*
Nitrazepam 10 mg*
Lormetazepam 2 mg
Zopiclone 7.5 mg*
BAC <0.05%
Oxazepam 50 mg*
Loprazolam 1 mg
Temazepam 20 mg
Lormetazepam 1 mg*
Nitrazepam 5 mg

BAC = 0%
Zolpidem 10 mg
Zaleplon 10 mg

Driving tests were performed 10-11 h after bedtime administration. The results are compared with blood alcohol concentration (BAC) that corresponds to the most common legal limits for driving a car. *Significantly different from placebo (P < 0.05; ANOVA).

The subjects who participated in these experimental studies enjoyed a full night of sleep. It is reasonable to assume that in sleep-deprived subjects or those experiencing disturbed sleep, the effects of psychoactive drugs are much more pronounced. Other medications that increase the risk of sleepiness-related crashes are long-acting hypnotics, sedating antihistamines (H1 class), and tricyclic antidepressants *(75)*. It appears that the risk is highest soon after the drug regimen is initiated and falls to near normal after several months *(76)*. Recreational drug use also may worsen sleepiness effects *(77)*.

Consumption of Alcohol

Many researchers have shown that sleepiness and alcohol interact with sleep restriction exacerbating the sedating effects of alcohol, and the combination adversely affects psychomotor skills to an extent greater than that of sleepiness or alcohol alone (78). WHTSA found that drivers had consumed some alcohol in nearly 20 percent of all sleepiness-related single-vehicle crashes (79). More than one in three New York State drivers surveyed in drowsy-driving crashes admitted to having consumed some alcohol (69).

Untreated Sleep Disorder

Untreated sleep apnea and narcolepsy increase the risk of automobile crashes (80). Persons with untreated sleep apnea syndrome usually have poor sleep quality which often leads to daytime sleepiness. Narcolepsy is a disorder of the sleep-wake mechanism that also causes excessive daytime sleepiness. These conditions are unrecognized and untreated in a substantial number of people.

Young People, Especially Young Men

Young people and males in particular are more likely to be involved in fall-asleep crashes (81). Definitions of "young people" among authors ranged from 16 to 29 years of age. Carsteaden offers a variety of age-specific reasons for the involvement of younger people, particularly adolescents (82). During this period, young people are learning to drive, experimenting and taking risks, and testing limits. At the same time, this age group is at risk for excessive sleepiness because of maturational changes that increase the need for sleep; changes in sleep patterns that reduce nighttime sleep or produce circadian disruptions; and cul-

tural and lifestyle factors which lead to insufficient sleep, especially a combination of schoolwork demands, part-time jobs, extracurricular activities, and late-night socializing. In one study, boys with greatest extracurricular time commitments were most likely to report falling asleep at the wheel (82). The subgroup at greatest risk comprised the brightest, most energetic hardworking teens. NHTSA data show that males are five times more likely than females to be involved in drowsy-driving crashes (74). The reasons young males have more crashes than do young females are not clear because both young men and young women are likely to be chronically sleep-deprived.

Sleepiness-Related Crashes and the Law

In the framework for prosecuting a sleepiness-related crash, three questions must be answered positively for a conviction to occur (83):
Question 1: There must be a law that the driver can be charged with.

Question 2: The prosecuting authorities must show that the driver did the acts voluntarily.

Question 3: The prosecuting authorities must show that the acts were done either intentionally or with reckless indifference to the consequences.

Question 1: There must be a law that the driver can be charged with.

As of 2008, the National Sleep Foundation reported that no state had a law that addresses non-fatal sleep-related motor vehicle crashes *(84)*. The only American state with a specific law under which a sleepy driver can be charged in a fatal crash is New Jersey. The statute adopted in 2004 allows law enforcement officials to charge individuals with vehicular homicide if, after going without sleep for more than 24 hours, they cause a fatal accident. This law, known as Maggie's law, was enacted in 2004 and named after Maggie McDonnell who was killed in a head-on collision in 1997 by a driver who had gone without sleep for thirty hours.

On the national level, and in response to a congressional mandate in 1995, a revised hours-of-service rule for commercial truck drivers was announced by the Department of Transportation's Federal Motor Carrier Safety Administration in April 2003 *(85)*. The legislation allows long-haul drivers to drive a maximum of 11 hours per day after 10 consecutive off-duty hours. Drivers are prohibited from driving after 14 hours on duty during a single shift.

Question 2: The prosecuting authorities have to show that the driver did the acts voluntarily.

One of the fundamental doctrines of criminal law is that a defendant (in this case, the driver) must act voluntarily to be convicted of a crime *(86)*. This voluntary act requirement has traditionally been encompassed by the term *actus reus*. The term "voluntarily" has a special meaning in the courts where it means that the person is engaged in some form of active control of the car. Conversely, if a driver falls asleep, that driver is acting involuntarily. Involuntariness and automatism are often used as synonyms. But, as this definition has been

questioned, the more general term of involuntariness is used *(83)*. Proof that a driver was asleep at the time of a crash is evidence of an involuntary act and is prima facie exculpatory.

Question 3: The prosecuting authorities have to show that the acts were done either intentionally or with reckless indifference to the consequences.

In conjunction with the *actus reus* voluntary act requirement, most serious crimes (such as those that carry a possible jail term) also require a fault or mental element known as the *mens rea*. This element requires that the person intended his/her actions, or that he/she was reckless to the consequences of those actions. In addition to potentially negating the voluntary element, sleepiness could also negate the mental element.

If it can be shown that the driver was awake at the time of the crash, and that the driver was aware of the consequences of his/her actions (*mens rea*), the driver could be found guilty. In the context of *mens rea*, if the driver could show that it was reasonable that he/she did not think that he/she would fall asleep, the driver would be able to defeat the imputation of recklessness. Negation of the *mens rea* question is normally considered in terms of mental capacity or insanity which is not applicable here. However, the general rule stands that if drivers had no idea that they were going to fall asleep or that they were dangerous on the road, they should be excused from responsibility for their actions *(83)*.

This is a difficult area of proof as, usually, the only persons present in these situations are the drivers themselves, and it is in their own interest to testify that they had no premonition of tiredness. It is in addressing this question that scientific research can have an important impact by evaluating how well people can assess their impairment, what factors they take into account in doing so, and how well they can use this information to engage in appropriate behaviors.

Research has shown that for situations involving a single period of acute sleep deprivation and the performance of laboratory tasks, individuals have an accurate perception of their performance (87). However, during simulated night shift work, the ability to predict performance was moderate at best (88). In a simulated driving experiment, Horne and Rayner found that while individuals may not be able to predict when they are about to fall asleep or report that they have fallen asleep, they consistently report the awareness of fighting sleep before they are objectively judged to have fallen asleep (89). The authors postulate that there are two reasons for this. The first is that sleep onset may be more rapid than people realize. The second is that the state of being sleepy interferes with the metacognitive ability to introspect about sleepiness. Nevertheless, even if people are not very good at predicting if they are about to fall asleep, research indicates that individuals can accurately assess how many hours of sleep they have had in the last 24 or 48 hours (90). This information may provide a basis for assessing whether the decision to drive was reasonable.

These conclusions are only supported by a limited number of studies and need to be more thoroughly investigated by further research. There is a question about whether the results obtained in the laboratory are generalizable to the fatigued driver as most of the studies involve single periods of sleep deprivation, whereas a typical sleepy driver would have either had some sleep or experienced a series of less than optimal sleep opportunities over the prior few days.

Industrial Accidents

Adequate sleep length and quality of sleep appear to be an important aspect of normal human functioning. The demands of society leave many individuals with too little or fragmented sleep and the need to work outside the typical daytime hours. Most (60 - 90%) industrial accidents are caused by human error although the proportion that is directly related to sleepiness is unknown (91). The 2009 Behavioral Risk Factor Surveillance System Survey of almost 75,000 Americans in 12 states found that, within the 30 days prior to taking the survey, 37.9% of respondents reported falling asleep unintentionally, 4.7% of respondents reported nodding off or falling asleep while driving, and 35.2% of respondents reported getting less than 7 hours of sleep (92). The Sleep in America Poll conducted in 2008 by the National Sleep Foundation found that 29% of respondents reported falling asleep or becoming very sleepy while at work.

There are many reasons why sleep/wake disorders go unrecognized. These include an underestimation of the clinical impact of sleep/wake disorders on the part of health-care providers; the tendency of clinicians to seek a "single diagnosis" even though sleep disorders are multifaceted (93); the relatively recent emergence of sleep medicine as a specialty; and the lack of adequate training in sleep medicine among healthcare providers (94).

Excessive sleepiness puts shift workers at high risk for accidents on the job or on their way home from work. Among shift workers, 19 to 29% (depending on the shift) report excessive sleepiness during times they are required to be awake, typically at night (94).

People working shifts have higher rates of accidents than those working fixed daytime schedules, and certain types of shift workers are at higher risk for accidents, including transportation workers, healthcare workers, and people with long commutes from work (95).

Night or rotating shift workers are two to four times as likely as daytime workers to fall asleep while driving, a result of cumulative effects of a number of factors such as impairment in concentration, reaction times, and alertness resulting from potential underlying causes including excessive sleepiness, sleep deprivation, and change in circadian rhythm.

Naps at work can enhance alertness and provide transient relief from sleepiness. However, sleep inertia refers to a period of impaired performance and reduced vigilance following awakening from the regular sleep episode or from a nap. Sleep inertia is a potent phenomenon resulting in impaired performance and vigilance averaging one hour and requiring two to four hours to dissipate in normal non-sleep-deprived individuals. This condition is worse following sleep deprivation.

Although industrial accidents can result in large-scale public losses, there have been no guilty verdicts attributed directly to sleep disorders or to sleep deprivation.

The accidents in Chernobyl and at Three Mile Island nuclear power plants and the grounding of the Exxon Valdez oil tanker are well publicized examples of large-scale industrial disasters partially attributed to sleep loss and shift work-related performance failures (91).

Health professionals can promote knowledge transfer to the public, to industry, and to policy makers which could promote behavior change and the formation of good public policy, and, as a result, improve public health.

SLEEP-RELATED DISABILITY

While many sleep disorders can be diagnosed adequately and treated with behavioral modification, medical therapy, or psychotherapy, there is a subset of patients who suffer from incapacitating symptoms despite aggressive and appropriate therapy. For example, despite optimal therapy with medications and behavioral modification, patients with narcolepsy may still experience severe daytime sleepiness.

The Americans with Disabilities Act (ADA) does not contain a list of medical conditions that constitute disabilities. Instead, the ADA has a general definition of disability that each person must meet (96). Therefore, some people with sleep disorders will have a disability under the ADA, and some will not.

Under the ADA, businesses with more than fifteen employees are required to accept requests from employees with disabilities for work accommodations (91). Patients with excessive daytime sleepiness who have difficulty waking up for the morning schedule are accommodated by moving them to the afternoon shift.

Patients with restless leg syndrome can often be 10-15 minutes late for work due to amount and quality of sleep. The employer can provide them with half an hour flexible start time. Depending on when the employee arrived, the time can be made up either in a break or at the end of the day. Laws to protect patients with disabilities are meant to defend the rights of those with temporary or permanent incapacities.

ADVANCING SLEEP MEDICINE

As sleep medicine advances, it will have a significant impact on the law. There is currently a large gap between neuroscience research and the realities of the day to day work of the justice system. Neuroscientists and legal professionals conduct their work in different geographical areas, using different methodologies and language. It is important that research and learning are shared among countries. In the United States, programs have brought together neuroscientists and lawyers for over a decade (50). For instance, the John D. and Catherine T. MacArthur Foundation funded the Law and Neuroscience project which has been underway since 2007. Other projects include the work of the National Academy of Sciences Committee on Science, Technology and the Law, the Federal Judicial Center, and the Bayler College of Medicine initiative on Neuroscience and the Law. This type of cooperation may lead to more effective efforts to manage the risks associated with violent behaviors and drowsiness as a matter of law and policy in the future.

KEY WORDS

- **EEG**
- **PSG**
- **RBD**
- **accident**
- **arousal**
- **automatism**
- **drowsy driving**
- **electroencephalography**
- **insanity**
- **Legal**
- **mens rea**
- **polysomnography**
- **sexsomnia**
- **sleepwalking**
- **somnumbulism**
- **violence**

REFERENCES

1. Ohayon MM, Caulet M, Priest RG. Violent behavior during sleep. J Clin Psychiatry **1997**; 58: 369–76; quiz 77.
2. Moldofsky H, Gilbert R, Lue FA, MacLean AW. Forensic sleep medicine: sleep-related violence. Sleep **1995**; 18: 731–39.
3. Guilleminault C, Moscovitch A, Leger D. Forensic sleep medicine: nocturnal wandering and violence. Sleep **1995**; 18: 740–48.
4. Guilleminault C, Leger D, Philip P, Ohayon MM. Nocturnal wandering and violence: review of a sleep clinic population. J Forensic Sci **1998**; 43: 158–63.
5. Schenck CH, Milner DM, Hurwitz TD, Bundlie SR, Mahowald MW. A polysomnographic and clinical report on sleep-related injury in 100 adult patients. Am J Psychiatry **1989a**; 146: 1166–73.
6. Rodin EA. Psychomotor epilepsy and aggressive behavior. Arch Gen Psychiatry **1973**; 28: 210–3.
7. Oudiette D, Leu S, Pottier M, Buzarre MA, Brion A, Arnulf I. Dreamlike mentations during

sleepwalking and sleep terrors in adults. Sleep **2009**; 32: 1621–7.

8. Mahowald MW, Schenck CH, Evolving concepts of human state dissociation archives Italiennes de Biologic 139: 2699-300. **2001**

9. Krueger, J. M., Obal Jr., F., Kapas, L. and Fang, J. Brain organization and sleep function Behav. Brain Res., 69: 177-185, **1995**

10. Hublin C, Kaprio J, Partinen M, Heikkila K, Koskenvuo M: Prevalence and genetics of sleepwalking: a population-based twin study. Neurology **1997**; 48:177–181.

11. Kales A, Soldatos C, Bixler E: Hereditary factors in sleepwalking and night terrors. Br J Psychiatry **1980**; 137:111–118.

12. Lecendreux M, Mayer G, Bassetti C, Neidhart E, Chappuis R,Tafti M: HLA class II association in sleep walking (abstract).Sleep **2000**; 23(suppl 2):A13.

13. Iranzo A, Molinuevo JL, Santamaria J, Serradell M, Marti MJ, Valldeoriola F, et al. Rapid-eye-movement sleep behaviour disorder as an early marker for a neurodegenerative disorder: a descriptive study. Lancet Neurol **2006** b; 5: 572–7.

14. Rosalind Cartwright, Ph.D. Sleepwalking Violence: A Sleep Disorder, a Legal Dilemma, and a Psychological Challenge Am J Psychiatry **2004**; 161:1149–1158

15. Ungerleider L, Mishkin M: Two cortical visual systems, in The Analysis of Visual Behavior. Edited by Ingle D, Mansfield R, Goodale M. Cambridge, Mass, MIT Press, **1982**, pp 549–586.

16. Broughton R: Sleep disorders: disorders of arousal? Science **1968**; 159:1070–1077.

17. Tassinari Ca, Tassi L, Calandra-Bounaura G, Stanzani-Maserati M, Fini N, Pizza F, et al Bit ing behavior, aggression and seizures. Epilepsia **2005**: 46:654-63

18. Oswald I, Evans J: On serious violence during sleep walking. Br J Psychiatry **1985**; 147:688–691

19. Bonkalo A: Impulsive acts and confusional states during incomplete arousal from sleep: criminological and forensic implications. Psychiatr Q **1974**; 48:400–409.

20. Brower Mc, Price BH. Neuropsychiatry of frontal lobe dysfunction in violent and criminal behavior: a critical review. J Neurol Neurosurg Psychiatr **2001**:71(6):720-6.

21. Best M, Williams JM, Coccaro EF. Evidence for a dysfunctional prefrontal circuit in patients with an impulsive aggressive disorder. Proc Natl Acad Sci USA **2002**; 99: 8448–53.

22. Coccaro EF, McCloskey MS, Fitzgerald DA, Luan Phan K. Amygdala and orbitofrontal reactivity to social threat in individuals with impulsive aggression. Biol Psychiatry **2007**; 62: 168–78.

23. Blair RJR. The roles of orbital frontal cortex in the modulation of antisocial behavior. Brain and Cognition **2004**; 55: 198–208.

24. Bassetti C, Vella S, Donati F, Wielepp P, Weder B. SPECT during sleepwalking. Lancet **2000**; 356: 484–5.

25. Terzaghi M, Sartori I, Tassi L, Didato G, Rustioni V, LoRusso G, et al.Evidence of dissociated arousal states during NREM parasomnia from an intracerebral neurophysiological study. Sleep 2009; 32: 409–12.

26. Siclari F, Khatami R, Urbaniok F, Nobili L, Mahowald MW, Schenck CH, Cramer Bornemann MA, Bassette, CL. Violence in sleep. Brain **2010**;133; 3494-3509

27. International Classification of Sleep Disorders. Third edition. American Academy of Sleep Medicine **2014 ISBN:0991543416**

28. Francesca Siclaria, Ramin Khatamia, Frank Urbaniokd, Claudio L. Bassettia: Violence in sleep Schweizer Archiv Für Neurologie Und Psychiatrie **2009**; 160(8):322–33 3030303030

29. Mahowald MW, Cramer-Bornemann MA. NREM sleep parasomnias. In: Kryger MH, Roth T, Dement W (eds) Principles and practice of sleep medicine. Philadelphia: Elsevier/Saunders; **2005**; pp 889–96.

30. Pressman MR. Disorders of arousal from sleep and violentbehavior: the role of physical contact and proximity.Sleep. **2007**; 30:1039–47 303030303

31. Broughton R, Billings R, Cartwright R, Doucette D, Edmeads J, Edwardh M, et al. Homicidal somnambulism:a case report. Sleep. **1994**; 17:253–64.

32. Howard C, D'Orban PT. Violence in sleep: medico-legal issues and two case reports. Psychol Med. **1987**; 17:915–25.

33. Broughton R. NREM arousal parasomnias. In: Kryger MH, Roth T. Dement WC, editors. Principles and practice of sleep medicine. 3rd ed. Philadelphia: Saunders; **2000**.p.1336.

34. Schenck CH, Arnulf I, Mahowald M. Sleep and sex: what can go wrong? A review of the literature on sleep related disorders and abnor-

mal sexual behaviors and experiences. Sleep. **2007**;30:683–702.

35. Guilleminault C., Moscovitch A., Yuen K., Poyares, D. A typical sexual behavior during sleep. Psychosomatic Med **2002** 64(2): 328-336.

36. Schenck CH, Lesa, Bornemann MA, Mahowald MW. Potential lethal behaviours associated with rapid eye movement and sleep behavior disorder: review of the literature and forensic implications. J Foreensic Sci **2009**: 54(6): 1475-1484

37. Fantani ML, Corona A, Cleric, S, Fexxini-Strambi L. Aggressive dream content without daytime aggressiveness in REM sleep behavior disorder. Neurology **2005**; 65(7): 1010-5

38. De Cock VC, Vidailhet M, Leu S, Texeira A, Apartis E, Elbaz A, et al. Restoration of normal motor control in Parkinson's disease during REM sleep. Brain. **2007**; 130:450–6

39. Iranzo A, Molinuevo JL, Santamaria J, Serradell M, Marti MJ, Valldeoriola F, et al. Rapid-eye-movement sleep behavior disorder as an early marker for a neurodegenerative disorder: a descriptive study. Lancet Neurol.**2006**; 5:572–7.

40. Olson EJ, Bradley FB, Boeve BF, Silber MH. Rapid eye movement sleep behavior disorder: demographic, clinical and laboratory findings in 93 cases. Brain. **2000**; 123:331–9.

41. American Psychiatric Association. Diagnostic and statistical manual of mental disorders. Washington DC: American Psychiatry Association; **2000**.

42. Morrison I, Rumbold John MM, Riha Renata L. Medicolegal aspects of complex behaviours arising from the sleep period: A review and guide for the practising sleep physician Sleep Medicine Reviews Volume 18, Issue 3, June **2014**, Pages 249–260.

43. Derry CP, Harvey AS, Walker MC, Duncan JS, Berkovic SF. NREM arousal parasomnias and their distinction from nocturnal frontal lobe epilepsy: a video EEG analysis. Sleep **2009**; 32: 1637–44.

44. Aldrich MS, Jahnke B. Diagnostic value of video-EEG polysomnography. Neurology **1991**: 41: 1060-6.

45. Pressman MR. Hypersynchronous delta sleep EEG activity and sudden arousals from slow-wave sleep in adults without a history of parasomnias:clinical and forensic implications. Sleep **2004**; 27: 706–10.

46. Guilleminault C, Poyares D, Aftab FA, Palombini L. sleep and wakefulness in somnambulism: a spectral analysis study. J Psychosom Res **2001**: 51: 411-6.

47. Guilleminault C, Kirisoglu C, Da Rosa AC, Lopes C, Chan A.Sleepwalking, a disorder of NREM sleep instability. Sleep Med **2006**; 7: 163–70.

48. Poldrack 2011 Mapping Mental Function to Brain Structure: How can Cognitive Neuroimaging Succeed? Perspectives on Psychological Science 5. 753-761

49. Pardo M.S., Patterson D; Philosophical Foundations of Law and Neuroscience **2010**

50. Brain Waves Module 4: Neuroscience and the Law Dec **2011 RS Policy document 05/11 IBSN 978-0-85403-932-6.** The Royal Society 2011

51. Tallis 2010 what neuroscience cannot tell us about ourselves. The New Altantis 29, 3-25

52. Criminal Liability: Insanity and Automatism A Discussion Paper 23 July 2013 Law Commission Discussion Paper (**July 2013**)

53. M'Naghton's Case (1843) 10 Clark and Finnelly 200, 210, (1843) 8 ER 718, [1843-60] All Er REsp 229 http://users.phhp.ufl.edu/rbauer/forensic_neuropsychology/mcnaghten.pdf

54. Broughton RJ, Shimizu T. Sleep-related violence: a medical and forensic challenge. Sleep **1995**; 18: 727–30.

55. Schenck C. Paradox lost: midnight in the battleground of sleep and dreams. 1 st Ed. Minneapolis, Minnesota, USA. Extreme-Nights, LLC, **2005**. ISBN -0-9763734-0-8

56. Raimundo Nonato Delgado-Rodrigues , Alexander N. Allen , Leandro Galuzzi dos Santos, Carlos H. Schenck: Sleep Forensics: a critical review of the literature and brief comments on the Brazilian legal situation. Medicina do sono Forense: uma revisão crítica da literatura e breves comentários sobre a situação brasileira Arq. Neuro-Psiquiatr. vol.72 no.2 São Paulo Feb. **2014**

57. Poldrack **2006** Can cognitive processes be inferred from neuroimaging data? Trends in cognitive sciences 10, 59-63.

58. National Transportation Safety Board. Safety Study: Fatigue Alcohol, Other Drugs, and Medical Factors in Fatal-to-the-Driver Heavy Truck Crashes. Washington, DC: National Transportation Safety Board; **1990**.

59. NCSDR/NHTSA Expert Panel on Driver Fatigue and Sleepiness 1996 http://www.nhtsa.gov/people/injury/drowsy_driving1/Drowsy.html#NCSDR/NHTSA

60. Dinges D: An overview of sleepiness and accidents. J Sleep Res **1995**; 4(2):4-114

61. Kribbs N, Dinges D, Barone N: Vigilance decrement and sleepiness. In: Harsh J, Ogilvie R, editors.(1994) Sleep onset: Normal and abnormal processes. Washington, DC: American Psychological Association; 1994. pp. 113-125.

62. Rosenthal L et al.: Level of sleepiness and total sleep time following various time in bed conditions. Sleep **1993**(a); 16:226-32.

63. Carskadon M, Roth T. Sleep restriction. In: Monk T, editor. Sleep, sleepiness and performance. Chichester: John Wiley & Sons; **1991**. pp. 155-67.

64. Naitoh P: Minimal sleep to maintain performance: The search for sleep quantum in sustained operations. In: Stampi C, editor. Why we nap: evolution, chronobiology, and functions of polyphasic and ultrashort sleep. Boston: Birkhauser; **1992**. pp. 199-216.

65. Carskadon M: Evaluation of excessive daytime sleepiness. Neurophysiol Clin **1993**(b);23:91-100

66. Carskadon M, Dement WC: Cumulative effects of sleep restriction on daytime sleepiness. Psychophysiology **1981**;18:107-13.

67. Dinges D et al.: Cumulative sleepiness, mood disturbance, and psychomotor vigilance performance -decrements during a week of sleep restricted to 4-5 hours per night. Sleep **1997**;20(4):267-77.

68. Dinges D et al.: Temporal placement of a nap for alertness: contributions of circadian phase and prior wakefulness. Sleep **1987**;10:313-29.

69. McCartt A et al.: The scope and nature of the drowsy driving problem in New York state. Accident Analysis New York GTSC Sleep Task Force, Public Information and Prevention **1996**;28(4):511-17

70. Maycock G: Sleepiness and driving: the experience strategiesof UK car drivers. JSleep Res **1996**; 5(220):220-37.

71. Marcus C, Loughlin G: Effect of sleep deprivation on driving safety in housestaff. Sleep **1996**;19(10):763-6.

72. Gold D et al.: Rotating shift work, sleep, and accidents related to sleepiness in hospital nurses. Am J Public Health **1992**;82(7):1011-4.

73. National Sleep Foundation. 2002 Sleep in America Poll. Washington: National Sleep Foundation; **2002**

74. Verster JC, Veldhuijzen DS, Volkerts ER. Residual effects of sleep medication on driving ability. Sleep Med Rev **2004**; 8: 309-325

75. Kozena L et al.: Vigilance impairment after a single dose of benzodiazepines. Psychopharmacology (Berl) **1995**; 119(1):39-45.

76. Ceutel C: Risk of traffic accident injury after a prescription for a benzodiazepine. Ann Epidemiol **1995**; 5(3):239-44.

77. Kerr et al.: Separate and combined effects of the social drugs on psychomotor performance. Psychopharmacology **1991**; 104:113-19.

78. Roehrs T et al.: Sleepiness and ethanol effects on simulated driving. Alcohol Clin Exp Res **1994**; 18(1):154-8.

79. Wang J, Knipling R, Goodman M: The role of driver in attention in crashes: new statistics from the 1995 crashworthiness data system. Fortieth Annual Proceedings of the Association for the Advancement of Automotive Medicine; **1996** Oct 7-9. pp. 377-92.

80. Aldrich M: Automobile accidents in patients with sleep disorders. Sleep **1989**;12(6):487-94.

81. Pack A et al.: Characteristics of crashes attributed to the driver having fallen asleep. AccidAnal Prev **1995**;27(6):769-75.

82. Carskadon M: Adolescent sleepiness: increased risk in a high-risk population. Alcohol Drugs and Driving **1990**; 5(4)/6(1):317-28.

83. Jones Christopher B, Dorrian Jillian and M.W. Rajaratnam, Shanthakumar: Fatigue and the Criminal Law Industrial Health **2005**, 43, 63-70

84. National Sleep Foundation. State of the States Report on Drowsy Driving: Summary of Findings. Washington, DC: National Sleep Foundation; **2008**.

85. Department of Transportation; Federal Motor Carrier Safety Administration . Hours of service of drivers; driver rest and sleep for safe operations. Fed Regist. **2003** ; 68 (81):22456 - 22517 .

86. American Law Institute (1962) Model Penal Code, ss

87. Dorrian J, Lamond N, Dawson D The ability to self monitor performance when fatigued. J Sleep Research 9(**2000**), 137–44.

88. Dorrian J, Lamond N, Holmes A, Roach G, Fletcher A, Dawson D: The ability to self-mon-

itor performance during a week of simulated night shifts. Sleep 26(**2003**), 871–7.

89. Horne JA, Reyner LA (**1995**) Driver sleepiness. J Sleep Res 4 (Suppl. 2), 23–9.

90. Rogers AE, Caruso CC, Aldrich MS: Reliability of sleep diaries for assessment of sleep/wake patterns. Nurs Res 42(**1993**), 368–72.

91. Vidya Krishnan, and Ziad Shaman: Legal Issues Encountered When Treating the Patient With a Sleep Disorder FCCP CHEST **2011**; 139(1): 200 – 207

92. Centers for Disease Control and Prevention. Unhealthy sleep-related behaviours-12 states 2009. MMWR. **2011**; 60:233-238.

93. McCarty D. E. Beyond Ockham's razor: redefining problem-solving in clinical sleep medicine using a "fire-finger" approach. J Clin Sleep Med. **2010**; 6: 292-296.

94. Rosen R, Mahowald M, Chesson A, et al. The Taskforce 2000 Survey: Medical education in sleep and sleep disorders. Sleep **1998**; 21; 235-238

95. Doghramji K, Advances in the management of shift-work disorder. Pharmacist supplement to U.S. December **2011**.Copyright 2011 by Jobson Medical Information LLC, New York.

96. US Department of Justice. Americans with Disabilities Act .http :// www . ada . gov . Accessed August 13, **2009** .

20

SLEEP AND DERMATOLOGY

Camila Hirotsu, Rachel Gimenes Albuquerque,
Sergio Tufik and Monica Levy Andersen

ABSTRACT

Sleep deprivation is a characteristic of the modern society. The effects of sleep loss on several systems of our body are widely known, however more recently studies also suggests that skin is affected too. Skin is the largest organ of our body, with an own complex immune system and tight connections to stress response system. For these reasons, it is possible that sleep plays an important role in the pathogenesis and exacerbation of skin diseases, as acne and psoriasis.

Camila Hirotsu
Department of Psychobiology, Universidade Federal de São Paulo, Rua Napoleão de Barros, Vila Clementino, São Paulo, Brazil

Rachel Gimenes Albuquerque
Department of Psychobiology, Universidade Federal de São Paulo, Rua Napoleão de Barros, Vila Clementino, São Paulo, Brazil

Sergio Tufik
Department of Psychobiology, Universidade Federal de São Paulo, Rua Napoleão de Barros, Vila Clementino, São Paulo, Brazil

Monica Levy Andersen
Department of Psychobiology, Universidade Federal de São Paulo, Rua Napoleão de Barros, Vila Clementino, São Paulo, Brazil

Corresponding author: Monica Levy Andersen
E-mail: ml.andersen12@gmail.com

20.1. INTRODUCTION

In the current society, the adjustments of the population to the habits of modern life has promoted daily interactions with challenging situations represented by changes in the work force and the increased social pressure, which contribute to the epidemic of sleep deprivation (Tufik et al., 2009). Indeed, the occurrence of stress factors, such as sleep deprivation, has increased in frequency among the global population and now affects nearly 30% of adults (Schoenborn et al., 2010).

This lifestyle imposes increasingly stressful situations to the organism, leading to serious damages to the body system through activation of the hypothalamic-pituitary-adrenal (HPA) axis (Smith et al., 2006). This may also reach the skin, since there are evidences of collagen loss, increased skin kallikrein-5 activity, impaired wound healing and decreased epidermal cell proliferation and corneodesmosomes density associated with stress conditions (Altemus et al., 2001; Choi et al., 2005;

Kahan et al., 2009; Sivamani et al., 2009; Hirotsu et al., 2012). Importantly, stress influences cellular and humoral immune responses by releasing glucocorticoid, catecholamine, corticotropin releasing hormone (CRH), and pro-opiomelanocortin (POMC), and altering cytokine profiles (Elenkov et al., 1999). On the other hand, the human skin expresses many elements of the HPA axis and present a local immune system regulated by cytokines and bioactive lipids that together provide defense against the stressors to reestablish tissue homeostasis (Slominski et al., 2007; Kendall et al., 2013). However, when immune responses are inadequate, these mediators contribute to skin lesions and pathologies (Kendall et al., 2013).

This important interface between sleep and skin has emerged more recently, through the development of the psychodermatology area, whose aim is to integrate both dermatology and psychiatry, due to the multidisciplinary investigation of the skin, the mind, and the environment. According to several studies, psychiatric disorders, skin diseases and stress may be tightly associated with a modulator factor in common: sleep. Although there is little scientific evidence about the direct relationship between sleep and skin, the common knowledge aware us of this connection a long time ago. After being sleep deprived in situations as for example studying for an exam or working until late of night, we can perceive that our facial appearance changes. Even more, it is commonly reported that during stress periods and consequently fragmented sleep, some skin diseases as acne and psoriasis can be triggered.

Skin has a complex physiology linked to the central nervous system (CNS). Thus,

it seems plausible to believe that sleep plays a role on skin components, such as extracellular matrix proteins, enzymes, cellular proliferation, and renovation, or even as a contributing factor to develop and/or modulate skin diseases. In this chapter, we will discuss how sleep can be involved with skin integrity, contributing to its homeostasis control and also being a contributing factor for dermatopathologies.

20.2. SLEEP DEPRIVATION AND SKIN

The skin is the largest organ in the human body and the primary barrier from the outside environment, protecting the host against physical, biological, and chemical stress through an organized system comprising elements of innate and adaptive immunity (Proksch et al., 2008). Skin is composed by three major layers: the epidermis, the dermis, and the hypodermis. The epidermis is a relatively thin and resistant layer; whose main cell population is the keratinocyte. The dermis is thicker than epidermis and mainly composed by mucopolysaccharides, collagen, elastin, and fibroblasts. The last one, hypodermis, is a layer of adipocytes, which controls temperature, serving as energy store, and mechanical protection.

Many stress-related changes, including sleep deprivation, can affect immune function and present further implications for skin regeneration and integrity (Kahan et al., 2009; Egydio et al., 2012). Other conditions, such as emotional or psychological stress can also pose adverse consequences on human health and contribute to cellular aging by loss of telomere length (Epel et al., 2004). In skin, stress can impair the barrier function, induce an inflammatory response, and trigger or exacerbate the course of skin disorders, such as psoriasis, acne, and atopic dermatitis (Theoharides et al., 1998; Chiu et al., 2003; Hanel et al., 2013).

It is well established that lack of sleep promotes stress (Andersen et al., 2005; Guyon et al., 2014) especially affecting our major stress response system, the HPA axis. After sleep deprivation, the HPA axis is responsible for a series of events, which results in glucocorticoids secretion. Excess glucocorticoids can accelerate the aging process (Oikarinen and Autio., 1991). The negative effects of glucocorticoids on the skin are mainly mediated by inadequate immuno-suppression and fibroblast function. *In vitro* and *in vivo* studies have demonstrated that variations in homeostasis cause changes in the extracellular matrix and collagen (Autio et al., 1994; Haapasaari et al., 1995; Oikarinen et al., 1998). For example, Oikarinen and collaborators (1992) demonstrated that the topical application of glucocorticoids in humans resulted in an 80% decrease in the levels of types I and III collagen. In cultures of human fibroblasts, treatment with glucocorticoids led to a 70% decrease in the fibers for types I and III collagen (Oikarinen et al., 1998).

Glucocorticoids are extensively used in inflammatory skin conditions and diseases (Schafer-Korting et al., 2005). However, it is well known that the excess of glucocorticoids promotes skin atrophy mainly due to its action on extracellular matrix proteins (Schoepe et al., 2006). Although most of studies were conducted around the side effects of topical glucocorticoids, it was also described that systemic therapy with glucocorticoids can suppresses collagen type I

and type III synthesis (Autio et al., 1994). Not only collagen, but other fundamental components as hyaluronan, involved in tissue homeostasis, are responsible for the repair and hydration processes of skin; and elastin, which plays a role in skin elasticity and resilience, is also decreased by glucocorticoids topical treatment (Kähäri et al., 1994; Gebhardt et al., 2010).

The effects of sleep loss lasting many hours on the integrity of human skin have also been evaluated. Following 42 h of sleep deprivation, there was a marked reduction in the recovery of skin barrier function, increased levels of interleukin-1 beta (IL-1β) and tumor necrosis factor alpha (TNF-α) and a greater number of natural killer (NK) cells (Altemus et al., 2001). From these results, the inhibition of collagen production was correlated with increased release of IL-1β.

Recently, Sundelin and colleagues (2013) showed that sleep deprivation may affect a number of facial characteristics related to fatigue. In this study, 40 observers rated 20 facial photographs with respect to fatigue, 10 facial cues, and sadness. The stimulus material consisted of 10 individuals photographed at 2 p.m. after normal sleep and after 31 h of sleep deprivation following a night with 5 h of sleep. It seems that many of the colloquial cues, such as droopy/hanging eyelids, red eyes, dark circles under the eyes, and pale skin, were indicative of both sleep deprivation and looking fatigued. The results showed that sleep deprivation affects features relating to the eyes, mouth, and skin, suggesting that sleep curtailment could be readily observable from a set of facial cues. Figure 20.1 illustrates the main mechanisms addressing the relationship between sleep deprivation and skin health.

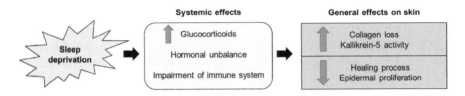

Figure 20.1. Systemic and general effects related to sleep deprivation and skin health.

20.3. SLEEP AND DERMATOPATHOLOGIES

20.3.1. Sleep and Acne

Acne is characterized by a chronic inflammation of the pilosebaceous unit leading to non-inflammatory and inflammatory lesions and even scarring. It is estimated that 9.4% of the global population is affected, making this disease the 8th most prevalent in the world (Hay et al., 2014). Acne tends to be more prevalent in boys at puberty, when they tend to have more severe acne than girls. At adult age, women are more affected than men (Tan and Bhate., 2015). The mechanism of acne development consists in four major processes: the release of inflammatory mediators on skin, altered keratinization, increased and altered sebum production (mediated by androgens) and *Propionibacterium acnes* colonization (William et al., 2012). Other factors as genetic predisposition, diet, sunlight exposure, smoking, skin hygiene and stress seems to play a role in acne, although this involvement remains unclear (William et al., 2012).

Sleep has been suggested as a contributing factor to acne development and triggering, especially due to be tied to the stress response system (Albuquerque et al., 2014). A community-based study conducted in Japan with 859 students between 13 and 19 years old demonstrated that 63.1 and 55.5% reported stress and lack of sleep, respectively, as factors believed to aggravate their acne (Kubota et al., 2010). Studies conducted with adult female acne patients demonstrated that most of women (50 to 71%) reported that their acne lesions got worsen during stress periods (Goulden et al., 1977; Poli et al., 2001).

Beyond these epidemiological data, experimental studies have demonstrated that the HPA axis activation is involved in the development of acne lesions, with particular involvement of the CRH. Sebaceous glands have a complete system composed by CRH, CRH-binding protein, and CRH receptors, which induce lipid synthesis and androgen production (Zouboulis and Böhm., 2004). According to Ganceviciene and colleagues (2009a, 2009b), acne-involved skin sebaceous glands demonstrated a very positive reaction for CRH in immunohistochemistry analysis in comparison to normal skin. Moreover, differentiating sebocytes from acne-involved skin demonstrated the strongest reaction for CHR binding protein and for type 2 CRH receptor. The authors suggested that the CRH system, which is abundant in acne-involved skin, possibly activates pathways leadings to inflammatory and immune process, which culminates with acne exacerbation on stress (Gancevíciene et al., 2009a, 2009b).

20.3.2. Sleep and Psoriasis

Psoriasis is a chronic inflammatory skin disease, whose degree of severity depends on genetics and environmental factors. Although the pathophysiology of psoriasis is still uncertain, several risk factors have been considered such as smoking, family history, stress, and obesity (Huerta et al., 2007). Psoriasis has a negative impact on quality of life (Martínez-García et al., 2014) and patients can present more traits of psychiatric disorders in comparison to general population as eating disorder (Crosta et al., 2014); anxiety and depression (Bilgic et al., 2010; Kurd et al., 2010; Kimball et al., 2012).

Several studies also reported that psoriasis impairs sleep. Among subjective trials, quality-of-life questionnaires have been used to measure sleep quality, and a number of hypotheses have been made about the factors involved in the relationship between sleep and psoriasis, including depression. Indeed, psoriasis is associated with depression, which has also been shown to alter sleep through modulation of substance P (Hirotsu et al., 2015). Psoriasis may cause itching that impairs sleep, commonly affecting the trunk and extremities, and negatively affecting quality of life (Savin et al., 1990). Scratching occurs predominantly during sleep stages N1 and N2 rather than N3, and precipitates arousals, leading to sleep fragmentation (Aoki et al., 1991). Callis Duffin and colleagues (2009) reported that 49.5% of respondents stated that their psoriasis negatively

affected sleep at least once per month. The same group revealed that psoriatic arthritis, lesion-related itch and pain, as well as emotional well-being were risk factors for the development of sleep disturbance in psoriasis patients.

Li and collaborators (2013) performed a prospective analysis that identified an association between rotating night shift work and increased risk of psoriasis. Recently, a link between psoriasis and obstructive sleep apnea (OSA) has also been proposed (Hirotsu et al., 2015). However, there are several comorbidities in common, observed in patients with psoriasis and patients with OSA, which make very difficult to establish a cause-effect relationship: both groups of patients present higher prevalence of hypertension, type 2 diabetes, obesity, and dyslipidemia (Parish et al., 2007; Love et al., 2011). Karaca and colleagues (2013) verified that in a sample of 33 patients, 18 presented mild OSA, whereas two had moderate OSA and five had severe OSA, that is, 54% of the patients with psoriasis presented OSA in some degree. In addition, a prospective study demonstrated that OSA patients have approximately 2-fold risk of developing psoriasis in comparison to general population (Yang et al., 2012). Shutty and colleagues (2013) also showed that patients with psoriasis are more likely to experience insomnia, however, in this study, sleep disturbances were more correlated with depression than to psoriasis by itself.

These data suggest that the population with psoriasis requires a greater attention to comorbidities, which in turn are related to cardiovascular diseases development and even metabolic syndromes, and to the presence of OSA. OSA is a very common sleep disturbance, but it is also underdiagnosed, and it is also underdiagnosed, but it is also underdiagnosed, and has a great impact in healthy and life quality of the patient.

20.4. CONCLUSIONS

Several negative effects of sleep loss have been documented regarding behavioral, hormonal, and immunological changes, in addition to a reduction in longevity. Recent evidence has also shown a close connection between sleep and skin along with the advance in the field of psychodermatology. Overall, it seems that there is a bidirectional relationship between sleep and skin. Sleep deprivation can impair skin integrity through HPA axis activation and immune modulation. In turn, some dermatopahologies such as psoriasis and acne are associated with poor sleep quality and high prevalence of sleep disorders (Fig. 20.2). Although many techniques and treatments for dealing with skin diseases and aging have been developed, which are especially important since many populations now have increased life expectancies, further studies about how sleep is related to skin health needs to be performed to improve the technology behind the skin health.

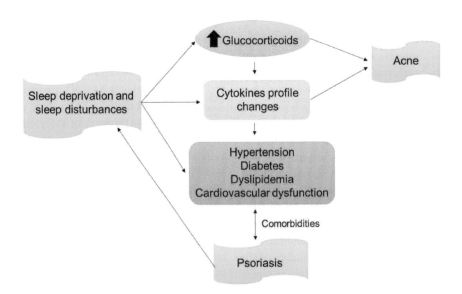

Figure 20.2. Scheme of the bidirectional relationship between sleep and skin.

REFERENCES

1. Albuquerque, R. G.; Rocha, M. A.; Bagatin, E.; Tufik, S.; Andersen, M. L. Could Adult Female Acne be Associated with Modern Life? *Arch. Dermatol. Res.* **2014,** *306 (8),* 683–688.

2. Altemus, M.; Rao, B.; Dhabhar, F. S.; Ding, W.; Granstein, R. D. Stress-Induced Changes in Skin Barrier Function in Healthy Women. *J. Invest Dermatol.* **2001,** *117 (2),* 309–317.

3. Andersen, M. L.; Martins, P. J.; D'Almeida, V.; Bignotto, M.; Tufik, S. Endocrinological and Catecholaminergic Alterations during Sleep Deprivation and Recovery in Male Rats. *J. Sleep Res.* **2005,** *14 (1),* 83–90.

4. Aoki, T.; Kushimoto, H.; Hishikawa, Y.; Savin, J. A. Nocturnal Scratching and its Relationship to the Disturbed Sleep of Itchy Subjects. *Clin. Exp. Dermatol.* **1991,** *16 (4),* 268–272.

5. Autio, P.; Oikarinen, A.; Melkko, J.; Risteli, J.; Risteli, L. Systemic Glucocorticoids Decrease the Synthesis of Type I and Type III Collagen in Human Skin in Vivo, Whereas Isotretinoin Treatment has Little Effect. *Br. J. Dermatol.* **1994,** *131 (5),* 660–663.

6. Bilgic, A.; Bilgic, Ö.; Akış, H. K.; Eskioğlu, F.; Kılıç, E. Z. Psychiatric Symptoms and Health-Related Quality of Life in Children and Adolescents with Psoriasis. *Pediatr. Dermatol.* **2010,** *27 (6),* 614–617.

7. Chiu, A.; Chon, S. Y.; Kimball, A. B. The Response of Skin Disease to Stress: Changes in the Severity of Acne Vulgaris as Affected by Examination Stress. *Arch. Dermatol.* **2003,** *139 (7),* 897–900.

8. Choi, E. H.; Brown, B. E.; Crumrine, D.; Chang, S.; Man, M. Q.; Elias, P. M.; Feingold, K. R. Mechanisms by Which Psychologic Stress Alters Cutaneous Permeability Barrier Homeostasis and Stratum Corneum Integrity. *J. Invest. Dermatol.* **2005,** *124 (3),* 587–595.

9. Crosta, M. L.; Caldarola, G.; Fraietta, S.; Craba, A.; Benedetti, C.; Coco, V.; Janiri, L.; Rinaldi, L.; De Simone, C. Psychopathology and Eating Disorders in Patients with Psoriasis. *G. Ital. Dermatol. Venereol.* **2014,** *149 (3),* 355–361.

10. Egydio, F.; Ruiz, F. S.; Tomimori, J.; Tufik, S.; Andersen, M. L. Can Morphine Interfere in the Healing Process during Chronic Stress? *Arch. Dermatol. Res.* **2012,** *304 (6),* 413–420.

11. Elenkov, I. J.; Chrousos, G. P. Stress Hormones, Th1/Th2 Patterns, Pro/Anti-Inflammatory Cytokines and Susceptibility to Disease. *Trends Endocrinol. Metab.* **1999,** *10 (9),* 359–368.

12. Epel, E. S.; Blackburn, E. H.; Lin, J.; Dhabhar, F. S.; Adler, N. E.; Morrow, J. D.; Cawthon, R. M. Accelerated Telomere Shortening in Response to Life Stress. *Proc Natl Acad Sci U S A.* **2004,** *101 (49),* 17312–17315.

13. Ganceviciene, R.; Graziene, V.; Fimmel, S.; Zouboulis, C. C. Involvement of the Corticotropin-Releasing Hormone System in the Pathogenesis of Acne Vulgaris. *Br. J. Dermatol.* **2009a,** *160 (2),* 345–352.

14. Ganceviciene, R.; Böhm, M.; Fimmel, S.; Zouboulis, C. C. The Role of Neuropeptides in the Multifactorial Pathogenesis of Acne Vulgaris. *Dermatoendocrinol.* **2009b,** *1 (3),* 170–176.

15. Gebhardt, C.; Averbeck, M.; Diedenhofen, N.; Willenberg, A.; Anderegg, U.; Sleeman, J. P.; Simon, J. C. Dermal Hyaluronan is Rapidly Reduced by Topical Treatment with Glucocorticoids. *J. Invest. Dermatol.* **2010,** *130 (1),* 141–149.

16. Goulden, V.; Clark, S. M.; Cunliffe, W. J. Post-Adolescent Acne: A Review of Clinical Features. *Br. J. Dermatol.* **1997,** *136 (1),* 66–70.

17. Guyon, A.; Balbo, M.; Morselli, L. L.; Tasali, E.; Leproult, R.; L'Hermite-Balériaux, M.; Van Cauter, E.; Spiegel, K. Adverse Effects of Two Nights of Sleep Restriction on the Hypothalamic-Pituitary-Adrenal Axis in Healthy Men. *J. Clin. Endocrinol. Metab.* **2014,** *99 (8),* 2861–2868.

18. Haapasaari, K. M.; Risteli, J.; Koivukangas, V.; Oikarinen, A. Comparison of the Effect of Hydrocortisone, Hydrocortisone-17-Butyrate and Betamethasone on Collagen Synthesis in Human Skin in Vivo. *Acta. Derm. Venereol. (Stockh).* **1995,** *75 (4),* 269–271.

19. Hanel, K. H.; Cornelissen, C.; Luscher, B.; Baron, J. M. Cytokines and the Skin Barrier. *Int. J. Mol. Sci.* **2013,** *14 (4),* 6720–6745.

20. Hay, R . J.; Johns, N. E.; Williams, H. C.; Bolliger, I. W.; Dellavalle, R. P.; Margolis, D. J.; Marks, R.; Naldi, L.; Weinstock, M. A.; Wulf, S. K.; Michaud, C.; Murray, J. L. C.; Naghavi, M. The Global Burden of Skin Disease in 2010: An Analysis of the Prevalence and Impact of Skin Conditions. *J. Invest. Dermatol.* **2014,** *134 (6),* 1527–1534.

21. Hirotsu, C.; Nogueira, H.; Albuquerque, R. G.; Tomimori, J.; Tufik, S.; Andersen, M. L. The Bidirectional Interactions between Psoriasis and

Obstructive Sleep Apnea. *Int. J. Dermatol.* **2015,** *54 (12),* 1352–1358.

22. Hirotsu, C.; Rydlewski, M.; Araujo, M. S.; Tufik, S.; Andersen, M. L. Sleep Loss and Cytokines Levels in an Experimental Model of Psoriasis. *PLoS One.* **2012,** *7 (11),* e51183.

23. Huerta, C.; Rivero, E.; Rodríguez, L. A. Incidence and Risk Factors for Psoriasis in the General Population. *Arch. Dermatol.* **2007,** *143 (12),* 1559–1565.

24. Kahan, V.; Andersen, M. L.; Tomimori, J.; Tufik, S. Stress, Immunity and Skin Collagen Integrity: Evidence from Animal Models and Clinical Conditions. *Brain Behav. Immun.* **2009,** *23 (8),* 1089–1095.

25. Kähäri, V. M. Dexamethasone Suppresses Elastin Gene Expression in Human Skin Fibroblasts in Culture. *Biochem. Biophys. Res. Commun.* **1994,** *201 (3),* 1189–1196.

26. Karaca, S.; Fidan, F.; Erkan, F.; Nural, S.; Pinarcı, T.; Gunay, E.; Unlu, M. Might Psoriasis be a Risk Factor for Obstructive Sleep Apnea Syndrome? *Sleep Breath.* **2013,** *17 (1),* 275–280.

27. Kendall, A. C.; Nicolaou, A. Bioactive Lipid Mediators in Skin Inflammation and Immunity. *Prog. Lipid. Res.* **2013,** *52 (1),* 141–164.

28. Kimball, A. B.; Wu, E. Q.; Guérin, A.; Yu, A. P.; Tsaneva, M.; Gupta, S. R.; Bao, Y.; Mulani, P. M. Risks of Developing Psychiatric Disorders in Pediatric Patients with Psoriasis. *J. Am. Acad. Dermatol.* **2012,** *67 (4),* 651–7.e1–2.

29. Kubota, Y.; Shirahige, Y.; Nakai, K.; Katsuura, J.; Moriue, T.; Yoneda, K. Community-Based Epidemiological Study of Psychosocial Effects of Acne in Japanese Adolescents. *J. Dermatol.* **2010,** *37 (7),* 617–622.

30. Kurd, S. K.; Troxel, A. B.; Crits-Christoph, P.; Gelfand, J. M. The Risk of Depression, Anxiety, and Suicidality in Patients with Psoriasis: A Population-Based Cohort Study. *Arch. Dermatol.* **2010,** *146 (8),* 891–895.

31. Li, W. Q.; Qureshi, A. A.; Schernhammer, E. S.; Han, J. Rotating Night-Shift Work And Risk of Psoriasis in US Women. *J Invest Dermatol.* **2013,** *133 (2),* 565–567.

32. Love, T. J.; Qureshi, A. A.; Karlson, E. W.; Gelfand, J. M.; Choi, H.K. Prevalence of the Metabolic Syndrome in Psoriasis: Results from the National Health and Nutrition Examination Survey, 2003–2006. *Arch. Dermatol.* **2011,** *147 (4),* 419–424.

33. Martínez-García, E.; Arias-Santiago, S.; Valenzuela-Salas, I.; Garrido-Colmenero, C.; García-Mellado, V.; Buendía-Eisman, A. Quality of Life in Persons Living with Psoriasis Patients. *J. Am. Acad. Dermatol.* **2014,** *71 (2),* 302–307.

34. Oikarinen, A.; Autio, P. New Aspects of the Mechanism of Corticosteroid-Induced Dermal Atrophy. *Clin. Exp. Dermatol.* **1991,** *16 (6),* 416–419.

35. Oikarinen, A.; Autio, P.; Kiistala, U.; Risteli, L.; Risteli, J. A New Method to Measure Type I and III Collagen Synthesis in Human Skin in Vivo: Demonstration of Decreased Collagen Synthesis after Topical Glucocorticoid Treatment. *J. Invest. Dermatol.* **1992,** *98 (2),* 220–225.

36. Oikarinen, A.; Haapasaari, KM.; Sutinen, M.; Tasanen, K. The Molecular Basis of Glucocorticoid-Induced Skin Atrophy: Topical Glucocorticoid Apparently Decreases both Collagen Synthesis and the Corresponding Collagen mRNA Level In Human Skin In Vivo. *Br. J. Dermatol.* **1998,** *139 (6),* 1106–1110.

37. Parish, J. M.; Adam, T.; Facchiano, L. Relationship of Metabolic Syndrome and Obstructive Sleep Apnea. *J. Clin. Sleep Med.* **2007,** *3 (5),* 467–472.

38. Poli, F.; Dreno, B.; Verschoore, M. An Epidemiological Study of Acne in Female Adults: Results of a Survey Conducted in France. *J. Eur. Acad. Dermatol. Venereol.* **2001,** *15 (6),* 541–545.

39. Proksch, E.; Brandner, J. M.; Jensen, J. M. The Skin: An Indispensable Barrier. *Exp. Dermatol.* **2008,** *17 (12),* 1063–1072.

40. Savin, J. A.; Adam, K.; Oswald, I.; Paterson W. D. Pruritus and Nocturnal Wakenings. *J. Am. Acad. Dermatol.* **1990,** *23,* 767–768.

41. Schafer-Korting, M.; Kleuser, B.; Ahmed, M.; Holtje, H. D.; Korting, H. C. Glucocorticoids for Human Skin: New Aspects of the Mechanism of Action. *Skin Pharmacol. Physiol.* **2005,** *18 (3),* 103–114.

42. Schoenborn, C. A.; Adams, P. F. Health Behaviors of Adults: United States, 2005–2007. National Center for Health Statistics. *Vital Health Stat.* **2010,** *10 (245),* 1-132.

43. Schoepe, S.; Schäcke, H.; May, E.; Asadullah, K. Glucocorticoid Therapy-Induced Skin Atrophy. *Exp. Dermatol.* **2006,** *15 (6),* 406–420.

44. Shutty, B. G.; West, C.; Huang, K. E.; Landis, E.; Dabade,T.; Browder, B.; O'Neill, J.; Kinney, M. A.; Feneran, A. N.; Taylor, S.; Yentzer, B.; Mc-

Call, W. V.; Fleischer, A. B. Jr.; Feldman, S. R. Sleep Disturbances in Psoriasis. *Dermatol. Online J.* **2013,** *19 (1),* 1.

45. Sivamani, R. K.; Pullar, C. E.; Manabat-Hidalgo, C. G.; Rocke, D. M.; Carlsen, R. C.; Greenhalgh, D. G.; Isseroff, R. R. Stress-Mediated Increases in Systemic and Local Epinephrine Impair Skin Wound Healing: Potential New Indication for Beta Blockers. *PLoS Med.* **2009,** *6 (1),* e12.

46. Slominski, A.; Wortsman, J.; Tuckey, R. C.; Paus, R. Differential Expression of HPA Axis Homolog in the Skin. *Mol. Cell Endocrinol.* **2007,** *265–266,* 143–149.

47. Smith, S. M.; Vale, W. W. The Role of the Hypothalamic-Pituitary-Adrenal Axis in Neuroendocrine Responses to Stress. *Dialogues Clin. Neurosci.* **2006,** *8 (4),* 383–395.

48. Sundelin, T.; Lekander, M.; Kecklund, G.; Van Someren, E. J.; Olsson, A.; Axelsson, J. Cues of Fatigue: Effects of Sleep Deprivation on Facial Appearance. *Sleep.* **2013,** *36 (9),* 1355–1360.

49. Tan, J. K.; Bhate, K. A Global Perspective on the Epidemiology of Acne. *Br. J. Dermatol.* [Online early access] DOI: 10.1111/bjd.13462. Published Online: January 17, **2015.** http://www.pubmed.com (accessed May 22, 2015).

50. Theoharides, T. C.; Singh, L. K.; Boucher, W.; Pang, X.; Letourneau, R.; Webster, E.; Chrousos, G. Corticotropin-Releasing Hormone Induces Skin Mast Cell Degranulation and Increased Vascular Permeability, a Possible Explanation for its Proinflammatory Effects. *Endocrinology.* **1998,** *139 (1),* 403–413

51. Tufik, S.; Andersen M. L.; Bittencourt L. R.; Mello M. T. Paradoxical Sleep Deprivation: Neurochemical, Hormonal and Behavioral Alterations. Evidence from 30 Years of Research. *An Acad. Bras. Cienc.* **2009,** *81 (3),* 521–538.

52. Williams, H. C.; Dellavalle, R. P.; Garner, S. Acne Vulgaris. *Lancet.* **2012,** *379 (9813),* 361–372.

53. Yang, Y. W.; Kang, J. H.; Lin, H. C. Increased Risk of Psoriasis Following Obstructive Sleep Apnea: A Longitudinal Population-Based Study. *Sleep Med.* **2012,** *13 (3),* 285–289.

54. Zouboulis, C. C.; Böhm, M. Neuroendocrine Regulation of Sebocytes -- A Pathogenetic Link between Stress and Acne. *Exp. Dermatol.* **2004,** *13 (4),* 31–35.

21

ASTHMA AND SLEEP

Nahid Sherbini, Abdullah Al-Harbi, Mohammad Khan,
and Hamdan Al-Jahdali

ABSTRACT

Asthma and sleep-disordered are common diseases that affect children and adults. Symptomatic asthma leads to sleep disturbances and sleep deprivation which may exaggerate asthma even more. Sleep breathing disorders also known risk factor for uncontrolled asthma. Patients with these disorders are commonly seen by general or subspecialties clinicians who should know about the interaction between these two conditions. There is increasing academic interest in asthma and sleep-disordered breathing. The purpose of this paper is to review the recent literature related to interaction between asthma and sleep disorders in a in a manner that is relevant for clinical practice.

Nahid Sherbini
Department of Medicine, Pulmonary Division, Sleep Disorders Center, King Saud University for Health Sciences, Riyadh, Saudi Arabia

Abdullah Al-Harbi
Department of Medicine, Pulmonary Division, Sleep Disorders Center, King Saud University for Health Sciences, Riyadh, Saudi Arabia

Mohammad Khan
Department of Medicine, Pulmonary Division, Sleep Disorders Center, King Saud University for Health Sciences, Riyadh, Saudi Arabia

Hamdan Al-Jahdali
Department of Medicine, Pulmonary Division, Sleep Disorders Center, King Saud University for Health Sciences, Riyadh, Saudi Arabia

Corresponding author: Hamdan Al-Jahdali. Adjunct Professor, McGill University; Professor Pulmonary and Sleep Medicine, King Saud University for Health Sciences; Head of Pulmonary Division Medical Director of Sleep Disorders Center, King Abdulaziz Medical City, PO Box 101830*11665, Riyadh, Saudi Arabia. Tel.: +966-12520088(Ext: 17597, 17531), Mobile: +966-505224271; E-mail: jahdali@yahoo.com

21.1. INTRODUCTION

Asthma is a chronic disease characterized by inflammatory reversible airflow obstruction and bronchial hyper responsiveness to a variety of stimuli. This inflammation causes recurrent episodes of wheezing, cough, shortness of breath and chest tightness (Table 21.1). The majority of Asthma patient's symptoms can be controlled with standard therapy. However, about 5–10% of them have severe Asthma which is persistently difficult to control. Global initiative for Asthma (GINA), Global Asthma Report 2015 recorded 334 million people have Asthma, 14% of the world's children suffer from Asthma symptoms, 8.6% of young adults (aged 18–45) experience Asthma symptoms and 4.5% of young adults have been diagnosed with Asthma and/or are taking treatment for Asthma.[1] The burden of Asthma is greatest for children aged 10–14 and the elderly aged 75–79. Asthma is considered the 14th most important disorder in the world in terms of the impact in life activates.[1]

Table 21.1. Asthma symptoms.

Cough
Recurrent Wheeze
Shortness of breath
Chest tightness
Frequent awakening from sleep or inability to sleep due to Asthma symptoms
Allergic rhinitis

The incidence of obstructive sleep apnoea (OSA) is reported to be 4% in men and 2% women as assessed using Level 1 Polysmnography (PSG).[2] National Sleep Foundation (USA) polled 1,506 adults in 2005 using Berlin Questionnaire which is a validated tool for OSA screening - 26% met high probability criteria of having OSA.[3]

OSA is classically manifested by Snoring, tiredness, observed apneas and excessive daytime sleepiness. Risk factors for OSA include male gender, age over 50, obesity with BMI > 35, nocturnal nasal congestion, high uncontrolled blood pressure, and neck circumference greater than 40 cm for male and 38 cm for female (Table 21.2).[4, 5]

Table 21.2. Risk factors and classical findings of OSA.

Male gender
Age more than 50
obesity
Snoring
Shortness of breath
Inability to sleep on supine position because of dyspnea
Frequent arousal from sleep
High blood pressure
Excessive daytime sleepiness
Unrefreshing sleep
Tiredness and Fatigue
Observed apnea
High Mallmpati score
Large size neck circumference
nocturnal nasal congestion

There are many risk factors affecting both diseases including allergic rhinitis, obesity, Gastro esophageal reflux disorder (GERD) and systemic inflammatory changes (Figure 21.1). The association between OSA and chronic obstructive pulmonary disease is well known as an overlap syndrome. However, the association of Asthma and OSA is less clear.[6, 7] Their coexistence has combined greater effects on patient symptoms, response to therapy, and outcomes.[8, 9]

Uncontrolled Asthma is an important concern and complex clinical problem with multiple risk factors such as disease phenotype, adherence issues, inhaler techniques, smoking, environmental triggers, and common associated comorbidities such as GERD and allergic rhinitis (AR) which can

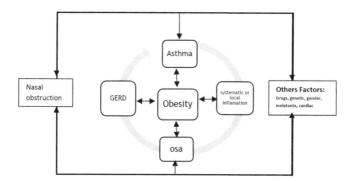

Figure 21.1. Interrelations between Asthma, Allergic rhinitis, GERD, Obesity and OSA modified.[14]

worsen Asthma. Moreover, OSA is associated with almost seven times greater likelihood of severe Asthma.[10] Both disorders are associated with histological changes due to release of inflammatory mediators such as cytokines. In addition, sleep deprivation, chronic upper airway edema, and inflammatory mediators release associated with OSA may further exacerbate nocturnal Asthma symptoms.[11, 12] Therefore, a high clinical suspicion of OSA in Asthma patients, who are refractory to standard treatment lines, is recommended.

The interactions between these conditions should highlight more specific mechanisms. The clinical, pathogenic, and therapeutic significance of the different phenotypes of the disease is still emerging and needs further evaluation. Asthma predisposes to OSA and OSA appears to affect Asthma control. Previous studies suggested some relationship between the two respiratory conditions, but researchers were not sure what the initial trigger is.[13] We present here a review on the relationships between Asthma and OSA, including their pathogenesis, clinical presentation, and therapeutic connections.

21.2. BURDEN OF THE DISEASE

21.2.1. Similarities

OSA and nocturnal Asthma are two distinct diseases that share similar risk factors and often overlapping symptoms, some used term "alternative overlap syndrome". [14] These disorders may result in fragmented sleep and frequent arousals, with significant excessive daytime sleepiness. Similarly, both may manifest with recurrent episodes of apneas, coughing, and breathlessness during the sleep. Both involve repetitive arousals, sleep stage changes and limitations in airflow and increasing respiratory efforts with a subsequent desaturation during sleep. This similarity further adds to the overlapping symptoms and pathogenesis.[15-17] Consequently, both disorders classically manifest with more symptoms during sleep. Given these similarities, nocturnal Asthma is considered by some to be another form of sleep-disordered breathing (SDB).[18, 19]

21.3. Pathogenesis and Proposed Mechanisms for Interaction Between OSA and Asthma

The exact mechanism of the interaction between Asthma and OSA is not known. However, we present several mechanisms postulated for interaction that might occur (Table 21.3):

Table 21.3. Proposed mechanisms for interaction between OSA and Asthma

---

A. Local inflammation and upper airway anatomy
B. Allergic rhinitis
C. Systemic inflammation
D. Circulating leptin
E. Neuromechanical reflex bronchoconstriction
F. Intermitted hypoxia
G. GERD
H. Obesity, weight gain
I. Asthma therapy

21.3.1 Local Airway Inflammation and Upper Airway Anatomy

Repeated mechanical injuries with the chronic persistent local inflammation in the respiratory system (upper and lower) caused by snoring and apnoea in OSA patients provokes chronic increase in bronchospasm and Asthma symptoms.[20] Snoring with or without OSA could also alter sympathetic and parasympathetic effects during sleep. Experimental studies suggest that pro-inflammatory biomarkers, such as TNF-α, IL-1, and macrophages, in the laryngeal and soft palate tissue was significantly high in rats exposed for sim-

ulated recurrent repetitive upper airway collapse and reopening, similar to what occurs in OSA.[21] Also, high level of IL-8 indicates trigger of inflammatory response was founded after exposure of the bronchial epithelial cells to prolong vibrations mimicking what happen with snoring and OSA after vibration.[22]

Supine position during sleep in OSA patient associated with reductions in lung volume, intrapulmonary pooling of blood, and sleep-associated upper airway narrowing leading to snoring and/or apnea.[23]

21.3.2. Allergic Rhinitis (AR)

AR is an independent risk factor for both Asthma and OSA. AR and Asthma are common diseases that frequently occur together and shared many pathophysiological features (Figure 21.2).[24, 25]

Both cause mucosal inflammation, consequently reducing nasal and upper airway cross-sectional area and may worsen OSA.[26] The increased nasal obstruction induces an increase in nasal resistance that cause increases the negative pressure in the upper airway during inspiration, and predispose to OSA. Kiely et al. noted that Apnea Hypoapnea index (AHI) improve in OSA patient with co-existing AR after four weeks of treatment with intranasal corticosteroid.[27] Kheirandish-Gozal et al. used intranasal budesonide for six weeks in children with moderate OSA and noted a significant improvement in the polysomnographic results, 54.1% of the children reported with a normal sleep test post treatment with budesonide.[28]

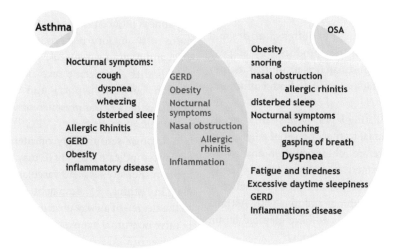

Figure 21.2. Similarities between OSA and asthma, common associated conditions, risk factors.

21.3.3. Systemic Inflammation

OSA by means of hypoxemia, hypercapnia, and sleep fragmentation, can aggravate pro-inflammatory states through effects on sympathetic hyperactivity, oxidative stress, or both. Oxidative stress secondary to repetitive oxygen desaturation in thought to be the reason for increased systemic inflammatory mediators.[29]

Many studies have reported high levels of systemic inflammatory markers in OSA patients.[30] These inflammatory markers include; C-reactive protein, tumor necrosis factor alpha (TNF-α), vascular cell adhesion molecule (VCAM), interleukin IL (IL-6 and 8) and intercellular adhesion molecule (ICAM).[30] These mediators cause endothelial cell damage and dysfunction. TNF-α is considered a marker of sleep-disordered breathing and a potent pro-inflammatory cytokine that interferes

with airway smooth muscle contractility and its secretion may play a role in circadian rhythm of sleep and an important role in the pathogenesis of Asthma. Inhibition for TNF-α can decrease the severity of OSA as examined and found after tonsillectomy.[31] These inflammatory responses in OSA patients provokes further Asthma symptoms and exacerbations.[32]

OSA has increase also lipid levels which promotes inflammatory responses. Gozal et al. noted significant improvements in lipid profiles, C-reactive protein, and apolipoprotein B after adenotonsillectomy.[31]

Other evidence showed that markers of chronic inflammation, like C-reactive protein, are related to severity of OSA and return to normal after CPAP treatment.[33] CPAP therapy results in an improvement in the levels of CRP, TNF-α, and IL-6.[33]

21.3.4. Circulating Leptin

Leptin is elevated in both asthmatics and OSA patients, it is produced by adipose tissue and postulated as pro-inflammatory marker which exacerbate both conditions. [20]

Circulating-leptin levels are directly proportional to the amount of adipose tissue; therefore, obese children and adults have elevated circulating-leptin levels. Leptin can be considered as a predictive factor for Asthma, children with Asthma are twice as likely to have elevated leptin levels as are those without. [34]

The treatment of OSA can reduce circulating-leptin levels as a result of the reduction in the AHI. As reported in one study that adults with OSA treated with CPAP showed a reduction leptin level that independently correlate with the reduction in the AHI. [35] These are some data supporting the theories behind the association of high leptin level and obesity in asthmatic patients which is also considered as a predictor for an Asthma severity. [36, 37]

21.3.5. Neuromechanical Reflex Bronchoconstriction

Irritation of upper airway neural receptors and altered nocturnal neuro-hormonal secretion due to sleep apnea, precipitate bronchoconstriction in asthmatic population causing Asthma symptoms and more exacerbations. [23, 29]

21.3.6. Hypoxia Induced Bronchoconstriction\Intermitted Hypoxia

Stimulation of carotid body by hypoxia triggered by repeated mechanical irritation of larynx during snoring and apnea causing apneas induced increase in vagal tone, more inflammatory responses worsening OSA and Asthma symptoms. [38]

Another important explanation of relation between hypoxia and worsening of Asthma is the presence of vascular endothelial growth factor (VEGF) which is a hypoxia-sensitive glycoprotein. The evidences indicate that VEGF may worsen hyper responsiveness and vascular remodeling in Asthma. [39, 40] It may be correlated to the degree of airway obstruction. [41] Repetitive nocturnal hypoxia is the most likely trigger of VEGF release in OSA patients. Studies also indicate that OSA patients have elevated concentrations of VEGF that correlate with the severity of the syndrome as reflected by the level of the AHI and to nocturnal oxygen desaturation level. [42]

21.3.7. Gastro Esophageal Reflux Disorder (GERD)

GERD is common in OSA, result in sleep disturbances that include difficulty in falling asleep, sleep fragmentation, and early morning awakenings. OSA also causes increase in negative pressure in the upper airway during inspiration and may induce GERD and micro aspiration. [43, 44]

Decreased intrathoracic pressure and increased transdiaphragmatic pressure during apnea are postulated to provoke Asthma by mean of vagal reflexes. This is elicited when the esophagus is exposed to acid and predispose to retrograde movement of gastric contents. [45] Guda et al. suggested that patients with GERD have more episodes of OSA than those without symptoms of GERD. [46]

GERD is considered a trigger for nocturnal Asthma, prevalence of GERD in Asthma population was reported to range from 36 to 80%.[46] It is more common in OSA patients than in the general population especially with uncontrolled Asthma. [44] The evidences supporting a link between GERD and the Asthma and OSA, treatment of GERD is helping to control both conditions. Studies showed OSA patients had GERD and increase arousals secondary to GERD which improve with CPAP therapy.[47] Treatment of GERD improves Asthma control, fundoplication was the most effective therapy in one study.[48]

21.3.8. Obesity

Obesity is associated with a specific Asthma phenotype; the exact mechanisms that underlie the interaction of obesity with Asthma remain unclear. However, obesity results in important changes to the mechanical properties of the respiratory system and these obesity-related factors appear to exert an additive effect to the Asthma-related changes seen in the airways.[49] OSA and obesity have a bi-interrelation which in turn predisposes to airway hyper-responsiveness causing again more Asthma symptoms and exacerbations.[50] Obese patients with OSA exhibit leptin levels approximately 50% higher than those of similarly obese men without OSA.[51, 52] Leptin is elevated in both asthmatics and OSA patients; its pro-inflammatory marker which may exacerbate both conditions.[20]

21.3.9. Asthma Therapy

Oral steroid and inhaled corticosteroid used to control Asthma may predispose OSA. Steroid may cause pharyngeal muscle myopathy, deposition of adipose tissue in the neck and parapharyngeal muscle. These changes lead to increase upper airway collapsibility and induce snoring and OSA.[53-55] Other medications used in Asthma such as theophylline and beta agonist may also cause insomnia, and sleep fragmentation.[56]

21.4. INTERRELATION BETWEEN ASTHMA AND INCIDENCE OF OSA

A negatively bi-directional relationship between OSA and Asthma is highlighted.[57] However, whether Asthma by itself is a risk for development of OSA only addressed recently. Teodorescu et al. examined the prospective relationship between patients diagnosed with Asthma and a new onset OSA in the Wisconsin sleep cohort study, which consists of randomly selected Wisconsin state employees that, at the time of enrollment in 1988, ranged between 30 and 60 years of age and were free of OSA.[13] Every 4 years, they did for them an overnight sleep study and completed health-related questionnaires. Asthma was associated with increased risk of new-onset OSA by almost 27% compared to 16% non-asthmatic patients over 4 years follow up, participants with a diagnosed Asthma were 40% more likely to develop OSA than those without pre-existing Asthma.[13]

A systematic review done to assess the relationship between Asthma and SDB and they found children with Asthma had a significantly higher risk for OSA: OR 1.9 [1.7–2.2]. Asthma seems to be a significant risk factor for developing OSA. However, the minority of the diagnosis of OSA studies in this review were based on polysomnography, considered the current gold standard for diagnosing OSA.[58] Both diseases are prevalent and whether such an association is causative or incidental is still not clear. This temporal relationships between both conditions should be addressed in future studies.

21.5. EFFECT OF OSA ON ASTHMA CONTROL

There is a higher prevalence of OSA symptoms in asthmatic patients independent of the severity of disease.[57, 59] A prospective cohort study showed a high prevalence of OSA in patients with difficult-to-control Asthma.[53] Julien et al. clearly demonstrated a higher prevalence of OSA among patients with severe compared to those with moderate Asthma patients.[60] Snoring and OSA were significantly associated with Asthma even after adjustment for other confounding factors and also associated with worse quality of life.[61, 62] Association of Asthma severity and OSA was studied also in a cross-sectional study of 167 patients by Byun et al. which investigated the other associating factors thought to increase predisposition for OSA, including the common known factors such as male gender, age, and obesity and the less studied factors including Asthma, GERD, postnasal drip syndrome, cardiovascular disease, diabetes, smoking, snoring, and COPD.[12] OSA defined at an AHI≥5/hr was strongly correlated with moderate to severe Asthma.[12]

OSA may cause nocturnal awakenings (arousal from apneas/hypopneas) confused with Asthma symptoms. Studies also have shown that recurrent Asthma exacerbations and OSA, rhinitis with snoring and apnea, rhinitis and sleepiness are all interrelated.[63, 64] Sleep deprivation, upper airway edema and inflammation associated with OSA may further exacerbate nocturnal Asthma symptoms.[50, 65] Observation suggests that OSA may contribute to the non-eosinophilic phenotype which recently more recognized between uncontrolled Asthma patients.[66]

Unrecognized OSA may lead to poor Asthma control despite optimal therapy and persistent Asthma symptoms during the day as well as the night. In Asthma patients, high OSA risk is significantly associated with uncontrolled Asthma independent of obesity and GERD.[57, 59]

The sleep-disordered breathing (SDB) in asthmatic children compared to non-asthmatic showed the prevalence of snoring was significantly high in asthmatic children (35.5%) than controls (15.7%) and the prevalence of poor sleep quality was significantly higher in asthmatic children (25.9%) than controls (10.6%) ($P < 0.001$).[67] There was a higher prevalence of SDB in asthmatic children compared to non-asthmatic children and the prevalence of SDB increased with increasing Asthma severity.[67] Insomnia in children with Asthma may reflect underlying OSA.[68]

Asthma patients have significant impairments in sleep quality that may be independent from GERD, OSA, and nocturnal Asthma manifestations. Poor sleep quality was associated with worse Asthma control and Asthma quality of life even after controlling for GERD and other covariates.[69]

The National Asthma Education and Prevention Program- NHLB - guidelines recommend evaluation for OSA in patients with Asthma with suboptimal control. NHLB also reported that difficult to treat Asthma need to be screened for OSA.[70]

Continuous positive airway pressure (CPAP) treatment of OSA in patients with Asthma improves outcomes, including Asthma symptoms, reliever bronchodilator use, lung functions, and quality of life.[7],[10] Another team found that two months of CPAP for OSA resulted in a significant reduction in Asthma symptoms.[8] Additionally, Chan et al. reported reduced Asthma symptoms and bronchodilator use, and improved PEFR after two weeks of CPAP in nine patients and found that cessation of CPAP returned PEFR to baseline level.[6] In another group of patients, CPAP improved both nocturnal and diurnal peak expiratory flow up to two weeks after starting therapy. When CPAP was applied in patients with nocturnal Asthma only, nocturnal expiratory airflow was not improved, and sleep quality worsened.[71]

21.6. EFFECT OF ASTHMA IN SLEEP

Seasonal variability of sleep breathing disordered was noticed in children which supporting the effect of Asthma in sleep.[72] Respiratory illnesses such as Asthma, which affects 7.5–10% of children, and atopy are reported to be risk factors for the development of OSA.[71]

Asthma can cause a spectrum of sleep breathing disorders by decreased sleep quality, snoring, daytime sleepiness, early awakening, difficulty maintaining sleep and end by OSA.

21.6.1. Daytime Sleepiness

Excessive daytime sleepiness (EDS) usually attributed to a direct effect of Asthma on nocturnal sleep quality. However, Teodorescu et al. studied 115 patients undergone a routine Asthma care, 55% perceived excessive daytime sleepiness and 47% had excessive daytime somnolence by Epworth sleeping scale (ESS) questionnaires independent from Asthma severity and he conclude that majority of sleepiness was due to undiagnosed OSA.[73]

21.6.2. Night Time Symptoms

Nocturnal Asthma can cause sleep disruption could cause periodic breathing and decreased upper airway muscle activity. Irritation of the upper airways may predispose to decreased lung function at night and hypoxemia may predispose to increased bronchial hyperresponsiveness. Nocturnal exacerbations affect about 2/3 Asthma patients, poor control and excessive daytime sleepiness, 90% of patients with Asthma report wheezing or cough at night.[74] Sleep disturbances as a consequence of nocturnal Asthma have been reported also in children.[69]

Understanding the mechanism of nocturnal Asthma and the factors that exacerbate Asthma during sleep would lead to better management of the condition. Many patients with Asthma experience nocturnal symptoms. In a large survey of 7729 patients with Asthma, 74% reported experiencing nocturnal cough and wheeze at least once a week.[16] The most studied causes and contributing factors to exacerbations of Asthma at night include circadian changes in airway hyperresponsiveness and inflammation, mucociliary clearance, ventilatory responses to hypercapnia and hypoxia, and hormone levels. Other possible explanations for refractory nocturnal

Asthma symptoms include non-adherence to drugs, nocturnal triggers, and OSA.

A greater proportion of Asthma patients tend to die at night than those in the general population. The occurrence of nocturnal Asthma symptoms is also reflected in increase mortality. In one study and over 1 year, 53% of Asthma deaths in one report occurred at night. In addition, 79% of these patients had complaints of Asthma affecting their sleep and occurring every night in 42%. Nocturnal Asthma symptoms suggesting poorly controlled Asthma which is worse at night, with nocturnal bronchospasm and cough inducing sleep disturbances.[6]

Although nocturnal exacerbations can indicate inadequate Asthma control and disturb sleep, poor sleep has been reported in patients with well-controlled Asthma suggesting that poor sleep may be independent from nocturnal Asthma symptoms.[75]

21.7. ALLERGIC RHINITIS, ASTHMA AND SLEEP

Evidences as explained earlier indicate that the majority of patients with Asthma have rhinitis. AR is associated with reduced upper airway cross-sectional area furthermore Asthma and AR increased airway mucosal inflammation causing sleep disruption worsening of OSA.[26]

Another possible aetiology for the high prevalence of OSA symptoms in asthmatic patients is the increased incidence of nasal obstruction in asthmatic patient due to concomitant allergic rhinitis. The nasal obstruction contributes to sleep-disordered breathing in predisposed individuals.[76]

Rhinitis and chronic sinusitis are common conditions that may cause nasal congestion and consequently contribute to upper airway obstruction in OSA. Nasal and nasopharyngeal polyps may also be associated with upper airway obstruction. [77] The increased nasal obstruction in asthmatic patients induces an increase in nasal resistance that cause increases the negative pressure in the upper airway during inspiration, increase airway collapsibility a predisposing factor for OSA.

Sinus disease and nasal congestion are common in patients with both GERD and Asthma and can manifest as snoring or OSA. Asthma is also a risk factor for increased respiratory complications after adenotonsillectomy in children with snoring and physical signs of upper airway obstruction.[78]

21.8. RELATION BETWEEN OSA–ASTHMA–OBESITY

Obesity is a risk factor for Asthma. It seems that the association between Asthma and OSA worsens the clinical picture of Asthma, given that OSA can stimulate weight gain, playing a significant role in the severity of Asthma.[59]

Proposed mechanisms linking the SBD to obesity are inflammation, insulin resistance, appetite regulating hormones, and sleep deprivation.[79] Other possible mechanisms may narrow the airways in patients with OSA and the repetitive Muller maneuver (inspiration against a closed glottis) which cause inspiratory collapse of sections of the airways. Increased vagal tone also may aggravate bronchoconstriction.[79] Thus there are numerous potential direct pathophysiological interactions between sleep apnea and Asthma.

OSA interferes with lipid homeostasis and systemic inflammation and, when associated with obesity, affects glycemic regulation, interfering with insulin sensitivity, independently of the BMI. Other team of investigators found an association of BMI, leptin levels, and adiponectin levels with decreased levels of exhaled nitric oxide in patients with Asthma.[31]

Obese asthmatic increased sputum neutrophils but not eosinophils. Neutrophilic Asthma defined as > 61% neutrophils. Sputum neutrophils positively correlated with BMI, 1% increase in neutrophils for every one unit increase of BMI.[80]

Obese individuals have functional residual capacity is close to residual volume and breathe at low lung volumes because of the mechanical effects of obesity on the respiratory system. In OSA, repetitive opening and closing of the airways is exacerbated by the increased tidal volume following an apnea. And also found that total lung resistance and elastance increased during the apneic episodes.[81] Influence of BMI, especially a BMI > 35 kg/m^2, was shown in other studies to have a significant correlation with symptoms of Asthma and OSA.[82]

There is under diagnosis of Asthma in morbidly obese.[83] Obese asthmatics often have poor Asthma control and respond poorly to therapy. It has been suggested that co-morbidities associated with obesity, such as reflux and obstructive sleep apnea, could be important factors contributing to poor Asthma control in obese patients.

The high prevalence of OSA in patients with Asthma appears to be associated with obesity. In a cross-sectional study involving 17,994 children, 14% of whom had a diagnosis of Asthma, the prevalence of Asthma was directly proportional to the BMI percentile. The prevalence of Asthma is higher in obese children and even higher in morbidly obese children. It has been suggested that, independent of a certain threshold of obesity, metabolic factors become involved in the pathophysiology of upper airway inflammation, as well as in bronchial hyperactivity, being able to interfere with the clinical manifestations of Asthma.[84]

Asthma, OSA and obesity are linked and associated with systemic inflammation worsen Asthma control. Treating obesity and OSA improves Asthma control and systemic inflammation.

21.9. CONCLUSIONS

When Asthma and OSA co-exist, successful treatment of one is dependent upon the effective treatment of the other. Both conditions separately required special education, close follow up and monitoring of the symptoms like shortness of breath, wheezing, snoring, worsening sleep quality, and/or daytime sleepiness.

Patient with uncontrolled Asthma should be screened for OSA before escalation of Asthma therapy. CPAP should be initiated in asthmatic patient with OSA and the need for warm air humidification in conjunction with the CPAP circuit should always be considered in the asthmatic patient.[14]

21.10. FINAL REMARKS

Future research may need to explore pathway to further confirm the hypothetical link in causality and management between the Asthma and OSA. We and other recommend that clinicians should understand the relationship between the two condition and how important is identifying the effect of each one on the other.[85]

KEY WORDS

- **asthma**
- **sleep disorders**
- **insomnia**
- **sleep apnea**
- **excessive sleepiness**

REFERENCES

1. GINA Global Strategy for Asthma Management and Prevention, http://www.ginasthma.org/local/uploads/files/GINA_Pocket_2015.pdf, accessdate 05-05-2015.
2. Young T, Palta M, Dempsey J, Skatrud J, Weber S, Badr S. The occurrence of sleep-disordered breathing among middle-aged adults. N Engl J Med. 1993 Apr 29;328(17):1230-5.
3. Hiestand DM, Britz P, Goldman M, Phillips B. Prevalence of symptoms and risk of sleep apnea in the US population: Results from the national sleep foundation sleep in America 2005 poll. Chest. 2006 Sep;130(3):780-6.
4. Chung F, Yegneswaran B, Liao P, Chung SA, Vairavanathan S, Islam S, et al. STOP questionnaire: a tool to screen patients for obstructive sleep apnea. Anesthesiology. 2008 May;108(5):812-21.
5. Tishler PV, Larkin EK, Schluchter MD, Redline S. Incidence of sleep-disordered breathing in an urban adult population: the relative importance of risk factors in the development of sleep-disordered breathing. JAMA. 2003 May 7;289(17):2230-7.
6. Chan CS, Woolcock AJ, Sullivan CE. Nocturnal asthma: role of snoring and obstructive sleep apnea. Am Rev Respir Dis. 1988 Jun;137(6):1502-4.
7. Guilleminault C, Quera-Salva MA, Powell N, Riley R, Romaker A, Partinen M, et al. Nocturnal asthma: snoring, small pharynx and nasal CPAP. Eur Respir J. 1988 Dec;1(10):902-7.
8. Ciftci TU, Ciftci B, Guven SF, Kokturk O, Turktas H. Effect of nasal continuous positive airway pressure in uncontrolled nocturnal asthmatic patients with obstructive sleep apnea syndrome. Respir Med. 2005 May;99(5):529-34.
9. Lafond C, Series F, Lemiere C. Impact of CPAP on asthmatic patients with obstructive sleep apnoea. Eur Respir J. 2007 Feb;29(2):307-11.
10. Teodorescu M, Polomis DA, Gangnon RE, Fedie JE, Consens FB, Chervin RD, et al. Asthma Control and Its Relationship with Obstructive Sleep Apnea (OSA) in Older Adults. Sleep Disord. 2013;2013:251567.
11. Bonekat HW, Hardin KA. Severe upper airway obstruction during sleep. Clin Rev Allergy Immunol. 2003 Oct;25(2):191-210.
12. Byun MK, Park SC, Chang YS, Kim YS, Kim SK, Kim HJ, et al. Associations of moderate to severe asthma with obstructive sleep apnea. Yonsei Med J. 2013 Jul;54(4):942-8.
13. Teodorescu M, Barnet JH, Hagen EW, Palta M, Young TB, Peppard PE. Association between asthma and risk of developing obstructive sleep apnea. JAMA. 2015 Jan 13;313(2):156-64.
14. Prasad B, Nyenhuis SM, Weaver TE. Obstructive sleep apnea and asthma: associations and treatment implications. Sleep Med Rev. 2014 Apr;18(2):165-71.
15. Guilleminault C, Stoohs R, Clerk A, Cetel M, Maistros P. A cause of excessive daytime sleepiness. The upper airway resistance syndrome. Chest. 1993 Sep;104(3):781-7.
16. Bohadana AB, Hannhart B, Teculescu DB. Nocturnal worsening of asthma and sleep-disordered breathing. J Asthma. 2002 Apr;39(2):85-100.
17. Larsson LG, Lindberg A, Franklin KA, Lundback B. Symptoms related to obstructive sleep apnoea are common in subjects with asthma, chronic bronchitis and rhinitis in a general population. Respir Med. 2001 May;95(5):423-9.
18. Sleep-related breathing disorders in adults: recommendations for syndrome definition and measurement techniques in clinical research. The Report of an American Academy of Sleep Medicine Task Force. Sleep. 1999 Aug 1;22(5):667-89.
19. Novak M, Shapiro CM. Drug-induced sleep disturbances. Focus on nonpsychotropic medications. Drug Saf. 1997 Feb;16(2):133-49.
20. Puthalapattu S, Ioachimescu OC. Asthma and obstructive sleep apnea: clinical and

pathogenic interactions. J Investig Med. 2014 Apr;62(4):665-75.

21. Almendros I, Carreras A, Ramirez J, Montserrat JM, Navajas D, Farre R. Upper airway collapse and reopening induce inflammation in a sleep apnoea model. Eur Respir J. 2008 Aug;32(2):399-404.

22. Puig F, Rico F, Almendros I, Montserrat JM, Navajas D, Farre R. Vibration enhances interleukin-8 release in a cell model of snoring-induced airway inflammation. Sleep. 2005 Oct;28(10):1312-6.

23. Ballard RD. Sleep, respiratory physiology, and nocturnal asthma. Chronobiol Int. 1999 Sep;16(5):565-80.

24. Khan DA. Allergic rhinitis and asthma: epidemiology and common pathophysiology. Allergy Asthma Proc. 2014 Sep-Oct;35(5):357-61.

25. Linneberg A, Henrik Nielsen N, Frolund L, Madsen F, Dirksen A, Jorgensen T. The link between allergic rhinitis and allergic asthma: a prospective population-based study. The Copenhagen Allergy Study. Allergy. 2002 Nov;57(11):1048-52.

26. Braido F, Baiardini I, Lacedonia D, Facchini FM, Fanfulla F, Molinengo G, et al. Sleep apnea risk in subjects with asthma with or without comorbid rhinitis. Respir Care. 2014 Dec;59(12):1851-6.

27. Kiely JL, Nolan P, McNicholas WT. Intranasal corticosteroid therapy for obstructive sleep apnoea in patients with co-existing rhinitis. Thorax. 2004 Jan;59(1):50-5.

28. Kheirandish-Gozal L, Gozal D. Intranasal budesonide treatment for children with mild obstructive sleep apnea syndrome. Pediatrics. 2008 Jul;122(1):e149-55.

29. Khan WH, Mohsenin V, D'Ambrosio CM. Sleep in asthma. Clin Chest Med. 2014 Sep;35(3):483-93.

30. Nadeem R, Molnar J, Madbouly EM, Nida M, Aggarwal S, Sajid H, et al. Serum inflammatory markers in obstructive sleep apnea: a meta-analysis. J Clin Sleep Med. 2013 Oct 15;9(10):1003-12.

31. Gozal D, Capdevila OS, Kheirandish-Gozal L. Metabolic alterations and systemic inflammation in obstructive sleep apnea among nonobese and obese prepubertal children. Am J Respir Crit Care Med. 2008 May 15;177(10):1142-9.

32. Arter JL, Chi DS, M G, Fitzgerald SM, Guha B, Krishnaswamy G. Obstructive sleep apnea, inflammation, and cardiopulmonary disease. Front Biosci. 2004 Sep 1;9:2892-900.

33. Baessler A, Nadeem R, Harvey M, Madbouly E, Younus A, Sajid H, et al. Treatment for sleep apnea by continuous positive airway pressure improves levels of inflammatory markers - a meta-analysis. J Inflamm (Lond). 2013;10:13.

34. Mai XM, Bottcher MF, Leijon I. Leptin and asthma in overweight children at 12 years of age. Pediatr Allergy Immunol. 2004 Dec;15(6):523-30.

35. Sanner BM, Kollhosser P, Buechner N, Zidek W, Tepel M. Influence of treatment on leptin levels in patients with obstructive sleep apnoea. Eur Respir J. 2004 Apr;23(4):601-4.

36. Juel CT, Ulrik CS. Obesity and asthma: impact on severity, asthma control, and response to therapy. Respir Care. 2013 May;58(5):867-73.

37. Nermin Guler EK, Ulker Ones, Zeynep Tamay, Nihal Salmayenli, Feyza Darendeliler. Leptin: Does it have any role in childhood asthma? J Allergy Clin Immunol. 2004;114(2):254-9.

38. Hudgel DW, Shucard DW. Coexistence of sleep apnea and asthma resulting in severe sleep hypoxemia. JAMA. 1979 Dec 21;242(25):2789-90.

39. Hoshino M, Takahashi M, Aoike N. Expression of vascular endothelial growth factor, basic fibroblast growth factor, and angiogenin immunoreactivity in asthmatic airways and its relationship to angiogenesis. J Allergy Clin Immunol. 2001 Feb;107(2):295-301.

40. Simcock DE, Kanabar V, Clarke GW, O'Connor BJ, Lee TH, Hirst SJ. Proangiogenic activity in bronchoalveolar lavage fluid from patients with asthma. Am J Respir Crit Care Med. 2007 Jul 15;176(2):146-53.

41. Asai K, Kanazawa H, Kamoi H, Shiraishi S, Hirata K, Yoshikawa J. Increased levels of vascular endothelial growth factor in induced sputum in asthmatic patients. Clin Exp Allergy. 2003 May;33(5):595-9.

42. Imagawa S, Yamaguchi Y, Higuchi M, Neichi T, Hasegawa Y, Mukai HY, et al. Levels of vascular endothelial growth factor are elevated in patients with obstructive sleep apnea--hypopnea syndrome. Blood. 2001 Aug 15;98(4):1255-7.

43. Fujiwara Y, Arakawa T, Fass R. Gastroesophageal reflux disease and sleep. Gastroenterol Clin North Am. 2013 Mar;42(1):57-70.

44. Shepherd KL, James AL, Musk AW, Hunter ML, Hillman DR, Eastwood PR. Gastro-oesopha-

geal reflux symptoms are related to the presence and severity of obstructive sleep apnoea. J Sleep Res. 2011 Mar;20(1 Pt 2):241-9.

45. Orr WC, Robert JJ, Houck JR, Giddens CL, Tawk MM. The effect of acid suppression on upper airway anatomy and obstruction in patients with sleep apnea and gastroesophageal reflux disease. J Clin Sleep Med. 2009 Aug 15;5(4):330-4.

46. Kiljander TO, Laitinen JO. The prevalence of gastroesophageal reflux disease in adult asthmatics. Chest. 2004 Nov;126(5):1490-4.

47. Ing AJ, Ngu MC, Breslin AB. Obstructive sleep apnea and gastroesophageal reflux. Am J Med. 2000 Mar 6;108 Suppl 4a:120S-5S.

48. Sontag SJ, O'Connell S, Khandelwal S, Greenlee H, Schnell T, Nemchausky B, et al. Asthmatics with gastroesophageal reflux: long term results of a randomized trial of medical and surgical antireflux therapies. Am J Gastroenterol. 2003 May;98(5):987-99.

49. Farah CS, Salome CM. Asthma and obesity: a known association but unknown mechanism. Respirology. 2012 Apr;17(3):412-21.

50. Kasasbeh A, Kasasbeh E, Krishnaswamy G. Potential mechanisms connecting asthma, esophageal reflux, and obesity/sleep apnea complex--a hypothetical review. Sleep Med Rev. 2007 Feb;11(1):47-58.

51. Phillips BG, Kato M, Narkiewicz K, Choe I, Somers VK. Increases in leptin levels, sympathetic drive, and weight gain in obstructive sleep apnea. Am J Physiol Heart Circ Physiol. 2000 Jul;279(1):H234-7.

52. Manzella D, Parillo M, Razzino T, Gnasso P, Buonanno S, Gargiulo A, et al. Soluble leptin receptor and insulin resistance as determinant of sleep apnea. Int J Obes Relat Metab Disord. 2002 Mar;26(3):370-5.

53. Yigla M, Tov N, Solomonov A, Rubin AH, Harlev D. Difficult-to-control asthma and obstructive sleep apnea. J Asthma. 2003 Dec;40(8):865-71.

54. Teodorescu M, Consens FB, Bria WF, Coffey MJ, McMorris MS, Weatherwax KJ, et al. Predictors of habitual snoring and obstructive sleep apnea risk in patients with asthma. Chest. 2009 May;135(5):1125-32.

55. Kakkar RK, Berry RB. Asthma and obstructive sleep apnea: at different ends of the same airway? Chest. 2009 May;135(5):1115-6.

56. Roehrs T, Merlotti L, Halpin D, Rosenthal L, Roth T. Effects of theophylline on nocturnal sleep and daytime sleepiness/alertness. Chest. 1995 Aug;108(2):382-7.

57. Teodorescu M, Polomis DA, Hall SV, Teodorescu MC, Gangnon RE, Peterson AG, et al. Association of obstructive sleep apnea risk with asthma control in adults. Chest. 2010 Sep;138(3):543-50.

58. Ross K. Sleep-disordered breathing and childhood asthma: clinical implications. Curr Opin Pulm Med. 2013 Jan;19(1):79-83.

59. Teodorescu M, Polomis DA, Teodorescu MC, Gangnon RE, Peterson AG, Consens FB, et al. Association of obstructive sleep apnea risk or diagnosis with daytime asthma in adults. J Asthma. 2012 Aug;49(6):620-8.

60. Julien JY, Martin JG, Ernst P, Olivenstein R, Hamid Q, Lemiere C, et al. Prevalence of obstructive sleep apnea-hypopnea in severe versus moderate asthma. J Allergy Clin Immunol. 2009 Aug;124(2):371-6.

61. Li L, Xu Z, Jin X, Yan C, Jiang F, Tong S, et al. Sleep-disordered breathing and asthma: evidence from a large multicentric epidemiological study in China. Respir Res. 2015 May 10;16(1):56.

62. Ekici A, Ekici M, Kurtipek E, Keles H, Kara T, Tunckol M, et al. Association of asthma-related symptoms with snoring and apnea and effect on health-related quality of life. Chest. 2005 Nov;128(5):3358-63.

63. ten Brinke A, Sterk PJ, Masclee AA, Spinhoven P, Schmidt JT, Zwinderman AH, et al. Risk factors of frequent exacerbations in difficult-to-treat asthma. Eur Respir J. 2005 Nov;26(5):812-8.

64. Desager KN, Nelen V, Weyler JJ, De Backer WA. Sleep disturbance and daytime symptoms in wheezing school-aged children. J Sleep Res. 2005 Mar;14(1):77-82.

65. Auckley D, Moallem M, Shaman Z, Mustafa M. Findings of a Berlin Questionnaire survey: comparison between patients seen in an asthma clinic versus internal medicine clinic. Sleep Med. 2008 Jul;9(5):494-9.

66. Foley SC, Hamid Q. Images in allergy and immunology: neutrophils in asthma. J Allergy Clin Immunol. 2007 May;119(5):1282-6.

67. Goldstein NA, Aronin C, Kantrowitz B, Hershcopf R, Fishkin S, Lee H, et al. The prevalence of sleep-disordered breathing in children with

asthma and its behavioral effects. Pediatr Pulmonol. 2014 Dec 2.

68. Levine B, Roehrs T, Stepanski E, Zorick F, Roth T. Fragmenting sleep diminishes its recuperative value. Sleep. 1987 Dec;10(6):590-9.

69. Luyster FS, Teodorescu M, Bleecker E, Busse W, Calhoun W, Castro M, et al. Sleep quality and asthma control and quality of life in non-severe and severe asthma. Sleep Breath. 2012 Dec;16(4):1129-37.

70. Simard B, Turcotte H, Marceau P, Biron S, Hould FS, Lebel S, et al. Asthma and sleep apnea in patients with morbid obesity: outcome after bariatric surgery. Obes Surg. 2004 Nov-Dec;14(10):1381-8.

71. Avital A, Steljes DG, Pasterkamp H, Kryger M, Sanchez I, Chernick V. Sleep quality in children with asthma treated with theophylline or cromolyn sodium. J Pediatr. 1991 Dec;119(6):979-84.

72. Greenfeld M, Sivan Y, Tauman R. The effect of seasonality on sleep-disordered breathing severity in children. Sleep Med. 2013 Oct;14(10):991-4.

73. Teodorescu M, Consens FB, Bria WF, Coffey MJ, McMorris MS, Weatherwax KJ, et al. Correlates of daytime sleepiness in patients with asthma. Sleep Med. 2006 Dec;7(8):607-13.

74. Greenberg H, Cohen RI. Nocturnal asthma. Curr Opin Pulm Med. 2012 Jan;18(1):57-62.

75. Brockmann PE, Bertrand P, Castro-Rodriguez JA. Influence of asthma on sleep disordered breathing in children: a systematic review. Sleep Med Rev. 2014 Oct;18(5):393-7.

76. Scharf MB, Cohen AP. Diagnostic and treatment implications of nasal obstruction in snoring and obstructive sleep apnea. Ann Allergy Asthma Immunol. 1998 Oct;81(4):279-87; quiz 87-90.

77. Corey JP, Houser SM, Ng BA. Nasal congestion: a review of its etiology, evaluation, and treatment. Ear Nose Throat J. 2000 Sep;79(9):690-3, 6, 8 passim.

78. Kalra M, Biagini J, Bernstein D, Stanforth S, Burkle J, Cohen A, et al. Effect of asthma on the risk of obstructive sleep apnea syndrome in atopic women. Ann Allergy Asthma Immunol. 2006 Aug;97(2):231-5.

79. Fletcher EC, Proctor M, Yu J, Zhang J, Guardiola JJ, Hornung C, et al. Pulmonary edema develops after recurrent obstructive apneas. Am J Respir Crit Care Med. 1999 Nov;160(5 Pt 1):1688-96.

80. Scott HA, Gibson PG, Garg ML, Wood LG. Airway inflammation is augmented by obesity and fatty acids in asthma. Eur Respir J. 2011 Sep;38(3):594-602.

81. Bijaoui EL, Champagne V, Baconnier PF, Kimoff RJ, Bates JH. Mechanical properties of the lung and upper airways in patients with sleep-disordered breathing. Am J Respir Crit Care Med. 2002 Apr 15;165(8):1055-61.

82. Alharbi M, Almutairi A, Alotaibi D, Alotaibi A, Shaikh S, Bahammam AS. The prevalence of asthma in patients with obstructive sleep apnoea. Prim Care Respir J. 2009 Dec;18(4):328-30.

83. van Maanen A, Wijga AH, Gehring U, Postma DS, Smit HA, Oort FJ, et al. Sleep in children with asthma: results of the PIAMA study. Eur Respir J. 2013 Apr;41(4):832-7.

84. Cottrell L, Neal WA, Ice C, Perez MK, Piedimonte G. Metabolic abnormalities in children with asthma. Am J Respir Crit Care Med. 2011 Feb 15;183(4):441-8.

85. Alkhalil M, Schulman ES, Getsy J. Obstructive sleep apnea syndrome and asthma: the role of continuous positive airway pressure treatment. Ann Allergy Asthma Immunol. 2008 Oct;101(4):350-7.

22

SENSORY PROCESSING IN SLEEP: AN APPROACH FROM ANIMAL TO HUMAN DATA

Marisa Pedemonte, Eduardo Medina-Ferret, and Ricardo A. Velluti

ABSTRACT

Men have developed devices in different ways based on sensory stimulation to induce sleep. It is a common experience that the sleep can be facilitated by reducing the amount of afferent sensory information to the brain during wakefulness, for example by closing the eyelids, etc. However, the common practices for centuries demonstrate that specific sensory stimulations (like proprioceptive and auditory) were effective in inducing sleep. There is now consensus that sleep is an active process

Marisa Pedemonte
Department of Physiology, Facultad de Medicina CLAEH, Punta del Este, Uruguay

Eduardo Medina-Ferret
EUTM, Hospital de Clínicas, Montevideo, Uruguay

Ricardo A. Velluti
Centro de Medicina del Sueño, Facultad de Medicina CLAEH, Punta del Este, Uruguay

Corresponding author: Marisa Pedemonte,
E-mail: marisa.pedemonte@gmail.com

implying sensorial networks that would actively participate in sleep modulation. Sleep means a change of networks, a new cooperative interaction among them with some functional purpose, considering that a single network may sub-serve several different functions, modulated by the sensory inputs. Information from the environment and the body influences the brain development both during wakefulness and also during sleep, in a context genetically determined, acting as epigenetic factors. All sensory systems present efferent controls allowing reciprocal influences (feedback) among the sensory inputs and the different states of the brain, asleep or awake.

We will explore the action of several sensory inputs on sleep, considering that the auditory system has special relevance because it has an important role in the philogenetic evolution, remaining open during sleep and acting as a continuous guard. The auditory processing never stop, however neuronal activity changed the pattern and frequency of discharges related to sleep state, studied in the auditory pathway from cochlear nucleus to the auditory primary cortex. We will also expose the relationship among sleep and other sensory systems as vision, proprioceptive system, termorreception and olfaction.

22.1. INTRODUCTION

It is possible that since the beginning of civilization men have developed devices in different ways based on sensory stimulation to induce sleep of newborns and toddlers. There is evidence that Greeks in the IV–V century BC (Museum Paestum, Italy), used cribs-swayed system that was successful because it lasted until today. These and other practices have claimed for centuries somatosensory and vestibular stimulation, probably completed by auditory stimulation of lullabies generating sensory changes, empirically proved effective in inducing sleep. On the other hand, it is a common experience that the sleep can be facilitated by reducing the amount of afferent sensory information to the brain during wakefulness, for example by closing the eyelids, avoiding movements that change the proprioceptive information and avoiding tactile stimuli, etc. However, the sleep probably depends not only on passive deafferentation processes (Hess, 1944). Bremer (1935) has emphasized the relevance of the sensory input in general and the reduction of it would be necessary to go into sleep: passive hypothesis. There are many evidences that sleep is an active process implied sensorial networks that would actively participate in sleep modulation (Moruzzi, 1972; Velluti, 1997).

Some sleep researchers were, perhaps unconsciously, looking for a "sleep center" that does not exist. A central nervous system (CNS) center may be real and useful for controlling functions such as the cardiovascular, the respiratory, etc., while, on the other hand, sleep is not a function *but a complete different CNS state*. This means different brains for the diverse waking conditions, for sleep stages N1 (somnolence), N2 (spindle sleep), N3 (slow wave sleep) and for Paradoxical Sleep or rapid eye movements (REM).

Hence, sleep means a whole change of networks/cell assemblies, a new cooperative interaction among them, considering that a single network may sub-serve several different functions (Kumar et al., 2010). The complex neuronal networks that are part of sleep and also wakefulness would be actively modulated by the sensory stimulation (Velluti and Pedemonte, 2012).

22.1.2. Conceptual Synthesis

Sleep is not a function but a different physiological state of wakefulness, during which the various changes occurring in both input information and its processing.

The cerebral cortex represents 80–85% of our brain, one of whose paradigms according to Fuster (2010) is the *Cognito* or cognitive network dependent on genetic factors, evolution and the presence of the environment through incoming sensory information "... so a *Cognito* is defined by a network of cortical neurons that are formed with environmental and educational experience of the individual", exist, according to this author, different levels of complexity in relation to a cognitive process, where is fundamentally important sensory input for the creation and modulation of these functional units. These concepts should also be introduced during the processes of sleep. The functionally different state of brain neural networks in different behavioral states determines the changes in the physiology of the body. It is recognized that there is no single, simple way to induce changes in a multitude of neural networks in the CNS and thus move from wakefulness *mode* to sleep *mode*. The hypothesis appears to argue that various neuronal assemblies would be involved, and that the process of change is dependent on multiple factors that would develop in a sequence that is still unknown.

Neural networks. The concept of neural networks is defined by the temporarily correlated neuronal activity with some functional purpose, acting cooperatively and integrating widely distributed neurons with different functional properties in the service of a state or physiological condition. Thus, the discharge of action potentials of a neuron in a network can process certain information, while sometime later, may be associated with another group or neural network with different functional purposes, for example, going from wakefulness to sleep (Velluti, 2008; Velluti and Pedemonte. 2010).

The sensory input is a huge spectrum of information reaching the CNS, which generate, after complex processing, expressions of the system in the form of motor acts, endocrine, autonomic, and behavioural responses and changes in the capabilities of the CNS, such as memory and learning. Information from two worlds, the outside and inside (the body) over life is a set of influences on brain development and in particular in the organization of sleep. The development of each brain is genetically determined, although conditioned to

a very significant event, as is the continued sensory input, a phenomenon that probably begin *in uterus* and continues throughout life. One consequence of this continuous sensory input to the CNS is that processing be differential according to the various states of the system, i.e, wakefulness and sleep, different levels and different stages (epigenetic factors). An important concept must be added: the CNS can influence the input, since it has efferent control (present in all sensory systems), which can exert influence on the receptors and the various nuclei of the afferent pathway to the same sensory cortices. Using this possibility of feedback sensory processing is complete, because the SNC can then select the information that runs a closed circuit (closed-loop), selecting entries.

22.2. THE AUDITORY SYSTEM DURING SLEEP

From several viewpoints, the auditory is a special system related to sleep neurophysiology, exhibiting a series of unique associated changes (Nakagawa et al., 1992; Velluti et al., 2000). The auditory system is a sensory organization that is relatively open during sleep. It has an *a priori* relevance, because is a telereceptor that can act as a continuous guard, to detect a predator or hear the gentle call of a little boy at night, processing that information and then deciding whether to continue sleeping or waking.

The auditory incoming signals to the CNS change the sleep characteristics. After destruction of both cochleae in hamsters and guinea pigs (Cutrera el al., 2000; Pedemonte et al., 1996), a significant enhancement of both paradoxical sleep and slow wave sleep together with decreased wakefulness, were observed for up to 45 days. The sleep augmentation consisted of an increment in the number of episodes of both slow wave and paradoxical sleep rather than in the duration of single episodes.

Conversely, the CNS can control by feedback mechanisms the auditory input carried out in close correlation with the sleep-wakefulness cycle. Receptor and auditory nerve action potentials exhibited amplitude changes when analyzed during quiet Wakefulness (W), Slow Wave Sleep (SWS) and Paradoxical Sleep (PS) in guinea pigs (Velluti et al., 1989). Both responses increased the amplitude on passing to slow wave sleep while decreased in paradoxical sleep to similar wakefulness' values. Analyzing the unitary activity in the auditory pathway from cochlear nucleus to the auditory primary cortex was observed changes in pattern and frequency of discharges (Fig. 22.1, Pedemonte and Velluti, 2011; Velluti and Pedemonte, 2002, 2010; Peña et al., 1999; Morales-Cobas et al., 1995). The units evoked activity was most marked during quiet W (~50%) and diminished during SWS; however, ~30% of the neuronal responses during SWS presented an equal or even greater firing than during W. During PS, the auditory responses were diminished in all the studied neurons; meanwhile some of them exhibited no evoked activity. A different proportion of auditory units firing were seen in the brainstem nuclei. In those *loci,* most of the units exhibited increasing and decreasing firing, while those units responding in sleep as during quiet W were present in a smaller number than in the auditory cortex.

The data from the guinea pigs auditory cortex (Peña et al., 1999) was recently con-firmed in primates (Issa and Wang, 2008). Around 50% of the auditory cortical (AI) units recorded during SWS and PS maintained a firing similar to the ones recorded during quiet W, postulated to continue monitoring the environment. Another set of cortical neurons were divided into those that increased and those that decreased their firing on passing from W to SWS or from SWS to PS. This latter group, although responding to the sound stimuli, is proposed to be engaged -associated to other neuronal network/cell assembly -in sleep-active processes (Fig. 22.2). Our hypothesis support the notion that these units, from the auditory cortex and the preoptic one (Suntsova et al., 2002) belong to the same sleep-related network, perhaps organized by a common *hub* neuron

This suggests that the auditory brain stem neurons that increase/decrease firing in sleep are postulated to be engaged in some sleep processes, particularly participating in sleep-active cell assemblies/networks.

22.3. HUMAN EVOKED BIOELECTRICAL POTENTIALS

In the 1960s it was shown an increase of human potential amplitude to go from wakefulness to sleep. Two types of evoked potentials, sensory and cognitive, were described. The first directly dependent on sensory input, whereas the latter are dependent on the cerebral cortex, for example, P300 and mismatch negativity (Bastuji and García-Larrea, 2005; Atienza el al., 2002). In all these ways response changes appear depending on the different stages of sleep. In Figure 22.3 evoked potentials and magneto-electroencephalography evoked

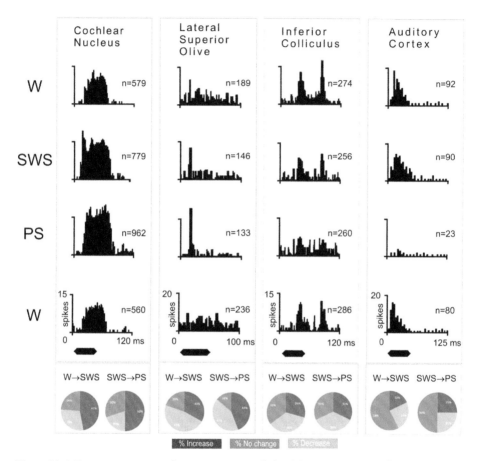

Figure 22.1. Four representative auditory neurons recorded at different auditory loci (cochlear nucleus, lateral superior olive, inferior colliculus and auditory cortex) during the sleep-waking cycle are shown. Post-stimulus time histograms exhibit changes in the pattern and/or in the number of discharge on passing from wakefulness to slow wave sleep and paradoxical sleep (PS). In these examples the cochlear nucleus recorded neuron increases the firing rate during sleep maintaining the same pattern of discharge; the lateral superior olive shows both a change in pattern and a decrease in firing during sleep. The inferior colliculus neuron exhibits a changed pattern but not significant variation in firing rate. The auditory cortex neuron significantly decreases its firing only during PS, recovering it in the following Wakefulness epoch. Stimuli: tone-burst (50 ms, 5 ms rise-decay time, at the unit characteristic frequency).

At the bottom, guinea pig percentages unitary evoked activity along the auditory pathway in the sleep-waking cycle. Pie charts show percentages (%) of neuronal firing shifts on passing from wakefulness to slow wave sleep and from slow wave sleep to paradoxical sleep. The subcortical nuclei, inferior colliculus and the lateral superior olive exhibited a higher percentage of increasing-decreasing firing neurons. Around 50% of the cortical neurons responded as during wakefulness. No silent auditory neurons were detected on passing to sleep or during sleep in any pathway level. (Data modified from Pedemonte and Velluti, 2011).

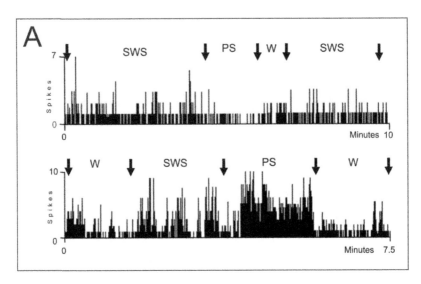

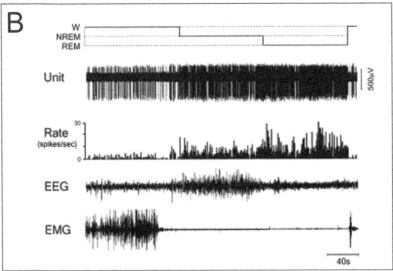

Figure 22.2. A. Two auditory cortical guinea pig neurons (A1) activity. Upper plot: spontaneous discharge as a function of time. After fluctuating during SWS, the firing rate markedly decreases during paradoxical sleep (PS). The number of spikes is quantified over 50 ms epochs, 10 min of continuous recording. Lower plot: discharge as a function of time. The firing rate shows peaks during SWS and a quasi-tonic increase, on passing to PS. The number of spikes is quantified over 450 ms epochs, 7.5 min of continuous recording of wakefulness (W), SWS, and PS, (Modified from Peña et al. 1999) ; B. Discharge of a "sleep-ON" or "sleep related" neuron during W, SWS and PS, recorded in the median preoptic nucleus of an unrestrained rat. Its firing-rate is low during waking, increases at sleep onset and during SWS, and reaches even higher levels in PS (Modified from Suntsova et al., 2002). Both units, the auditory cortex and the preoptic one, belong to the same sleep-related network, perhaps organized by a common *hub* neuron/s.

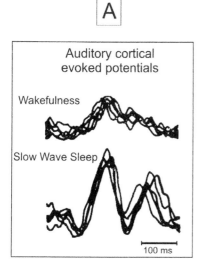

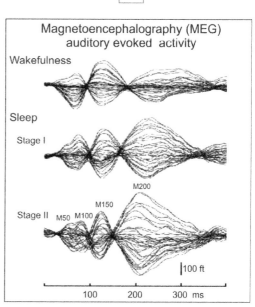

Figure 22.3. A. Auditory evoked potentials in waking and slow-wave sleep in human. Five overlapping responses have net differences with increased amplitude, breadth and complexity of waves passing from wakefulness to slow wave sleep (stimulus, click 1/s).

B. Magnetoencephalography (MEG), magnetic field caused in the human auditory cortex, by a 250 Hz pure tone was presented to the right ear, and a record was made in the left hemisphere. Several averaged wave fields showed variations in comparison between waking and sleep stages with M 200 significantly larger during sleep. Modified from Vanzulli et al., 1961; Kakigi et al, 2003).

activity changes during sleep are shown (Vanzulli et al., 1961; Kakigi et al., 2003).

Human Brainstem Auditory Evoked Potentials (BAEP) shifts during sleep are still under discussion. Early research reported sleep-related latency delays of wave V (Amadeo and Shagass, 1973; Osterhammel et al., 1985); although no significant effect of sleep was observed in later studies (Campbell and Bartoli, 1986; Bastuji et al., 1988). On the other hand, Bastuji et al. (1988) excluded the possibility of a direct sleep effects. Small BAEP latency changes during nocturnal sleep were found

to correlate with the very small circadian night fall of body temperature rather than with sleep stages (Bastuji et al., 1988, Bastuji and García-Larrea, 2005).

Besides, Kräuchi et al. (2006) reported that proximal as well as distal increases in skin temperature are present during nap time. These reports are indicative of a great difference in temperature control during both night and nap sleep. Then, we began to study the BAEP during *nap-sleep-like* in human subjects. Volunteers without any pathology were stimulated with alternating clicks (80 dB SPL, 10/s) and BAEP were recorded together with on-line sleep polysomnographic control, under Chloral Hydrate (CH) hypnotic dose. The experimental design was to record during the afternoon nap (from 1:00 to 4:00 p.m.) with monitored skin temperature and evoked responses studied, showed a significant increase in wave V latency not related to body temperature but to sleep N2 (Fig. 22.4). Waves I to V (10 ms) were also studied in the frequency domain (FFT) showed sig-

nificant decreases in the power spectra of 400, 500 and 600 Hz during N2 *sleep-like* (Medina-Ferret et al., 2009; Picton, 2010).

22.3.1. Noise and Human Sleep

Human sleep organization is extremely sensitive to acoustic stimuli, and noise generally exerts an arousing influence on it. A noisy night-time ambiance leads to a decrease in total sleep time and in delta

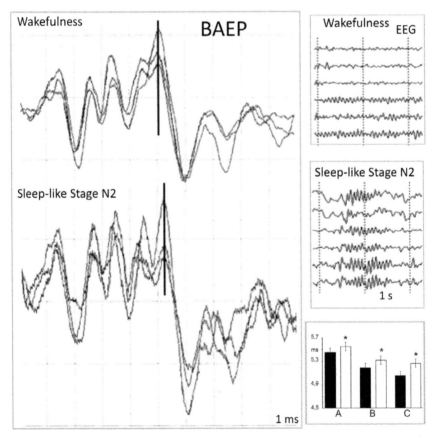

Figure 22.4. Brain auditory evoked potentials (BAEP), super-positions of 3 different subjects (A, B, C). Right side EEG showing a wakefulness recording (upper), while in the middle one a sleep-like N2 stage, under chloral hydrate, is depicted. The lower right corner shows the corresponding latencies change significance ($p < 0.05$; Student t test). Black bars, wakefulness; open bars, sleep-like N2 (Medina-Ferret et al, 2009).

wave sleep (Stage IV, N3) and REM, with the consequent increase in the time spent in Stage II, (or N2) and W. Moreover, the remarkable sleep improvement after noise abatement, suggests that the environment is continuously scanned by the auditory system, notion also supported by the unitary analysis in sleeping animals.

22.3.1.1. Late Auditory Evoked Potentials (LAEP)

Different waveforms, varying with the sleep stages, were observed (Osterhammel et al., 1973). Later, Campbel and Colrain (2000) demonstrated a progressive increase in amplitude on entering sleep, from N1 to N3, while the amplitude diminished, almost equal to wakefulness, in a REM sleep condition.

22.3.1.2. Deaf Human Intracochlear Implanted

Velluti et al. (2010) demonstrated that the auditory input in humans can introduce changes in central nervous system activity leading to shifts in sleep characteristics (Fig. 22.5). To properly demonstrate the effect of auditory input on sleep of intracochlear implanted patients, they were recorded during four nights: two nights with the implant "off", with no auditory input, and two nights with the implant "on", that is, with normal auditory input, being only the common night sounds present, without any additional auditory stimuli delivered. On comparison of the night recordings with the implant "on" and "off", a new sleep organization was observed for the recordings with the implant "on", suggesting that brain plasticity may produce changes

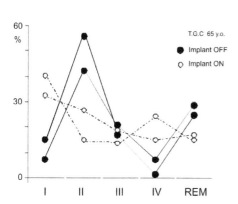
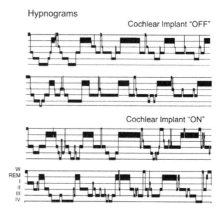

Figure 22.5. Example of sleep stages organization in a deaf implanted patient. At the left, filled circles show control recording nights with the implant "off", that is, recorded as a deaf patient; and the open circles, implant "on", that is, as a hearing patient. The decrease of stage II and paradoxical sleep (REM), together with the stage IV percentage increase are seen. At the right, hypnograms display the whole night's sleep ultradian cycle recordings (8 h), two nights with the implant "off" and two nights with the implant "on". No shifts in the sleep ultradian cycles were observed. Sleep stages are I, II, III, IV (equivalent to SWS or non-REM sleep), and REM (Modified from Velluti et al, 2010).

in the sleep stage percentages while maintaining the ultradian rhythm. During sleep with the implant "on", the analysis of the electroencephalographic delta, theta and alpha bands in the frequency domain, using the Fast Fourier Transform, revealed a diversity of changes in the power originated in the contralateral cortical temporal region. Authors are postulating that an intracochlear-implanted deaf patient may have a better recovery if the implant is maintained "on" during the night, that is, during sleep.

22.3.2. Vision and Sleep

Centrifugal fibers that reach the retina were described by Ramón y Cajal (1935) and were later shown that the efferent system has direct effects on retinal ganglion cells. However, it was not until recently that Galambos et al. (1994) demonstrated that rat electroretinography was modified during the sleep-wake cycle, with increased amplitude during SWS diminishing similar to waking in the paradoxical (or REM) sleep values. These results mimicked those changes obtained simultaneously in evoked potentials in the visual cortex, where they recorded the highest amplitudes during slow sleep.

Continuous exposure to bright light in human subjects introduced changes in the organization of sleep. Unlike with the dim light, bright light induced a significant increase in slow wave sleep and decreased wakefulness, while increased levels of alert (Kohsaka et al., 1995; Livingtone and Hubel, 1981). Intact human visual system is essential to synchronize the circadian system; therefore blind subjects whose lesions include retino-hipothalamic-fibers will show circadian alterations.

22.3.2.1. Cortical Visual Units

The spontaneous discharge of neurons in the primary visual cortex of cats was recorded during sleep by Evarts (1963) who showed that most of neurons discharged at higher frequencies during REM sleep than during slow sleep.

22.3.2.2. Magnetoencephalography

Visual induced magnetic fields show variations when passing from wakefulness to N1 and N2 in human sleep (Kakigi et al., 2003).

22.3.2.3. Lateral Geniculate Nucleus

The configuration of the spontaneous discharge cells in the lateral geniculate nucleus is different in waking and sleep. It was described a depression in the responses during slow sleep. It was demonstrated that upon awakening, the neuronal firing changed to either decrease or increase, depending on the stimulus applied. Other changes described in the configuration of neuronal discharges included in the paradoxical sleep analysis as well as cross with hippocampal theta rhythm, changing correlation in the different behavioral stages (Gambini et al., 2002). Pedemonte et al. (2005) demonstrated that during wakefulness, a change in light flash stimulation pattern (stimuli frequency shift and stimuli "on" and "off") caused an increment in the theta band power in 100% of the cases and a phase locking of the spikes in 53% of the recorded neurons. During slow wave sleep, there were no consistent changes in the theta power notwithstanding 13% of the neurons exhibited phase locking, i.e., novelty may induce changes in the temporal correlation of visual neuronal activity with the hippocampal theta rhythm in sleep. Those results suggest that visual processing in slow wave sleep exists, while auditory information and learning were reported during slow wave sleep in animals and newborn humans (Cheour, et al., 2002). The changes in the theta power as well as in the neuronal phase locking amount indicate that in slow wave sleep, the ability of the hippocampus to detect/process novelty, although present, may be decreased. This is consistent with the noticeable decrease in awareness of the environment during sleep (Fig. 22.6).

22.3.2.4. Cortical Visual Evoked Potentials

Human averaged visual evoked potential records showed increased amplitude during sleep at the expense mainly of increased positive waves in the response. The directional coherence between head, eye, and limb movements in REM sleep behavior disorder suggest that REMs imitate the scanning of the dream scene, the concordance can be extended to normal REMs (Leclair-Vissoneau, et al., 2010).

Visual imagery during sleep has long been a topic of persistent speculation, but its private nature has hampered objective analysis. A neural decoding approach were described in which machine-learning models predict the contents of visual imagery during the sleep-onset period, given measured brain activity, by discovering links between human functional magnetic resonance imaging patterns and verbal reports with the assistance of lexical and image databases. Decoding models trained on stimulus-induced brain activity in vi-

sual cortical areas showed accurate classification, detection, and identification of contents. Our findings demonstrate that specific visual experience during sleep is represented by brain activity patterns shared by stimulus perception, providing a means to uncover subjective contents of dreaming using objective neural measurement (Horikawa et al., 2013).

22.3.3. Proprioceptive System

Different objective studies were carried out to evaluate the impact of sleep deprivation on proprioception. The balance perturbations tests, objective posturography measurements, and stability were analyzed to investigate whether postural stability and adaptation differed after a normal night of sleep comparing to after 24 and 36 h of sleep deprivation. Sleep deprivation might affect postural stability through reduced adaptation ability and lapses in attention (Patel et al., 2008).

22.3.4. Termorrecepción and Sleep

Information about body temperature, both increases and decreases, comes from two main sources: a) the receptors of skin temperature and b) the central receptors concentrated mainly in the anterior hypothalamus. The hypothalamus integrates information relevant to the regulation of peripheral and core temperature, particularly in view of the mechanisms underlying the powerful influences of temperature changes in the brain over sleep. Exposing animals to variations in ambient temperature, a decrease in SWS and REM sleep were shown (Fig. 22.7). Thermal stress changes because it promotes sleep thermoregulatory responses that are

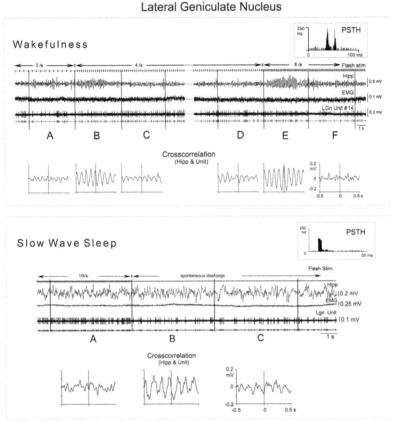

Figure 22.6. Lateral geniculate nucleus visual units discharge and hippocampal theta rhythm pre- and post-flashes of light frequency stimulation changes during wakefulness (upper panel) and SWS (lower panel). Both panels show: the insert with post-stimulus time histogram (PSTH) to characterize the unit as a visual one and traces showing from top to bottom: flash stimulation changing from 2/s to 4/s and, after five minutes of recording, shifting from 4/s to 8/s (at the top box) and changing from 16/s to no stimulation (spontaneous activity, bottom panel); Hippocampal field activity (Hipp); electromyogram (EMG); extracellular unitary discharge (digitized below).

At the top panel, it is shown a quiet wakefulness characteristic recording. Six epochs (5 s each) were selected for processing (A to F, divided by vertical lines). A and D are immediately prior to the frequency stimulation changes, whereas B-C and E-F are successive windows after the light rate stimulation changes. Bottom, each column corresponds to the temporal window shown at the top. Cross-correlation between hippocampal field activity and spikes was calculated by spike-triggered averaging. Temporal correlation (phase locking) between unitary activity and theta rhythm appears after changing the flash rate (windows B and E) and disappears ~5 s later (windows C and F).

Bottom panel, other example during SWS. This recording is characteristic of SWS because of the high amplitude and low frequency in the Hipp field activity and the EMG low activity. Three epochs (5 s each) were selected for processing (A to C, divided by vertical lines); A is immediately previous to the end of flash stimulation while B-C is successive windows without flashes (spontaneous firing). Temporal correlation (phase locking) between unitary discharge and theta rhythm appears after changing the flash stimulation (window B) and disappears some seconds (~5 s) later (window C). (Modified from Pedemonte et al., 2005).

Human hypnograms

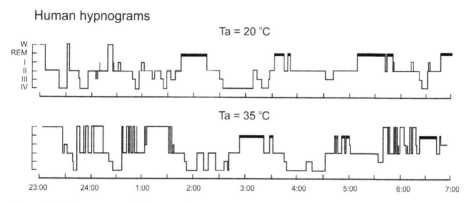

Figure 22.7. Variations in the human hypnograms caused by changes in ambient temperature (Ta) from 20° C to 35° C. When Ta increases there are repeated awakenings and disorganization of the sleep architecture (lower hypnogram). W, wakefulness; REM and stages I, II, III, IV of SWS. Seven hours of registration. (Modified from Libert and Bach, 2005).

not compatible with the normal development of sleep, for example, long exposure to cold produce paradoxical sleep deprivation. Extrahypothalamic neurons respond to transient changes in temperature applied to the skin showing differences in the sleep-wake cycle. Neurons were seen in structures thermo sensitive striatal-thalamic-limbic located dorsally in relation to classical anterior hypothalamic area; these neurons discharged during wakefulness and slow sleep, being depressed or altered during REM sleep. It has been shown that during slow sleep cats develop thermoregulatory mechanisms in response to cold and heat; in contrast, these mechanisms are suppressed during paradoxical sleep. The amount of time animals and humans spend in REM sleep depends on the ambient temperature; maximum durations are obtained in neutral temperatures, reducing both the heat and the cold. Variations in ambient temperature cause changes in the sleep architecture (Parmeggiani, 2005; Parmeggiani, 2011; Libert and Bach, 2005)

22.4. OLFACTORY SYSTEM

In the genome of mammals a large percentage of all genes are involved in the detection of odors. The huge amount of genetic information related to smell reflects the significance of this sensory system. For most animals smell is the primary sense used to identify food, predators, colleagues, etc., while, from the human point of view, smell is mainly an aesthetic sense, without recognizing the many capabilities that this sensory input can have . Little information is available from the olfactory influences on sleep and vice versa. The electrical activity of the olfactory bulb is modulated by the sleep-wake cycle.

Pioneering works demonstrated large waves in the olfactory bulb during W in cats characteristics waves in bursts of 40–50 Hz simultaneous with respiration increase during sniffing and decrease during slow sleep, disappearing in the paradoxical sleep, in cat (Hernández-Peón, 1963; Velluti and Hernández-Peón, 1963). The bilateral olfactory bulbectomy in rats significantly reduced the number of episodes of paradoxical sleep, causing an overall reduction in this stage, while not induce changes in slow wave sleep. Furthermore, episodes of paradoxical sleep returned to normal 15 days after the lesion (Araki et al., 1980). Different results were presented: injury to the lateral portion of the olfactory bulbs in cats caused suppression of sleep. When the olfactory bulbs were electrically stimulated an increase of slow wave sleep appears, while decreased wakefulness. Specific modulation of olfactory input involves the action of centrifugal fibers. The central effects on sensory input are mainly exerted on the mitral cells. Changes in the excitability of mitral cells related to food, effective during wakefulness, disappeared in slow wave sleep (Affanni and Cervino, 2005).

22.4.1. Effects of Odors on Sleep

Cats have been reported in two main groups of fibers from the olfactory bulb. It is suggested that both are necessary for normal functioning during wakefulness, emphasizing the importance of the medial forebrain bundle in mediating responses during slow sleep, having shown it takes at least one medial path to have the neocortical awakening in response to the smell of food.

Preliminary results in guinea pigs have shown changes in the percentage amounts of sleep stages when income of olfactory information is suppressed by injury receptor itself (Fig. 22.8, López et al., 2007). In humans the information about olfaction

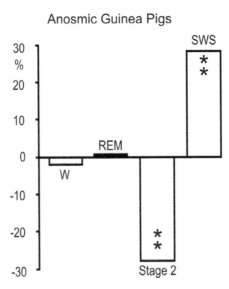

Figure 22.8. Average of percent of changes in sleep stages of anosmic guinea pigs. Olfactory receptors were destroyed through bilateral intranasal instillation of zinc chloride (100–200 mM). Male guinea pigs ($n = 8$) with chronic implants for the diagnosis of sleep-wake cycle were used. Histological monitoring showed total epithelial injury. The results show a statistically significant increment in the percentage of SWS (~30% increase) while sleep spindles decreased (~30%) ($p < 0.001$, Mann-Whitney's test, 95% confidence intervals). Decrease in the percentage of total wake time, and increment in the REM were not consistent (López et al., 2007).

and sleep is limited. Available data indicate the existence of reactions awakening when an olfactory stimulation occurs during sleep.

22.5. SUMMARY

Sensory input and subsequent processing are definitely present in sleep, but show different characteristics than during wakefulness.

The almost complete deafferentation in cats caused a state of drowsiness in these animals.

The interaction between sleep and sensory physiology is an important factor because any sufficiently intense sensory stimulation always produces an awakening, from any stage of sleep.

Interestingly enough, each sensory system has an efferent pathway, with centrifugal projections ending in virtually all cores afferents and on the receiver itself. Therefore, incoming sensory information can alter the physiology of sleep and wakefulness, and these states modulate incoming information.

Normal sleep depends on many aspects of sensory input. Neural networks that command sleep and wakefulness are modulated by many sensory inputs, a proportion of the 'passive' effects must be associated with active mechanisms of sleep. Gains or losses sensory inputs produce imbalances in neuronal networks involved in the sleep-wake cycle, changing their relative proportions of active and not being mere passive processes.

KEY WORDS

- **sensory systems**
- **vision**
- **auditory system**
- **olfaction**
- **termorregulation**
- **proprioception**
- **sleep**
- **theta rhythm**

REFERENCES

Affanni, M.; Cervino, J. Interactions Between Sleep, Wakefulness and the Olfactory System. In *The Physiologic Nature of Sleep*; Parmeggiani, P. L., Velluti, R. A., Eds., Imperial College Press:London, 2005; pp 571–599.

Amadeo, M.; Shagass, C. Brief Latency with Evoked Potentials During Wake and Sleep in Man. *Psychophysiol.* **1973**, *10*, 244–250

Araki, H.; Yamamoto. T.; Watanabe. S.; Ueki, S. Changes in Sleep Wakefulness Pattern Following Bilateral Olfactory Bulbectomy in Rats. *Physiol. Behav. 1980*, *24*, 73–78.

Atienza, M.; Cantero, J. L.; Gomez, C. M. Mismatch Negativity (MMN): An Objective Measure of Sensory Memory and Long-Lasting Memory During Sleep. *Int. J. Phsychophysiol.* **2002**, *46*, 215–225.

Bastuji, H.; Garcia-Larrea, L.; Bertrand, O.; Mauguiere, F. BAEP Latency Changes During Nocturnal Sleep are not Correlated with Sleep Stages but with Body Temperature. *Electroenceph. Clin. Neurophysiol.* **1988**, *70*, 9–15

Bastuji, H.; García-Larrea, L. Human Auditory Information Processing During Sleep Assessed with Evoked Potentials. In *The Physiologic Nature of Sleep*; Parmeggiani, P. L., Velluti, R. A. Eds., Imperial College Press:London, 2005; pp 509–34.

Bremer, F. Cerveau "isole" et Physiologie du Sommeil. *CR. Soc. Biol.* **1935**, *118*, 1235–1241.

Campbell, Colrain. Information Processing During Sleep. *Canadian. J. Experimental. Psycology.* **2000**, *54*, 208–218.

Campbell, K.; Bartoli, E. Human Auditory Evoked Potentials During Natural Sleep: The Early Components. *Electroenceph. Clin. Neurophysiol.* **1986**, 65, 142–149

Cheour, M.; Martynova, O.; Naatanen, R.; Erkkola, R.; Sillanpaa, M.; Kero, et al. Speech Sounds Learned by Sleeping Newborns. *Nature.* **2002**, 415, 599–600.

Cutrera, R.; Pedemonte, M.; Vanini, G.; Goldstein, N.; Savorini, D.; Cardinali, D. P.; Velluti, R. A. Auditory Deprivation Modifies Biological Rhythms in the Golden Hamster. *Arch. Ital. Biol.* **2000**, 138, 285–293.

Evarts, E. V. Photically Evoked Responses in Visual Cortex Units During Sleep and Waking. *J. Neurophysiol.* **1963**, 26, 229–248.

Fuster, J. M. El Paradigma Reticular de la Memoria Cortical. *Rev. Neurol.* **2010**, 50(Supl 3), S3–10.

Galambos, R.; Juhász, G.; Kékesi, A. K.; Nyitrai, G.; Szilágy, N. Natural Sleep Modifies the Rat Electroretinogram. *Proc. Natl. Acad. Sci. USA.* **1994**, 91, 5153–5157.

Gambini, J. P.; Velluti, R. A.; Pedemonte, M. Hippocampal Theta Rhythm Synchronizes Visual Neurons in Sleep and Waking. *Brain. Res.* **2002**, 926, 37–141.

Hernández-Peón, R.; Chávez-Ibarra, G.; Morgane, J. P.; Timo-Iaria, C. Limbic Cholinergic Pathways Involved in Sleep and Emotional Behavior. *Exp. Neurol.* **1963**, 8, 93–111.

Hess, W. R. Das Schlafsyndrom als Folge Dienzephaler Reizung. *Helv. Physiol. Pharamcol. Acta.* **1944**, 2, 305–344.

Horikawa. T.; Tamaki, M.; Miyawaki, Y.; Kamitani1, Y. Neural Decoding of Visual Imagery During Sleep. *Science.* **2013**, 340, 639–642.

Issa, E. B.; Wang, X. Sensory Responses During Sleep in Primate Primary and Secondary Auditory Cortex. J. Neurosci. **2008**, 28,14467–14480.

Kakigi, R.; Naka, D.; Okusa, T.; Wang, X.; Inui, K.; Qiu, Y. et al. Sensory Perception During Sleep in Humans: A Magnetoencephalograhic Study. *Sleep. Med.* **2003**, 4, 493–507.

Kohsaka, M.; Honma, H.; Fukuda, N. et al. Effect of Daytime Bright Light on Sleep Structure and Alertness. *Sleep. Res.* **1995**, 24A, 33.

Kräuchi, K.; Knoblauch, V.; Wirz-Justice, A.; Cajochen, C. Challenging the Sleep Homeostat does not Influence the Thermoregulatory System in Men: Evidence from a Nap vs. Sleep-Deprivation Study. *Am. J. Physiol.- Regul.Integr. Comp. Physiol.* **2006**, 290, 1052–1061

Kumar, A.; Rotter, S.; Aertsen, A. Spiking Activity Propagation in Neuronal Networks: Reconciling Different Perspectives on Neural Coding. *Nature. Rev. Neurosci.* **2010**, 11, 615–627.

Leclair-Visonneau, L.; Oudiette, D.; Gaymard, B.; Leu-Semenescu, S.; Arnulf, I. Do the Eyes Scan Dream Images During Rapid Eye Movement Sleep? Evidence from the Rapid Eye Movement Sleep Behaviour Disorder Model. *Brain.* **2010**, 133, 1737–1746.

Libert, J. P.; Bach, V. Thermoregulation and Sleep in the Human. In *The Physiologic Nature of Sleep*; Parmeggiani, P. L., Velluti, R. A., Eds.Imperial College Press:London, 2005; pp 407–429.

Livingstone, M. S.; Hubel, D. H. Effects of Sleep and Arousal on the Processing of Visual Information in the Cat. *Nature.* **1981**,291, 554–561.

López, C.; Rodríguez-Servetti, Z.; Velluti, R. A.; Pedemonte, M. Influence of the Olfactory System on the Wakefulness-Sleep Cycle. World Federation of Sleep Medicine Sleep Research Societies, Cairns, Australia. *Sleep. Biol. Rhythms.* **2007**, 5, 54.

Medina-Ferret, E.; Peña-Rehbein, J. L.; Bentancor, C.; Pedemonte, M.; Schiavo, L.; Velluti, R. A. In *Brainstem AuditoryEevoked Potentials During Sleep-Like Cloral Hydrate Effect*, XXXI Congress of Association for Research in Otolaryngology (ARO) in Baltimore, USA, 2009.

Morales-Cobas, G.; Ferreira, M. I.; Velluti, R. A. Sleep and Waking Firing of Inferior Colliculus Meurons in Response to Low Frequency Sound Stimulation. *J. Sleep. Res.* **1995**, 4, 242–251.

Moruzzi, G. The Sleep-Waking Cycle. Ergeb. Physiol. **1972**, 64, 1–165.

Nakagawa, H.; Sack, R. L.; Lewy, A, J. Sleep Propensity Free-Runs with Temperature, Melatonin and Cortisol Rhythms in a Totally Blind Person. *Sleep.* **1992**, 15, 330–336.

Osterhammel, P .A.; Davis, H.; Wier, C. C.; Hirsh, S. K. Adult Evoked Potentials in Sleep. *Audiology.* **1973**, 12, 116–128.

Osterhammel, P. A.; Shallop, J. K.; Terkildsen, K. The Effect of Sleep on the Auditory Brainstem Response (ABR) and the Middle Latency Response (MLR). *Scand. Audiol.* **1985**, 14, 47–50

Parmeggiani, P. L. Sleep Behavior and Temperature. In *The Physiologic Nature of Sleep*; Parmeggiani, P. L., Velluti, R. A., Eds.; Imperial College Press:London, 2005; pp 387–405.

Parmeggiani, P. L. *Systemic Homeostasis and Poikilostasis in Sleep*; Imperial Collegge Press:London, 2011, pp 39–78.

Patel, M.; Gomez, S.; Berg, S.; Almbladh, P.; Lindblad, J.; Petersen, H.; Magnusson, M.; Johansson, R.; Fransson, P. A. Effects of 24-h and 36-h Sleep Deprivation on Human Postural Control and Adaptation. *J. Exp. Brain. Res.* **2008**, *185(2)*, 165–173.

Pedemonte, M.; Gambini, J. P.; Velluti, R. A. Novelty-Induced Correlation Between Visual Neurons and the Hippocampal Theta Rhythm in Sleep and Wakefulness. *Brain. Res.* **2005**, *1062*, 9–15.

Pedemonte, M.; Peña, J. L.; Torterolo. P.; Velluti, R. Auditory Deprivation Modifies Sleep in the Guinea-Pig. Neurosci. Lett.**1996**, *233*,1–4.

Pedemonte, M.; Velluti, R. A. The Brain Auditory Processing During Behaviour Seen from the Unitary Activity. In *Horizons in Neuroscience*; Costa, A., Villalba, E., Eds.; Nova Science Publishers:New York, 2011; Vol 6, 123–146

Peña, J. L.; Pérez-Perera, L.; Bouvier, M.; Velluti, R. A. Sleep and Wakefulness Modulation of the Neuronal Firing in the Auditory Cortex of the Guinea Pig. *Brain. Res.* **1999**, *816*, 463–470.

Picton, P. W. *Human Auditory Evoked Potentials*; Plural publishing:California, 2010, pp 161–166.

Ramón y Cajal, S. Traveaux du Laboratoire de Recherche de l' Université de Madrid, **1935**, *30*, 1–210.

Suntsova, N.; Szymusiak, R.; Alam, M. N.; Guzman-Marin, R.; McGinty, D. Sleep-Waking Discharge Patterns of Median Preoptic Nucleus Neurons in Rats. *J. Physiol.* **2002**, *543*(Pt 2), 665–677.

Vanzulli, A.; Bogacz, J. García-Austt, E. Evoked Responses in Man. III. Auditory Response. *Acta. Neurol. Latinoam.* **1961**, *7*, 303–308.

Velluti, R. A.; Hernández-Peón, R. Atropine Blockade Within a Cholinergic Hypnogenic Circuit. *Exp. Neurol.* **1963**, *8*, 20–29.

Velluti, R. A.; Pedemonte, M.; García-Austt, E. Correlative Changes of Auditory Nerve and Microphonic Potentials Throughout Sleep. *Hearing. Res.* **1989**, *39*, 203–208.

Velluti, R. A.; Pedemonte, M.; Suárez, H.; Bentancor, C.; Rodríguez-Servetti, Z. Auditory Input Modulates Sleep: An Intra-Cochlear-Implanted Human Model. *J. Sleep. Res.* **2010**, *19*, 585–590.

Velluti, R. A.; Pedemonte, M. Auditory Neuronal Networks in Sleep and Wakefulness. *Int. J. Bifurcat. Chaos.* **2010**, *20*, 403–407.

Velluti, R. A.; Pedemonte, M. In Vivo Approach to the Cellular Mechanisms for Sensory Processing in Sleep and Wakefulness. *Cell. Mol. Neurobiology.* **2002**, *22*, 501–515.

Velluti, R. A.; Pedemonte, M. Sensory Neurophysiologic Functions Participating in Active Sleep Processes. *Sleep. Sci.* **2012**, *5*, 103–106.

Velluti, R. A.; Peña, J. L.; Pedemonte, M. Reciprocal Actions Between Sensory Signals and Sleep. *Biol. Signals. Recept.* **2000**, *9*, 297–308.

Velluti, R. A. Interactions Between Sleep and Sensory Physiology. A review Paper. *J. Sleep. Res.* **1997**, *6*, 61–77.

Velluti, R. A. *The Auditory System in Sleep.* Elsevier/Academic Press: Amsterdam, 2008.

23

SYNOPSIS OF SLEEP IN INFANTS, CHILDREN, AND ADOLESCENTS

Margeaux M. Schade, Christopher E. Bauer, Colleen N. Warren, Hawley E. Montgomery-Downs

ABSTRACT

Pediatric sleep has played a pivotal role in the advancement of modern sleep science. In fact, the celebrated discovery and naming of rapid eye movement (REM) sleep was first described in infants (Aser-

Margeaux M. Schade
Department of Psychology, West Virginia University, Morgantown, WV, USA. Address: 1124 Life Sciences Building, 53 Campus Drive, Morgantown, WV 26506

Christopher E. Bauer
Center for Neuroscience, West Virginia University, Morgantown, WV, USA. Address: Center for Neuroscience, 1 Medical Center Drive, Morgantown, WV 26505

Colleen N. Warren
Department of Psychology, West Virginia University, Morgantown, WV, USA. Address: 1124 Life Sciences Building, 53 Campus Drive, Morgantown, WV 26506

Hawley E. Montgomery-Downs
Department of Psychology, West Virginia University, Morgantown, WV, USA. Address: 1124 Life Sciences Building, 53 Campus Drive, Morgantown, WV 26506

Corresponding author: Hawley E. Montgomery-Downs, E-mail: Hawley. Montgomery-Downs@mail.wvu.edu

insky and Kleitman, 1953a; Aserinsky and Kleitman, 1953b; Aserinsky and Kleitman, 1955b; Datta and MacLean, 2007) although eye motility, body twitching and reports of dreaming had been previously documented (Berrien, 1930). The contemporary study of pediatric sleep is moving the field toward an understanding of sleep's broader developmental impact and ability to adaptively respond to environmental insult across the lifespan. Our goal with this chapter is to provide a synopsis of pediatric sleep, including recording methodology, normative parameters, clinical considerations and current issues.

We begin with a general overview of pediatric sleep monitoring techniques, their benefits, and limitations. Subsequently, the chapter is organized into chronological age-related sections: infancy (birth to ~2 years), early school-age (~3 to 5 years), school-age (~6 to 12 years) and adolescence (~13 to 18 years). For each age period, we provide an overview of developmentally typical sleep; the most prevalent sleep disorders and their known consequences; and recommendations for healthy sleep practices. We hope this broad and general summary of historic and current pediatric sleep research will inspire the reader to pursue a more detailed understanding and make their own contributions to the field. Toward those ends, we have provided additional resources and future directions at the conclusion of the chapter.

23.1 PEDIATRIC SLEEP MONITORING TECHNIQUES

Since the 1950s, technology for assessing sleep has advanced markedly. Although we now possess the ability to monitor and evaluate sleep in much greater detail than was possible 70 years ago, our standard laboratory tools still include electroencephalography (EEG), electrooculography (EOG), and electromyography (EMG) (Fig. 23.1). Intensive instrumentation of pediatric patients or research participants for sleep recording is now standardized and age-specific modifications have been recommended (Iber et al., 2007; Queensland Department of Health, 2013). Specialized expansion of the EEG to include more channels is also recommended for cases where differentiation between pediatric parasomnias and nocturnal seizures is necessary (Aurora et al., 2012).

Nevertheless, direct observation of behavioral state remains the gold-standard for studying young infants (Thoman, 1975). Videosomnography is also a particularly valuable tool for coding behavioral state in the normal home environment (Anders and Sostek, 1976; Burnham, et al., 2002).

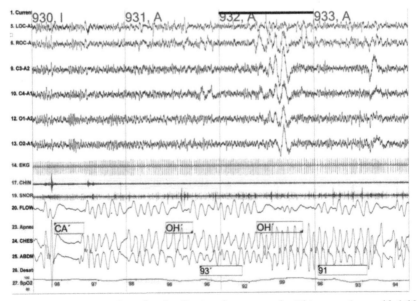

Figure 23.1. Four 30-second epochs of pediatric polysomnography. This was a 6 year-old child with obstructive sleep apnea and no other remarkable medical history. Central apnea (CA) marked is a compensatory pause and would not be marked as a respiratory event. The obstructive hypopnea (OH) events were associated with desaturations and, in the second event, cortical arousal.

With this vulnerable population, it is even more important to choose the least intrusive tools needed to answer our clinical or research questions. In that theme, actigraphy has become quite popular. Actigraphy is a class of small (wrist- or ankle-worn) devices that use recorded activity patterns and specific algorithms to estimate when the wearer is asleep. Though these devices have generally been shown valid for use with adults (American Academy of Sleep Medicine [AASM], 2001; Morgenthaler, et al., 2007; De Souza et al., 2003; Marino et al., 2013) there are significant issues that still need to be overcome for its use with infants and children (Gnidovec et al., 2002; Insana, et al., 2010; Sitnick et al., 2008; Galland et al., 2014; Meltzer et al., 2012). Actigraphy may be a viable resource for preclinical estimates, but the current technology does not appear consistent enough for exclusive use in clinical evaluation or diagnosis. The AASM supports use of pediatric actigraphy, "... for delineating sleep patterns, and to document treatment responses in normal infants and children..." (Morgenthaler et al., 2007).[1] Figure 23.2 is an example of data from a pediatric actigram.

The utility of subjective sleep diaries and surveys are obviously dependent upon the age of the pediatric subject and/or how aware the parent completing the diary or survey is of their child's sleep. Survey value also depends upon its intended use (i.e. research or clinical) and ability to measure the construct of interest based on its performance in systematic, psycho-

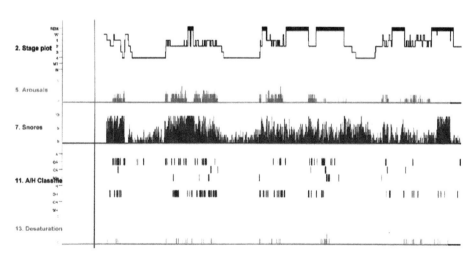

Figure 23.2: Pre-adenotonsillectomy summary overnight polysomnography data from a 4 year-old child with severe obstructive sleep apnea.

---

[1]Morgenthaler, T.; Alessi, C.; Friedman, L.; Owens, J.; Kapur, V.; Boehlecke, B.; Brown, T.; Chesson, A.; Coleman, J.; Lee-Chiong, T.; Pancer, J.; Swick, T. J.; Standards of Practice Committee; American Academy of Sleep Medicine. Practice Parameters for the Use of Actigraphy in the Assessment of Sleep and Sleep Disorders: An Update for 2007. *Sleep* **2007**, 519-529.

metric evaluation (Spruyt and Gozal et al., 2011a). Surveys with thoroughly vetted, High-qualify psychometrics permit large-scale data collection without labor and cost prohibitive instrumentation and are able to capture subjective information such as attitudes or mood that may be useful in identifying unique contributions of subjectively perceived sleep. Only a handful of pediatric sleep surveys meet or approach the most rigorous criteria for psychometric evaluation, notably the child and adolescent versions of the sleep disorders inventory for students (Luginbuehl et al., 2008; Spruyt and Gozal, 2011b).

23.2 INFANCY

23.2.1 Normative Infant Sleep

Infant sleep changes quickly. These changes can be observed behaviorally, electroencephalographically, organizationally, and temporally across the 24 h day. Watching a sleeping neonate reveals a remarkable diversity of expression including sucking, twitches, startles, joint-spanning movements, brief cries, smiles, grimaces, stretching, REMs, and sighs (Prechtl, 1964; Thoman, 1975). The reason these behaviors are so clear and facilitate identification of sleep state is that early infant sleep is characterized by a lack of muscle atonia (a brainstem-mediated mechanism of motor output suppression; Kohyama, 1996). Among young infants, sleep is classified as either active (generally considered analogous to REM), quiet (analogous to non-rapid eye movement [NREM]), or indeterminate (containing mixed features; Anders et al., 1971). The AASM pediatric task force has suggested that the scoring of infant sleep according to traditional adult criteria not begin until two months (Grigg-Damberger et al., 2007; Iber et al., 2007). During and after that time, behavioral observation is recommended to complement EEG scoring because sleep stage-defining features are still developing (Grigg-Damberger et al., 2007).

Infant sleep states are organized into roughly 70 min cycles and as they develop, an increasing proportion of infants' sleep occurs within the cycle structure (Scher et al., 2005). This and an increase in the organization of transitions between stages (Groome et al., 1997) reflect increasing physical maturity (Ficca et al., 2000). Through at least two weeks of age active and quiet sleep occur in roughly equal proportions (Sankupellay et al., 2011) after which the proportion of active sleep steadily decreases, and quiet sleep becomes more concentrated at the beginning of the nocturnal sleep period (Fagioli and Salzarulo, 1982; Ficca, et al., 2000). By six months, when NREM and REM are consistently used to describe infant sleep, NREM occupies about 65% and REM about 35% of sleep time (Sankupellay et al., 2011). These values continue to change gradually until they are close to adult-like proportions of 70 and 30%, respectively, around 12 months (Sankupellay et al., 2011).

Additional information about infant sleep comes from the times when it normally occurs. Neonatal sleep is initially organized in three to four hour bouts around the clock (Peirano et al., 2003). Nocturnal sleep consolidation occurs gradually (Iglowstein et al., 2003), and this process has been used as an index of developmental maturity (Hoppenbrouwers, 1982). Light exposure (Harrison, 2004), other environmental cues, and genetic and behavioral influences (Fisher et al., 2012) all likely contribute to development of an increasingly diurnal sleep rhythm (Jenni et al., 2006). About 60% of infant sleep occurs at night by six months and about 71% by 12 months (Harrison, 2004). Commensurate with an increase in nocturnal sleep time is a decrease in infant daytime napping, from two to three naps during the first six months to a single (typically afternoon) nap that may persist into early childhood (Montgomery-Downs and Gozal, 2006).

Nocturnal awakenings are another infant sleep characteristic that has received considerable attention, largely because of the consternation that this causes parents (Meltzer and Mindell, 2007). More nocturnal awakenings and reduced nocturnal sleep duration are also predictive of more parent-reported infant sleep problems within the first two years (Byars et al., 2011). Up to two nocturnal awakenings are typical through four months (about 93% of infants; Meijer and Van Den Wittenboer, 2007), after which one or fewer nocturnal awakenings are the norm through the remainder of infancy (Montgomery-Downs and Gozal, 2006; O'Connor et al., 2007; Weinraub et al., 2012). Only about 4% of infants still have frequent nocturnal awakenings at 12 months (Meijer and Van Den Wittenboer, 2007).

23.2.2 Infant Sleep Disorders and Consequences

Infant sleep may serve as a window onto neonatal adjustment to the extrauterine environment and as an index of maturity among older infants. For example, sleep may indicate periods of developmental sensitivity (Chu et al., 2014), achievement of developmental milestones or delayed maturation (Sankupellay et al., 2011) and may even play a direct role in the development of functional cortical networks (Kurth et al., 2013; Piantoni et al., 2013). Less indeterminate relative to active and quiet sleep, for example, has been interpreted as indicating greater sleep maturity (Anders et al., 1971). Distinctive EEG waveforms and power spectral changes in the theta, sigma, and alpha ranges may also provide developmental insight (Sankupellay et al., 2011; Chu et al., 2014). REM sleep in particular may be critically important for brain development (Marks et al., 1995; Mirmiran et al., 2003; Roffwarg et al., 1966).

The immediate effects of poor sleep on infant behavior include reduced performance on attention tasks and elevated distractibility among infants 4–18 months old (Geva et al., 2013). Unfortunately, the most compelling evidence is from infants whose sleep is insufficient or disturbed.

23.2.2.1 Behavioral Insomnia

Though, as described above, nocturnal awakenings are normal in infancy, parental reports suggest that 10 to 30% of six months old infants have sleep problems (Blunden, 2012; Byars et al., 2011; Morgenthaler et al., 2006), a proportion that persists though 36 months (Byars et al., 2011). Earlier sleep problems (i.e. 6–24 months) appear largely related to nocturnal awakenings and overall reduced nocturnal sleep duration (Byars et al., 2011).

Routines at bedtime are generally considered good practice, but a dependent association between an environmental factor and sleep onset can lead to difficulty re-initiating sleep after nocturnal awakening (AASM international classification of sleep disorders (ICSD-2), 2001). That is, to fall back asleep after normal awakenings between sleep cycles the infant may require return of the parent or other stimuli that were present when the infant was falling asleep but that are absent when they awaken at night. To avoid such sleep-onset associations, clinical experts recommend that infants fall asleep at the beginning of the night in the same environment in which they will later awaken and specifically that they be put to bed awake but drowsy (Mindell et al., 2006).

Treatments such as sleep training, which may include a set bedtime, scheduled awakenings, positive sleep hygiene and systematic extinction of sleep associations and other unwanted behaviors can be effective at correcting behavioral insomnia (Adachi et al., 2009; Mindell, 1999; Mindell et al., 2009a; Mindell and Meltzer, 2008; Meltzer, 2010). At present, there is scarce evidence (France and Hudson, 1990) that intervention effects can last up to 24 months, so the long-term efficacy of behavioral intervention remains a concern (Mindell et al. 2006; Mindell et al., 2009a; Owens and Mindell, 2011).

23.2.2.2 Sleep-Disordered Breathing

Other sleep disruptions may go unnoticed by parents though they may have more deleterious and long-term consequences. Anywhere from 12 to 20% of infants snore multiple times per week, a primary sign of sleep-disordered breathing (SDB) (Byars et al., 2011). From 2 to 4% of infants suffer the most severe form of SDB, obstructive sleep apnea (OSA), which causes reduction

or cessation of breathing, transient reductions in blood oxygenation and frequent awakenings that fragment sleep (Robison et al., 2013). OSA has been associated with behavioral deficits in infants (Robison et al., 2013). Even mild SDB may carry risk. Infants who snore from shortly after birth through the first year score significantly lower on developmental assessments than controls (Grigg-Damberger and Ralls, 2012; Piteo et al., 2011). If SDB is not treated before preschool, the risks may be compounded; a great deal more is known about the deleterious impact of childhood SDB (Barnes et al., 2009; Key et al., 2009; Montgomery-Downs et al., 2003). Therefore, there are both immediate and long-term developmental reasons to ensure that infant sleep problems are treated effectively.

23.2.3 Infant Healthy Sleep Recommendations

23.2.3.1 Sudden Infant Death Syndrome (SIDS)

SIDS is "the sudden death of an infant less than 1 year of age that cannot be explained after a thorough investigation is conducted, including a complete autopsy, examination of the death scene, and a review of the clinical history" (Centers for Disease Control and Prevention, 2015a).[2] SIDS deaths have decreased in the U.S. since the American Academy of Pediatrics (AAP) began a campaign to promote infants being placed to sleep on their backs (AAP, 2011). However, the rate of sudden unexpected infant death syndrome (SUID)

---

[2]Centers for Disease Control and Prevention. Sudden Unexpected Infant Death and Sudden Infant Death Syndrome: About SUID and SIDS. http://www.cdc.gov/sids/aboutsuidandsids.htm (accessed Apr 3, 2015a).

due to unsafe sleeping environments and accidental strangulation and suffocation while sleeping has increased (Centers for Disease Control and Prevention, 2015b). Recent reports have shown that nearly 55% of infants in the U.S. sleep with unsafe bedding (National Institutes of Health, NIH, 2015b). In some locations, SIDS cases have decreased more significantly, not because of a decrease in infant deaths but because of improved investigations and more accurately determined causes of deaths that can include SUID (Stanley, 2015) or even deaths related to Munchausen's Syndrome by Proxy (Southall et al., 1997).

While many aspects of SIDS remain a mystery, the most recent policy statement and technical report released by the AAP (2011) lists evidence-based SIDS risk factors: prone sleeping position, overheating during sleep, premature birth, exposure to tobacco smoke (either during gestation or after birth), unsafe sleeping environments, failure to receive immunizations, and failure to breastfeed. Additionally, the effectiveness of infant monitors or other products that claim to prevent SIDS are either untested or invalid and should not be trusted as SIDS prevention measures (AAP, 2011).

23.2.3.2 Parental Practices

Safe infant sleep practice has been guided by SIDS prevention recommendations. These recommendations include: no loose bedding, pillows, stuffed animals, crib bumpers or anything that might pose a suffocation, and/or strangulation hazard in an infant's sleep area. Cribs and mattresses should meet Consumer Product Safety Commission (CPSC) guidelines, and it is recommended that infants never sleep on couches, stuffed chairs, or adult mattresses (CPSC, 2010). Though co-sleeping is highly prevalent (Montgomery-Downs and Gozal, 2006) and recommendations doing so safely are needed, the AAP unilaterally recommends room-sharing without bed-sharing (AAP, 2011). In addition, the NIH recommends: to avoid overheating, an infant should sleep in a room that is a comfortable temperature for an adult without excessive covering; a thin blanket may be tucked in at the bottom and sides of the crib mattress with the infant's feet at the foot of the crib; and infants should be placed in the supine position for sleep, with adequate supervised time in the prone position while awake to prevent positional plageocephaly (NIH, 2015b).

23.2.3.3 Infant Feeding Methods

Breastfeeding may have beneficial effects on infant sleep. It may help to alleviate infantile colic (Engler et al., 2012), a condition which may reduce sleep duration (Thiedke, 2001). Melatonin, a hormone known to induce sleepiness, may also occur in greater quantities in breast milk compared to infant formulas and may aid in circadian regulation (Engler et al., 2012). Perhaps most significantly, there is some evidence that breastfeeding may have a long-term effect on sleep promotion by reducing coughing, snoring, and breathing problems during sleep (Brew et al., 2014; Galbally et al., 2013; Montgomery-Downs et al., 2007). Children who snore habitually have significantly reduced SDB severity if they were breastfed as infants (Brew et al., 2014; Montgomery-Downs et al., 2007). Though the mechanism is unknown, these studies suggest that breast-feeding may help to ameliorate or prevent SDB. Given that the beneficial cognitive impacts of breastfeeding (Anderson et al., 1999; Deoni et al., 2013; Issacs et al., 2010; Kafouri et al., 2012; Kramer et al., 2008; McCrory and Murray, 2013; Mortensen et al., 2002) seem to affect the same faculties as pediatric SDB (Bourke, 2011a), evaluating the potential overlap between these is provocative.

These benefits notwithstanding, the relationship between breastfeeding and infant sleep is complex. Although breastfeeding is often considered the best for the child, there is equivocal evidence based largely on parent report that breastfed infants may also have an increased number and duration of nocturnal awakenings compared to non-breastfed infants (Demirci et al., 2012; Engler et al., 2012; Galbally et al., 2013; Mindell et al., 2012; Ramamurthy et al., 2012). There is also some evidence that breastfed infants have more difficulty sleeping alone (Galbally et al., 2013), have more sleep fragmentation (Mindell et al., 2012), and less consolidated sleep overall(Galbally et al., 2013; Mindell et al., 2012; Ramamurthy et al., 2012). A potential mechanism for these findings, in cases of co-sleeping, may be that parents' movements inadvertently awaken the infant (Ramamurthy et al., 2012). Although breastfeeding is a predictor of more nocturnal awakenings in young infants, by six months parental presence at sleep onset becomes the strongest predictor of awakenings and by nine months differences between breast- and formula-fed infants disappear (Mindell et al., 2012).

23.3 EARLY CHILDHOOD

23.3.1 Normative Early Childhood Sleep

Early childhood sleep duration, distribution, and EEG continue to develop with age, albeit less rapidly than during infancy. There is also evidence of developmental continuity; for example, sleep duration during infancy is predictive of early child sleep duration (Blair et al., 2012). The amount of parental involvement required at bedtime at four years is also related to the child's number of awakenings when they were 12 months (Tikotzky and Shaashua, 2012).

Average nocturnal sleep duration from three to five years is a fairly consistent 11 h, but there are considerable individual differences (Blair et al., 2012; Iglowstein et al., 2003; Teng et al., 2012). During early childhood fewer than two additional hours of sleep take place during daytime naps (Blair et al., 2012; Iglowstein et al., 2003; Teng et al., 2012). At five years, the roughly 8% of children who nap do so only occasionally, a reduction from about 25% at 3.5 years (Blair et al., 2012; Iglowstein et al., 2003). Thus, most sleep during early childhood (in the cultures assessed) is consolidated to the nighttime so that the highest proportions of nocturnal sleep across the lifespan occur during this developmental period (Iglowstein et al., 2003; Williams et al., 2013). By three years, only 10% of parents report that their children have ongoing nocturnal awakenings (Weinraub et al., 2012). Of a child's time in bed, the proportion spent asleep (a measure of sleep quality) is close to 90% (Biggs et al., 2012; Tikotzky and Shaashua, 2012).

Proportions of sleep spent in REM and NREM during this period are similar to adults (20–25% and 70–75%, respectively), but young children spend 25–30% of their sleep time in NREM stage three (Biggs et al., 2012), in contrast to young adults' roughly 15% (Ehlers and Kupfer, 1997). NREM stage three is characterized by high power, low frequency delta wave EEG (Chu et al., 2014; Biggs et al., 2012) and corresponds to increased growth hormone release (Takahashi et al., 1968). The gradual decrease in delta power across a night of sleep provides evidence of a developed homeostatic sleep drive at this age (Biggs et al., 2012). During early childhood the posterior (occipital region) EEG rhythm also comes to resemble adult-like alpha – a waveform that precedes and aids in polysomnographic identification of sleep onset (Iber et al., 2007; Scraggs, 2012).

23.3.2 Early Childhood Sleep Disorders and Consequences

23.3.2.1 Sleep-Disordered Breathing

Conservative estimates are that 1–4% of children suffer from OSA (Lumeng and Chervin, 2008), the most severe manifestation of SDB. However, this disorder is under-diagnosed in young children, as the prevalence of habitual snoring, the cardinal symptom of SDB, may be at least three times higher than the prevalence of OSA diagnosis (Bonuck et al., 2011; Lumeng and Chervin, 2008). This is especially concerning given the immediate (Chervin et al., 2002; Bourke et al., 2011a; Perfect et al., 2013) and long-term (Kohler et al.,

2009) cognitive, behavioral, and academic problems of untreated SDB.

Adenotonsillectomy is the most common treatment for pediatric OSA (Archbold and Parthasarathy, 2009; Guilleminault et al., 2005; Kohler et al., 2009). Removal of these airway-obstructing soft tissues often leads to cognitive, behavioral, and academic improvements (Friedman et al., 2003; Kohler et al., 2009; Marcus et al., 2013), but such findings are not always consistent (Kohler et al., 2009; Marcus et al., 2013). Continuous positive airway pressure (CPAP) is also used at this age (Archbold and Parthasarathy, 2009; Guilleminault et al., 2005) though it requires a unique approach to be effective (Archbold, 2013; Archbold and Parthasarathy, 2009). Figures 23.3 and 23.4 illustrate the summary effects of child OSA and effective treatment.

23.3.2.2 Sleep Terrors and Confusional Arousals

Both sleep terrors and confusional arousals are parasomnias that can occur during NREM sleep, usually upon arousing from NREM stage three (AASM ICSD-2, 2001; Thorpy et al., 1988). Most notably, these conditions involve abnormal arousal from sleep (AASM ICSD-2, 2001) that is postulated to be incomplete, sharing qualities of both sleep and wake (Modi et al., 2014). Both confusional arousals and sleep terrors have typical presentation during early childhood (overall prevalence 6.9 and 2.7%, respectively) (Bjorvatn et al.,

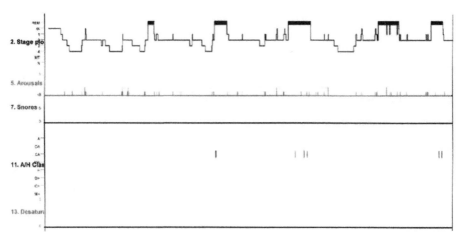

Figure 23.3. Post-adenotonsillectomy summary overnight polysomnography data from the same 4 year-old child three months post-surgery.

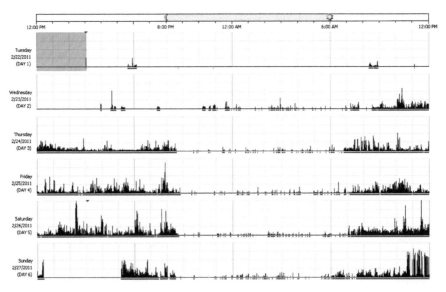

Figure 23.4. Six days of continuous actigraphy data from an 11 year-old child. Note that the first day the child did not wear the actigraph. Periods of wake during sleep are clearly visible but it would difficult to identify precise sleep onset and offset periods without behavioral corroboration from the parent. In addition, note that the period of quiescence on the last day from ~12:30 pm to ~4:05pm may be a daytime nap or a period when the watch was off – again, it is not possible to determine without additional information from the family.

2010) with a high parent-reported prevalence of sleep terrors among the youngest in this age group (near 40% at 2.5 years) (Petit et al., 2007). Confusional arousals are aptly named, featuring disorientation after awakening but are not usually dangerous (ICSD-2, 2001). Sleep terrors have a more violent manifestation with fear-like defensive behaviors and inconsolable crying and/or screaming that can be particularly troubling for parents who are usually unable to calm their child for sometimes prolonged periods of 10 min or more (Kotagal, 2009). Though the child has no recollection of these events, there is some potential for injury (ICSD-2, 2001) and children with terrors also suffer increased awakenings, reduced sleep efficiency, and have less NREM stage three continuity (Bruni et al., 2008). Nevertheless, terrors and confusional arousals usually resolve with age and do not appear to indicate risk for other psychopathology (ICSD-2, 2001; Howell, 2012).

23.3.3 Early Childhood Healthy Sleep Recommendations

Best sleep practices during this developmental period center on family routines and behavioral dynamics. Certainly there are physiological contributions to sleep timing, but early work suggests that over three quarters of a child's sleep problems can be addressed behaviorally (Jones and Verduyn, 1983).

Because failure to effectively impose limits may manifest in childhood behavioral insomnia, limit-setting behavioral techniques include parents' reinforcing compliance with bedtime practices (ICSD-2, 2001). It is important to set limits that are developmentally (related to needed hours of sleep) and temporally (related to the circadian timing of bed and rise times) appropriate. For example, child bedtimes are sometimes notably earlier than their physiological sleep onset facilitated by dim-light melatonin-onset (LeBourgeois et al., 2013a); greater misalignment between these factors is associated with longer time to sleep onset and greater incidence of child bedtime resistance (LeBourgeois et al., 2013b).

Early childhood is a transitional period for napping marked by significant individual differences in napping cessation with some continuing napping through early school age (Blair et al., 2012; Iglowstein et al., 2003). Preschoolers who nap regularly perform better after a nap on certain learning tasks and do not recover performance deficits after a skipped nap, indicating a unique benefit of daytime sleep for some children (Kurdziel et al., 2013). Behavioral regulation also appears to be lower after nap deprivation (Miller et al., 2014). Afternoon nap time up to 1.5 h should be adequate to accommodate most children who are still napping (Blair et al., 2012). While it may be tempting to schedule obligatory nap time in light of these data, it is also important to keep in mind that increased nap duration is associated with later bedtime (Komada et al., 2012) and that there is wide individual variability in nap need with age (Blair et al., 2012; Iglowstein et al., 2003). Recall that misaligned bedtime and sleep onset time for children are related to nocturnal sleep issues (LeBourgeois et al., 2013b), which pose a challenging scenario for using translational research to meet the individual sleep needs of young children.

23.4 SCHOOL-AGE CHILDREN

23.4.1 Normative School-Age Sleep

The ongoing developmental decreases in total sleep time, overall EEG power (Chu et al., 2014; Scraggs, 2012), and amount of NREM deep sleep continue throughout childhood and adolescence, ultimately stabilizing around 16 years (Iglowstein et al., 2003). Changes in sleep quantity and organization are more modest across the school-age span than during infancy and early childhood. The typical adult 90 min sleep cycle is established at the beginning of childhood (Sheldon, 1996) and total sleep time decreases from an average of 11.9 h in early childhood to an average of 9.2 h through about 12 years (Galland et al., 2012). It should be noted, however, that the individual variability in sleep time that was true at earlier ages also applies to child sleep (Galland et al., 2012).

23.4.2 School-Age Sleep Disorders and Consequences

23.4.2.1 Sleep-Disordered Breathing

The implications of SDB extend beyond child health and into the classroom for school-age children. There is an association between reduced academic performance and increased behavioral problems in children between six and 12 years who have SDB (Bourke et al., 2011a; Bourke et al., 2011b; Perfect et al., 2013). SDB has also been linked to decreased pediatric neurocognitive performance (Bourke et al., 2011a), executive function, attention, and intellectual ability even in mild cases (Kohler et al., 2009). The entire range of SDB severity has been associated with deficits (Bourke et al., 2011a; Bourke et al., 2011b; Perfect et al., 2013; Grigg-Damberger and Ralls, 2012). Ten to 20% of children under seven years (Bonuck et al., 2011) are reported to snore habitually, while 1–4% (Lumeng and Chervin, 2008) of all children are diagnosed with OSA.

Hyperactivity is also more common in children who have SDB (Chervin et al., 2002; Perfect et al., 2013). In children between two and 13 years, increased hyperactive behavior has been associated with habitual and loud snoring (Chervin et al., 2005). Habitual snoring is also associated with daytime sleepiness among children (Chervin et al., 2002). Further, the deficits observed among children with SDB may persist even after successful treatment. Wide-ranging neurocognitive deficits and an average 10 point IQ deficit have been reported among children with SDB compared to matched controls with effects persisting 6 months after treatment (Kohler et al., 2009).

23.4.2.2 Nocturnal Enuresis

The American Psychological Association (APA, 2015) presents nocturnal enuresis, or bedwetting, as concerning if it still occurs at five years and at least twice per week for three months or more. Up to 9.5% of children still have nocturnal enuresis at nine years (Butler and Heron, 2008). A critical aspect of treatment is avoiding shaming and seeking medical expertise since causes can include diabetes and urinary tract infection (Bansal and Sheldon,

2008). Conditioning therapies may be used to accustom the child to awakening to urinate. A moisture-sensitive bed alarm that awakens the child when urination begins is an example of such a conditioning therapy and one of the most effective treatments for this disorder (Houts et al., 1994).

23.4.2.3 Somnambulism

Up to 40% of children may experience sleepwalking at some point during their childhood, but only around 3–4% will experience these episodes frequently (Mason and Pack, 2007). The most common ages of onset for sleepwalking are between five and ten years (Kotagal, 2009). Sleepwalking episodes may include standing, walking, and in some cases, eating or urinating (Kotagal, 2009). Episodes usually last between 10 and 20 min during the first third of the night and may be present from a few times a week to a couple times a month (Kotagal, 2009). A sleepwalking child can usually be led back to bed fairly easily, even without awakening (Rosen, 2014). If there is concern that sleepwalking might cause injury, measures should be taken to protect the child from falls or other hazards and to alert the parent that the child is out of bed (Kuhn and Floress, 2008).

23.4.3 School-Age Healthy Sleep Recommendations

Though there is scarce evidence to support it, the National Sleep Foundation (NSF, 2015a) recommends that school-age children be allowed 9–11 h of sleep and have consistent bedtime routines. Exercise should be completed at least 3 h before bedtime (NSF, 2015b). Caffeine should not be consumed nor should naps be taken in the late afternoon or later as these can both reduce the ability to fall asleep at night or negatively affect sleep quality (Calamaro et al., 2009; Drapeau et al., 2006; Mindell et al., 2009b). Because both hunger and satiation can affect the ability to fall asleep, a light snack in the evening is recommended (NSF, 2015b). The bedroom should be cool and dark without electronics or other sources of artificial light (NSF, 2015b). Social media, games, television, computers, phones, and other activities should not be used in the bedroom as their presence can delay sleep onset and cause awakenings (Calamaro et al., 2009; Mindell et al., 2009b).

23.5 ADOLESCENCE

23.5.1 Normative Adolescent Sleep

Ongoing increases in neuronal myelination, fluctuations in grey matter and synaptic pruning throughout adolescence may be reflected in the EEG (Giedd et al., 1999). The slowest EEG activity, delta, decreases in power between early adolescence and about 16 years (Chu et al., 2014; Feinberg et al., 2006; Feinberg and Campbell, 2010). Other primary developmental changes during this period are circadian and behavioral.

The marked shift toward a delay in circadian timing during adolescence is associated with the delta activity change noted above (Campbell et al., 2007) and puberty (Carskadon et al., 1993; Feinberg et al., 2006). Though normative (Millman et al., 2005), adolescent sleep phase delay is at odds with school times that require early awakenings. For example, there are decreases in adolescent sleep on weekdays, including a decrease in REM sleep (Thorpy et al., 1988). Weekday sleep truncation may manifest in weekend sleep time increases by about 3 h (Crowley et al., 2007). Our normative section about adolescents is brief because regretfully, adolescent sleep duration is generally inadequate (Mitchell et al., 2013; Owens et al., 2014; Stallones et al., 2006) and little is known about how adolescents would sleep were their schedules more compatible with their circadian clock.

23.5.2 Adolescent Sleep Disorders and Consequences

23.5.2.1 Insufficient Sleep

Adolescents' sleep is often of insufficient quantity and their chronic sleep deprivation can have problematic effects both immediately and later in development. Adolescents in the U.S. are generally sleep deprived, averaging about 1.5 h less per night than recommended (Mitchell et al., 2013; Stallones et al., 2006). Reduced sleep duration or quality in adolescents is correlated with an increased risk of accidental injury (Danner and Phillips, 2008; Stallones et al., 2006) and risky behavior (O'Brien and Mindell, 2005). Suicidal ideation and self-harm behaviors are also associated with difficulty sleeping (Wong et al., 2011).

With an estimated prevalence of 11–30%, adolescent behavioral sleep problems are regretfully common (Blunden, 2012; Wolfson and Carskadon, 2003). Inadequate sleep hygiene and insufficient sleep are two of the most common problems in this category (Blunden, 2012). These problems, which primarily decrease sleep quantity, can affect school performance and other cognitive tasks such as memory, attention, and other learning (Blunden, 2012; Wolfson and Carskadon, 2003).

23.5.2.2 Narcolepsy

Narcolepsy is a hypersomnia of central origin sleep disorder defined by daytime hypersomnolence, sleep attacks that may be related to emotional reaction or cataplexy and sleep-onset REM periods that can be accompanied by hypnogogic hallucinations (AASM ICSD-2, 2001). The first age period for usual clinical presentation is roughly 13 to 15 years (Dauvilliers et al., 2001; Rao et al., 2012). Pediatric onset narcolepsy is associated with a family history of the disorder and more severe symptoms such as cataplexy (Dauvilliers et al., 2001). Pediatric onset of narcolepsy is also associated with puberty; Poli et al. (2013) have reported that 17% of narcolepsy cases with cataplexy also had medical history of precocious puberty and 41% had accelerated pubertal development. Children diagnosed with narcolepsy have shorter sleep onset and REM latencies (Rao et al., 2012). Early detection of pediatric onset narcolepsy may be improved by screening for human leukocyte antigen DQB1*0602 (Hood and Harbord, 2004) or by carefully attending to children at risk due to family history (Dauvilliers et al., 2001) for onset symptoms to minimize impact of excessive daytime sleepiness on school performance (Millman et al., 2005).

23.5.3 Adolescent Healthy Sleep Recommendations

Because the healthy sleep practices and recommendations for adolescents are similar to those for school-age children above, we focus here on the additional recommendations unique to adolescents. The NSF (2015a and b) recommends that teens should reserve 8–10 h for sleep each night. However, there is overwhelming evidence of adolescents' delayed phase preference, or biological shift to later sleep and wake times. Since this is incompatible with standard school-start schedules, sleep researchers have called for later school-start times (Carskadon, et al., 1980; Carskadon et al., 1993). Though there has been positive momentum in this, full implementation of this schedule still needs to occur.

23.6 CONCLUSION

In a relatively short historical timeframe, ongoing advances in sleep science, encouraged by technological breakthroughs, have quickly improved our understanding of the ontogeny of sleep. We have also learned a great deal about physiological, neurological, and developmental influences of pediatric sleep disturbance. In this overview we hope to have effectively touched on evidence of key techniques for monitoring pediatric sleep and normative age-related sleep features, age-specific disorders and recommendations for treatment, and prevention of sleep problems. However, many issues remain for further investigation. Contemporary imaging technologies, capacity for longitudinal and big data approaches, and translational application are currently contributing some of the most significant breakthroughs. For example, evaluating parent–infant sleep dynamics to improve family bedtime practices based on an ecological systems perspective may lead to sleep improvements for the whole family. Effective evidence-based translation of child and adolescent circadian sleep demands have led to policy shifts modifying school schedules toward better enabling adolescent success. We are also edging closer to determining whether the general recommendations for sleep times meet individual sleep needs. It is our hope that the methods, basic physiology, clinical considerations, and current issues in pediatric sleep summarized here will encourage the reader to include sleep in their future training, research, and/or clinical programs.

23.7 ADDITIONAL RESOURCES

This review – by definition a synopsis of a relatively vast developmental period – is necessarily incomplete and with regret we have neglected much of our colleagues' important work. We hope the inspired reader will consider these additional resources.

BOOKS AND REVIEWS

Principles and Practice of Pediatric Sleep Medicine, 2nd Ed. Sheldon, S. H.; Ferber, R.; Kryger, M. H.; Gozal, D. London: Elsevier 2014.

Meltzer, L.; Montgomery-Downs, H. E.; Walsh, C.; Insana, S. P. Use of Actigraphy in Pediatric Sleep Research; Review. *Sleep Medicine Reviews*, **2012**, 16(5), 463–475.

Wolfson, A., Montgomery-Downs, H. E., Eds.; *Oxford Handbook of Infant, Child, and Adolescent Sleep and Behavior*; Oxford University Press: Online, 2013.

WEB INFORMATION

American Academy of Sleep Medicine: www.aasmnet.org

National Institute of Child Health and Human Development, Infant Sleep Safety: www.nichd.nih.gov/sts/Pages/default.aspx

National Sleep Foundation: www.sleepfoundation.org

Sleep Research Society: www.sleepresearchsociety.org

KEY WORDS

- pediatric
- infant
- child
- adolescent
- development

REFERENCES

Adachi, Y.; Sato, C.; Nishino, N.; Ohryoji, F.; Hayama, J.; Yamagami, T. A Brief Parental Education for Shaping Sleep Habits in 4-Month-Old Infants. *J.Clin. Med. & Res.* **2009**, 7(3), 85–92.

American Academy of Pediatrics. Policy Statement: SIDS and Other Sleep-Related Infant Deaths: Expansion of Recommendations for a Safe Infant Sleeping Environment. *American Academy of Pediatrics.***2011**, DOI: 10.1542/peds.2011–2284. www.pediatrics.org/cgi/doi/10.1542/peds.2011–2284

American Academy of Sleep Medicine.*The International Classification of Sleep Disorders, Revised: Diagnostic and Coding Manual*; American Academy of Sleep Medicine: Chicago, 2001.

American Psychological Association. Fact Sheet: Enuresis in Children and Adolescents. http://www.apadivisions.org/division-54/evidence-based/enuresis.aspx?__utma=12968039.1120 231989.1421203650.1428542757.1428688106 .6&__utmb=12968039.3.10.1428688106&__ utmc=12968039&__utmx=-&__utmz=1296803 9.1428688106.6.4.utmcsr=yahoo|utmccn=%28or ganic%29|utmcmd=organic|utmctr=%28not%20 provided%29&__utmv=-&__utmk=169751457 (accessed Apr 14, 2015).

Anders, T. F.; Emde, R. N.; Parmelee, A. H. *A manual of standardized terminology, techniques and criteria for scoring of states of sleep and wakefulness in newborn infants.* UCLA Brain Information Service Publications Office **1971**, NINDS Neurological Information Network.

Anders, T. F.; Sostek, A. M. Use of Time Lapse Video Recording of Sleep-Wake Behavior in Human Infants. *Psychophysiology* **1976**, 15(2), 155–157.

Anderson, J. W.; Johnstone, B. M.; Remley, D. T. Breast-feeding and cognitive development: a meta-analysis. *Am. J.Clin.Nutr.***1999**, 70, 525–535.

Archbold, K. H. Pediatric Sleep Apnea and Adherence to Positive Airway Pressure (PAP) Therapy. In *The Oxford Handbook of Infant, Child, and Adolescent Sleep and Behavior*; Wolfson, A. R.; Montgomery-Downs, H. E., Eds.; Oxford University Press: New York, USA, 2013; pp 362.

Archbold, K. H.; Parthasarathy, S. Adherence to Positive Airway Pressure Therapy in Adults and Children. *Curr.Opin.Pulm. Med.* **2009**, 16(6), 585–590.

Aserinsky, E.; Kleitman, N. Regularly Occurring Periods of Eye Motility, and Concomitant Phenomena, During Sleep. *Science* **1953a,** 118, 273–274.

Aserinsky, E.; Kleitman, N. Eye Movements During Sleep. *Fed. Proc.* **1953b,** 12, 6–7.

Aserinsky, E.; Kleitman, N. Two Types of Ocular Motility Occurring in Sleep. *J. Appl. Physiol.* **1955a,** 8(1), 1–10.

Aserinsky, E.; Kleitman, N. A Motility Cycle in Sleeping Infants as Manifested by Ocular and Gross Body Activity. *J. Appl. Physiol.***1955b,** 8(1), 11–18.

Aurora, R. N.; Lamm, C. I.; Zak, R. S.; Kristo, D. A.; Bista, S. R.; Rowley, J. A.; Casey, K. R. Practice Parameters for the Non-Respiratory Indications for Polysomnography and Multiple Sleep Latency Testing for Children. *Sleep* **2012**, 35(11), 1467–1473.

Bansal, R.; Sheldon, S.H. Parasomnias in Childhood. In *Sleep and Psychiatric Disorders in Children and Adolescents*; Ivanenko, A., Ed.; Informa Healthcare: New York, 2008; Chap. 22; pp 312.

Barnes, M. E.; Huss, E. A.; Garrod, K. N.; Raay, E. V.; Dayyat, E.; Gozal, D.; Molfese, D. L. Impairments in Attention in Occasionally Snoring Children: An Event-Related Potential Study. *Dev.Neuropsychol.* **2009**, 34(5), 629–649.

Berrien, F. K. Recall of Dreams During the Sleep Period. *J. Abn. Soc. Psych.* **1930**, 25(2), 110–114. DOI: 10.1037/h0071688

Biggs, S. N.; Walter, L. M.; Nisbet, L. C.; Jackman, A. R.; Anderson, V.; Nixon, G. M.; Davey, M. J.; Trinder, J.; Hoffman, R.; Armitage, R.; Horne, R. S. C. Time Course of EEG Slow-Wave Activity in Pre-School Children With Sleep Disordered Breathing: A Possible Mechanism for Daytime Deficits? *Sleep Med.* **2012**, 13, 999–1005.

Bjorvatn, B.; Gronli, J.; Pallesen, S. Prevalence of Different Parasomnias in the General Population. *Sleep Med.* **2010**, 11(10), 1031–1034.

Blair, P. S.; Humphreys, J. S.; Gringras, P.; Taheri, S.; Scott, N.; Emond, A.; Henderson, J.; Fleming, P. J. Childhood Sleep Duration and Associated Demographic Characteristics in an English Cohort. *Sleep* **2012**, 35(3), 353–360.

Blunden, S. L. Behavioral Sleep Disorders Across the Developmental Age Span: An Overview of Causes, Consequences, and Treatment Modalities. *Psych.* **2012**, 3, 249–256.

Bonuck, K. A.; Chervin, R. D.; Cole, T. J.; Emond, A.; Henderson, J.; Xu, L.; Freeman, K. Prevalence and Persistence of Sleep Disordered Breathing Symptoms in Young Children: A 6-Year Population-Based Cohort Study. *Sleep* **2011**, 34(7), 875–844. DOI: 10.5665/SLEEP.1118

Bourke, R. S.; Anderson, V.; Yang, J. S. C.; Jackman, A. R.; Killedar, A.; Nixon, G. M.; Davey, M. J.; Walker, A. M.; Trinder, J.; Horne, R. S. C. Cognitive and Academic Functions are Impaired in Children with all Severities of Sleep-Disordered Breathing. *Sleep Med.* **2011a**, 12, 489–496.

Bourke, R. S.; Anderson, V.; Yang, J. S. C.; Jackman, A. R.; Killedar, A.; Nixon, G. M.; Davey, M. J.; Walker, A. M.; Trinder, J.; Horne, R. S. C. Neurobehavioral Function is Impaired in Children with all Severities of Sleep disordered breathing. *Sleep Med.* **2011b**, 12, 222–229.

Brew, B. K.; Marks, G. B.; Almqvist, C.; Cistulli, P. A.; Webb, K.; Marshall, N. S. Breastfeeding and snoring: a birth cohort study. *PLOS ONE* **2014**, 9(1), e84956.

Bruni, O.; Ferri, R.; Novelli, L.; Finotti, E.; Miano, S.; Guilleminault, C. NREM Sleep Instability in Children with Sleep Terrors: The Role of Slow Wave Activity Interruptions. *Clin.Neurophysiol.* **2008**, 119, 985–992.

Burnham, M. M.; Goodlin-Jones, B. L.; Gaylor, E. E.; Anders, T. F. Nighttime Sleep-Wake Patterns and Self-Soothing from Birth to One Year of Age: A Longitudinal Intervention Study. *J. Child Psychol. Psychiatry* **2002**, 43(6), 713–725.

Butler, R.; Heron, J. The Prevalence of Infrequent Bedwetting and Nocturnal Enuresis in Childhood: A Large British Cohort. *Scand. J. Urol. Nephrol.***2008**, 2, 257–264.

Byars, K. C.;Yolton, K.; Rausch, J.; Lanphear, B.; Beebe, D. W. Prevalence, Patterns, and Persistence of Sleep Problems in the First 3 Years of Life. *Pediatr.* **2011**, 129 (2), e276-e284.

Calamaro, C. J.; Mason, T. B. A.; Ratcliffe, S. J. Adolescents Living the 24/7 Lifestyle: Effects of Caffeine and Technology on Sleep Duration and Daytime Functioning. *Pediatr.* **2009**, 123(6), e1005-e1010. DOI: 10.1542/peds.2008–3641.

Campbell, I. G.; Higgins, L. M.; Trinidad, J. M.; Richardson, P.; Feinberg, I. The Increase in Longitudinally Measured Sleepiness Across Adolescence is Related to the Maturational Decline in Low-Frequency EEG Power. *Sleep* **2007**, 30(12), 1677–1687.

Carskadon, M.A.; Harvey, K.; Duke, P.; Anders, T.F.; Litt, I.F.; Dement, W.C. Pubertal Changes in Daytime Sleepiness. *Sleep* **1980**, 2, 453–460.

Carskadon, M.A.; Vieira, C.; Acebo, C. Association Between Puberty and Delayed Phase Preference, *Sleep* **1993**, 16(3), 258–262.

Centers for Disease Control and Prevention. Sudden Unexpected Infant Death and Sudden Infant Death Syndrome: About SUID and SIDS. http://www.cdc.gov/sids/aboutsuidandsids.htm (accessed Apr 3, 2015a).

Centers for Disease Control and Prevention. Sudden Unexpected Infant Death and Sudden Infant Death Syndrome: Data and Statistics. http://www.cdc.gov/sids/data.htm. (accessed Apr 3, 2015b).

Chervin, R. D.; Archbold, K. H.; Dillon, J. E.; Panahi, P.; Pituch, K. J.; Dahl, R. E.; Guilleminault, C. Inattention, Hyperactivity, and Symptoms of Sleep-Disordered Breathing. *Pediatr.* **2002**, 109(3), 449–456.

Chervin, R. D.; Ruzicka, D. L.; Archbold, K. H.; Dillon, J. E. Snoring Predicts Hyperactivity Four Years Later. *Sleep* **2005**, 28(7), 885–890.

Chu, C. J.; Leahy, J.; Pathmanathan, J.; Kramer, M. A.; Cash, S. S. The Maturation of Cortical Sleep Rhythms and Networks over Early Development. *Clin.Neurophysiol.* **2014**, 125, 1360–1370.

Consumer Product Safety Commission. Safety Standards for Full-Size Baby Cribs and Non-Full-Size Baby Cribs; Notice of Proposed Rulemaking; Proposed Rule. *Fed. Reg.* .**2010**, 75(141), 43307–

43327. http://www.cpsc.gov//PageFiles/99324/cribstd.pdf

Crowley, S. J.; Acebo, C.; Carskadon, M. A. Sleep, Circadian Rhythms, and Delayed Phase in Adolescence. *Sleep Med.*, **2007**, 8, 602–612.

Danner, F.; Phillips, B. Adolescent Sleep, School Start Times, and Teen Motor Vehicle Crashes. *J.Clin. Sleep Med.* **2008**, 4(6), 533–535.

Datta, S.; MacLean, R. R. Neurobiological Mechanisms for the Regulation of Mammalian Sleep-Wake Behavior: Reinterpretation of Historical Evidence and Inclusion of Contemporary Cellular and Molecular Evidence. *Neurosci.Biobehav. Rev.* **2007**, 31, 775–824.

Dauvilliers, Y.; Montplaisir, J.; Molinari, N.; Carlander, B.; Ondze, B.; Besset, A.; Billiard, M. Age at Onset of Narcolepsy in Two Large Populations of Patients in France and Quebec. *Neurology* **2001**, 57, 2029–2033.

De Souza, L.; Benedito-Silva, A. A.; Pires, M. N.; Poyares, D.; Tufik, S.; Calil, H. M. Further Validation of Actigraphy for Sleep Studies. *Sleep* **2003**, 26(1), 81–85.

Demirci, J. R.; Braxter, B. J.; Chasens, E. R. Breastfeeding and Short Sleep Duration in Mothers and 6 to 11 Month Old Infants. *Infant Behav. Dev.* **2012**, 35(4), 884–886.

Deoni, S. C. L.; Dean III, D. C.; Piryatinsky, I.; O'Muircheartaigh, J.; Waskiewicz, N.; Lehman, K.; Han, M.; Dirks, H. Breastfeeding and Early White Matter Development: A Cross-Sectional Study. *NeuroImage* **2013**, 82, 77–86.

Doi, Y.; Ishihara, K.; Uchiyama, M. Sleep/Wake Patterns and Circadian Typology in Preschool Children Based on Standardized Parental Self-Reports. *Chronobiol. Int.* **2014**, 31(3), 328–336.

Drapeau, C.; Hamel-Hebert, I.; Robillard, R.; Selmaoui, B.; Filipini, D.; Carrier, J. Challenging Sleep in Aging: The Effects of 200 mg of Caffeine During the Evening in Young and Middle-Aged Moderate Caffeine Consumers. *J. Sleep Res.* **2006**, 15, 133–141. DOI: 10.1111/j.1365-2869.2006.00518.x

Ehlers, C. L.; Kupfer, D. J. Slow-Wave Sleep: Do Young Adult Men and Women Age Differently? *J. Sleep Res.* **1997**, 6, 211–215.

Emde, R. N.; Walker S. Longitudinal Study of Infant Sleep: Results of 14 Subjects Studied at Monthly Intervals. *Psychophysiology* **1976**, 13(5), 456–461.

Engler, A. C.; Hadash, A.; Shehadeh, N.; Pillar, G. Breastfeeding May Improve Nocturnal Sleep and Reduce Infantile Colic: Potential Role of Breast Milk Melatonin. *Eur. J. Pediatr.* **2012**, 171, 729–732.

Fagioli, I; Salzarulo, P. Sleep States Development in the First Year of Life Assessed Through 24-h Recordings. *Early Hum. Dev.* **1982**, 6(2), 215–228.

Feinberg, I.; Campbell, I. G. Sleep EEG Changes During Adolescence: An Index of a Fundamental Brain Reorganization. *Brain Cogn.* **2010**, 72(1), 56–65.

Feinberg, I.; Higgins, L. M.; Khaw, W. Y.; Campbell, I. G. The Adolescent Decline of NREM Delta, an Indicator of Brain Maturation, is Linked to Age and Sex but Not to Pubertal Stage. *Am. J. Physiol. Regul. Integr. Comp. Physiol.* **2006**, 291(6), r1724-r1729.

Ficca, G.; Fagioli, I.; Salzarulo, P. Sleep Organization in the First Year of Life: Developmental Trends in the Quiet Sleep-Paradoxical Sleep Cycle. *J. Sleep Res.* **2000**, 9(1), 1–4.

Fisher, A.; vanJaarsveld, C. H. M.; Llewellyn, C. H.; Wardle, J. Genetic and Environmental Influences on Infant Sleep. *Pediatrics* **2012**, 129, 1091–1096.

France, K. G. Behavior Characteristics and Security in Sleep-Disturbed Infants Treated with Extinction. *J. Pediatr. Psychol.* **1992**, 17, 467–475.

France, K. G.; Blampied, N. M.; Wilkinson, P. Treatment of Infant Sleep Disturbance by Trimeprazine in Combination with Extinction. *J. Dev. Behav. Pediatr.* **1991**, 12, 308–314.

France, K. G.; Hudson, S, M. Behavior Management of Infant Sleep Disturbance. *J. Appl. Behav. Anal.* **1990**, 23, 91–98.

Friedman, B. C.; Hendeles-Amitai, A.; Kozminsky, E.; Leiberman, A.; Friger, M.; Tarasiuk, A.; Tal, A. Adenotonsillectomy Improves Neurocognitive Function in Children with Obstructive Sleep Apnea Syndrome. *Sleep* **2003**, 26(8), 999–1005.

Galbally, M.; Lewis, A. J.; McEgan, K.; Scalzo, K.; Islam, F. A. Breastfeeding and Infant Sleep Patterns: An Australian Population Study. *J. Paediatr. Child Health* **2013**, 49, e147-e152.

Galland, B.; Meredith-Jones, K.; Terrill, P.; Taylor, R. Challenges and Emerging Technologies Within the Field of Pediatric Actigraphy. *Front. Psychiatry* **2014**, 5(99), 1–5.

Galland, B. C.; Taylor, B. J.; Elder, D. E.; Herbison, P. Normal Sleep Patterns in Infants and Children: A Systematic Review of Observational Studies. *Sleep Med. Rev.* **2012**, 16, 213–222.

Geva, R.; Yaron, H.; Kuint, J. Neonatal Sleep Predicts Attention Orienting and Distractibility. *J. Atten. Disord.* **2013**, 1–13. DOI: 10.1177/1087054713491493

Giedd, J. N.; Blumenthal, J.; Jeffries, N. O.; Castellanos, F. X.; Liu, H.; Zijdenbos, A.; Paus, T.; Evans, A. C.; Rapaport, J. L. Brain Development During Childhood and Adolescence: A Longitudinal MRI Study. *Nat. Neurosci.* **1999**, 2, 861–863.

Gnidovec, B.; Neubauer, D.; Zidar, J. Actigraphic Assessment of Sleep-Wake Rhythm During the First 6 Months of Life. *Clin. Neurophysiol.* **2002**, 113, 1815–1821.

Grigg-Damberger M.; Gozal, D.; Marcus, C. L.; Quan, S. F.; Rosen, C. L.; Chervin, R. D.; Wise, M.; Picchietti, D. L.; Sheldon, S. H.; Iber, C. The Visual Scoring of Sleep and Arousal in Infants and Children. *J. Clin. Sleep Med.* **2007**, 3(2), 201–240.

Grigg-Damberger, M.; Ralls, F. Cognitive Dysfunction and Obstructive Sleep Apnea: From Cradle to Tomb. *Curr. Opin. Pulm. Med.* **2012**, 18(6), 580–587.

Groome, L.; Swiber, M.; Atterbury, J.; Bentz, L. Similarities and Differences in Behavioral State Organization During Sleep Periods in the Perinatal Infant Before and After Birth. *Child Dev.* **1997**, 68(1), 1–11.

Guilleminault, C.; Lee, J. H.; Chan, A. Pediatric Obstructive Sleep Apnea Syndrome. *Arch. Pediatr. Adolesc. Med.* **2005**, 159, 775–785.

Harrison, Y. The Relationship Between Daytime Exposure to Light and Night-Time Sleep in 6–12 Week-Old Infants. *J. Sleep Res.* **2004**, 13(4), 345–352.

Hood, B. M.; Harbord, M. G. Narcolepsy: Diagnosis and Management in Early Childhood. *J. Pediatr. Neurol.* **2004**, 2(2), 65–71.

Hoppenbrouwers, T.; Hodgman, J. E.; Harper, R. M.; Sterman, M. B. Temporal Distribution of Sleep States, Somatic Activity, and Autonomic Activity During the First Half Year of Life. *Sleep* **1982**, 5(2), 131–144.

Houts, A. C.; Berman, J. S.; Abramson, H. Effectiveness of Psychological and Pharmacological Treatments for Nocturnal Enuresis. *J. Consult. Clin. Psychol.* **1994**, 62(4), 737–745. http://dx.doi.org/10.1037/0022-006X.62.4.737

Howell, M. J. Parasomnias: An Updated Review. *Neurotherapeutics* **2012**, 9(4), 753–775.

Iber, C.; Ancoli-Israel, S.; Chesson, A.; Quan, S. F. *AASM Manual for the Scoring of Sleep and Associated Events: Rules, Terminology, and Technical Specifications*; Westchester, IL: American Academy of Sleep Medicine, **2007**, 32–33.

Insana, S. P.; Gozal, D.; Montgomery-Downs, H. E. Invalidity of One Actigraphy Brand for Identifying Sleep and Wake Among Infants. *Sleep Med.* **2010**, 11, 191–196.

Iglowstein, I.; Jenni, O. G.; Molinari, L.; Largo, R. H. Sleep Duration from Infancy to Adolescence: Reference Values and General Trends. *Pediatrics* **2003**, 111, 302–307.

Issacs, E. B.; Fischl, B. R.; Quinn, B. T.; Chong, W. K.; Gadian, D. G.; Lucas, A. Impact of Breast Milk on Intelligent Quotient, Brain Size, and White Matter Development. *Pediatr. Res.* **2010**, 67(4), 357–362.

Jenni, O. G.; Deboer, T.; Achermann, P. Development of the 24-h Rest-Activity Pattern in Human Infants. *Infant Behav. Dev.* **2006**, 29, 143–152.

Jones, D. P. H.; Verduyn, C. M. Behavioural Management of Sleep Problems. *Arch. Dis. Child.* **1983**, 58, 442–444.

Kafouri, S.; Kramer, M.; Leonard, G.; Perron, M.; Pike, B.; Richer, L.; Toro, R.; Veillette, S.; Pausova, Z.; Paus, T. Breastfeeding and Brain Structure in Adolescence. *Int. J. Epidemiol.* **2012**, 42(1), 150–159.

Kataria, S.; Swanson, M. S.; Trevathon, G. E. Persistence of Sleep Disturbances in Preschool Children. *J. Pediatr.* **1987**, 110, 642–646.

Key, A.P.F.; Molfese, D. L.; O'Brien, L.; Gozal, D. Sleep-Disordered Breathing Affects Auditory Processing in 5–7 Year-Old Children: Evidence From Brain Recordings. *Dev.Neuropsychol.* **2009**, 34(5), 615–628.

Kohler, M. J.; Lushington, K.; van den Heuvel, C. J.; Martin, J.; Pamula, Y.; Kennedy, D. Adenotonsillectomy and Neurocognitive Deficits in Children with Sleep Disordered Breathing. *PLOS ONE* **2009**, 4(10), e7343. DOI: 10.1371/journal.pone.0007343

Kohyama, J. A Quantitative Assessment of the Maturation of Phasic Motor Inhibition During REM Sleep. *J. Neurol. Sci.* **1996**, 143 (1–2), 150–155.

Komada, Y.; Asakoa, S.; Abe, T.; Matsuura, N.; Kagimura, T.; Shirakawa, S.; Inoue, Y. Relationship Between Napping Pattern and Nocturnal Sleep among Japanese Nursery School Children. *Sleep Med.* **2012**, 13, 107–110.

Kotagal, S. Parasomnias in Childhood. *Sleep Med. Rev.* **2009**, 13(2), 157–168. DOI:10.1016/j.smrv.2008.09.005

Kramer, M. S.; Aboud, F.; Mironova, E.; Vanilovich, I.; Platt, R. W.; Matush, L.; Igumnov, S.; Fombonne, E.; Bogdanovich, N.; Ducret, T.; Collet, J. P.; Chalmers, B.; Hodnett, E.; Davidovsky, S.;

Skugarevsky, O.; Trofimovich, O.; Kozlova, L.; Shapiro, S. Breastfeeding and Child Cognitive Development; New Evidence from a Large Randomized Trial. *Arch. Gen. Psychiatry* **2008,** 65(5), 578–584.

Kuhn, B.R.; Floress, M.T. Nonpharmacological Interventions for Sleep Disorders in Children. In *Sleep and Psychiatric Disorders in Children and Adolescents;* Ivanenko, A., Ed.; Informa Healthcare: New York, 2008; Chap. 19; pp 269.

Kurdziel, L.; Duclos, K.; Spencer, R. M. C. Sleep Spindles in Midday Naps Enhance Learning in Preschool Children. *Proc. Natl. Acad. Sci. U.S.A.* **2013,** 110 (43), 17267–17272.

Kurth, S.; Achermann, P.; Rusterholz, T.; LeBourgeois, M. K. Development of Brain EEG Connectivity Across Early Childhood: Does Sleep Play a Role? *Brain Sci.* **2013,** 3(4), 1445–1460.

LeBourgeois, M. K.; Wright, K. P. Jr.; LeBourgeois, H. B.; Jenni, O. G. Dissonance Between Parent-Selected Bedtimes and Young Children's Circadian Physiology Influences Nighttime Settling Difficulties. *Mind Brain Educ.* **2013a,** 7 (4), 234–242.

LeBourgeois, M. K.; Carskadon, M. A.; Akacem, L. D.; Simpkin, C. T.; Wright, K. P. Jr.; Achermann, P.; Jenni, O. G. Circadian Phase and its Relationship to Nighttime Sleep in Toddlers. *Mind Brain Educ.* **2013b,** 7 (4), 234–242.

Luginbuehl, M.; Bradley-Klug, K. L.; Ferron, J.; Anderson, W. M.; Benbadis, S. R. Pediatric Sleep Disorders: Validation of the Sleep Disorders Inventory for Students. *School Psychol. Rev.* **2008,** 37(3), 409–431.

Lumeng, J. C.; Chervin, R. D. Epidemiology of Pediatric Obstructive Sleep Apnea. *Proc. Am. Thorac. Soc.* **2008,** 5(2), 242–252.

Marcus, C. L.; Moore, R. H.; Rosen, C. L.; Giordani, B.; Garetz, S. L.; Taylor, H. G.; Mitchell, R. B.; Amin, R.; Katz, E. S.; Arens, R.; Paruthi, S.;Muzumdar, H.; Gozal, D.; Thomas, N. H.; Ware, J.; Beebe, D.; Snyder, K.; Elden, L.; Sprecher, R. C.; Willging, P.; Jones, D.; Bent, J. P.; Hoban, T.; Chervin, R. D.; Ellenberg, S. S.; Redline, S. A Randomized Trial of Adenotonsillectomy for Childhood Sleep Apnea. *N. Engl. J. Med.* **2013,** 368(25), 2366–2376.

Marino, M.; Li, Y.; Rueschman, M. N.; Winkelman, J. W.; Ellenbogen, J. M.; Solet, J. M.; Dulin, H.; Berkman, L. F.; Buxton, O. M. Measuring Sleep: Accuracy, Sensitivity, and Specificity of Wrist Actigraphy Compared to Polysomnography. *Sleep* **2013,** 36(11), 1747–1755.

Marks, G. A.; Shaffery, J. P.; Oksenberg, A.; Speciale, S. G.; Roffwarg, H. P. A Functional Role for REM Sleep in Brain Maturation. *Behav. Brain Res.* **1995,** 69, 1–11.

Mason, T. B. A.; Pack, A. I. Pediatric Parasomnias. *Sleep* **2007,** 30(2), 141–151.

McCrory, C.; Murray, A.The Effect of Breastfeeding on Neuro-Development in Infancy. *Matern. Child Health J.* **2013,** 17, 1680–1688.

Meijer, A. M.; van den Wittenboer, G. L. H. Contribution of Infants' Sleep and Crying to Marital Relationship of First-Time Parent Couples in the 1st Year After Childbirth. *J. Fam. Psychol.* **2007,** 21(1), 49–57.

Meltzer, L. J. Clinical Management of Behavioral Insomnia of Childhood: Treatment of Bedtime Problems and Night Wakings in Young Children. *Behav. Sleep Med.* **2010,** 8(3), 172–189.

Meltzer, L. J.; Mindell, J. A. Relationship Between Child Sleep Disturbances and Maternal Sleep, Mood, and Parenting Stress: A Pilot Study. *J. Fam. Psychol.* **2007,** 21(1), 67–73.

Meltzer, L.; Montgomery-Downs, H. E.; Walsh, C.; Insana, S. P. Use of Actigraphy in Pediatric Sleep Research; Review. *Sleep Med. Rev.* **2012,** 16(5), 463–475.

Miller, A. L.; Seifer, R.; Crossin, R.; LeBourgeois, M. K. Toddler's Self-Regulation Strategies in a Challenge Context are Nap-Dependent. *J. Sleep Res.* **2014.** DOI:10.1111/jsr.12260

Millman, R. P.; Working Group on Sleepiness in Adolescents/Young Adults; AAP Committee on Adolescents. Excessive Sleepiness in Adolescents and Young Adults: Causes, Consequences, and Treatment Strategies. *Pediatrics* **2005,** 115(6), 1774–1786.

Mindell, J. Empirically Supported Treatments in Pediatric Psychology: Bedtime Refusal and Night Wakings in Young Children. *J. Pediatr. Psychol.* **1999,** 24(6), 465–481.

Mindell, J. A.; Du Mond, C.; Tanenbaum, J. B.; Gunn, E. Long-Term Relationship Between Breastfeeding and Sleep. *Child. Health Care* **2012,** 41, 190–203.

Mindell, J. A.; Kuhn, B; Lewin, D. S.; Meltzer, L. J.; Sadeh, A. Behavioral Treatment of Bedtime Problems and Night Wakings in Infants and Young Children. *Sleep* **2006,** 29(10), 1263–1276.

Mindell, J. A.; Meltzer, L. J. Behavioural Sleep Disorders in Children and Adolescents. *Ann. Acad. Med. Singap.* **2008,** 37(8), 722–728.

Mindell, J. A.; Telofski, L. S.; Wiegand, B.; Kurtz, E. S. A Nightly Bedtime Routine: Impact on Sleep in

Young Children and Maternal Mood. *Sleep* **2009a**, 32(5), 599–606.

Mindell, J. A.; Meltzer, L. J.; Carskadon, M. A.; Chervind, R. D. Developmental Aspects of Sleep Hygiene: Findings from the 2004 National Sleep Foundation Sleep in America Poll. *Sleep Med.* **2009b**, 10(7), 771–779.

Mirmiran, M.; Maas, Y. G. H.; Ariagno, R. L. Development of Fetal and Neonatal Sleep and Circadian Rhythms. *Sleep Med. Rev.* **2003**, 7(4), 321–334.

Mitchell, J. A.; Rodriguez, D.; Schmitz, K. H.; Audrain-McGovern, J. Sleep Duration and Adolescent Obesity. *Pediatr.* **2013**, 131(5), e1428-e1434.

Modi, R. R.; Camacho, M.; Valerio, J. Confusional Arousals, Sleep Terrors, and Sleepwalking. *Sleep Med. Clin.* **2014**, 9(4), 537–551.

Montgomery-Downs, H. E.; Crabtree, V. M.; Capdevila O. S.; Gozal D. Infant-Feeding Methods and Childhood Sleep-Disordered Breathing. *Pediatr.* **2007**, 120, 1030–1035.

Montgomery-Downs, H. E.; Gozal, D. Sleep Habits and Risk Factors for Sleep-Disordered Breathing in Infants and Young Toddlers in Louisville, Kentucky. *Sleep Med.* **2006**, 7, 211–219.

Montgomery-Downs, H. E.; Jones, V. F.; Molfese, V. J.; Gozal, D. Snoring in Preschoolers: Associations with Sleepiness, Ethnicity, and Learning. *Clin. Pediatr.* **2003**, 42, 719–725.

Montgomery-Downs, H. E.; Meltzer, L. J. Actigraphy. In *Sleep Disordered Breathing in Children: A Comprehensive Guide to Clinical Evaluation and Treatment*; Kheirandish-Gozal, L.; Gozal, D., Ed.; Springer Science and Business Media: New York, 2012; pp 177–185.

Morgenthaler, T.; Alessi, C.; Friedman, L.; Owens, J.; Kapur, V.; Boehlecke, B.; Brown, T.; Chesson, A.; Coleman, J.; Lee-Chiong, T.; Pancer, J.; Swick, T. J.; Standards of Practice Committee; American Academy of Sleep Medicine. Practice Parameters for the Use of Actigraphy in the Assessment of Sleep and Sleep Disorders: An Update for 2007. *Sleep* **2007**, 519–529.

Morgenthaler, T. I.; Owens, J.; Alessi, C.; Boehlecke, B.; Brown, T. M.; Coleman, J.; Friedman, L.; Kapur, V. K.; Lee-Chiong, T.; Pancer, J.; Swick, T. J. Practice Parameters for Behavioral Treatment of Bedtime Problems and Night Wakings in Infants and Young Children. *Sleep* **2006**, 29(10), 1277–1281.

Mortensen, E. L.; Michaelsen, K. F.; Sanders, S. A.; Reinisch, J. M. The Association Between Duration of Breastfeeding and Adult Intelligence. *JAMA* **2002**, 287(18), 2365–2371.

National Institutes of Health. Nearly 55 Percent of U.S. Infants Sleep with Potentially Unsafe Bedding. http://www.nichd.nih.gov/news/releases/Pages/120114-SIDS-bedding.aspx. (accessed Apr 3, 2015a).

National Institutes of Health. Babies Need Tummy Time! http://www.nichd.nih.gov/sts/about/Pages/tummytime.aspx. (accessed Apr 8, 2015b).

National Sleep Foundation. Children and Sleep. http://sleepfoundation.org/sleep-topics/children-and-sleep?page=0%2C2 (accessed Apr 3, 2015a).

National Sleep Foundation. Healthy Sleep Tips. http://sleepfoundation.org/sleep-tools-tips/healthy-sleep-tips (accessed Apr 3, 2015).

National Sleep Foundation. Teens and Sleep. http://sleepfoundation.org/sleep-topics/teens-and-sleep (accessed Apr 3, 2015).

O'Brien, E. M.; Mindell, J. A. Sleep and Risk-Taking Behavior in Adolescents. *Behav. Sleep Med.* **2005**, 3(3), 113–133.

O'Connor, T. G.; Caprariello, P.; Blackmore, E. R.; Gregory, A. M.; Glover, V.; Fleming, P.; ALSPAC Study Team. Prenatal Mood Disturbance Predicts Sleep Problems in Infancy and Toddlerhood. *Early Hum. Dev.* **2007**, 83, 451–458.

Owens, J.; Adolescent Sleep Working Group; Committee on Adolescence. Insufficient Sleep in Adolescents and Young Adults: An Update on Causes and Consequences. *Pediatrics* **2014**, 134 (3), e921-e932. DOI: 10.1542/peds.2014-1696

Owens, J. A.; Mindell, J. A. Pediatric Insomnia. *Pediatr. Clin. North Am.* **2011**, 58, 555–569.

Paavonen, E. J.; Fjallberg, M.; Steenari, M. R.; Aronen F. T. Actigraph Placement and Sleep Estimation in Children. *Sleep* **2002**, 25, 235–237.

Pack, A. I.; Pien, G. W. Update on Sleep and Its Disorders. *Annu. Rev. Med.* **2011**, 62, 447–460.

Peirano, P.; Algarin, C.; Uauy, C. Sleep-Wake States and their Regulatory Mechanisms Throughout Early Human Development. *J. Pediatr.* **2003**, 143 (4), 70–79.

Perfect, M. M.; Archbold, K.; Goodwin, J. L.; Levine-Donnerstein, D.; Quan, S. F. Risk of Behavioral and Adaptive Functioning Difficulties in Youth with Previous and Current Sleep Disordered Breathing. *Sleep* **2013**, 36(4), 517–525.

Petit D.; Touchette, E.; Tremblay, R. E.; Boivin, M.; Montplaisir, J. Dyssomnias and Parasomnias in

Early Childhood. *Pediatrics* **2007**, 119 (5) e1016–1025.

Piantoni, G.; Poli, S. S.; Linkenkaer-Hansen, K.; Verweij, I. M.; Ramautar, J. R.; Van Someren, E. J.; Van Der Werf, Y. D. Individual Differences in White Matter Diffusion Affect Sleep Oscillations. *J. Neurosci.* **2013**, 33(1), 227–233.

Piteo, A. M.; Lushington, K.; Roberts, R. M.; Martin, A. J.; Nettelbeck, T.; Kohler, M. J.; Kennedy, J. D. Parental-Reported Snoring from the First Month of Life and Cognitive Development at 12 Months of Age. *Sleep Med.* **2011**, 12(10), 975–980.

Poli, F.; Pizza, F.; Mignot, E.; Ferri, R.; Pagotto, U.; Teheri, S.; Finotti, E.; Bernardi, F.; Pirazzoli, P.; Cicognani, A.; Balsamo, A.; Nobili, L.; Bruni, O.; Plazzi, G. High Prevalence of Precocious Puberty and Obesity in Childhood Narcolepsy with Cataplexy. *Sleep* **2013**, 36(2), 175–181.

Prechtl, H. F. R. The Behavioural States of the Newborn Infant. *Brain Res.* **1964**, 76(2), 185–212.

Pritchard, A. A.; Appleton, P. Management of Sleep Problems in Pre-School Children: Effects of a Behavioural Programme on Sleep Routines, Maternal Depression and Perceived Control. *Early Child Dev. Care* **1988**, 34(1), 227–240.

Queensland Department of Health Guideline: Polysomnography Set-up (Paediatric Patients); Queensland Government: Queensland, AU, 2013. http://www.google.com/url?sa=t&rct=j&q=&esrc=s&source=web&cd=1&ved=0CB8QFjAA&url=http%3A%2F%2Fwww.health.qld.gov.au%2Fqhpolicy%2Fdocs%2Fgdl%2Fqh-gdl-405.pdf&ei=7TArVa-ABIqmgwTjzoGoDg&usg=AFQjCNGfdNnXJ2Tc2W8l4yF3gW1LHSQi3g&bvm=bv.90491159,d.eXY (Accessed April 10, 2015).

Ramamurthy, M. B.; Sekartini, R.; Ruangdaraganon, N.; Huynh, D. H. T.; Sadeh, A.; Mindell, J. A. Effect of Current Breastfeeding on Sleep Patterns in Infants from Asia-Pacific Region. *J. Paediatr. Child Health* **2012**, 48, 669–674.

Rao, S. C.; Mansukhani, M. P.; Lloyd, R. M.; Slocumb, N. L.; Kotagal, S. Nocturnal Polysomnographic Characteristics of Pediatric Narcolepsy. *Sleep Biol. Rhythms* **2012**, 10, 69–71.

Robison, J. G.; Wilson, C.; Otteson, T. D.; Chakravorty, S. S.; Mehta, D. K. Analysis of Outcomes in Treatment of Obstructive Sleep Apnea in Infants. *Laryngoscope* **2013**, 123, 2306–2314.

Roffwarg, H. P.; Muzio, J. N.; Dement, W. C. Ontogenetic Development of the Human Sleep-Dream Cycle. *Science* **1966**, 152, 605–619.

Rosen, G. M.; Mahowald, M. W. Disorders of Arousal. In *Principles and Practice of Pediatric Sleep Medicine, 2nd Ed;* Sheldon S. H. et al., Eds.; Elsevier: London, 2014, pp. 293–304.

Sadeh, A. Activity-Based Assessment of Sleep-Wake Patterns During the 1st Year of Life. *Infant Behav. Dev.* **1995**, 18(3), 329–337.

Sankupellay, M.; Wilson, S.; Heussler, H. S.; Parsley, S.; Yuill, M.; Dakin, C. Characteristics of Sleep EEG Power Spectra in Healthy Infants in the First Two Years of Life. *Clin.Neurophysiol.* **2011**, 122, 236–243.

Sazonov, E.; Sazonova, N.; Schuckers, S.; Neuman, M.; CHIME Study Group.Activity-Based Sleep-Wake Identification in Infants. *Physiol. Meas.* **2004**, 25(5), 1291.

Scher, M. S.; Johnson, M. W.; Holditch-Davis, D. Cyclicity of Neonatal Sleep Behaviors at 25 to 30 Weeks' Postconceptional Age. *Pediatr. Res.* **2005**, 57(6), 879–882.

Scraggs, T. L. EEG Maturation: Viability Through Adolescence. *Neurodiagn. J.* **2012**, 52, 176–203.

Sheldon, S. H. Evaluating Sleep in Infants and Children; Lippincott-Raven: Philadelphia, 1996.

Sitnick, S. L.; Goodlin-Jones, B. L.; Anders, T. F. The Use of Actigraphy to Study Sleep Disorders in Preschoolers: Some Concerns about Detection of Nighttime Awakenings. *Sleep* **2008**, 31(3) 395.

Southall D. P.; Plunkett, M. C., Banks, M. W.; Falkov, A. F.; Samuels, M. P. Covert Video Recordings of Life-Threatening Child Abuse: Lessons for Child Protection. *Pediatrics* **1997**, 100, 735–60.

Spruyt, K.; Gozal, D. Development of Pediatric Sleep Questionnaires as Diagnostic or Epidemiological Tools: A Brief Review of Do's and Don'ts. *Sleep Med. Rev.* **2011a**, 15, 7–17.

Spruyt, K.; Gozal, D. Pediatric Sleep Questionnaires as Diagnostic or Epidemiological Tools: A Review of Currently Available Instruments. *Sleep Med. Rev.* **2011b**, 15, 19–32.

Stallones, L.; Beseler, C.; Chen, P. Sleep Patterns and Risk of Injury among Adolescent Farm Residents. *Am. J. Prev. Med.* **2006**, 30(4), 300–304.

Stanley, K. SIDS deaths in Florida have plummeted: Why? *Tampa Bay Times*, Feb. 6, 2015. http://www.tampabay.com/news/health/sids-deaths-in-florida-have-plummeted-why/2216719 (accessed Apr. 8, 2015).

Staton, S. L.; Smith, S. S.; Thorpe, K. J. "Do I Really Need a Nap?": The Role of Sleep Science in Informing Sleep Practices in Early Childhood Education and Care Settings. *Trans. Issues. Psychol. Sci.* **2015,** 1(1), 32–44.

Takahashi, Y.; Kipnis, D. M.; Daughaday, W. H. Growth Hormone Secretion During Sleep. *J. Clin. Invest.* **1968,** 47(9), 2079–2090.

Teng, A.; Bartle, A.; Sadeh, A.; Mindell, J. Infant and Toddler Sleep in Australia and New Zealand. *J. Paediatr. Child Health* **2012,** 268–273.

Thiedke, C. C. Sleep Disorders and Sleep Problems in Childhood. *Am. Fam. Phys.* **2001,** 63(2), 277–287.

Thorpy, M. J.; Korman, E.; Spielman, A. J.; Glovinsky, P. B. Delayed Sleep Phase Syndrome in Adolescents. *J. Adolesc. Health Care* **1988,** 9(1), 22–27.

Tikotzky, L.; Shaashua, L. Infant Sleep and Early Parental Sleep-Related Cognitions Predict Sleep in Pre-School Children. *Sleep Med.* **2012,** 13, 185–192.

Thoman, E. B. Sleep and Wake Behaviors in Neonates: Consistencies and Consequences. *Merrill Palmer Q.,* **1975,** 21 (4), 295–314.

Weinraub, M.; Bender, R. H.; Friedman, S. L.; Susman, E. J.; Knoke, B.; Bradley, R.; Houts, R.;

Williams, J. Patterns of Developmental Change in Infants' Nighttime Sleep Awakenings from 6 Through 36 Months of Age. *Dev. Psychol.* **2012,** 48(6), 1511–1528.

Werner, H.; Molinari, L.; Guyer, C.; Jenni, O. G. Agreement Rates Between Actigraphy, Diary, and Questionnaire for Children's Sleep Patterns. *Arch. Pediatr. Adolesc. Med.* **2008,** 162(4), 350–358.

Williams, J. A.; Zimmerman, F. J.; Bell, J. F. Norms and Trends of Sleep Time Among US Children and Adolescents. *JAMA Pediatr.* **2013,** 167(1), 55–60.

Wolfson, A. R.; Carskadon, M. A. Understanding Adolescents' Sleep Patterns and School Performance: A Critical Appraisal. *Sleep Med. Rev.* **2003,** 7(6), 491–506.

Wong, M. W.; Brower, K. J.; Zucker, R. A. Sleep Problems, Suicidal Ideation, and Self-Harm Behaviors in Adolescents. *J. Psychiatr. Res.* **2011,** 45(4), 505–511.

24

Sleep in the Elderly: Normal and Abnormal

Adam B. Hernandez and Steven H. Feinsilver

ABSTRACT

Sleep complaints increase with aging, to the degree that it is often difficult to know what normal sleep is in the elderly. In most respects, sleep quality appears to decline, with increased awakenings and reduced slow wave sleep. Circadian rhythm changes may occur with a tendency toward advanced sleep phase. Many medical illnesses are increasingly common with aging and lead to worsened sleep. For all of these reasons, insomnia is more common and often more challenging in older patients. REM behavior disorder is a disease seen largely in the elderly, may be more common than

Adam B. Hernandez
Department of Medicine, Pulmonary, Critical Care and Sleep Medicine, 17 East 102 Street Floor 6th Floor Room West Tower, New York, NY 10029, USA

Steven H. Feinsilver
Icahn School of Medicine at Mount Sinai, Division of Pulmonary, Critical Care and Sleep Medicine, 1 Gustave Levy Place, New York, NY 10029, USA

Corresponding author: Steven H. Feinsilver, E-mail: steven.feinsilver@mssm.edu

generally recognized, and in some is the earliest manifestation of progressive neurologic disease.

The fastest growing segment of the world's population is the elderly, defined as people aged 65 or above. In the United States, this population is expected to increase from about 48 million in 2015 to about 79 million in 2035.[1] Sleep problems are even more common in the elderly than in the general population. The elderly have more complaints of daytime sleepiness and more complaints of difficulty initiating and maintaining sleep. Some specific problems such as REM sleep behavior disorder are nearly unique to the elderly. This chapter will review normal sleep in the healthy elderly, and specific problems in the elderly: advanced sleep phase disorder, REM behavior disorder, insomnia, and sleep disordered breathing.

24.1. NORMAL SLEEP IN THE HEALTHY ELDERLY

Defining normal sleep in the elderly is problematic. Changes occur in both sleep timing and quality. Two popular assumptions are that older people need less sleep and are more likely to be sleepy during the day. Neither may be true, at least in the healthy elderly. It does appear to be true that with aging subjects get less sleep.

An expected feature of geriatric sleep is a decline in sleep efficiency: the ratio of time asleep to time in bed. This is mostly because of multiple nocturnal awakenings, which are a frequent complaint. An increase in time needed to fall asleep (sleep latency) is a less common chief complaint. In addition to changes with normal aging, disease and medication may cause decreased sleep efficiency. As examples, any disease causing chronic pain is likely to cause multiple awakenings, and diuretics may cause nocturnal wakes.

Age-related changes in polysomno-graphic parameters have been somewhat inconsistent in the literature. Most studies have shown reductions in total sleep time, sleep efficiency and slow wave sleep, with an increase in wake after sleep onset. Changes in REM sleep and sleep latency have been less consistent. The sleep heart health study (2004) provided sleep architecture values for 2685 participants between age 37 and 92. Patients with excessive alcohol intake, psychotropic medication use, restless leg syndrome symptoms, or systemic pain were excluded.[2] In this study there was a notable decline in slow wave sleep with aging in men, but this was not seen in women. REM sleep declined modestly in both men and women.

In contrast, a meta-analysis done by Ohayon et al. of 65 studies showed a decline in total sleep time, sleep efficiency, slow wave sleep, and the percentage of REM sleep from young adulthood to elderly subjects in both genders.[3] Wake after sleep onset was increased and the percentage of stage I and stage to sleep was increased. Interestingly, sleep latency was not changed.

A recent study in Brazil sampled 1024 random individuals without significant "mental and physical disturbances" from age 20–80.[4] There was a fairly linear increase in wake after sleep onset with aging. Smaller effect sizes were seen for increases in sleep latency, REM latency, and percentages of stages 1 and 2. Total sleep time, sleep efficiency, and slow wave sleep are reduced with aging. The effects were not gender specific.

The reduction in slow wave sleep seen with aging is mostly seen from adulthood to middle age. This may be related to a decrease in amplitude of the EEG signal because of increasing electrical resistance of the skull and scalp with aging. If the requirement for a 75 μv amplitude for Delta waves is eliminated, slow wave sleep may appear more normal in elderly subjects.[5] However, a parallel decrease in growth hormone secretion in men is seen, suggesting this does have physiologic significance (Figure 24.1).[6]

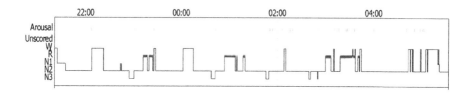

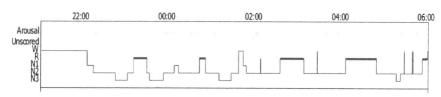

Figure 24.1. Hypnograms from normal subjects. Top healthy 85-year-old. Note early sleep onset and early morning wakefulness suggesting advanced sleep phase, several wakes, and less slow wave sleep compared with bottom hypnogram from healthy 33-year-old.

24.1.1. Are Normal Elderly Subjects Sleepier?

The significance of changes in sleep in the normal elderly might best be assessed by measuring daytime functioning and daytime sleepiness. The best validated measure of daytime sleepiness is the multiple sleep latency test (MSLT), although this has been criticized as testing the ability to fall asleep rather than the ability to maintain wakefulness. Earlier studies suggested that older people were sleepier.[7] However, in a study of 110 healthy subjects, there was an age-related reduction in daytime sleep propensity on MSLT, despite an increase in wake after sleep onset, reduction in slow wave sleep and sleep continuity in the elderly.[8] This study did not address sleep propensity during evening hours when it is possible that older adults are sleepier than younger adults. Another study of 26 young versus 11 elderly subjects showed that healthy elderly subjects were less sleepy and did better after sleep deprivation than younger subjects.[9]

In the cardiovascular health study, with 4578 adults older than 65, 20% of participants reported being "usually sleepy in the daytime." However, by design this was not a selected population of elderly subjects; comorbidities were allowed, and sleepiness was more common in those with depression, loud snoring, heart failure, and sedentary lifestyle, among other factors.[10]

Circadian rhythms also change with aging, generally becoming weaker, and more desynchronized, and losing amplitude.[11] There is deterioration of the suprachiasmatic nucleus, with decreased response to external time cues and decreased melatonin secretion. At the same time exposure to external time cues, particularly bright light, may be reduced in the elderly, especially in those who are institutionalized. Circadian phase advance is common, leading to earlier sleep onset and offset. Social pressure to maintain sleep times more typical of the general population may cause many elderly patients to go to bed later than their optimal circadian time leading to sleep loss.

Many patients will compensate for decreased sleep at night by napping during the day. It is unclear whether this is beneficial. In one study of 455 70-year-old subjects followed for 12 years survival was significantly reduced in those who reported napping. In fact napping appeared to be a significant independent predictor of mortality.[12] However, napping may improve cognitive performance in the elderly. In a study of 33 healthy subjects between 55 and 85 years of age, the opportunity to nap in the midafternoon improved cognitive performance with little effect on subsequent nighttime sleep quality or duration.[13] It therefore remains unclear whether elderly patients complaining of insufficient

nocturnal sleep should be encouraged to nap, or discouraged from napping.

Duffy et al. compared 26 younger (mean age 21.9) to 11 older (mean age 68.2) subjects in their response to 26 h of enforced wakefulness.[14] Older subjects exhibited less impairment and fewer unintentional sleep episodes. This suggests that excessive sleepiness is not the norm in healthy elderly subjects. As in the rest of the population, those with excessive daytime somnolence need to be evaluated for sleep deprivation, mood disorders, medical illness, or primary sleep disorders.

A summary of changes in sleep in healthy elderly individuals is shown in Table 24.1.

24.2. ADVANCED SLEEP–WAKE PHASE DISORDER

Advanced sleep–wake phase disorder (ASWPD) is a circadian rhythm disorder characterized by early habitual sleep and wake times and associated daytime sleepiness or early morning awakening when the habitual schedule is not followed. The

Table 24.1. Sleep in healthy elderly.

Characteristics of normal sleep in the healthy elderly	
Parameter	Characteristic
Total sleep time (TST)	Reduced
Sleep efficiency (SE)	Reduced
Wake after sleep onset (WASO)	Increased
REM sleep	Slightly reduced
Sleep latency	Increased
REM latency	Probably reduced
Slow wave sleep (SWS)	Reduced
Sleep latency (MSLT)	Possibly increased

terms advanced sleep phase syndrome and advanced sleep phase type are applied to individuals with a phase advanced sleep schedule but without any accompanying symptoms. ASWPD likely has higher prevalence in the elderly population.

24.2.1. Clinical Characteristics

The habitual bedtime in ASWPD patients is typically between 6 and 9 p.m. and the habitual wake time is usually between 2 and 5 a.m. Attempts to either delay bedtime or stay in bed in an effort to delay wake time may result in sleep restriction, sleepiness and insomnia (early morning awakening). The delay in bedtime is often sought for social reasons, sometimes in an effort to synchronize with the sleep–wake schedule of a spouse or partner. Sleepiness is usually maximal in the late afternoon and early evening but may be present to some extent throughout the day, especially if significant sleep restriction is present.[15]

24.2.2. Epidemiology

The prevalence of ASWPD is unknown but it is likely rare, perhaps because affected individuals may willingly adopt their habitual sleep schedule and their circadian phase is less likely to interfere with conventional work times.[16,17] In addition, elderly patients may be retired or partially retired and more able to conform to their habitual sleep times. Prevalence increases with age and the diagnosis may be more common in males.[15, 17, 18]

24.2.3. Pathophysiology

The etiology of ASWPD is not well defined; however, there are several possible mechanisms of note. A shortened intrinsic circadian period (tau) has been documented in case reports.[19,20] Familial clustering with an autosomal dominant pattern has been reported.[20,21] In addition, within one family a genetic polymorphism was identified in the clock gene, human Period2 (hPer2).[22]

24.2.4. Diagnosis

The diagnosis of ASWPD is most often made clinically. The primary characteristics are a reported difficulty staying awake until the desired or socially acceptable bed time and an inability to remain asleep until the desired or socially acceptable wake time (ICSD-3).[23] If patients do sleep in phase with their habitual sleep–wake times they will have normal sleep duration and quality and no symptoms of sleepiness or insomnia. This schedule should be confirmed by sleep logs or actigraphy, and their symptoms should not be better explained by an alternative diagnosis.[23] Importantly, early morning awakening is a common symptom of depression in the elderly, so this diagnosis should be ruled out with care.[24] The use of a morningness–eveningness questionnaire may also be helpful in establishing the diagnosis.[25,26]

Core body temperature and dim light melatonin onset (DLMO) are advanced by several hours compared to normal in ASWPD patients.[20, 21, 27–30] These parameters are most useful in the context of clinical and research studies, and are not generally available in clinical practice (Table 24.2).

Table 24.2. ICSD-3 criteria for advanced sleep–wake phase disorder.

Diagnostic criteria for advanced sleep–wake phase disorder

- Difficulty remaining awake until a desired conventional bed time or inability to remain asleep until a desired conventional wake time
- Normal sleep quality and duration when sleeping in phase with advanced sleep schedule
- Sleep log or actigraphy documenting advanced sleep schedule
- Symptoms not better explained by another disorder or substance

24.2.5. Treatment

Data on treatment options for ASWPD is limited likely due to the rarity of the disorder in clinical practice. As with other circadian disorders, the optimal treatment involves overlapping the patient's circadian phase with their socially preferred sleep–wake schedule. Thus, patients may choose to simply "comply" with their advanced habitual bed and wake times; though if this were a palatable option for the patient they would likely adopt this schedule and not be seeking medical advice. Evening bright light therapy (between 7 and 9 p.m.) has been shown to delay circadian phase and reduce awakenings in multiple studies including one of elderly subjects, although positive results have not been universal.[27, 29–32] Evening bright light therapy is supported by American Academy of Sleep Medicine practice parameters.[26] Other potential options include chronotherapy (progressive advancement of bed time) and morning melatonin although neither is a proven therapy.[33,34] In addition, daytime administration of melatonin may result in sedation which is of particular concern in the elderly.

24.3. RAPID EYE MOVEMENT SLEEP BEHAVIOR DISORDER

Rapid eye movement (REM) sleep behavior disorder (RBD) is a type of REM parasomnia characterized by the loss of normal REM sleep atonia. The clinical hallmark of RBD is dream-enactment behavior (DEB) and the polysomnographic correlate is REM sleep without atonia (RSWA). This disorder has an increased prevalence in elderly subjects and may coexist with or portend diagnosis of Parkinson's disease (PD) or other neurodegenerative disorders.[35]

24.3.1. Clinical Characteristics

Patients often present with concerns over DEB which may range in manifestation from brief jerking movements and vocalizations to violent thrashing, hitting, kicking, yelling, and/or falling out of bed.[36] Episodes may occur up to several times per week and are more frequent during the latter half of the sleep period. Patients infrequently sleep–walk or leave the bedroom and these activities are more likely to be related to a non-REM (NREM) parasomnia. DEB may result in injury to the patient (most often due to falling out of bed) or bed partner. Vocalizations may loud, emotional, and contain language and expletives which may be uncharacteristic of the patient during wakefulness.[37] If patients awaken they often remember the content of their dreams and the bed partner may be able to correlate this with the specific DEB. Importantly, patients are alert and do not have sustained confusion upon awakening, which may aid in distinguishing the disorder from a NREM parasomnia. Symptoms of RBD usually manifest in late adulthood and the diagnosis is often delayed for several years.[35]

RBD may manifest in patients with a known neurodegenerative disorders or may pre-date symptoms and diagnosis by months to decades. Among patients with idiopathic RBD, approximately 40–75% will develop PD, multiple system atrophy (MSA), or dementia with Lewy bodies (DLB) in 10 years, and up to 90% at 14 years.[38,39] The mean time to development of symptoms was 7.5 years in one study.[39] In addition, patients may demonstrate subtle or subclinical signs of impaired motor or cognitive function.[36] Age, family history of dementia, and autonomic and motor symptoms may predict a higher risk of development of an alpha-synucleinopathy (PD, MSA, DLB).[40]

24.3.2. Epidemiology

RBD is uncommon (overall prevalence 0.5%) but has increased prevalence in the elderly (possibly as high as 2%) with a strong male predominance.[35,41–43] RBD is very common in patients with movement disorders with prevalence of approximately 50% in PD, 70–90% in MSA, and 80% in DLB.[44–46] RBD may also occur less

commonly in variety of other neurologic disorders including Alzheimer disease, amyotrophic lateral sclerosis, progressive supranuclear palsy, and Huntington disease.[36,47]

24.3.3 Pathophysiology

Normal REM sleep is characterized by atonia and skeletal muscle paralysis precluding DEB. Loss of REM atonia appears to be related to dysfunction in the REM-on and REM-off nuclei of the pons.[36] RBD may occur in a variety of clinical contexts including neurodegenerative diseases, narcolepsy, pontine lesions, and medications.[48–50]

In the case of neurodegenerative diseases, specifically the alpha-synucleinopathies, RBD symptoms manifest when the degenerative process involves the pontine nuclei which control REM sleep.[36] It is thought that the majority if not all cases of idiopathic RBD may be an early manifestation of preclinical PD, MSA, or DLB.[36]

Medication-associated (toxic) RBD is most commonly associated with antidepressants (selective serotonin reuptake inhibitors, tricyclic antidepressant, and monoamine oxidase inhibitors) and may occur in up to 6% of patients using them.[35,36] Of note, medication-associated RBD may predict future development and an alpha-synucleinopathy although not as strongly as idiopathic RBD.[51] In some cases antidepressant use may result in RSWA without symptoms of RBD (i.e., DEB).[52]

24.3.4. Diagnosis

Diagnosis of RBD requires DEB, RSWA, and that the symptoms are not better explained by another disorder or seizures (ICSD-3).[53] DEB is most often documented during the patient's history but may also be observed during polysomnography (PSG) with video. Importantly, obstructive sleep apnea (OSA), which is common in the elderly, can mimic RBD (pseudo-RBD) and should be ruled out with PSG.[54]

RSWA is defined as a phasic or tonic increase in EMG tone above the minimum amplitude observed in NREM (AASM 2007 manual).[55] RSWA may also be observed in patients without a history of DEB, though it is unknown what percentage of these patients will subsequently develop clinical RBD or an alpha-synucleinopathy (Figure 24.2, Table 24.3).[56]

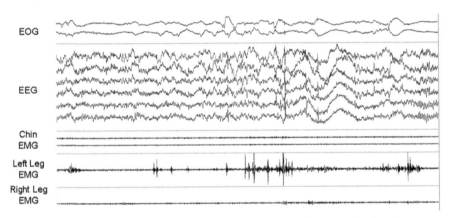

Figure 24.2. A 30 s epoch of REM sleep in a patient with REM sleep behavior disorder. The patient is a 73-year-old male with a history of a tremor and episodes of violent thrashing, punching, kicking, and falling out of bed at night. The left leg EMG demonstrates excessive phasic activity. The chin EMG and right leg EMG show normal atonia during REM.

Table 24.3. ICSD-3 criteria for Rapid eye movement sleep behavior disorder.

Diagnostic criteria for Rapid eye movement (REM) sleep behavior disorder

---

- REM sleep without atonia on EMG
- Disruptive or injurious behavior (dream enactment behavior) by history or observed druing REM sleep on PSG with video
- Absence of seizure as a cause
- Symptoms not better explained by another disorder or substance

---

24.3.5. Treatment

Treatment is symptom-focused, and the primary goal is to prevent injury to the patient and/or bed partner. Interventions may include environmental safety precautions and pharmacotherapy. In addition, discontinuation of any potential medications which may be causing or contributing to RBD should be considered.[36]

Sleeping environment interventions may include moving dangerous items out of the bedroom, placing a mattress next to the bed, bed rails or other form of barrier to prevent falling out of the bed, barriers placed between the patient and spouse, and/or use of a sleeping bag to restrict movement.[36, 57]

For patients in whom environmental interventions are insufficient, pharmacotherapy can be considered. Although randomized-controlled trials are lacking, the most evidence-based effective treatment for RBD is the benzodiazepine, clonazepam. Low doses (0.5–2 mg) at bedtime have been shown to reduce DEB in the majority of patients.[35, 58] Although, of particular concern in the elderly, side effects of clonazepam are not uncommon and include daytime sedation, cognitive impairment, and falls; thus, this agent should be used with caution in this population.[59,60]

Alternatively, melatonin, an over-the-counter agent, may be useful in patients who develop side effects on clonazepam or who are at high risk of side effects. Relatively high doses of melatonin (6–15 mg) have been shown in case series to restore REM atonia and reduce DEB.[61,62] Importantly, melatonin has a less noxious side effect profile than clonazepam and is associated with few medication-related adverse effects; thus melatonin may be a more optimal first-line therapy in the elderly.[63] A variety of other medications have been used in the treatment of RBD but overall evidence for use of these agents is scant and/or conflicting.[57,60,64]

Finally, patients with idiopathic RBD should be informed of the risk of future development of neurologic disease and appropriate counseling and/or neurology referral should be provided.[65]

24.4. INSOMNIA

Insomnia is a common complaint in the general population, but it is especially prevalent in the elderly and treatment poses specific challenges in this population. In particular, the elderly suffer from higher rates of comorbid disease and may be at higher risk of adverse effects of hypnotic medications.

24.4.1. Clinical Characteristics

Insomnia is defined as difficulty initiating or maintaining sleep or early awakening in combination with daytime symptoms. While previous classifications have differentiated primary and comorbid insomnia and sub-classified primary insomnias (i.e., psychophysiological, idiopathic, paradoxical), the recent ICSD-3 has grouped the majority of these previous classifications under the term "chronic insomnia" (ICSD-3).[66] While causality has not been ascertained, insomnia is associated with poorer physical function, cognitive decline, lower quality of life, depression, falls, and mortality.[67-72]

24.4.2. Epidemiology

Insomnia increases in prevalence with age and may affect up to one third of older adults.[73,74] Amongst older adults, one large study showed that women had a higher rate of subjective sleep complaints while men had greater abnormalities in sleep architecture. In addition, older age may be a risk factor for persistence of symptoms.[75] Insomnia is more common in older adults with higher rates of chronic medical and psychiatric disease.[76]

24.4.3. Pathophysiology

Certain etiologic factors are specific to or particularly important in the elderly. Older adults have a higher rate of chronic, comorbid diseases (e.g., depression, memory impairment, chronic pain, heart disease), and poor overall health which correlates with an increased risk of insomnia.[76-78] Polypharmacy is common in the elderly and medication side effects may be a factor in insomnia.[79] Psychosocial factors common to the elderly may also play a role. Both lower levels of physical activity and caregiver status have been identified as risk factors for insomnia in this population.[78,80,81]

Age-related circadian factors which may play a role in insomnia include both a decrease in the amplitude of circadian pacemaker output, which may lead to irregularity in sleep schedule, and a circadian phase advance, which may result in early morning awakening.[82]

24.4.4. Diagnosis

Insomnia is a clinical diagnosis based upon reported difficulty initiating or maintaining sleep in combination with daytime complaints and in the setting of adequate sleep opportunity and environment (ICSD-3).[83] PSG is not routinely indicated but may be needed to rule out OSA as a factor.

24.4.5. Treatment

Treatment options include pharmacotherapy, behavioral therapies, and psychosocial interventions.

Behavioral treatments (including cognitive behavioral therapy for insomnia (CBTI)) have proven effective for treating insomnia in older adults in randomized controlled trials.[84-86] One study investigated the efficacy of individual components of behavioral therapy (stimulus control therapy (SCT) and sleep restriction therapy (SRT)) compared to a multiple component intervention (SCT and SRT) and a wait-list control.[87] All three treatment arms proved equally efficacious in sleep latency, sleep efficiency, and other measures of sleep quality. The multicomponent therapy was more likely to result in sustained remission of symptoms. Compared to hypnotic medications, behavioral therapies demonstrate similar (or better) short-term efficacy and improved long-term efficacy.[87,88] Consensus guideline recommendations from experts in the fields of sleep and geriatrics advise behavioral therapies as a first-line approach to insomnia in the elderly (Table 24.4).[79]

Table 24.4. Specific characteristics and treatment issues in insomnia in elderly subjects.

Insomnia in elderly subjects compared to the general population

- Higher prevalence of insomnia and sleep complaints
- High prevalence of comorbid chronic medical and psychiatric disease
- Increased risk of side effects from hypnotic medications
- Circadian (advanced sleep–wake phase) and psychosocial factors may play a significant role

---

As pharmacotherapy opions, the non-benzodiazepine, benzodiazepine receptor agonists, and low dose doxepin have been shown to be effective at improving sleep in the elderly in randomized trials.[89–92] In addition, ramelteon, a melatonin receptor agonist, and prolonged release melatonin have been specifically studied in the elderly. Randomized controlled trials have shown modest benefit on sleep parameters with minimal to no adverse effects.[93–96]

It is important to note that multiple studies have documented an increased risk of adverse events in elderly patients taking hypnotics for treatment of insomnia. Both benzodiazepines and non-benzodiazepine benzodiazepine receptor agonists are associated with an increased risk of hip fractures in the elderly, although one study correlated falls with sleep disturbance and not with hypnotic use.[70, 97–100] A large meta-analysis showed a significantly increased risk of adverse cognitive events, adverse psychomotor events, and daytime fatigue in older adults taking hypnotics for insomnia.[101] The American Geriatrics Society has released recommendations to avoid the use of benzodiazepine medications for insomnia in the elderly (and to avoid us-

ing non-benzodiazepine benzodiazepine receptor agonists for over 90 days) due to the risk of adverse effects.[102] Of note, CBTI may be beneficial when attempting to taper older patients off of chronic hypnotic therapy (Table 24.5).[103]

Frequent physical activity may protect against insomnia in the elderly.[104] One small study of elderly men and women showed improved subjective (but not objective) sleep quality with scheduled daily physical and social activities.[105] Another study of subjects living in a retirement facility showed an increase in slow wave sleep with schedule daily physical and social activities.[106] A randomized controlled trial of a moderate intensity exercise regimen in elderly patients with sleep complaints showed improvements in multiple sleep parameters.[107] A recent randomized trial of a cognitive intervention showed improvements in sleep quality in older adults.[108] In sum, the evidence suggests that structured exercise and possibly social activities may be beneficial in older patients with insomnia. In light of the minimal risk incurred by these interventions and potential other benefits, these options should be considered in addition to behavioral and possibly pharmacologic therapies.

24.5. SLEEP DISORDERED BREATHING

It has been suggested that in older adults sleep apnea is a different disorder with a distinct phenotype.[109,110] With aging, there is an increased tendency for upper airway collapse with lengthening of the soft palate and increased upper airway fat.[111] Some, but not all, studies show increased airway collapsability with aging, although there

Table 24.5. Non-benzodiazepine hypnotic options and dosing in the elderly.[1-4]

Medication	Dose in the elderly	Class	Side effects	Notes
Melatonin	5–10 mg	Over-the-counter	Sedation	Very modest hypnotic effect
Ramelteon	8 mg	Melatonin receptor agonist	Sedation, dizziness, headache	Not a controlled substance
Doxepin	3–6 mg	Antidepressant	Minimal side effects with low dose (at higher doses: drowsiness, dizziness, confusion, blurred vision, dry mouth, constipation, urinary retention, arrhythmias, orthostatic hypotension)	Not a controlled substance
Zolpidem	5 mg	Non-benzodiazepine benzodiazepine receptor agonist	Sedation, gait instability, falls, sleepwalking, dizziness, headache	Use with caution in the elderly. Associated with an increased risk of side effects. Avoid using in patients with impaired mobility, gait instability, dementia, and/or impaired cognition. Avoid long-term use.
Eszopiclone	1–2 mg	Non-benzodiazepine benzodiazepine receptor agonist	Sedation, gait instability, falls, sleepwalking, unpleasant taste, dizziness, headache	Use with caution in the elderly. Associated with an increased risk of side effects. Avoid using in patients with impaired mobility, gait instability, dementia, and/or impaired cognition. Avoid long-term use.
Zaleplon	5–10 mg	Non-benzodiazepine benzodiazepine receptor agonist	Sedation, gait instability, falls, sleepwalking, nausea, myalgia	Use with caution in the elderly. Associated with an increased risk of side effects. Avoid using in patients with impaired mobility, gait instability, dementia, and/or impaired cognition. Avoid long-term use.
Suvorexant	10–20 mg	Orexin antagonist	Sedation, memory loss, sleep paralysis, depression	Contraindicated with liver disease, strong CYP3A inhibitors. May have advantages related to novel mechanism of action

may also be increased upper airway size.[112–114] Respiratory instability may be caused by age-related changes in ventilator control and loop gain (of particular importance in central apnea) but it appears the increase in obstructive sleep apnea with aging is primarily because of changes in the anatomy and physiology of the upper airway.[110]

24.5.1. Clinical Characteristics and Epidemiology

The prevalence of obstructive sleep apnea clearly increases with age. In the Sleep Heart Health Study, the prevalence of sleep apnea in the general population was 1.7 times higher in those over 60 compared to 40 to 60-year-old adults, using an apnea hypopnea index cutoff of greater than or equal to 15.[115] Nineteen percent of those between 60 and 69, and 21% of those between 70 and 79 had an AHI greater than or equal to 15. In a study of community-dwelling adults 62–91 years old, nearly one third had an AHI of 15 or more.[116] The prevalence of sleep apnea increases in patients who are generally less healthy. The prevalence of sleep apnea is higher in institutionalized elderly compared to the community.[117]

Sleep apnea may present differently in the elderly. In the Sleep Heart Health Study, it was found that elderly subjects were much less likely to report witnessed apneas.[115] The male predominance of obstructive sleep apnea appears to disappear after about age 50, probably because of the effects of menopause.[118] In an observational study of 389 consecutive patients with sleep apnea, patients greater than or equal to 65 years of age did not differ significantly in terms of disease severity, poly-

somnographic findings, or the therapeutic implications of their diagnostic study when compared with younger patients.[119]

At least when severe, untreated sleep apnea is clearly associated with increased mortality and multiple morbidities. However, these associations are less well established in the elderly. In a prospective cohort study following 6441 adults for an average of 8 years, untreated severe sleep apnea defined as an apnea hypopnea index greater than 30 was associated with increased mortality compared to those with an apnea hypopnea index less than 5.[120] However, in the subgroup of 110 adults greater than 70 years old, untreated severe sleep apnea was not associated with increased mortality (relative risk 1.27, 95% CI 0.86–1.86 among men; relative risk 1.14, 95% CI 0.65–2.01 among women).

Untreated sleep apnea has been linked to increased risk for stroke.[121] In a six-year follow-up of a cohort of 400 adults aged 72–100, patients with untreated severe sleep apnea had an increased risk of developing and ischemic stroke compared with those without apnea (adjusted hazard ratio 2.52, 95% CI 1.04–6.01).[122]

Dementia and milder forms of cognitive impairment are of extreme importance in older adults. A number of observational studies have found an association between sleep apnea and cognitive defects. In one prospective study 298 women with a mean age of 82 years were followed for 4 years with cognitive testing.[123] Women with sleep apnea as defined as AHI greater than or equal to 15 were more likely to develop mild cognitive defects or dementia compared with women without sleep apnea (45 vs. 31%, adjusted odds ratio 1.85, 95% CI 1.11–3.08).

Untreated sleep apnea is also associated with hypertension. Data from the Sleep Heart Health Study show a closer association with time spent below 90% oxygen saturation and hypertension in those >65 years old than in younger subjects.[124] Apnea hypopnea index did not predict hypertension. Other studies that have used apnea hypopnea index alone have shown a reduced association between sleep apnea and systemic hypertension with increasing age.[125,126] in a prospective cohort of the Sleep Heart Health Study of 4422 adults followed for almost 9 years, untreated severe sleep apnea(AHI greater than or equal to 30) was associated with an increased risk of coronary artery disease in men between 40 and 70 years of age. This effect was not seen in older men or in women.[127] Other studies have had too few events to confirm this association in the elderly.[120] Heart failure may also be associated with untreated severe sleep apnea in both older and younger men, but the data from the previously cited study shows only a trend (adjusted hazard ratio 1.58, 95% CI 0.93–2.66).[127]

There is some reason to believe that the effects of sleep apnea might be reduced in older adults; in a study in rats, older animals show less evidence of oxidative stress than younger animals in response to repetitive airway obstructions.[128] In general, however, it is likely that the consequences of obstructive sleep apnea are similar in the elderly but more difficult to prove because of comorbid illnesses.

24.5.2. Diagnosis

As in younger adults, the most common symptoms of sleep apnea are excessive daytime sleepiness and snoring. Cognitive deficits and nocturia may be more frequent presentations of sleep apnea in the elderly, and may be easily attributed to other etiologies in this age group. Sleepiness and attention problems may be recognized later in patients who are not working or otherwise required to keep a consistent schedule. As the normal apnea hypopnea index may be different in the elderly, it remains controversial at what point elderly patients should be treated, particularly in those who do not have significant symptoms or comorbidities.

24.5.3. Treatment

Treatment with positive airway pressure remains the most important modality for sleep apnea but may be more difficult in older adults. Older adults experience more nocturnal awakenings, and may have more difficulty achieving a good mask fit especially if they are edentulous. An oral appliance may be useful for patients with mild disease, but this may be a less satisfactory option as it generally requires adequate dentition. Elderly patients may be poor candidates for a surgical approach to treatment.

As always, avoiding alcohol and any sedative medications is important if possible, as their use can decrease respiratory drive, worsen upper airway dysfunction during sleep and severely worsen sleep apnea. Older patients may be more sensitive to these effects.

Data on treatment outcomes are limited. In a recent multicenter trial including 278 adults greater than or equal to 65 years of age with newly diagnosed sleep apnea, patients receiving CPAP plus best supportive care had improvement in daytime sleepiness compared with best supportive care alone.[130] Benefits of CPAP were present at 3 and 12 month time points and correlated with greater CPAP usage. Efficacy may have been limited by a median nightly CPAP usage of only 2 h despite a high self-reported CPAP adherence rate.

A prospective study of 939 patients 65 or older showed reduced cardiovascular mortality in patients with an AHI greater than or equal to 30 who used CPAP for at least 4 h a night.[131] A small study has shown improved outcome in stroke patients with sleep apnea treated with CPAP (mean age about 62).[132] There has been a great deal of interest in the possibility that sleep apnea diagnosis and treatment may greatly impact stroke prevention and therapy.[132]

Although not unexpected, only recently has there been evidence that sleep apnea may increase chances of cognitive impairment and Alzheimer's disease. The Apnea Positive Pressure Long-Term Efficacy Study (APPLES) failed to show much long-term improvement in neurocognitive tests in a six-month randomized trial in a mixed age group.[124] There has been some recent evidence that treating sleep apnea may decrease cognitive decline in Alzheimer's disease patients; in a recent study of 23 patients with severe apnea (AHI greater than or equal to 30), decline in mini mental status exam was slower in those treated with CPAP.[133] A recent study by Osorio et al. from the Alzheimer's Disease Neuroimaging Initiative did show that the presence of sleep apnea was associated with an earlier age of cognitive decline which was delayed by treatment with CPAP.[134]

24.6. SUMMARY

Some features of sleep even in the completely healthy elderly make it difficult to know when normal aging becomes disease. Nocturnal sleep quality is reduced by most measures, and napping may be appropriate. Advanced sleep phase is common and may not necessarily need addressing. REM sleep behavior disorder can be dangerous, can generally be controlled with medication, but may be the first manifestation of a neurodegenerative disease. Insomnia is common, and should generally be treated behaviorally. The observation of sleep apnea is common in the elderly, but may not always indicate disease requiring treatment. It is hoped that more data about outcomes will provide further proof that treatment of sleep disorders can significantly improve quality of life, morbidity, and mortality in the expanding aging population.

KEY WORDS

- **elderly**
- **sleep apnea**
- **insomnia**
- **circadian rhythm**
- **REM behavior disorder**

REFERENCES

1. US Census Bureau. Population Estimates and Projections Https://www.census.gov/population/projections/data/national/2014.html

2. Redline, S.; Kirchner, H. L.; Quan, S. F.; et al. The Effects of Age, Sex, Ethnicity, and Sleep Disordered Breathing on Sleep Architecture. *Arch. Int. Med.* **2004,** *164,* 406–418.

3. Ohayon, M. M.; Carskadon, M. A.; Guilleminault, C.; Vitiello, M. V. Meta-Analysis of Quantitative Sleep Parameters from Childhood

to Old Age in Healthy Individuals: Developing Normative Sleep Values across the Human Lifespans. *Sleep*. **2004**, *27*, 1255–1273.

4. Moraes, W.; Piovezan, R.; Poyares, D.; Bittencourt, L. R.; Santos-Silva, R.; Tufik, S. Effects of Aging on Sleep Structure through Adulthood: A Population-Based Study. *Sleep Med.* **2014**, *15*, 401–409.

5. Webb, W. B.; Dreblow, L. M. A Modified Method for Scoring Slow Wave Sleep of Older Subjects. *Sleep*. **1982**, *5*, 195–199.

6. Van Cauter, E.; Leproult, R.; Plat, L. Age-Related Changes in Slow Wave Sleep and REM Sleep and Relationship with Growth Hormone and Cortisol Levels in Healthy Men. *JAMA* **2000**, *284*, 861–868.

7. Carskadon, M. A.; Van den Hoed, J.; Dement, W. C. Insomnia and Sleep Disturbances in the Aged. Sleep and Daytime Sleepiness in the Elderly. *J. Geriatr. Psychiatry*. **1980**, *13*, 135–151.

8. Dijk, D. J.; Groeger, J. A.; Stanley, N.; Deacon, S. Age-Related Reduction in Daytime Sleep Propensity and Nocturnal Slow Wave Sleep. *Sleep*. **2010**, *33*, 211–223.

9. Duffy, J. F.; Wilson, H. J.; Wang, W.; Czeisler, C. A. Healthy Older Adults Better Tolerate Sleep Deprivation than Young Adults. *J. Am. Geriatr. Soc.* **2009**, *57*, 1245–1251.

10. Whitney, C. W.; Enright, P. L.; Newman, A. B.;et al. Correlates of Daytime Sleepiness in 4578 Elderly Persons: The Cardiovascular Health Study. *Sleep*. **1998**, *21*, 27–36.

11. Pandi-Perumal, S. R.; Spence, D. W.; Sharma, V. K. Aging and Circadian Rhythms: General Trends. In *Principles and Practice of Geriatric Sleep Medicine*, Pandi-Perumal, S. R., Monti, J. M., Monjan, A. A., Eds; Cambridge University Press: New York, **2010**, 3–11.

12. Bursztyn, M.; Stressman, J. The Siesta and Mortality: Twelve Years of Prospective Observations in 70-year-olds. *Sleep*. **2005**, *28*, 345–357.

13. Campbell, S. S.; Murphy, P. J.; Stauble, T. N. Effects of a Nap on Nighttime Sleep and Waking Function in Older Subjects. *J. Am. Geriatr. Soc.* **2005**, *53*, 48–53.

14. Duffy, J. F.; Willson, H. J.; Wang, W.; Czeisler, C. A. Healthy Older Adults Better Tolerate Sleep Deprivation than Young Adults. *J. Am. Geriatr. Soc.* **2009**, *57*, 1245–1251<References 9 and 14 are same>

15. Ando, K.; Kripke, D. F., Ancoli-Israel, S. Delayed and Advanced Sleep Phase Symptoms. *Isr. J. Psychiatry Relat. Sci.* **2002**, *39 (1)*, 11–18.

16. Schrader, H.; Bovim, G.; Sand, T. The Prevalence of Delayed and Advanced Sleep Phase Syndromes. *J. Sleep. Res.* **1993**, *2 (1)*, 51–55.

17. Paine, S. J.; et al. Identifying Advanced and Delayed Sleep Phase Disorders in the General Population: A National Survey of New Zealand Adults. *Chronobiol. Int.* **2014**, *31 (5)*, 627–636.

18. Carrier, J.; et al. Sleep and Morningness-Eveningness in the 'middle' Years of Life (20–59 y). *J. Sleep Res.* **1997**, *6 (4)*, 230–237.

19. Czeisler, C. A.; et al. Bright Light Resets the Human Circadian Pacemaker Independent of the Timing of the Sleep-Wake Cycle. *Science*. **1986**, *233 (4764)*, 667–671.

20. Jones, C. R., et al., Familial Advanced Sleep-Phase Syndrome: A Short-Period Circadian Rhythm Variant in Humans. *Nat. Med.* **1999**, *5 (9)*, 1062–1065.

21. Reid, K. J; et al. Familial Advanced Sleep Phase Syndrome. *Arch. Neurol.* **2001**, *58 (7)*, 1089–1094.

22. Toh, K. L.; et al. An hper2 Phosphorylation Site Mutation in Familial Advanced Sleep Phase Syndrome. *Science*, **2001**, *291 (5506)*, 1040–1043.

23. American Academy of Sleep Medicine. International classification of sleep disorders, 3rd ed. Darien, IL: American Academy of Sleep Medicine, 2014. P 198–203.

24. Leblanc, M. F.; Desjardins, S.; Desgagne, A. Sleep Problems in Anxious and Depressive Older Adults. *Psychol. Res. Behav. Manag.* **2015**, *8*, 161–169.

25. Horne, J. A.; Ostberg, O. A Self-Assessment Questionnaire to Determine Morningness-Eveningness in Human Circadian Rhythms. *Int. J. Chronobiol.*, **1976**, *4 (2)*, 97–110.

26. Morgenthaler, T. I.; et al. Practice Parameters for the Clinical Evaluation and Treatment of Circadian Rhythm Sleep Disorders. An American Academy of Sleep Medicine Report. *Sleep*. **2007**, *30 (11)*, 1445–1459.

27. Lack, L.; et al. The Treatment of Early-Morning Awakening Insomnia with 2 Evenings of Bright Light. *Sleep*. **2005**, *28 (5)*, 616–623.

28. Satoh, K.; et al. Two Pedigrees of Familial Advanced Sleep Phase Syndrome in Japan. *Sleep*. **2003**, *26 (4)*, 416–417.

29. Suhner, A. G.; Murphy, P. J.; Campbell, S. S. Failure of Timed Bright Light Exposure to Alleviate Age-Related Sleep Maintenance Insomnia. *J. Am. Geriatr. Soc.* **2002,** *50 (4),* 617–623.

30. Campbell, S. S.; Dawson, D.; Anderson, M. W. Alleviation of Sleep Maintenance Insomnia with Timed Exposure to Bright Light. *J. Am. Geriatr. Soc.* **1993,** *41 (8),* 829–836.

31. Lack, L.; Wright, H. The Effect of Evening Bright Light in Delaying the Circadian Rhythms and Lengthening the Sleep of Early Morning Awakening Insomniacs. *Sleep.* **1993,** *16 (5),* 436–443.

32. Pallesen, S.; et al. Bright Light Treatment has Limited Effect in Subjects over 55 years with Mild Early Morning Awakening. *Percept. Mot. Skills* **2005,** *101 (3),* 759–770.

33. Moldofsky, H.; Musisi, S.; Phillipson, E. A. Treatment of a Case of Advanced Sleep Phase Syndrome by Phase Advance Chronotherapy. *Sleep.* **1986,** *9 (1),* 61–65.

34. Zee, P. C. Melantonin for the Treatment of Advanced Sleep Phase Disorder. *Sleep.* **2008,** *31 (7),* 923; author reply 925.

35. Olson, E. J.; Boeve, B. F.; Silber, M. H. Rapid Eye Movement Sleep Behaviour Disorder: Demographic, Clinical and Laboratory Findings in 93 Cases. *Brain,* **2000,** *123 (2),* 331–339.

36. Boeve, B. F. REM Sleep Behavior Disorder: Updated Review of the Core Features, the REM Sleep Behavior Disorder-Neurodegenerative Disease Association, Evolving Concepts, Controversies, and Future Directions. *Ann. N. Y. Acad. Sci.,* **2010,** *1184,* 15–54.

37. Sasai, T.; Inoue, Y.; Matsuura, M. Do Patients with Rapid Eye Movement Sleep Behavior Disorder have a Disease-Specific Personality? *Parkinsonism Relat. Disord.* **2012,** *18 (5),* 616–618.

38. Postuma, R. B.; et al. Quantifying the Risk of Neurodegenerative Disease in Idiopathic REM Sleep Behavior Disorder. *Neurology* **2009,** *72 (15),* 1296–1300.

39. Iranzo, A.; et al. Neurodegenerative Disorder Risk in Idiopathic REM Sleep Behavior Disorder: Study in 174 Patients. PLoS One. **2014,** *9 (2),* e89741.

40. Postuma, R. B.; et al. Risk Factors for Neurodegeneration in Idiopathic Rapid Eye Movement Sleep Behavior Disorder: A Multicenter Study. *Ann. Neurol.* **2015,** *77 (5),* 830–839.

41. Kang, S. H.; et al. REM Sleep Behavior Disorder in the Korean Elderly Population: Prevalence

and Clinical Characteristics. *Sleep.* **2013,** *36 (8),* 1147–1152.

42. Bodkin, C. L.; Schenck, C. H. Rapid Eye Movement Sleep Behavior Disorder in Women: Relevance to General and Specialty Medical Practice. *J. Womens Health (Larchmont).* **2009,** *18 (12),* 1955–1963.

43. Ohayon, M. M.; Caulet, M.; Priest R. G. Violent Behavior During Sleep. *J. Clin. Psychiatry.* **1997,** *58 (8),* 369–736; quiz 377.

44. Poryazova, R.; et al. REM Sleep Behavior Disorder in Parkinson's Disease: A Questionnaire-Based Survey. *J. Clin. Sleep Med.* **2013,** *9 (1),* 55–59.

45. Plazzi, G.; et al. REM Sleep Behavior Disorders in Multiple System Atrophy. *Neurology.* **1997,** *48 (4),* 1094–1097.

46. Boeve, B. F.; Silber, M. H.; Ferman, T. J. REM Sleep Behavior Disorder in Parkinson's Disease and Dementia with Lewy Bodies. *J. Geriatr. Psychiatry Neurol.* **2004,** *17 (3),* 146–157.

47. Wang, P.; et al. Rapid Eye Movement Sleep Behavior Disorder in Patients with Probable Alzheimer's Disease. *Aging Clin. Exp. Res.* [Epub] **2015,** May 29.

48. Dauvilliers, Y.; Jennum, P.; Plazzi, G. Rapid Eye Movement Sleep Behavior Disorder and Rapid Eye Movement Sleep Without Atonia in Narcolepsy. *Sleep Med.,* **2013,** *14 (8),* 775–781.

49. Jianhua, C.; et al. Rapid Eye Movement Sleep Behavior Disorder in a Patient with Brainstem Lymphoma. *Intern. Med.* **2013,** *52 (5),* 617–621.

50. Teman, P. T.; et al. Idiopathic Rapid-Eye-Movement Sleep Disorder: Associations with Antidepressants, Psychiatric Diagnoses, and Other Factors, in Relation to Age of Onset. *Sleep Med.* **2009,** *10 (1),* 60–65.

51. Postuma, R. B.; et al. Antidepressants and REM Sleep Behavior Disorder: Isolated Side Effect or Neurodegenerative Signal? *Sleep.* **2013,** *36 (11),* 1579–1585.

52. Winkelman, J. W.; James, L. Serotonergic Antidepressants are Associated with REM Sleep Without Atonia. *Sleep.* **2004,** *27 (2),* 317–321.

53. American Academy of Sleep Medicine. International classification of sleep disorders, 3rd ed. Darien, IL: American Academy of Sleep Medicine, 2014. p. 246–253.

54. Iranzo, A.; Santamaria, J. Severe Obstructive Sleep Apnea/Hypopnea Mimicking REM Sleep

Behavior Disorder. *Sleep.* **2005,** *28 (2),* 203–206.

55. Iber, C.; Ancoli-Israel, S.; Chesson. A.; Quan, S. F. for the American Academy of Sleep Medicine. The AASM Manual for the Scoring of Sleep and Associated Events: Rules, Terminology and Technical Specifications, 1st ed.: Westchester, IL: American Academy of Sleep Medicine, **2007,** 42–43.

56. Sasai-Sakuma, T.; et al. Quantitative Assessment of Isolated Rapid Eye Movement (REM) Sleep without Atonia without Clinical REM Sleep Behavior Disorder: Clinical and Research Implications. *Sleep Med.* 2014. *15 (9),* 1009–1015.

57. Devnani, P.; Fernandes, R. Management of REM Sleep Behavior Disorder: An Evidence Based Review. *Ann. Indian Acad. Neurol.* **2015,** *18 (1),* 1–5.

58. Schenck, C. H.; Hurwitz, T. D.; Mahowald, M. W. Symposium: Normal and Abnormal REM Sleep Regulation: REM Sleep Behaviour Disorder: An Update on a Series of 96 Patients and a Review of the World Literature. *J. Sleep Res.* **1993,** 2 *(4),* 224–231.

59. Gagnon, J. F.; Postuma, R. B.; Montplaisir, J. Update on the Pharmacology of REM Sleep Behavior Disorder. *Neurology.* **2006,** *67 (5),* 742–747.

60. Anderson, K. N.; Shneerson, J. M. Drug Treatment of REM Sleep Behavior Disorder: The Use of Drug Therapies other than Clonazepam. *J. Clin. Sleep Med.* **2009,** *5 (3),* 235–239.

61. Boeve, B. F.; Silber, M. H.; Ferman, T. J. Melatonin for Treatment of REM Sleep Behavior Disorder in Neurologic Disorders: Results in 14 Patients. Sleep Med. **2003,** *4 (4),* 281–284.

62. Kunz, D.; Mahlberg, R. A Two-Part, Double-Blind, Placebo-Controlled Trial of Exogenous Melatonin in REM Sleep Behaviour Disorder. *J. Sleep Res.* **2010,** *19 (4),* 591–596.

63. McCarter, S. J.; et al. Treatment outcomes in REM sleep behavior disorder. *Sleep Med.* **2013,** *14 (3),* pp 237–242.

64. Trotti, L. M. REM sleep behaviour disorder in older individuals: epidemiology, pathophysiology and management. *Drug Aging.* **2010,** *27 (6),* pp 457–470.

65. Vertrees, S.; Greenough, G. P. Ethical considerations in REM sleep behavior disorder. *Continuum (Minneap Minn).* **2013,** *19 (1 Sleep Disorders),* pp 199–203.

66. American Academy of Sleep Medicine. International classification of sleep disorders, 3rd ed. Darien, I. L: American academy of sleep medicine, **2014,** P.21

67. Dam, T. T.; et al. Association between sleep and physical function in older men: the osteoporotic fractures in men sleep study. *J. Am. Geriatr. Soc.* **2008,** *56 (9),* pp 1665–1673.

68. Tworoger, S. S.; et al. The association of self-reported sleep duration, difficulty sleeping, and snoring with cognitive function in older women. *Alzheimer. Dis. Assoc. Disord.* **2006,** *20 (1),* pp 41–48.

69. Cricco, M.; Simonsick, E. M.; Foley, D. J. The impact of insomnia on cognitive functioning in older adults. *J. Am. Geriatr. Soc.* **2001,** *49 (9),* pp 1185–1189.

70. Avidan, A. Y.; et al. Insomnia and hypnotic use, recorded in the minimum data set, as predictors of falls and hip fractures in Michigan nursing homes. *J. Am. Geriatr. Soc.* **2005,** *53 (6),* pp 955–962.

71. Stone, K. L.; et al., Self-reported sleep and nap habits and risk of mortality in a large cohort of older women. *J. Am. Geriatr. Soc.* **2009,** *57 (4),* pp 604–611.

72. Jaussent, I.; et al. Insomnia and daytime sleepiness are risk factors for depressive symptoms in the elderly. *Sleep.* **2011,** *34 (8),* pp 1103–1110.

73. Foley, D. J.; et al. Sleep complaints among elderly persons: an epidemiologic study of three communities. *Sleep.* **1995,** *18 (6),* pp 425–432.

74. Unruh, M. L.; et al. Subjective and objective sleep quality and aging in the sleep heart health study. *J. Am. Geriatr. Soc.* **2008,** *56 (7),* pp 1218–1227.

75. Morphy, H.; et al. Epidemiology of insomnia: a longitudinal study in a UK population. *Sleep.* **2007,** *30 (3),* pp 274–280.

76. Vitiello, M. V.; Moe, K. E.; Prinz, P. N. Sleep complaints cosegregate with illness in older adults: clinical research informed by and informing epidemiological studies of sleep. *J. Psychosom. Res.* **2002,** *53 (1),* pp 555–559.

77. Foley, D.; et al. Sleep disturbances and chronic disease in older adults: results of the 2003 National Sleep Foundation Sleep in America Survey. *J. Psychosom. Res.* **2004,** *56 (5),* pp 497–502.

78. Morgan, K. Daytime activity and risk factors for late-life insomnia. *J. Sleep Res.* **2003,** *12 (3),* pp 231–238.

79. Bloom, H. G.; et al. Evidence-based recommendations for the assessment and management of sleep disorders in older persons. *J. Am. Geriatr. Soc.* **2009**, *57 (5)*, pp 761–789.

80. Castro, C. M.; et al. Sleep patterns and sleep related factors between caregiving and non-caregiving women. *Behav. Sleep Med.* **2009**, *7 (3)*, pp 164–179.

81. Wilcox, S.; King, A. C. Sleep complaints in older women who are family caregivers. *J. Gerontol. B Psychol. Sci. Soc. Sci.* **1999**, *54 (3)*, pp P189–198.

82. Czeisler, C. A.; et al., Association of sleep-wake habits in older people with changes in output of circadian pacemaker. Lancet. **1992**, *340 (8825)*, pp 933–936.

83. American Academy of Sleep Medicine. International classification of sleep disorders, 3rd ed. Darien, IL: American Academy of Sleep Medicine, 2014. P. 19–39.

84. Epstein, D. R.; et al. Dismantling multicomponent behavioral treatment for insomnia in older adults: a randomized controlled trial. *Sleep.* **2012**, *35 (6)*, pp 797–805.

85. Mitchell, M. D.; et al. Comparative effectiveness of cognitive behavioral therapy for insomnia: a systematic review. *BMC Fam. Pract.* **2012**, *13*, p. 40.

86. Irwin, M. R.; Cole, J. C.; Nicassio, P. M. Comparative meta-analysis of behavioral interventions for insomnia and their efficacy in middle-aged adults and in older adults 55+ years of age. *Health Psychol.* **2006**, *25 (1)*, pp 3–14.

87. Morin, C. M.; et al. Behavioral and pharmacological therapies for late-life insomnia: a randomized controlled trial. *JAMA.* **1999**, *281 (11)*, pp 991–999.

88. Sivertsen, B.; et al. Cognitive behavioral therapy vs zopiclone for treatment of chronic primary insomnia in older adults: a randomized controlled trial. *JAMA.* **2006**, *295 (24)*, pp 2851–2858.

89. Ancoli-Israel, S.; et al. Zaleplon, a novel non-benzodiazepine hypnotic, effectively treats insomnia in elderly patients without causing rebound effects. *Prim. Care Companion J. Clin. Psychiatry* **1999**, *1 (4)*, pp 114–120.

90. Scharf, M.; et al. A 2-week efficacy and safety study of eszopiclone in elderly patients with primary insomnia. *Sleep.* **2005**, *28 (6)*, pp 720–727.

91. Scharf, M.; et al. Efficacy and safety of doxepin 1 mg, 3 mg, and 6 mg in elderly patients with primary insomnia: a randomized, double-blind, placebo-controlled crossover study. *J. Clin. Psychiatry.* **2008**, *69 (10)*, pp 1557–1564.

92. Krystal, A. D.; et al. Efficacy and safety of doxepin 1 mg and 3 mg in a 12-week sleep laboratory and outpatient trial of elderly subjects with chronic primary insomnia. *Sleep.* **2010**, *33 (11)*, pp 1553–1561.

93. Roth, T.; et al. A 2-night, 3-period, crossover study of ramelteon's efficacy and safety in older adults with chronic insomnia. *Curr. Med. Res. Opin.* **2007**, *23 (5)*, pp 1005–1014.

94. Roth, T.; et al. Effects of ramelteon on patient-reported sleep latency in older adults with chronic insomnia. *Sleep Med.* **2006**, *7 (4)*, pp 312–318.

95. Lemoine, P.; et al. Prolonged-release melatonin improves sleep quality and morning alertness in insomnia patients aged 55 years and older and has no withdrawal effects. *J. Sleep Res.* **2007**, *16 (4)*, pp 372–380.

96. Wade, A. G.; et al. Efficacy of prolonged release melatonin in insomnia patients aged 55–80 years: quality of sleep and next-day alertness outcomes. *Curr. Med. Res. Opin.* **2007**, *23 (10)*, pp 2597–2605.

97. Wang, P. S.; et al. Zolpidem use and hip fractures in older people. *J. Am. Geriatr. Soc.* **2001**, *49 (12)*, pp 1685–1690.

98. Cumming, R. G.; Le Couteur, D. G. Benzodiazepines and risk of hip fractures in older people: a review of the evidence. *CNS Drugs.* **2003**, *17 (11)*, pp 825–837.

99. Berry, S. D.; et al. Nonbenzodiazepine sleep medication use and hip fractures in nursing home residents. *JAMA Intern. Med.* **2013**, *173 (9)*, pp 754–761.

100. Kang, D. Y.; et al. Zolpidem use and risk of fracture in elderly insomnia patients. *J. Prev. Med. Public Health.* 2012, *45 (4)*, pp 219–226.

101. Glass, J.; et al. Sedative hypnotics in older people with insomnia: meta-analysis of risks and benefits. *BMJ.* **2005**, *331 (7526)*, pp 1169.

102. American Geriatrics Society Beers Criteria Update Expert, P., American Geriatrics Society updated Beers Criteria for potentially inappropriate medication use in older adults. *J. Am. Geriatr. Soc.* **2012**, *60 (4)*, pp 616–631.

103. Morin, C. M.; et al. Randomized clinical trial of supervised tapering and cognitive behavior therapy to facilitate benzodiazepine discontinu-

ation in older adults with chronic insomnia. *Am. J. Psychiatry.* **2004,** *161 (2),* pp 332–342.

104. Inoue, S.; et al. Does habitual physical activity prevent insomnia? A cross-sectional and longitudinal study of elderly Japanese. *J. Aging Phys. Act.* **2013,** *21 (2),* pp 119–139.

105. Benloucif, S.; et al. Morning or evening activity improves neuropsychological performance and subjective sleep quality in older adults. *Sleep.* **2004,** *27 (8),* pp 1542–1551.

106. Naylor, E.; et al. Daily social and physical activity increases slow-wave sleep and daytime neuropsychological performance in the elderly. *Sleep.* **2000,** *23 (1),* pp 87–95.

107. King, A. C.; et al., Moderate-intensity exercise and self-rated quality of sleep in older adults. A randomized controlled trial. *JAMA.* **1997,** *277 (1),* pp 32–37.

108. Haimov, I.; Shatil, E. Cognitive training improves sleep quality and cognitive function among older adults with insomnia. *PLoS One.* **2013,** *8 (4),* pp e61390.

109. Launois, S. H.; Pepin, J. L.; Levy, P. Sleep apnea in the elderly: a specific entity? *Sleep Med. Rev.* **2007,** *11,* 87–97.

110. Edwards, B. E.; Wellman, A.; Sands, S. A.: et al. Obstructive sleep apnea in older adults is a distinctly different physiological phenotype. *Sleep.* **2014,** *37 (7),* 1227–1236.

111. Malhotra, A.; Huang, Y.; Fogel, R.; et al. Aging influences on pharyngeal anatomy and physiology: the predisposition to pharyngeal collapse. *Am. J. Med.* **2006,** *119,* e19–e14.

112. Eikermann, M.; Jordann, A. S.; Chamberlin, N. L. et al. The influence of aging on pharyngeal collapsibility during sleep. *Chest.* **2007,** *131,* 1702–1709.

113. Mayer, P.; Pepin, J. L.; Bettega, G.; et al. Relationship between body mass index, age and upper airway measurements in snorers and sleep apnoea patients. *Eur. Respir. J.* **1996,** *9,* 1801–1809.

114. Burger, C. D.; Stanson, A. W.; Sheedy, P. F., II; Daniels, B. K.; Shepard, J. W., Jr. Fast-computed tomography evaluation of age-related changes in upper airway structure and function in normal men. *Am. Rev. Respir. Dis.* **1992,** *145,* 846–852.

115. Young, T.; Shahar, E.; Nieto, F. J. et al. Predictors of sleep disordered breathing in community dwelling adults: the sleep heart health study. *Arch. Intern. Med.* **2002,** *162,* 893–900.

116. Endeshaw, Y. Clinical characteristics of obstructive sleep apnea in community-dwelling older adults. *J. Am. Geriatr. Soc.* **2006,** *54,* 1740–1744.

117. Ancoli-Israel, S. Epidemiology of sleep disorders. Clin. Geriatr. Med. **1989,** *5,* 347–362.

118. Young. T.; Finn, L.; Austin, D.; et al. Menopausal status and sleep-disordered breathing in the Wisconsin sleep cohort study. *Am. J. Respir. Crit. Care Med.* **2003,** *167,* 1181–1185.

119. Levy, P.; Pepin, J. L.; Malauzat, D.; et al. Is sleep apnea syndrome in the elderly a specific entity? *Sleep.* **1996,** *19,* S29.

120. Punjabi NM, Caffo BS, Goodwin JL, et al. Sleep disordered breathing and mortality: a prospective cohort study. *PLoS Med.* **2009,** *6,* e1000132.

121. Yaggi, H. K.; Concato, J.; Kernan, W. N.; et al. Obstructive sleep apnea as a risk factor for stroke and death. *N. Engl. J. Med.* **2005,** *353,* 2034–2041.

122. Munoz, R.; Duran-Cantolla, J.; Martinez-Vila, E.; et al. Severe sleep apnea and risk of ischemic stroke in the elderly. *Stroke.* **2006,** *37,* 2317–2321.

123. Yaffe, K.; Laffan, A. M.; Harrison, S. L.; et al. Sleep disordered breathing, hypoxia, and risk of mild cognitive impairment and dementia in older women. *JAMA.* **2011,** *306,* 613–619.

124. Nieto, F. J.; Young, T. B.; Lind, B. K. et al. Association of sleep disordered breathing, sleep apnea, and hypertension in a large community-based study. Sleep heart health study. *JAMA.* **2000,** *283,* 1829–1836.

125. Bixler, E. O.; Vgontzas, A. N.; Lin, H. M.; et al. Association of hypertension and sleep disordered breathing. *Arch. Intern. Med.* **2000,** *160,* 2289–2295.

126. Grote, L.; Ploch, T.; Heitmann, J.; et al. Sleep-related breathing disorder is an independent risk factor for systemic hypertension. *Am. J. Resp. Crit. Care Med.* **1999,** *160,* 1875–1882.

127. Gottlieb, D. J.; Yenokyan, G.; Newman, A. B. et al. Prospective study of obstructive sleep apnea and incident coronary heart disease and heart failure: the sleep heart health study. *Circulation.* **2010,** *122,* 352–360.

128. Dalmases, M.; Torres, M.; Marquez-Kisinousky, L.; et al. Brain tissue hypoxia and oxidative stress induced by obstructive apneas is different in young and aged rats. *Sleep.* **2014,** *37,* 1249–1256.

129. McMillan, A.; Bratton, D. J.; Faria, R.; et al. Continuous positive airway pressure in older people with obstructive sleep apnoea syndrome (PREDICT): a 12 month multicentre, randomised trial. *Lancet Respir. Med.* **2014,** 2, 804–812.

130. Martinez-Garcia, M. A.; Campos-Rodriquez, F.; Catalan-Serra, P.; et al. Cardiovascular mortality in obstructive sleep apnea in the elderly: role of long-term continuous positive airway pressure treatment: a prospective observational study. Am. J. Respir. Crit. Care Med. **2012,** *186,* 909–916.

131. Ryan, C. M.; Bayley, M.; Green, R.; et al. Influence of CPAP on outcomes of rehabilitation in stroke patients with OSA. *Stroke.* **2011,** *42,* 1062–1067.

132. Kushida, C. A.; Nichols, D. A.; Holmes, T. H.; et al. Effects of continuous positive airway pressure on neurocognitive function in obstructive sleep apnea patients: the apnea positive pressure long-term efficacy study (APPLES). *Sleep.* **2012,** *35,* 1593–1602.

133. Troussiere, A. C.; Charley, C. M.; Salleron, J.; et al. Treatment of sleep apnoea syndrome decreases cognitive decline In patients with Alzheimer's disease. *J. Neurol. Neurosurg. Psychiatry* **2014,** *85 (12),* 1405–1408.

134. Osorio, R. S.; Gumb, T.; Pirraglia, M. A.; et al. Sleep-disordered breathing advances cognitive decline in the elderly. *Neurology.* **2015,** *84,* 1964–1971.

1. Morin, C. M.; Benca, R. M. Insomnia nature, diagnosis, and treatment. *Handb. Clin. Neurol.* 2011, 99, pp 723–746.

2. Bloom, H. G.; et al. *Evidence-based recommendations for the assessment and management of sleep disorders in older persons. J. Am. Geriatr. Soc.* 2009, 57 (5), pp 761–789.

3. Scharf, M.; et al. *A 2-week efficacy and safety study of eszopiclone in elderly patients with primary insomnia. Sleep.* 2005, 28 (6), pp 720–727.

4. Scharf, M.; et al. Efficacy and safety of doxepin 1 mg, 3 mg, and 6 mg in elderly patients with primary insomnia: a randomized, double-blind, placebo-controlled crossover study. *J. Clin. Psychiatry.* 2008, 69 (10), pp 1557–1564.

5. Herring, W.; Connor, K.; Snavely, D.; et al. Clinical profile of suvorexant over 3 months in elderly patients with insomnia. *Neurology.* 2015, 84, S 53.

25

POLYSOMNOGRAPHY I: PROCEDURE AND TECHNOLOGY

Ahmed S. BaHammam, Divinagracia E. Gacuan,
Smitha George, Karen Lorraine Acosta,
Seithikurippu Ratnas Pandi-Perumal, and Ravi Gupta

Ahmed S. BaHammam, MD, FACP

University Sleep Disorders Center, College of Medicine, King Saud University, Riyadh, Saudi Arabia

The Strategic Technologies Program of the National Plan for Sciences and Technology and Innovation in the Kingdom of Saudi Arabia

Divinagracia E. Gacuan, RPSGT

University Sleep Disorders Center, College of Medicine, King Saud University, Riyadh, Saudi Arabia

Smitha George, RPSGT

University Sleep Disorders Center, College of Medicine, King Saud University, Riyadh, Saudi Arabia

Karen Lorraine Acosta, RPSGT

University Sleep Disorders Center, College of Medicine, King Saud University, Riyadh, Saudi Arabia

Seithikurippu Ratnas Pandi-Perumal

Somnogen Canada Inc., College Street, Toronto, ON M6H 1C5, Canada

Ravi Gupta, MD

Department of Psychiatry & Sleep Clinic, Himalayan Institute of Medical Sciences, Swami Ram Nagar, Doiwala, Dehradun, India

Corresponding author: Ahmed S. BaHammam, E-mails: ashammam2@gmail.com; ashammam@ksu.edu.sa

ABSTRACT

Polysomnography (PSG) is the gold standard diagnostic test for sleep disorders. During PSG, several physiological parameters are monitored while the patient is asleep. These parameters include brain waves, the oxygen level in the blood, heart rate and breathing, body position as well as eye and leg movements, along with synchronized audiovisual monitoring. In this chapter, we will take the reader in a journey to demonstrate PSG technology and how PSG is performed from the time patient arrives until his discharge.

25.1. INTRODUCTION

Polysomnography (PSG) is the gold standard diagnostic test for several sleep disorders. The test is usually performed at a sleep disorders unit within a hospital or at a sleep disorders center. During PSG, several physiological parameters are monitored while the patient is asleep. These parameters include brain waves, the oxygen level in the blood, heart rate and breathing, body position, as well as eye and leg movements, along with synchronized audiovisual monitoring. Moreover, in certain conditions, additional parameters may be included such as esophageal pH monitoring, esophageal manometry, and overnight blood pressure monitoring. In addition to its important diagnostic role, PSG can be used to adjust treatment plan and to reach the optimal positive airway pressure (PAP) setting for patients with sleep-disordered breathing (SDB).

Sleep architecture is largely divided into non-rapid eye movement (NREM) and rapid eye movement (REM) sleep. NREM is further divided into three stages: N1, N2, and N3, with N3 being the deepest stage of sleep. REM sleep alternates with NREM sleep approximately every 90 min and a normal person who sleeps for 7–8 h usually has 4–6 cycles of REM and NREM sleep (Fig. 25.1). Monitoring of the different sleep stages, sleep interruptions, movements, and the other respiratory and cardiac signals are clinically helpful for identifying the nature of patient's sleep

problems and assessing response to treatment.

In this chapter, we will take the reader in a journey to demonstrate PSG technology and how PSG is performed and interpreted to help reaching the accurate diagnosis and prescribing the correct treatment.

25.2. DATA COLLECTION

25.2.1. Pre-study Patient Evaluation

PSG is typically performed at night. A request for sleep study is usually completed by a sleep medicine specialist indicating the clinical diagnosis, the type of sleep study needed, and instructions to the sleep technologist. The patient may be asked to visit the sleep disorders unit a day before undergoing a sleep study to be familiar with the new sleep environment. Patients are usually asked to complete a pre-designed questionnaire that includes some validated questionnaires to assess daytime sleepiness and the risk of certain sleep disorders as a part of a comprehensive assessment of the patient before the overnight sleep study.

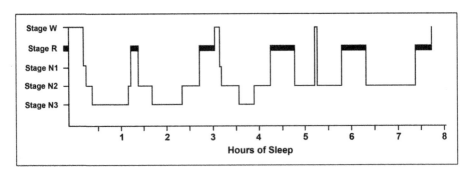

Figure 25.1. A hypnogram showing normal distribution of sleep stages.

During the pre-study evaluation, patients are given information about sleep study and patients are usually instructed to avoid naps and stimulants during the day of the sleep study. Table 25.1 gives an example of the instructions given to the patient in preparation for a sleep study.

25.2.2. Performing Sleep Study

Applying the electrodes and sensors is the most important part of the sleep study to ensure quality data. Electrodes and sensors represent the link between the patient and the data output. The care with which the electrodes and sensors are applied is the most important factor for the quality and accuracy of the recorded data. There are suggested sequence for applying electrodes and sensors (first is the scalp and face electrodes, followed by the snore sensor, ECG electrodes, leg electrodes, respiratory bands, airflow sensors and lastly the oximeter probe). However, every technologist

Table 25.1. An example of the instructions given to the patient prior to PSG

Instructions
• Maintain your regular daytime schedule.
• Avoid napping on the day of the study.
• Avoid alcohol, caffeine (coffee, tea, cocoa, or chocolate), sedatives, and stimulants for 24 h, unless otherwise directed by your physician.
• Wash and dry your hair and do not apply hair sprays, oils or gels. Oils and moisturizers on your skin and hair may interfere with our ability to attach electrodes to your skin or scalp.
• For female patients, please remove any nail polish.
• Take your regularly scheduled medications as you normally would unless your physician instructs otherwise. Please inform your sleep technologist of any medications you have already taken or plan to take on the night of your study.
• Eat your "regular" **meals** – including dinner before you arrive to the sleep disorders center. Notify the medical team in advance of any special dietary needs. You may bring a small snack and decaffeinated beverage.
• Bring comfortable sleep attire (preferably pajamas) to wear during your sleep tests.
• You may also want to bring some reading materials. There is a television in each bedroom.
• For patients under 18 years of age and those patients who may need special assistance, a parent or guardian is required to stay in the sleep disorders center for the entire duration of testing.
• It is important that you arrive on time. A technologist will greet you and will show you your bedroom. If you have special needs or concerns, please tell the sleep center staff ahead of time.

may develop his/her own routine for applying these electrodes and sensors. Figure 25.2 illustrates the location of the various electrodes and sensors used for monitoring sleep. During electrodes and sensors application, the technologist needs to reassure the patient and explain what each electrode and sensor measures.

25.2.2.1. Surface Electrodes Application

Surface electrodes measure very small potentials caused by synchronized activity in very large numbers of synapses in the cerebral cortex (electroencephalography [EEG]), electrical activity from the chin and leg muscles (electromyography [EMG]), electrical potentials generated by eye movements (electrooculography [EOG]), and the conducting system of the heart (electrocardiography [ECG]). These potentials arising from the body are called biopotentials, which are extremely small and require amplification in order to be seen while monitoring. Good preparation of the site of the electrode and proper ap-

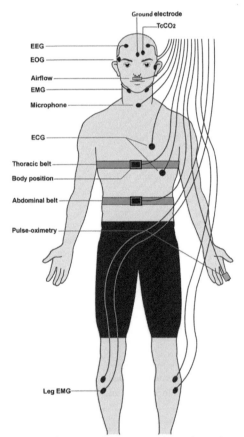

Figure 25.2. An illustration of the location of the various electrodes and sensors used for monitoring sleep. It includes the EEG, EOG, electromyogram (chin EMG), airflow (nasal and/or oral), ECG, pulse oximetry, respiratory effort (thoracic/abdominal), snore microphone, and body position sensor.

plication of the electrode are essential for a good quality, noise-free signal. The patient should be given a brief explanation about the hook-up procedure. To prepare the site for electrode placement, the site should be prepared by cleaning the area with alcohol swab to eliminate oil on the skin. Then, using a small cotton tip applicator, scrub (using abrasive skin preparation such as Nu-Prep) the area where the electrode will be placed. Some patients feel slight burning sensation with this and they need to be reassured. If excess prep material remains, it should be gently cleaned with the rubbing alcohol as it interferes with the sticking of the adhesive tape, which is used to fix the sensor.

25.2.2.2. EEG

25.2.2.2.1. International 10–20 System

The proper placement of the EEG electrodes is essential for good EEG signal. The placement of the EEG electrodes on the scalp follows an international system known as the 10–20 system of electrode placement (Fig. 25.3). This is a standardized method of identifying equally spaced electrode positions on the scalp, based on four identifiable skull landmark (nasion: the bridge of the nose or lowest indentation point between nose and forehead; inion: the bony ridge at the base of the back of the skull; pre-auricular points: indentations just above the cartilage that covers

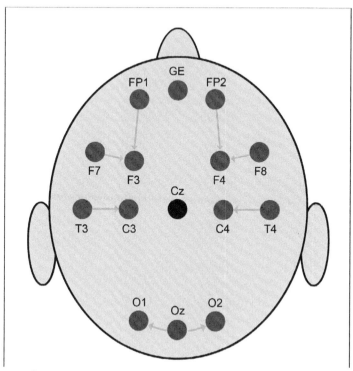

Figure 25.3. An illustration showing the location of the EEG electrodes on the scalp according to the 10–20 system of electrode placement.

the external ear openings; and the mastoid process: bony area located just behind the ear). The "10" and "20" refers to the 10 or 20% inter-electrode distance. The displayed EEG signal depends on the electrode derivation, which means from where the signals are being derived. A differential amplifier will measure and amplify the difference between the electro-potentials of two input sites (input 1 and input 2). The American Academy of Sleep Medicine (AASM) recommends three derivations for EEG recording; F4-M1, C4-M1, and O2-M1 (F: frontal; C: central; O: occipital; and M: mastoid).[1] This is called the recording montage. Alternative montages and additional electrodes may be used as directed by the treating physician for the evaluation of a patient with a possible nocturnal seizure disorder.

25.2.2.2.2. EEG Waves

EEG signal is represented by waveforms classified according to their amplitude and frequency. Table 25.2 shows different sleep, EEG waves, frequencies, and the derivations that show them clearly. These waveforms are observed in different stages of sleep which help the scorer to identify different sleep stages. EEG waves are defined based on their frequency which refers to the number of waveforms appearing within the span of one second expressed as cycles per second (cps) or hertz (Hz); morphology, which refers to the shape or structure of certain waveforms or groups of waveforms (Fig. 25.4 and amplitude which refers to the vertical size (height) of waveform from peak to peak and reflects the voltage of the incoming signal and is expressed in microvolts (μV). Figure 25.4 shows examples of different EEG wave morphology.

25.2.2.3. EOG

EOG picks up movements of the eye balls; based on recording the electro-potential difference between the cornea and the

Table 25.2. Different sleep EEG waves frequencies and the derivations that show them clearly

Waves	Frequency	Derivation
Alpha	8–13 Hz	Occipital
Beta	More than 13 Hz	
Theta	4–7.99 Hz	
Delta	0–3.99 Hz	Frontal
Vertex		Central
Spindles	12–14 Hz	Central
K–Complex	Less than 2 Hz	Frontal
Sawtooth	2–6 Hz	Central

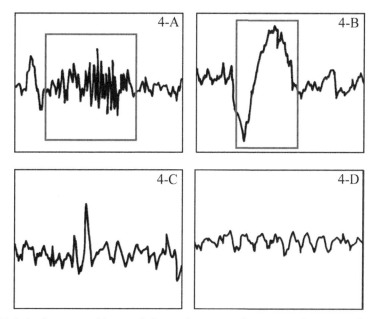

Figure 25.4. An illustration of the morphology and structure of certain waveforms. A: A sleep spindle, which is a burst of oscillatory short synchronized brain activity visible on an EEG that occurs during stage N2 sleep. It consists of 12–14 Hz waves that occur for at least 0.5 s; B: K complex defined as an EEG event composed of well-defined negative sharp wave immediately followed by a positive component with a total duration of ≥0.05 s seen maximally over the frontal region; C: Vertex sharp wave: It is a negative (upward) sharp wave lasting <0.5 s that stands out of the EEG background activity and usually seen in the central leads. It is seen in stage N1; D: Sawtooth EEG pattern is characteristic variant of theta activity (containing waveforms with a notched or sawtooth-shaped appearance) and seen frequently during REM stage sleep.

retina (the cornea has a positive voltage output, while the retina has a negative polarity). There are two reasons for recording EOG. The first is to record rapid eye movements of REM sleep, and the second is to assess sleep onset, which is associated with slow rolling eye movements.

25.2.2.4. Chin EMG

EMG records chin muscle tone at the mentalis and submentalis muscles. It is a mandatory recording parameter for staging REM sleep and is essential to determine the onset of REM sleep. In general, muscle tone decreases during sleep with maximal reduction occurring during REM sleep. For men who have beard, the use of collodion or EC2 adhesive paste may be needed unless they arrive clean-shaven. Figure 25.5 shows EEG, EOG, and EMG patterns for wakefulness, REM sleep, and NREM (N2) sleep.

25.2.2.5. Leg EMG

The monitoring of the anterior tibialis muscle is needed to diagnose periodic limb

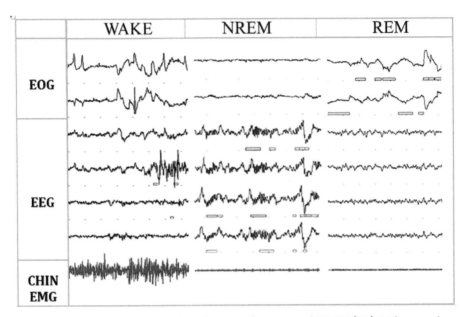

Figure 25.5. Examples of PSG recording of stages wake, REM, and NREM (N2). Wake stage shows eye blinking (EOG), mixed frequency EEG waves are seen with attenuated alpha rhythm, and high chin EMG tone. REM shows rapid eye movements (EOG), mixed frequency EEG waves and absent chin EMG tone. NREM (N2) shows no eye movements (EOG), sleep spindles and K complex (EEG), and reduced chin EMG tone.

movements in sleep (PLMS). Two electrodes are properly placed about 2 to 4 cm apart longitudinally, along the belly of the anterior tibialis muscle of each leg.

25.2.2.6. ECG

ECG is the record of the electrical potential changes in the conducting system of the heart. Recording the electrical activity of the heart during PSG is essential in order to measure the heart rate evolution and notice any signs of cardiac arrhythmias, which may be linked to sleep disorders. Moreover, ECG helps in calculating the pulse transit time (PTT).

25.2.2.7. Respiratory Monitoring

Respiration during sleep is monitored by different sensors. These sensors measure the respiratory flow, thoraco–abdominal movements associated with breathing, oxygen saturation, and carbon dioxide (CO_2) output. Together these signals provide an idea about SDB and its subtypes.

25.2.2.7.1. Respiratory Airflow

The recognition of SDB requires the recording of both airflow and respiratory effort. Airflow can be monitored by thermal sensors or pressure sensing devices. The AASM guidelines recommend that both should be used simultaneously during monitoring. Thermal sensors (thermistors or thermocouples) are commonly used to record airflow.[2] Their non-quantitative signals are based on the variation between the temperature of inhaled and exhaled air. Pressure sensing device uses a nasal cannula connected to a very sensitive pressure transducer and produces a qualitative signal quite comparable to a pneumotachograph (the most accurate quantitative way of measuring airflow but unrealistic to use in a clinical setting). For the identification of an apnea during a diagnostic study, the AASM recommends the use of an oronasal, thermal airflow sensor to monitor airflow.[1] For the identification of a hypopnea during a diagnostic study, the AASM recommends the use of a nasal pressure transducer to monitor airflow.[1]

25.2.2.7.2. Respiratory Effort

Breathing effort and breathing pattern can be assessed by recording chest and abdominal wall movements. Belts around the chest and abdomen are used to monitor respiratory movements. Proper placement of the belts is essential for good signal. The thoracic belt is positioned just below the armpits while the abdominal belt is secured around the waist, at the level of the navel. The belts need to be fit so they are snug, but not too tight or too loose. For obese patients, a location that will provide a good fit and allow the belt to stay in place throughout the night should be chosen. The essential role of this signal is to distinguish obstructive apnea (presence of respiratory effect) from central apnea (lack of respiratory effort). Respiratory inductance plethysmography (RIP) and piezo technology

are the most common methods used to monitor respiratory effort.[2] Piezo technology uses piezo-electric crystals that produce an output charge when compressed, flexed, or stretched. The output signal of piezo technology is not linear; therefore, it cannot be used to assess hypopnea. Moreover, piezo technology can produce false paradoxical breathing signal when tension is applied to the belt during patient movements. RIP technology utilizes a belt with a wire interweaved in a zigzag pattern along its length. A battery passes a weak current though the wire, which creates a weak magnetic field. Breathing movements modify the magnetic field, which is converted into voltage output and a waveform recorded in the PSG. The output signal of the RIP technology is linear and hence more accurate than the piezo technology. If calibrated appropriately, RIP provides accurate information regarding paradoxical breathing and flow volume loops.

25.2.2.7.3. Snore Sensor

This sensor records the low frequency snoring sound produced by the vibrations of the upper airway during sleep. Snoring can be monitored using different sensors: (1) Microphone, which simply records the sounds produced when the patient is snoring; (2) Piezo snoring sensor, which allows the recording of snoring vibrations from the side of the neck by converting these vibrations into different voltages; and (3) Nasal pressure cannula, where snoring signal can be seen superimposed on the airflow waveform signal when the signal from airflow sensor cannula is unfiltered.

25.2.2.7.4. Pulse Oximetry

It records both pulse and oxygen saturation. The pulse oximeter probe is usually placed on a finger, preferably the index finger. Alternative sites include the ear lobe or the toe; however, the sensor tends to dislodge more easily from these sites during sleep. For a good signal, it is important to prepare the site by removing artificial nails and nail polish. For pulse oximetry used during sleep study, a fast sampling rate oximeter is recommended (shorter interval e.g., 3 s or less) to improve sensitivity as patients with SDB usually have short-lasting intermittent hypoxemia. Overnight pulse oximetry is an important parameter for the evaluation of respiratory disturbances during sleep particularly for scoring hypopneas for which desaturation is one of the criteria. In addition, it indicates the severity of SDB and the need to supplement PAP device with oxygen therapy.

25.2.2.7.5. Carbon Dioxide Monitoring (PCO₂)

In pediatrics, end-tidal carbon dioxide (E_TCO_2) is considered as a standard practice during PSG monitoring. On the other hand, in adult patients, the AASM recommends the use of arterial PCO_2, E_TCO_2, or transcutaneous PCO_2 for the detection of hypoventilation during a diagnostic sleep study, and the use of arterial PCO_2 or transcutaneous PCO_2 for the detection of hypoventilation during PAP titration.[1] Both E_TCO_2 and transcutaneous PCO_2 are noninvasive, validated, indirect methods to predict arterial PCO_2.[3] E_TCO_2 reflects exhaled CO_2 at end-tidal sample of exhaled gas. Infrared spectroscopy is the technique usually used to assess E_TCO_2 in the exhaled gas. When the patient is being ventilated using a closed circuit (e.g., endotracheal tube), E_TCO_2 can be measured directly with good accuracy. However, during sleep study, E_TCO_2 is measured using a nasal cannula in a spontaneously breathing patient; therefore, side-stream sampling often occurs. When the patient is on a noninvasive PAP device, the increased flow within the open circuit results in dilution of the exhaled gas. Therefore, the numerical value or the displayed waveform of E_TCO_2 may not accurately reflect arterial PCO_2. Mouth breathing may influence the displayed signal too. Transcutaneous PCO_2 is obtained through the skin where the electrode warms the skin surface, increasing local capillary perfusion, and measures the CO_2 gas as it diffuses from the dermis across a gas permeable membrane. Currently available devices require less heating (42°C for adults and 41°C for neonates); therefore, resulting in less discomfort and less potential for skin damage. Nevertheless, the sensor may need to be repositioned at least once during sleep study. Transcutaneous PCO_2 absolute value is affected by skin thickness and capillary density. Therefore, it is important to place the electrode at a site of high capillary density and thin skin. This presents no problem in the newborn, in which these conditions are usually present. However, in adults, there is greater variation from site to site. The suggested locations for best transcutaneous measurements are the forehead, forearm, chest, or abdomen. Transcutaneous PCO_2 provides a good alternative to E_TCO_2 during PAP titration to assess the response to treatment.

25.2.2.7.6. Position Sensor

Body position monitoring is important during PSG due to the sleep-dependent nature of SDB. Body position monitoring allows the technician to capture changes in SDB breathing in different positions (supine, lateral, and prone) and to accurately gauge the true severity and appropriate PAP titration. Body position sensor reports the position of the sensor rather than the position of the human body; therefore, it is essential to assure that the sensor is oriented correctly to the patient.

25.2.2.8. Cardiac Monitoring

25.2.2.8.1. ECG

ECG recording is an essential component of PSG recording. It usually includes recording from the right and left subclavicular areas (essentially lead I) or the right subclavicular area and the lower left thorax (essentially lead II), which is the recommended derivation by the AASM. The ECG quality must be carefully observed prior to "lights out." To ensure a quality recording, ECG electrodes should be applied with the same care and preparation as with any other electrodes used in the sleep study. Viewing the ECG and arrhythmia recognition is facilitated by "spreading the recording out," utilizing a 10 s display or less.

25.2.2.8.2. Bio-calibration

Bio-calibration is a series of actions that the patient is asked to perform prior to initiating a sleep study, to ensure that the sensors are measuring what they are supposed to measure and to verify signal quality. However, the importance of this procedure to the quality of PSG recording is often unrecognized. Bio-calibration allows the sleep technologist to verify that the monitoring equipment, electrodes, and sensors are working properly before starting a sleep study. The patient is asked to relax and lie still, and perform certain maneuvers. Noting the different signals when the patient is awake, particularly EEG, EMG, and EOG patterns of eyes-open and eyes-closed help the sleep study scorer to recognize sleep stages and abnormalities during sleep. Table 25.3 shows the routine bio-calibration procedure, instructions, and quality signal acquired from each test. The impedance of the head electrodes should also be checked prior to starting the study. An ideal impedance should not exceed 5 kΩ, although 10 kΩ or less is acceptable. Electrodes with higher impedances should be changed. Bio-calibration is to be done before starting every sleep study, after hooking up the patients with all sensors.

25.2.2.8.3. Filters

During PSG recording, there are signals that are undesirable even though they are physiological in nature. Because there are plenty of competing signals of varying amplitudes and frequencies that can obscure the record, filters are needed to eliminate or reduce these unwanted signals in order to obtain noise-free monitoring signals. Signals are filtered with high frequency filter (HFF), low frequency filter (LFF), and 60 Hz notch filter. HFF which is also known as a "low-pass" filter attenuates all frequencies above the cut-off frequency of

Table 25.3. Routine biocalibration tests, instructions and quality signal acquired from each test

Instructions	Electrode/Sensor	Expected signal
Relax and keep your eyes open.	EEG	Mixed frequency EEG waves are seen (attenuated alpha rhythm).
	EOG	Slow eye movements.
Close your eyes and relax without sleeping (30 s).	EEG	Alpha waves are typically prominent in occipital channels with eyes closed.
Without moving your head, look to the right, look left (repeat 3 times each direction).	EOG	Detects eye movements.
Slowly and distinctly blink your eyes 5 times.	EOG	Blinking eye movements.
Clench your jaw, smile or grit your teeth (5 times); relax.	Chin EMG	Burst of activity on chin EMG channel or EMG amplitude should increase compared in the relax state.
Take slow breaths, inhale then exhale (5 breaths).	Respiratory channels (airflow and effort sensor)	With normal breathing, all the airflow and respiratory effort channels should be in phase (sinusoidal wave) (polarity of inspiration is usually upward).
	Pulse oximetry	With normal breathing, the signal should be flat line with oxygen saturation >90%.
While remaining still, take a breath and hold it for 10 s.	Respiratory channels (airflow and effort ensor)	Respiratory effort and airflow channels should show a flat line (helps in apnea detection).
Count from one to five.	Snore sensor	There is an oscillation or activity in the snore sensor signal.
Flex and extend your right foot and then left foot (5 times).	Leg EMG	Burst of activity while recording each leg (reference to evaluate leg movements).

the filter and allows all frequencies below the cut-off frequency to pass unchanged. LFF which is also known as a "high-pass" filter, attenuates all frequencies below the cut-off frequency of the filter and allows all frequencies above the cut-off frequency to pass unchanged. 60 Hz notch filter is provided by equipment manufacturers as a temporary means of eliminating 50/60 Hz

frequency artifact from electrical sources without affecting other frequencies.

25.2.2.8.4. Amplifiers

Acquired signals during PSG monitoring are tiny and need amplification to a viewable size. Therefore, amplifiers are used during PSG recording, which include both

AC and DC amplifiers. AC amplifiers can amplify only AC signals, whereas DC amplifiers can amplify both AC and DC signals.[4] AC amplifiers are designed to process fast signal frequencies (such as EEG, EOG, EMG, and ECG waves). AC amplifiers have controls to set the sensitivity, polarity, and filters in order to condition the signal. Its distinguishing feature is the presence of both LLF and HFF. On the other hand, as DC amplifiers are designed to record relatively constant or slowly fluctuating voltages (such as respiration, pressure, oximetry, and respiratory effort), a LFF is not included although a HFF is present.

25.2.2.9. Audiovisual Recording

Synchronized video and audio recording is an important component of a sleep study as it allows the scorer to observe the patient's behavior during the study, evaluate parasomnias, nocturnal seizures and other motor events during sleep, and monitor snoring and vocalization during sleep. Most sleep centers will have an intercom audio system between the patient's room and the control room to allow the patient and the sleep technologists to communicate.

25.3. DOCUMENTATION

One of the essential parts of every sleep study is the proper documentation by sleep technologists. Sleep technologist's responsibility is to maintain a good quality sleep study that will assist the treating physician to reach the accurate diagnosis. Included in the technologist's documentation are all the events, especially unusual sleep related events that are noted during the PSG, as well as, any medication taken by or given to the patient during the time of the study. Table 25.4 demonstrates examples of subjective and objective events that need documentation during PSG.

25.4. PATIENT DISCHARGE

Once sleep study is completed, the sleep technologist will remove electrodes and

Table 25.4. Examples of the subjective and objective events that need documentation during PSG

Subjective	Objective
Patient's perception regarding sleep quality during the study	Sleep–wake patterns of patient's sleep
Any unusual complaints i.e., headache, pain, nightmares etc.	Unusual events i.e., short REM sleep latencies, leg movements, hypopneas, apneas, sleep walking, parasomnias, sleep talking etc.
The subjective feeling when a PAP device titration was implemented.	Artifacts
Patient's perception of PAP device therapy	Response to PAP device titration

sensors and patients are asked to complete the post-sleep study questionnaires and evaluation forms. A follow-up appointment with the treating physician to discuss the results of the PSG and the management plan should be given to the patient.

- EMG
- EOG
- snore
- airflow
- apnea

ACKNOWLEDGEMENT

This project was partially funded by the Strategic Technologies Program of the National Plan for Sciences and Technology and Innovation in the Kingdom of Saudi Arabia (King Saud University and King Abdulaziz City for Science and Technology).

The authors would like to thank Mrs. Samar Nashwan and Mrs. Flordelita Gatdula for their help in preparing the illustrations.

KEY WORDS

- polysomnography
- EEG

REFERENCES

1. Berry, R. B.; Brooks, R.; Gamaldo, C. E.; Harding, S. M.; Lloyd, R. M.; Marcus, C. L.; et al. The AASM manual for the scoring of sleep and associated events: rules, terminology and technical specifications, Version 2.1. www.aasmnet.org, Darien, Illinois: American Academy of Sleep Medicine, 2014.
2. Farre, R.; Montserrat, J. M.; Navajas, D. Non-invasive monitoring of respiratory mechanics during sleep. *Eur. Respir. J.* **2004,** *24(6),* 1052–1060.
3. Kirk, V. G.; Batuyong, E. D.; Bohn, S. G. Transcutaneous carbon dioxide monitoring and capnography during pediatric polysomnography. *Sleep* **2006,** *29(12),* 1601–1608.
4. Thomas, S. J. Basic principles of polysomnography including electrical concepts. *Respir. Care. Clin. N. Am.* **2005,** *11(4),* 587–595.

26

POLYSOMNOGRAPHY II: SCORING

Ahmed S. BaHammam, Divinagracia E. Gacuan,
Smitha George, Karen Lorraine Acosta,
Seithikurippu Ratnas Pandi-Perumal, and Ravi Gupta

Ahmed S. BaHammam, MD, FACP
University Sleep Disorders Center, College of
Medicine, King Saud University, Riyadh, SAUDI
Arabia

The Strategic T echnologies Program of the
Na-tional Plan for Sciences and T echnology
and Innovation in the Kingdom of Saudi Arabia

Divinagracia E. Gacuan, RPSGT
University Sleep Disorders Center, College of
Medicine, King Saud University, Riyadh, SAUDI
Arabia

Smitha George, RPSGT
University Sleep Disorders Center, College of
Medicine, King Saud University, Riyadh, SAUDI
Arabia

Karen Lorraine Acosta, RPSGT
University Sleep Disorders Center, College of
Medicine, King Saud University, Riyadh, SAUDI
Arabia

Seithikurippu Ratnas Pandi-Perumal
Somnogen Canada Inc., College Street,
Toronto, ON M6H 1C5, Canada

Ravi Gupta, MD
Department of Psychiatry & Sleep Clinic,
Himalayan Institute of Medical Sciences,
Swami Ram Nagar, Doiwala, Dehradun, India.

Corresponding author: Prof. Ahmed
BaHammam, Professor of Medicine Director,
Sleep Disorders Center, College of Medicine,
King Saud University Box 225503, Riyadh
11324, Saudi Arabia Telephone: 966-1-467-1521;
Fax: 966-1-467-2558; E-mails: ashammam2@
gmail.com; ashammam@ksu.edu.sa

ABSTRACT

After polysomnography (PSG) is completed, the recording is analyzed. Scoring an
overnight sleep study requires mastering of
the scoring rules and good experience. In
this chapter, we cover the scoring of sleep
stages; different respiratory events such as
apnea, hypopnea and hypoventilation; and
periodic limb movements. Several illustrations are discussed to elucidate the scoring
rules.

INTRODUCTION

After the test is completed, the recording
is analyzed. A PSG recording of 7 hours
will be displayed in approximately 840 epochs (one epoch represents 30 seconds).
These epochs are scored manually by an
expert scorer. PSG is scored according to
the AASM scoring rules.[1] The scorer may
modify the duration of the epoch depending on the data to be scored. While EEG
signals are scored in 30 seconds epoch,
respiratory parameters and movements are
scored in 2 minutes epochs. On the other
hand, Cheyne-Stokes breathing may be

scored in 5 minutes epochs and ECG is usually scored in 10-15 seconds epochs.

SCORING SLEEP STAGES

Although sleep may seem like a steady state, it actually consists of several stages that cycle throughout the night. Three parameters are being used in order to properly recognize different sleep stages; these are the EEG, EOG and the EMG (muscle tone particularly the sub-mentalis or chin muscles activity). EEG waves help to define different sleep stages. Sleep Stages are divided into: wake (W), non-rapid eye movements (NREM) 1 (N1), NREM 2 (N2), NREM 3 (N3) and rapid eye movements (REM) (R).

Stage W

The first several epochs of the record will usually be stage W, although occasionally a patient is so sleepy and is asleep when the recording is started. During normal wakefulness with eyes closed, the posterior dominant rhythm (PDR) is detected over the occipital leads and gets attenuated when the eyes are open. PDR is characterized by a sinusoidal rhythm with a frequency of 8-13 Hz, roughly the range of alpha frequency, and thus is also called the alpha rhythm. Stage W is scored when there is alpha rhythm in greater than 50% of the epoch. In patients without detectable alpha rhythm on EEG (approximately 10% of all recordings), stage W may still be scored if any one of the following markers of alertness is detected: Eye blinks that appears as conjugate vertical eye movements at a frequency of 0.5-2 Hz, reading eye movements consisting of a slow phase followed by a rapid movement in the opposite direction, and the presence of irregular conjugate eye movements with normal or high chin muscle tone suggesting that the subject is awake and looking around. Figure 1 shows an epoch of stage W.

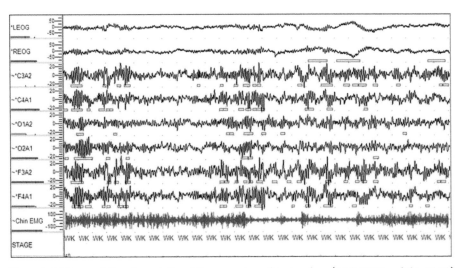

Figure 1: A 30 seconds Epoch consisting of the parameters of staging sleep (EEG, EOG, and chin EMG) showing stage wake (W). The record shows alpha rhythm and increased chin EMG tone.

Stage N1

This is the lightest stage of sleep. It is defined by the presence of slow eye movements (SEM), which is conjugate, reasonably regular and sinusoidal eye movements. The EEG shows low amplitude, mixed frequency activity, predominantly of 4–7 Hz (theta waves) encompassing more than 50% of the epoch. Vertex sharp waves can be seen, which are sharply contoured waves with duration <0.5 seconds, seen mostly over the central region (Figure 4-C in PSG I chapter). The term "Sleep Onset" is defined as the beginning of the first epoch scored as any stage other than stage W. Figure 2 shows an epoch of stage N1.

Stage N2

Stage N2 sleep constitutes the largest percentage of total sleep time in a normal adult. It is defined by the presence of sleep spindles and/or K complexes (Figure 4-A and 4-B in PSG I chapter). Sleep Spindles are short rhythmic waveform clusters of 11–16 Hz (most commonly 12–14 Hz), often showing a waxing and waning appearance with a duration ≥ 0.5 seconds, seen maximally over the central derivations. This stage is also characterized by the appearance of K complexes, which are negative sharp waves immediately followed by a slower positive component with total duration ≥ 0.5 seconds, seen maximally over the frontal derivations. Figure 3 demonstrates an epoch of stage N2.

Stage N3

Stage N3 sleep is frequently referred to as "deep sleep" or "slow wave sleep". It is defined when 20% or more of a 30-seconds

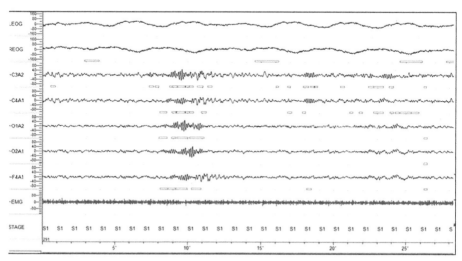

Figure 2: A 30 seconds Epoch consisting of the parameters of staging sleep (EEG, EOG, and chin EMG) showing stage N1. The EOG shows presence of slow eye movements (SEM) which is conjugate, reasonably regular and sinusoidal eye movements and the EEG shows low amplitude, mixed frequency activity, predominantly of 4–7 Hz (theta waves).

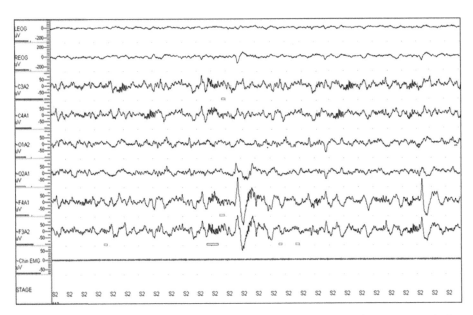

Figure 3: A 30 seconds Epoch consisting of the parameters of staging sleep (EEG, EOG, and Chin EMG) showing stage N2. EEG demonstrates sleep spindles (short rhythmic waveform clusters of 11–16 Hz, often showing a waxing and waning appearance with a duration ≥0.5 seconds and K complexes (negative sharp waves immediately followed by a slower positive component with total duration ≥0.5 seconds shown maximally in the frontal leads).

epoch contains slow wave activity, which are waves of a frequency of 0.5–2 Hz with a peak-to-peak amplitude >75 µV as measured over the frontal regions. Stage N3 tends to occur more in the first half of the night. It is often more difficult to awaken sleepers during stage N3 sleep compared with stages N1 and N2. NREM parasomnias typically occur during stage N3. Figure 4 shows an epoch of stage N3.

Stage R

REM sleep also called Stage R is defined by REMs (rapid eye movements), which are conjugate, irregular and sharply peaked eye movements. Low or absent chin EMG

tone is the hallmark of this stage. Sawtooth waves are seen, which are often serrated waves of 2–6 Hz and seen best over the central region (Figure 4-D in PSG I chapter). Transient muscle activity (short irregular bursts of EMG activity usually of duration < 0.25 seconds superimposed on low EMG tone) may be present and this activity may be seen in the chin or anterior tibialis EMG derivations. Figure 5 shows an epoch of stage R.

Sleep Efficiency (SE)

After scoring sleep stages, SE can be calculated. SE is defined as the number of minutes of sleep divided by the number of min-

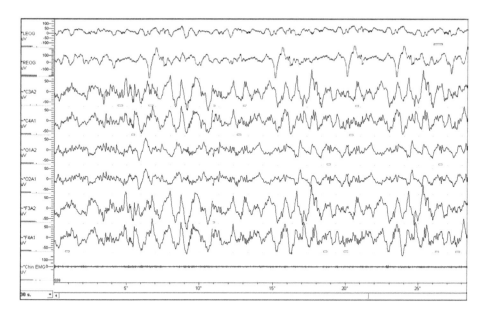

Figure 4: A 30 seconds Epoch consisting of the parameters of staging sleep (EEG, EOG, and chin EMG) showing stage N3. The EEG shows slow wave activity, which are waves of a frequency of 0.5–2 Hz with a peak-to-peak amplitude >75 μV.

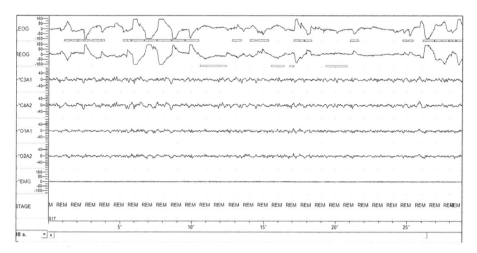

Figure 5: A 30 seconds Epoch consisting of the parameters of staging sleep (EEG, EOG, and Chin EMG) showing stage R. The EOG shows rapid eye movements and the EEG shows mixed frequency low amplitudes waves. Chin EMG is absent.

utes in bed. Normal SE is approximately 85 to 90% or higher.

Sleep Onset

Sleep onset is defined as the start of the first epoch scored as any stage other than stage W.[1] In most subjects, stage N1 appears at sleep onset.

Sleep Onset Latency

During PSG recording, sleep onset latency is defined as the time period from "lights out," or bedtime, to the onset of sleep.

SCORING RESPIRATORY EVENTS

Recognizing common respiratory events encountered during an overnight sleep study and understanding their scoring rules will lead to accurate diagnosis and proper treatment of SDB. Apneas (obstructive and mixed) and hypopneas are the two classical breathing abnormalities seen in obstructive sleep apnea (OSA) patients. Normal breathing appears as a series of fairly equal "sinusoidal" waveforms. The inspiratory phase of the breathing cycle is negative or has an upward position.

Apnea

The AASM defines an apnea as a drop in the airflow peak signal excursion by ≥ 90% of pre-event baseline using an oronasal thermal sensor for at least 10 seconds. Apneas are divided into obstructive, central and mixed apneas.[1]

Obstructive Apnea

In obstructive apnea, there is no airflow in the presence of continued effort to breathe. It is caused by closure of the upper airway while the patient is trying to breathe. The event lasts at least 10 seconds in adults and 2 missed breaths in children. Figure 6 demonstrates an example of an obstructive apnea during PSG recording.

Central Apnea

Central Apnea occurs when respiratory effort ceases, leading to absence of airflow. The cessation of breathing in this case is not due to obstruction of the upper airway but secondary to a central cause. Figure 7 shows an example of a central sleep apnea.

Mixed Apnea

Mixed Apnea occurs when the respiratory effort ceases, the airway collapses, and then respiratory effort resumes against a closed airway. Most mixed apneas appear as a central apnea followed by an obstructive apnea because effort is absent initially but resumes in the second portion of the event. Figure 8 shows an example of a mixed apnea.

Hypopnea

Hypopnea indicates reduction in airflow during sleep usually due to partial closure of the upper airway. The new AASM scoring guidelines of hypopneas define hypopnea as a reduction in peak signal of airflow by ≥ 30% of pre-event baseline using nasal pressure, PAP device flow, or an alternative hypopnea sensor for at least 10 seconds followed by desaturation or associated with an arousal.[1] Hypopneas are also classified into obstructive hypopnea and central hypopnea. In general, obstructive hypopnea is scored if the event is associated with snoring, flattening of the inspiratory portion of the nasal pressure waveform or thoraco-abdominal paradox. Figure 9 shows an example of an obstructive hypopnea.

Obesity Hypoventilation (OHS)

The ICSD-3 defines OHS as the combined presence of obesity (BMI >30 kg/m^2) with awake arterial hypercapnia (PaCO$_2$ >45 mmHg) and SDB in the absence of other causes of alveolar hypoventilation such as lung diseases, medication use, neurologic disorder, muscle weakness, or a known

congenital or idiopathic central alveolar hypoventilation syndrome.[2] The AASM scoring manual Version 2.3 defines sleep hypoventilation as an increase in end-tidal PCO$_2$ to a value >55 mmHg for ≥10 minutes or as an increase in end-tidal PCO$_2$ of ≥10 mmHg during sleep (compared with an awake supine value) to a value exceeding 50 mmHg for ≥10 minutes.[1] Figure 10 demonstrates an example of sleep hypoventilation.

Cheyne-Stokes Breathing (CSB)

CSB is a breathing pattern characterized by regular "crescendo-decrescendo" fluctuations in respiratory rate and tidal volume. It is more common among patients with heart failure and low ejection fraction. CSB is scored if both criteria are met, there

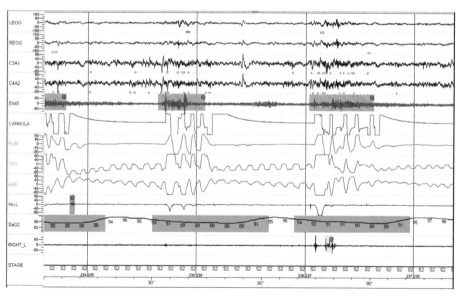

Figure 6: A zoomed 2 minutes Epoch showing examples of obstructive apneas. Airflow is absent despite persisting respiratory paradoxical effort. The event is followed by arousal and desaturation.

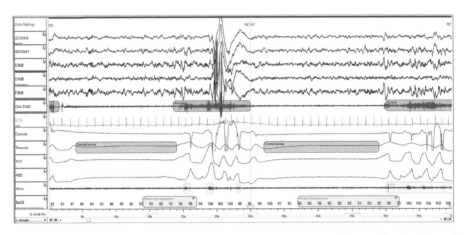

Figure 7: A zoomed 1 minute Epoch showing two central apneas. There is absence of both airflow and respiratory efforts.

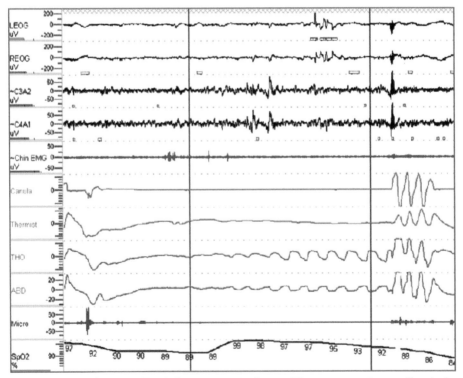

Figure 8: A zoomed 1 minute Epoch showing one mixed apnea. There are no respiratory efforts in the first part of the apnea. In the second part of the apnea, respiratory efforts are present.

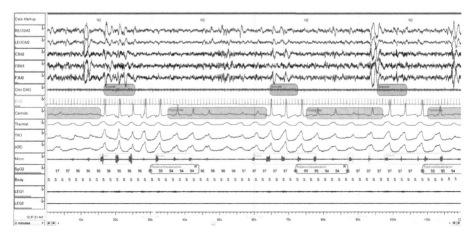

Figure 9: A zoomed 2 minutes Epoch showing obstructive hypopneas. The record demonstrates reduction in airflow by ≥ 30% of pre-event baseline associated with snoring, desaturation and arousals.

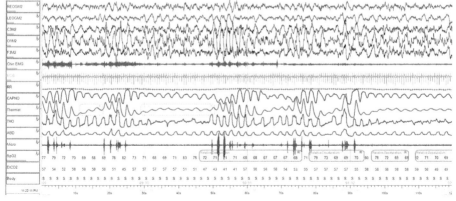

Figure 10: A zoomed 2 minutes Epoch showing sustained hypoxemia and an increase in end-tidal PCO_2 to a value >55 mmHg indicative of sleep hypoventilation.

are episodes of more than 3 consecutive central apneas and/or central hypopneas separated by a crescendo and decrescendo change in breathing amplitude with a cycle length of at least 40 seconds (typically 45 to 90 seconds), and there are more than 5 central apneas and/or central hypopneas per hour of sleep associated with the crescendo/decrescendo breathing pattern recorded over more than 2 hours of monitoring.[1] Figure 11 shows an example of CSB.

SCORING LEG MOVEMENTS

Periodic Limb Movements in Sleep (PLMS)

Periodic limb movements are characterized by repetitive movements of the limbs during sleep. These movements may be jerking and kicking motions, an upward flexing of the feet or brief twitches. It can be asymptomatic or produce a complaint of either insomnia or excessive sleepiness. The patient is usually not aware of the movements which are more typically a bed partner complaint. Table 5 presents the AASM scoring criteria for leg movements event and PLM series.[1] Figure 12 shows an example of PLMS.

REPORT GENERATION

The organization of a sleep study report is dependent on the needs of the end user. Reports do come with superfluous amounts of data, which fortunately, can be customized according to the needs of the users. For in depth analysis and research purposes, a report may contain multiple tables, statistics, graphs, and charts. On the other hand, if the report is mainly for clinical use, it usually includes patient's demographic data, the main findings of the sleep study, sleep study hypnogram that reflects the summary of the sleep study and the recommendations of the treating physician. Important sleep parameters that must appear in the report include, summary of sleep architecture, respiratory events, oxygen saturation levels, limb movements, arousals, and heart rate. If PAP device was used, the report should summarize the

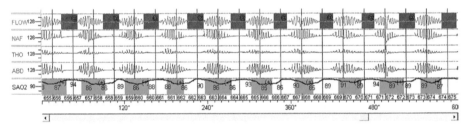

Figure 11: A zoomed window of flow, respiratory effort and oxygen saturation. The records shows the crescendo/decrescendo breathing pattern indicative of Cheyne-Stokes breathing associated with desaturation.

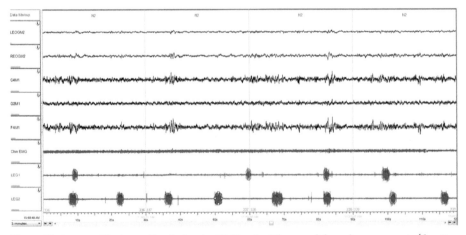

Figure 12: A zoomed 2 minutes Epoch showing PLM series. The record shows leg movements (duration >0.5 seconds and <10 seconds) separated from each other by more than 5 seconds and less than 90 seconds.

Table 5: The AASM scoring criteria for periodic limb movements in sleep (PLMS) scoring rules.[1]

Leg Movement (LM) Event	PLM Series
The following define a significant (LM) event:	The following define a PLM series:
1. The minimum duration of a LM event is 0.5 seconds	1. The minimum number of consecutive LM events is 4 LMs
2. The maximum duration of a LM event is 10 seconds	2. The minimum period length between LMs (define as the time between onsets of consecutive LMs) is 5 seconds
3. The minimum amplitude of a LM event is an 8 μV increase in EMG voltage above resting EMG	3. The maximum period length between LMs (define as the time between onsets of consecutive LMs) is 90 seconds
4. The timing of the onset of a LM event is defined as the point at which there is an 8 μV increase in EMG voltage above resting EMG	4. Leg movements on 2 different legs separated by less than 5 seconds between movement onsets are counted as a single leg movement
5. The timing of the ending of a LM event is defined as the start of a period lasting at least 0.5 seconds during which the EMG does not exceed 2 μV above resting EMG	

subjective and objective response to therapy and the optimal setting of the device. The sleep technologists' comments are important and can add further insight into the study report as they stay around the patient during monitoring. Examples of important technologists' comments include an account of the patient's physical and emotional status during the study, unusual sleep related behavior or atypical findings that are not evident in the monitored sleep parameters.

ACKNOWLEDGEMENT

This project was partially funded by The Strategic Technologies Program of the National Plan for Sciences and Technology and Innovation in the Kingdom of Saudi Arabia (King Saud University and King Abdulaziz City for Science and Technology).

The authors would like to thank Mrs. Samar Nashwan and Mrs. Flordelita Gatdula for their help in preparing the illustrations.

KEYWORDS

- **sleep stages**
- **N1**
- **N2**
- **N3**
- **R**
- **apnea**
- **hypopnea**
- **hypoventilation**
- **periodic limb movements**
- **Cheyne stokes breathing**

REFERENCES

1. Berry RB, Brooks R, Gamaldo CE, Harding SM, Lloyd RM, Marcus CL, et al. The AASM Manual for the Scoring of Sleep and Associated Events: Rules, Terminology and Technical Specifications, Version 2.1. www.aasmnet.org, Darien, Illinois: American Academy of Sleep Medicine, 2014.

2. American Academy of Sleep Medicine. International classification of sleep disorders (ICSD), 3rd ed. Darien, IL: American Academy of Sleep Medicine, 2014.

INDEX

O

Printed and bound by CPI Group (UK) Ltd, Croydon, CR0 4YY

23/10/2024

01777702-0016